Errata Sheet

LNS 140 *Case Studies in Bayesian Statistics Volume IV*
Constantine Gatsonis Bradley Carlin Andrew Gelman Mike West
Robert E. Kass Alicia Carriquiry Isabella Verdinelli (Editors)

On page 289, line -12 Figure ?? should read Figure 1

On page 291, line 18, Section ?? should read Section 3.2

On page 292, line -1 Figure ?? should read Figure 2

On page 294, line -12 Figure ?? should read Figure 2

On page 296, line 3 Figure ?? should read Figure 3

On page 296, line 15 Figure ?? should read Figure 4

On page 296, line 17 Figure ?? should read Figure 5

On page 296, line -11 Figure ?? should read Figure 6

On page 297, line -5 Figure ?? should read Figure 7

On page 297, line -4 Figure ??(a) should read Figure 7(a)

On page 299, line 1 Figure ??(c) should read Figure 7(c)

On page 299, line 9 Figure ??(b) should read Figure 7(b)

On page 299, line 10 Figure ??(d) should read Figure 7(d)

On page 299, line 20 Figure ??(d) should read Figure 7(d)

On page 299, line 20 Figure ??(b) should read Figure 7(b)

Lecture Notes in Statistics **140**

Springer
New York
Berlin
Heidelberg
Barcelona
Hong Kong
London
Milan
Paris
Singapore
Tokyo

Constantine Gatsonis
Robert E. Kass
Bradley Carlin
Alicia Carriquiry
Andrew Gelman
Isabella Verdinelli
Mike West (Editors)

Case Studies in Bayesian Statistics

Volume IV

Springer

Constantine Gatsonis
Center for Statistical Sciences
Brown University
Box G-A416
Providence, RI 02912

Robert E. Kass
Department of Statistics
Carnegie Mellon University
Baker Hall 229
Pittsburgh, PA 15213

Bradley Carlin
Division of Biostatistics
University of Minnesota
Box 303 Mayo Building
Minneapolis, MN 55455

Alicia Carriquiry
Department of Statistics
Iowa State University
222 Snedecor Hall
Ames, IA 50011-1210

Andrew Gelman
Department of Statistics
Columbia University
618 Mathematics Bldg.
New York, NY 10027

Isabella Verdinelli
Department of Statistics
Carnegie Mellon University
232 Baker Hall
Pittsburgh, PA 15213

Mike West
Institute of Statistics and Decision Sciences
Duke University
214A Old Chemistry Building
Durham, NC 27708-0251

CIP data available.
Printed on acid-free paper.

Camera ready copy provided by the editors.
Printed and bound by Braun-Brumfield, Ann Arbor, MI.
Printed in the United States of America.

9 8 7 6 5 4 3 2 1

ISBN 0-387-98640-5 Springer-Verlag New York Berlin Heidelberg SPIN 10695467

Preface

The 4th Workshop on Case Studies in Bayesian Statistics was held at the Carnegie Mellon University campus on September 27-28, 1997. As in the past, the workshop featured both invited and contributed case studies. The former were presented and discussed in detail while the latter were presented in poster format. This volume contains the four invited case studies with the accompanying discussion as well as nine contributed papers selected by a refereeing process. While most of the case studies in the volume come from biomedical research the reader will also find studies in environmental science and marketing research.

INVITED PAPERS

In *Modeling Customer Survey Data*, **Linda A. Clark, William S. Cleveland, Lorraine Denby,** and **Chuanhai Liu** use hierarchical modeling with time series components in for customer value analysis (CVA) data from Lucent Technologies. The data were derived from surveys of customers of the company and its competitors, designed to assess relative performance on a spectrum of issues including product and service quality and pricing. The model provides a full description of the CVA data, with random location and scale effects for survey respondents and longitudinal company effects for each attribute. In addition to assessing the performance of specific companies, the model allows the empirical exploration of the conceptual basis of consumer value analysis. The authors place special emphasis on graphical displays for this complex, multivariate set of data and include a wealth of such plots in the paper.

The advent of functional neuroimaging has revolutionized the study of the brain in recent years. The new technology generates enormous sets of data with intricate spatio-temporal patterns. The significant statistical challenges presented by such data are discussed in *Functional Connectivity in the Cortical Circuits Subserving Eye Movements* by **Chris Genovese** and **John Sweeney**. Since eye movement abnormalities provide a commonly used marker for several neurologic and psychiatric disorders, the study of how the brain controls eye movements is important for acquiring an understanding of brain abnormalities related to these disorders. The authors analyze functional Magnetic Resonance Imaging (fMRI) data to study the human eye-movement system. Through the use of hierarchical modeling, they examine the functional relationship between subserving rapid eye repositioning movements and smooth visual pursuit of a target. The discussion of the specific analysis of the eye movement data is used to showcase a broad spectrum of modeling and computational issues confronting investigators in functional imaging.

The recent explosive growth in genetic research and in our understanding of the genetic susceptibility to various diseases provides the background for the next invited case study in this volume. In *Modeling Risk of Breast Cancer and Decisions about Genetic Testing*, **Giovanni Parmigiani, Donald A. Berry, Edwin Iversen, Jr., Peter Müller, Joellen Schildkraut** and **Eric P. Winer** make use of the extensive capabilities of modern Bayesian modeling and computing to study clinical and health policy issues related to genetic testing for susceptibility to breast cancer. Their paper discusses aspects of an integrated set of research projects on breast cancer at Duke University. The main questions include the prediction of genetic susceptibility to breast cancer (presence of BRCA1 and BRCA2 mutations), the assessment of the risk for developing breast cancer, and the incorporation of this knowledge in the construction of decision models for breast cancer prevention and treatment.

The last of the invited case studies discusses the use of Bayesian methods in another key area of biomedical research, drug development. In *The Bayesian approach to population pharmacokinetic/pharmacodynamic modeling*, **Jon Wakefield, Leon Aarons** and **Amy Racine-Poon** take a population perspective of drug development, aiming at characterizing the variability in drug absorption, distribution and elimination, and drug efficacy and toxicity across individuals. The analysis incorporates informative prior distributions to capture information from previous studies or from biological and clinical considerations. The authors emphasize the strengths of Bayesian methods for problems in pharmacokinetics and pharmacodynamics, particularly with respect to the ability of these methods to handle large numbers of parameters and non-linear, subject-specific models.

CONTRIBUTED PAPERS

The study of toxicity resulting from radiation treatment for cancer is addressed in *Longitudinal Modeling of the Side Effects of Radiation Therapy* by **Sudeshna Adak** and **Abhinanda Sarkar**. The authors analyze data from a recent study of lung cancer patients who were observed over a four week period. The focus of the paper is on Bayesian modeling for the prediction of the longitudinal course of toxicity in these patients using pre-treatment patient data, such as age and stage of the disease.

In *Analysis of hospital quality monitors using hierarchical time series models*, **Omar Aguilar** and**Mike West** analyze longitudinal quality of care measures from Veterans Administration hospitals. Their approach is designed to assess the relationship among monitors of the various aspect of quality of care and to examine the variability of quality indicators across hospitals both cross-sectionally and over time.

The link between ambient ozone and pediatric Emergency Room visits for asthma attacks is the the focus of *Spatio-Temporal Hierarchical Models for Analyzing Atlanta Pediatric Asthma ER Visit Rates* by **Bradley P. Carlin, Hong Xia, Owen Devine, Paige Tolbert,** and **James Mulholland**. The models include demographic and meteorologic covariates, and allow for spatial and spatio-temporal autocorrelation, errors in the ozone estimates. In addition to deriving numerical estimates of the risk of asthma attacks due to ozone exposure, the authors discuss methods for drawing smoothed maps for relative risks, which can be used in a variety of epidemiologic studies.

The topic of *Validating Bayesian Prediction Models: a Case Study in Genetic Susceptibility to Breast Cancer* by **Edwin Iversen Jr, Giovanni Parmigiani,** and **Donald Berry** is closely linked to the work presented earlier in this volume by Parmigiani et al. This paper is focused on the validation of a probability model for risk of BRCA1 and BRCA2 mutations. The approach allows joint assessment of test sensitivity and specificity and carrier score error and treats genetic status as a latent variable. Receiver operating characteristic (ROC) curves are used to assess diagnostic performance of the model for a range of thresholds.

In *Population Models for Hematologic Data*, **J.Lynn Palmer** and **Peter Müller** study the distribution of blood stem cells collected longitudinally from cancer patients during the time period prior to undergoing high-dose therapy. The resulting models are used as the basis for designing optimal blood cell collection schedules for patients scheduled for intensive therapy.

The application of Bayesian structured mixture models in the process of drug screening is discussed by **Susan Paddock, Mike West, S. Stanley Young** and **Merlise Clyde** in *Mixture Models in the Exploration of Structure-Activity Relationships in Drug Design.* The authors use mixture models to handle missing information on binding configurations in drug screening experiments. They report their experience with both simulated and real data sets and discuss ways in which the analysis of the latter led them to enhancements of their their original model.

The next two contributed case studies come from research in environmental science. In *A Hierarchical Spatial Model for Constructing Wind Fields from Scatterometer Data in the Labrador Sea*, **J. A. Royle, L. M. Berliner, C. K. Wikle**and **R. Milliff** discuss the analysis of high resolution wind data using a model incorporating spatial structure components. The model leads to realistic reproductions of both wind and pressure fields, holding promise of generalizability outside the domain of the original data collection.

In *Redesigning a Network of Rainfall Stations*, **Bruno Sansó** and **Peter Müller** use a decision model to evaluate alternative designs for a system of rainfall monitoring stations in Venezuela. The goal of their analysis is to reduce the overall number of stations with minimal loss of information about local rainfall.

The final paper in the volume discusses the use of change-point modeling to screening for prostate cancer. In *Using PSA to detect prostate cancer onset: An application of Bayesian retrospective and prospective change-point identification*, **Elizabeth H. Slate** and **Larry C. Clark** analyze an extensive set of longitudinal prostate-specific antigen (PSA) measurements on participants in the Nutritional Prevention of Cancer Trial. Through the use of hierarchical modeling the authors study the natural history of PSA levels and assess the diagnostic performance of rules based on PSA data.

ACKNOWLEDGMENTS

A number of individuals provided tremendous help for the Workshop and this volume. We would like to thank Norene Mears and Heidi Sestrich for their work on the arrangements for the Workshop, Heidi Sestrich for editorial support, Ashish Sanil and Alix Gitelman for compiling the author and subject index of the volume, and several referees for their reviews of contributed case studies. Our thanks also go to the National Science Foundation and the National Institutes of Health for financial support for the Workshop . Above all we would like to thank the workshop participants for their contribution to the continued success of this series.

Providence, RI	**Constantine Gatsonis**
Minneapolis, MN	**Bradley Carlin**
Ames, IA	**Alicia Carriquiry**
New York, NY	**Andrew Gelman**
Pittsburgh, PA	**Robert E. Kass**
Pittsburgh, PA	**Isabella Verdinelli**
Durham, NC	**Mike West**

Contents

Invited Papers

Contributed Papers

List of Contributors

Leon Aarons, University of Manchester, UK

Sudeshna Adak, Harvard School of Public Health & Dana-Farber Cancer Institute, Boston, MA

Omar Aguilar, Duke University, Durham, NC

L.M. Berliner, National Center for Atmospheric Research, Boulder, CO

Donald A. Berry, Duke University, Durham, NC

N.G. Best, Imperial College, London

Frédéric Y. Bois, Lawrence Berkeley National Laboratory, Berkeley, CA

Eric T. Bradlow, Wharton School, Philadelphia, PA

Bradley P. Carlin, University of Minnesota, Minneapolis, MN

Larry C. Clark, Arizona Cancer Center, Tucson, AZ

Linda Clark, Bell Labs, Murray Hill, NJ

William Cleveland, Bell Labs, Murray Hill, NJ

Merlise Clyde, Duke University, Durham, NC

Marie Davidian, North Carolina State University, Raleigh, NC

Lorraine Denby, Bell Labs, Murray Hill, NJ

Owen Devine, Centers for Disease Control and Prevention, Atlanta, GA

Christopher Genovese, Carnegie Mellon University, Pittsburgh, PA

Samuel Greenhouse, George Washington University, Washington, DC

Edwin S. Iversen, Jr., Duke University, Durham, NC

Kirthi Kalyanam, Santa Clara University, Santa Clara, CA

Larry Kessler, FDA/CDRH, Rockville, MD

Changying Liu, University of Michigan, Ann Arbor, MI

Chuauhai Liu, Bell Labs, Murray Hill, NJ

R. Milliff, National Center for Atmospheric Research, Boulder, CO

James Mulholland, Georgia Institute of Technology, Atlanta, GA

Peter Müller, Duke University, Durham, NC

Susan Paddock, Duke University, Durham, NC

J. Lynn Palmer, M.D. Anderson Cancer Center, Durham, NC

Giovanni Parmigiani, Duke University, Durham, NC

Amy Racine-Poon, Novartis, Basel, Switzerland

Jonathan Raz, University of Michigan, Ann Arbor, MI

Peter E. Rossi, University of Chicago, Chicago, IL

J.A. Royle, National Center for Atmospheric Research, Boulder, CO

Bruno Sanso, Universidad Simon Bolivar

Abhinanda Sarkar, M.I.T., Cambridge, MA

Joellen M. Schildkraut, Duke University, Durham, NC

Nozer Singpurwalla, George Washington University, Washington, DC

Steven J. Skates, Massachusetts General Hospital, Boston, MA

Elizabeth H. Slate, Cornell University, Ithaca, NY

John A. Sweeney, Western Psychiatric Institute and Clinic, Pittsburgh, PA

Paige Tolbert, Emory University, Atlanta, GA

Jon Wakefield, Imperial College School of Medicine, London, UK

Mike West, Duke University, Durham, NC

C.K. Wikle, National Center for Atmospheric Research, Boulder, CO

Eric P. Winer, Dana Farber Cancer Institute, Boston, MA

Ying Nian Wu, University of Michigan, Ann Arbor, MI

Hong Xia, Minneapolis Medical Research Foundation, Minneapolis, MN

Stan Young, Glaxo Wellcome Inc., Research Triangle Park, NC

INVITED PAPERS

Modeling Customer Survey Data

Linda A. Clark
William S. Cleveland
Lorraine Denby
Chuanhai Liu

ABSTRACT In customer value analysis (CVA), a company conducts sample surveys of its customers and of its competitors' customers to determine the relative performance of the company on many attributes ranging from product quality and technology to pricing and sales support. The data discussed in this paper are from a quarterly survey run at Lucent Technologies.

We have built a Bayesian model for the data that is partly hierarchical and has a time series component. By "model" we mean the full specification of information that allows the computation of posterior distributions of the data — sharp specifications such as independent errors with normal distributions and diffuse specifications such as probability distributions on parameters arising from sharp specifications. The model includes the following: (1) survey respondent effects are modeled by random location and scale effects, a t-distribution for the location and a Weibull distribution for the scale; (2) company effects for each attribute through time are modeled by integrated sum-difference processes; (3) error effects are modeled by a normal distribution whose variance depends on the attribute; in the model, the errors are multiplied by the respondent scale effects.

The model is the first full description of CVA data; it provides both a characterization of the performance of the specific companies in the survey as well as a mechanism for studying some of the basic notions of CVA theory.

Building the model and using it to form conclusions about CVA, stimulated work on statistical theory, models, and methods: (1) a *Bayesian theory of data exploration* that provides an overall guide for methods used to explore data for the purpose of making decisions about model specifications; (2) an approach to *modeling random location and scale effects* in the presence of explanatory variables; (3) a reformulation of integrated moving-average processes into *integrated sum-difference models*, which enhances interpretation, model building, and computation of posterior distributions; (4) *post-posterior modeling* to combine certain specific exogenous information — information from sources outside of the data — with the information in a posterior distribution that does not incorporate the exogenous information; (5) *trellis display*, a framework for the display of multivariable data.

1 Introduction

1.1 CVA

In customer value analysis (CVA), a company conducts sample surveys of its customers and of its competitors' customers to determine the relative performance of the company on many *attributes* ranging from product quality and technology to pricing and sales support (Gale, 1994; Naumann and Kordupleski, 1995). The goal is to use CVA to look across the business to prioritize where resources would best be committed to improve performance. The basic premise of CVA is that the market share of a company is directly linked to the company's performance on the polled attributes.

The concept of CVA goes back to the 1970s and is often attributed to Sidney Schoeffler at General Electric. But the practice of CVA by major corporations is relatively new and is still in the stage of gaining acceptance. It represents a broadening of the quality movement which for the last few decades has focused largely on process improvement to meet engineering specifications. CVA adds information about customers' opinions so that in addition to quality as conformance to specifications, there is customer-perceived quality. And CVA introduces the practice of measuring competitors' customers.

1.2 Data

The data discussed in this paper are from a quarterly survey of supplier companies in the information technology industry. Customers are polled and rate companies on attributes on a scale of 1 to 10. The data of the paper cover 9 attributes and there are 3505 pollings. If there were no missing data we would have 3505 × 9 = 31545 ratings. But because of missing data there are 25309 ratings. In all, 19 suppliers are rated over a period of 9 quarters. Most respondents are polled just once but a small fraction are polled more than once.

These data are a subset of a much larger data set; 7 of the 9 attributes studied here have sub-attributes, questions that clarify issues of the main attribute. In the interest of space and readability we study only the 9 main attributes here, rather than the full set of 45 attributes. Also, not all of the companies in the survey are included here; for example, Lucent Technologies is not included. Nor are all of the currently available quarters included.

The information in our survey is highly sensitive; individual companies compete, but they also form many partnerships — for example, partnerships to develop and market certain product lines, and industry-wide partnerships to establish standards. For this reason we must regard the company names as confidential. We will rename the companies with three letter words such as "Cat" and "Jet".

1.3 The Two Goals of the Data Analysis

Ambitious surveys of customers such as ours are instituted to characterize the relative performance of companies on the polled attributes; this provides a basis for actions to improve performance from a customer perspective. Thus one goal of our data analysis is to provide a rigorous mechanism for going from the survey data to characterizations of performance. The mechanism involves the combination of information from the data with the basic tenets of customer-value-analysis theory. The tenets amount to premises about the perceptions of customers on a variety of attributes and how these perceptions are structured. Gale (1994) and Naumann and Kordupleski (1995) present the current status of CVA theory that is adopted by most practitioners. A second goal of our analysis is to study CVA theory both to test old tenets and to develop new ones. Thus the outcome of our analysis is (1) characterizations of relative performance and (2) statements about CVA theory. While most of this paper is devoted to model development, estimation, and computation, we briefly illustrate the two-part outcome of our work at the end of the paper.

2 Overviews

We cannot in this one paper describe in full detail our analysis of the survey data. In this section we will summarize (1) the contents of papers on theory, models, and methods that were stimulated by work on the CVA data; (2) the contents of this paper; (3) aspects of the data analysis not conveyed here.

By the word "model" we mean the full specification of information that allows the computation of posterior distributions of the data — sharp specifications such as independent errors with normal distributions and diffuse specifications such as probability distributions on parameters arising from sharp specifications.

By the phrase "exogenous information" we mean information about the subject under study from sources outside of the data. Some might use the term "prior information" but we do not simply to avoid the unfortunate connotation that prior information is formally specified prior to the exploration and study of the data.

2.1 Theory, Models, and Methods

Philosophies of Model Building

We did not simply write down a model for the customer data and then carry out diagnostic checks based on a complete model to see if it appeared to be reasonable. The sources of variability in our data are too varied and complex to allow such an approach to succeed. Instead, we carried out a process of building up a model by a detailed exploration of the structure of the data, starting with a minimum of specifications, and ending with a complete model. At each stage of our model building there were specifications in place that guided the exploration, and at the end there was exploration based on a tentative complete model.

Since data visualization methods are the single most powerful approach to studying the structure of data, visualization played a fundamental role in the model building.

The origins of this philosophy of searching the data for structure as a part of model building go back to the emerging ideas on diagnostic checking of models (e.g., Anscombe and Tukey, 1961; Box and Hunter, 1965; Daniel and Wood, 1971).

But recent evolutions of this philosophy have had significant changes (Cleveland, 1993). In recent decades computing technology has changed so profoundly that it calls for revised fundamentals for approaching data. There has been an enormous increase in the power of visualization tools for exploration. This has led to a significant increase in the potential reliance we can place on visualization as a basis of modeling decisions. With this increase has been a corresponding decrease in the amount of reliance that we need to put on using more formal tests of sharp hypotheses. Rather it means that the myriad decisions that must be made in building a model for data are often made more judiciously and expeditiously via data exploration and less formal methods. There has been an enormous increase in the speed of numerical computations and in the ease and flexibility of specifying and fitting models. Consequently, data exploration can be more than simply a look at the raw data. It can involve exploration based on many tentative structures for the data, some partial and some complete.

A Theory of Data Exploration for Model Building

One result that we hope will excite some is that it is possible to see model building via data exploration quite explicitly as part of the overall Bayesian paradigm.

Cleveland and Liu (1998a) have developed a theory of data exploration that invokes Bayesian principles for its rationale. The theory begins with a number of premises about data analysis in practice, both what is actually done and what is desirable. Many of these premises have been expressed in a number of quite disparate works, both Bayesian and frequentist (Box, 1980; Draper, 1970; Dempster, 1970; Draper, Hodges, Mallows, and Pregibon, 1993; Edwards, Lindman, and Savage, 1963; Gelman, Meng, and Stern, 1996; Good, 1957; Hill, 1986; Hill, 1990; Kass and Raftery, 1995; Rubin, 1984; Savage, 1961). From these premises comes a key concept of the theory: exploration methods allow us to reason about the likelihood of the data given specifications. We combine our judgment of likelihood based on the data exploration with our judgment of likelihood based on exogenous information to form a new likelihood that is the basis for decisions about specifications. The combination is analogous to the use of Bayes theorem when we have exogenous information and likelihood described mathematically. But, for data exploration, in place of the combining by mathematical computation, we use a process of direct, intuitive reasoning. While we lose the precision of mathematically described inductive inferences, we gain an enormous breadth of coverage of competing specifications that is simply not feasible to capture mathematically.

Casting data exploration as a vehicle for judging the likelihood of the data given specifications has important implications for how we carry out exploration. Here are four of many: (1) The theory encourages accompanying exploratory methods with assessments of the variability of the displayed functions of the data, either by simulating the displays with data generated from credible specifications, or by simple mathematical probabilistic calculations based on these specifications. This may seem like a break with data exploration as it is practiced by many, because we invoke notions of probability, but we believe it is only a break with how data exploration is often portrayed. (2) The theory encourages the use of visualization methods to explore the data because this often provides powerful assessments of likelihood. Visualization methods can reveal much information in the data and thus allow us to judge the likelihood of the data given a wide variety of specifications seen to be credible based on exogenous information. (3) The theory discourages the use of test statistics that reduce the vast information in the data to a small amount of information because this provides exceedingly limited assessments of likelihood. For example, the theory explains why a normal probability plot is a far more powerful tool for deciding on a sharp specification of normality than a chi-squared test (even a chi-squared test calibrated by a posterior predictive distribution). (4) The theory encourages the fitting of many alternative models, some to the full data, and others to functions of the data, as a vehicle for assessing likelihood.

Building Models with Random Locations and Scales

Many sets of data consist of cases, each of which has an associated set of measurements of a response variable. The response depends on explanatory variables, and the goal is to describe the dependence, but the cases have an effect as well, which complicates the study of dependence.

For our survey data, the response is the ratings, the cases are the respondents, and the explanatory variables are attribute, company, and time.

It is common to model a case effect by a random location effect. But for many case effects, scales vary as well. For example, it is widely acknowledged that for rater data, respondents have varying locations and scales. But it is uncommon to model such scale effects by random effects, in part because model building, model fitting, and diagnostic checking are complex. Often, the issue is ignored, or effects are estimated as fixed which uses up too many degrees of freedom unless each rater does ratings of many cases (Longford, 1995).

Cleveland, Denby, and Liu (1998) have developed an approach to modeling random location and scale effects in the presence of explanatory variables. An overall framework is posited that can be verified by diagnostic checking. Within this framework, they have developed procedures for exploring the data to get insight into the form of the location, scale, and error distributions as well as the dependence of the response on the explanatory variables. The problem is that the functions of the data used in the exploration convolve the population location and scale distributions with error distributions and with sampling distributions. So their methods, which have an empirical Bayes flavor, amount to deconvolution procedures.

Integrated Sum-Difference Models

We employ an integrated moving-average time series model (Box and Jenkins, 1970) to describe changes through time in company-attribute effects. To do this we developed a class of models — integrated sum-difference models, or ISD(d,q) — which are IMA(d,q) models but are structured differently; the new structure enhances interpretation, model building, and computation of posterior distributions (Cleveland and Liu, 1998b). The new structure models a differenced series by a sum of orthogonal series, each the output of applying, to white noise, filters that are made up of products of powers of first-difference filters and first-sum filters.

Post-Posterior Exploration, Analysis, and Modeling

The posterior distribution of a complicated model such as ours is a complex numerical object. It is in many ways like the data themselves. Just as with data, it can require extensive exploratory study, for example by visualization methods, to comprehend its structure. We can find ourselves, simply as an exploratory device, fitting simplified mathematical functions to the posterior information as a way of summarizing the information. Even more, it can be judicious to impose even further model specifications, using exogenous information not incorporated at the outset and combining the posterior and exogenous information directly rather than returning to the original model specification. Cleveland and Liu (1998a) discuss in detail these issues of *post-posterior exploration, analysis, and modeling*.

Trellis Display

Our model building, our model diagnostics, and our post-posterior exploration, analysis, and modeling depend heavily on trellis display, a framework for the visualization of multivariable data (Becker, Cleveland, and Shyu, 1996; Becker and Cleveland, 1996). Its most prominent aspect is an overall visual design, reminiscent of a garden trelliswork, in which panels are laid out into rows, columns, and pages. On each panel of the trellis, a subset of the data is graphed by a display method such as a scatterplot, curve plot, boxplot, 3-D wireframe, normal quantile plot, or dot plot. Each panel shows the relationship of certain variables conditional on the values of other variables.

2.2 *Contents of the Paper*

Section 3 describes the data. Section 4 attacks the scale of measurement of the data. The respondent ratings are integers from 1 to 10. We had to make an important decision: build a discrete model whose rating scale allows only integer values from 1 to 10, or build a model whose ratings emanate from an underlying continuous scale that is rounded to discrete values by the measurement process. Section 4 describes the decision and the basis for it.

Section 5 briefly describes the model building process. The result of the process is a Bayesian model: sharp specifications and prior distributions on parameters arising from the sharp specifications. The model is partly hierarchical and has a time series component. The model specifications include the following: (1) Survey respondent effects are modeled by random location and scale effects, a t-distribution for the location and a Weibull distribution for the scale. (2) Company effects for each attribute through time are modeled by integrated sum-difference processes. (3) Error effects are modeled by a normal distribution whose variance depends on the attribute; in the model, the errors are multiplied by the respondent scale effects. Section 6 describes the full model.

Earlier we stated two goals in analyzing the data: (1) characterizing the relative performances of the companies on the attributes, and (2) studying the basic tenets of CVA. In Section 7 we briefly describe how information in the posterior distribution is used to achieve these two goals.

2.3 *What is Not in the Paper*

Our actual model building was a highly iterative process in which we accumulated specifications for a complete model, fitted the complete model, carried out a variety of diagnostic checks (including posterior predictive checks), found inadequacies, altered the model, carried out more diagnostics, and so forth. The actual process was far too complicated to remember let alone recount. But by the end of our process we had discovered a logical path of accumulating assumptions that serves as one form of validation of our model; as we move along the path, we are able to use previous assumptions together with exploration of the data and our exogenous knowledge to form additional assumptions. The logical path, a series of steps, does begin with a specification of overall structure, but this was checked by a series of diagnostics. Section 5 describes the logical path rather than the actual model building process.

We carried out posterior predictive checking (Rubin, 1984; Gelman, Meng, and Stern, 1996). Because of the extensive model building process we did not need to rely fully on such checking to justify our model, but we did generate data from the posterior predictive distribution and plot it in a variety of ways. This is not discussed in the paper.

Computation of the posterior was a challenging matter because of the complexity of the model. We report briefly on computational matters in Section 6 but far more care than described there needed to be given. For example, convergence was a matter that needed much attention but is not discussed here.

3 The Data Studied Here

Our survey is administered on a quarterly basis. For the data studied here, 19 supplier companies are rated over a period of 9 quarters. Altogether there are 3505 rating sessions, or pollings; in each polling, a person rates one company on 9 attributes on a scale of 1 to 10 where 1 = poor and 10 = excellent.

A small number of people participated more than once. There are 3385 people. 3271 people were polled once, 106 were polled twice, and 6 were polled three times, which results in

$$3505 = 3271 + 2 \times 106 + 3 \times 6$$

pollings.

The 9 polled attributes cover quality, technology, price, and the interaction between the customer and the company. The attributes and their abbreviated names are the following:

prod-qual product quality

over-qual overall quality of company processes

delivery delivery of the product

cost cost of the product to the customer and the process of working with the customer to establish the cost

features technological excellence of the product

pre-sup providing product information before delivery to support customer processes

response responsiveness of the company to customer needs

value value of the product relative to the cost of the product

service service provided by the sales team

The survey is designed with a skip pattern that results in no respondent rating all nine attributes; only purchasing agents rate `delivery` and only product and process designers rate `pre-sup`. The remaining 7 attributes are intended to be rated by all respondents, but some respondents choose not to rate all attributes about which they are asked. Table 1 shows counts of the number of attributes rated in the 3505 pollings:

Number of Attributes Rated	1	2	3	4	5	6	7	8
Number of Pollings	5	9	26	64	150	350	1106	1795

TABLE 1. Counts of Pollings According to Number of Attributes Rated

The number of respondents selected for polling for each company in the survey is determined by a number of factors but a chief one is the market share. Companies with a larger share have a larger number of customers.

Table 2 shows the number of ratings for all combinations of companies and attributes. The rows are ordered so that the totals for the companies increase from top to bottom. The columns are ordered so that the totals for the attributes increase from left to right.

	deli	pre-	prod	serv	cost	resp	valu	feat	over
Nut	15	15	25	29	28	32	33	33	32
Gas	17	25	26	34	40	43	42	44	46
Ham	13	35	38	43	42	47	46	50	50
Pub	32	32	56	59	61	67	68	70	71
Key	35	41	49	70	72	76	77	77	76
Cab	25	74	65	85	88	104	107	113	111
Ear	28	74	75	92	94	102	106	106	107
Bee	38	76	87	114	119	121	128	131	132
Mug	58	71	106	116	120	132	136	138	137
Jet	26	128	102	138	147	153	158	162	162
Rug	54	100	124	148	163	169	174	167	174
Log	63	97	121	156	165	167	170	172	171
Toy	47	110	132	145	152	170	177	178	181
Oak	84	110	169	174	197	205	218	219	216
Ace	77	174	186	254	240	270	267	273	275
Inn	83	187	180	255	260	283	287	287	285
Ski	112	189	250	282	300	321	334	328	335
Duo	121	292	309	408	389	420	420	431	430
Fan	132	303	320	400	403	431	455	465	463

TABLE 2. Counts of Pollings According to Number of Questions Answered

4 Modeling: The Measurement Scale

The survey response scale is an integer scale from 1 to 10. We must make a critical decision here, at the onset, that will set the strategy of the ensuing model development. Should we build an ordinal-scale model that allows only integer values from 1 to 10, or should we suppose respondents have an underlying continuous scale that is rounded to an integer by the measurement process?

With a continuous underlying scale, we could divide the measurement scale into disjoint rounding intervals and estimate the interval endpoints (Albert and Chib (1996), Bradlow and Zaslavsky (1997), Johnson (1997)). Or, we could ignore the rounding with a supposition that a continuous scale is a good approximation to the measurement scale. And, in the continuous case, we would search for a transformation of the data to make the model simpler.

For a continuous model it is likely that the analysis would be simpler and results would be easier to comprehend. The rating scale would then, implicitly, be given a stronger metric interpretation. Model development, model diagnostics, and model interpretation would focus on a comparison of the relative values of ratings on a continuous scale, rather than focusing on relative distributions of probabilities of ratings on an ordinal scale, which would likely lead to more easily fathomed results. And computations for the continuous scale would likely be simpler, or at the very least, involve better understood methods.

Figures 1 to 3 are a trellis display of the ratings. Each panel has a histogram of all ratings of one attribute for one supplier company; the company and attribute labels are in the strip labels at the tops of the panels. Each block of panels has the 19 company histograms for one attribute; there are three blocks on each page.

The histograms show substantial regularity in the ratings. The mode is typically at 7 or 8. Because the scale ranges from 1 to 10, more of the scale is available below the mode than above. This results in skewness to the left. But there does not appear to be an undue build-up at the high end of the scale, in particular, at 10.

The regularity of the ratings and the lack of build-up at either end of the scale are a positive signal that a continuous model is at least worth attempting. The regularity and lack of build-up also suggest that we can take the simple route of supposing that each response k arises from rounding a continuous value in an interval extending from $k - 0.5$ to $k + .5$. The rounding process has a variance of 1/12, exceedingly small compared with the variability that we observe in the histograms, so we will attempt a first model that treats the ratings as if they were continuous variables from 1 to 10 and ignore the rounding. But this approach would rest on a more solid foundation if the scale of measurement resulted in symmetric distributions rather than skewed.

Next, we attack the skewness problem. The histograms of Figures 1 to 3 provide less incisive assessments of distributional shapes and we move to normal quantile plots in Figure 4; in the interest of space we show the plots only for the `prod-qual` attribute, but the shapes for other attributes are quite similar. As with all of the quantile plots to come, we use the following procedure. Suppose the data for a particular quantile plot are x_i for $i = 1$ to n. Suppose F is a theoretical distribution. To check whether the empirical distribution of the data is well approximated by F we plot $x_{(i)}$, the i-th order statistic, against $F^{-1}((i-0.5)/n)$.

The skewness of the data is apparent in Figure 4 but the discontinuity of the patterns arising from the discreteness of the data does interfere with our ability to visually judge the adequacy of an underlying continuous distribution. To assist our visual processes we will smooth the distribution function of the data by the following procedure and then display the smoothed data quantiles. Consider the ratings of one attribute for one company. Suppose, for $k = 1$ to 10, that there are $n(k)$ values equal to k, then we replace the $n(k)$ values of k by $k - 0.5 + (i - 0.5)/n(k)$ for $i = 1$ to $n(k)$. The result is shown in Figure 5. In addition, each panel has lines on it. The oblique line is drawn through two points: the lower

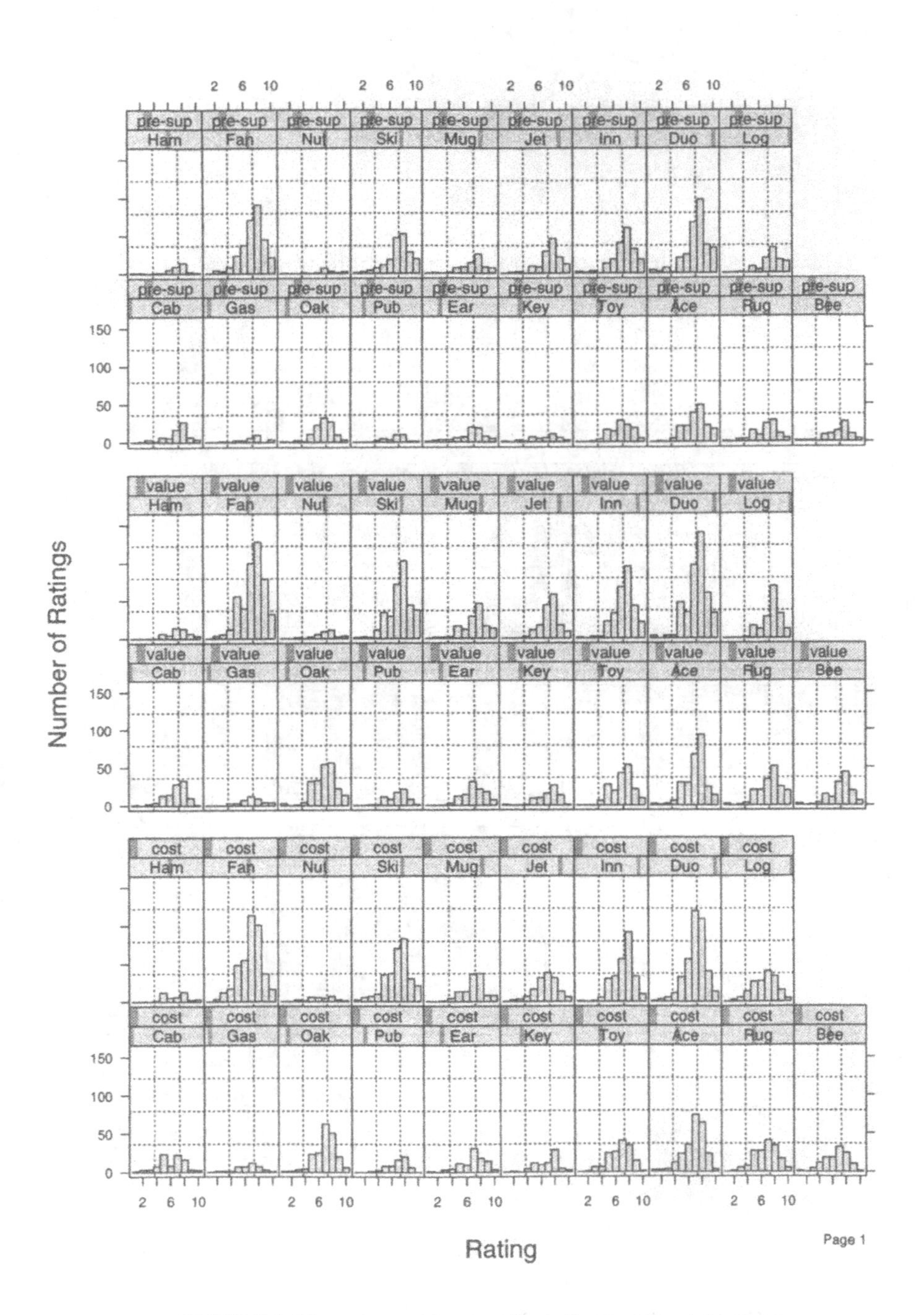

FIGURE 1. Histograms of Rating Given Company and Attribute

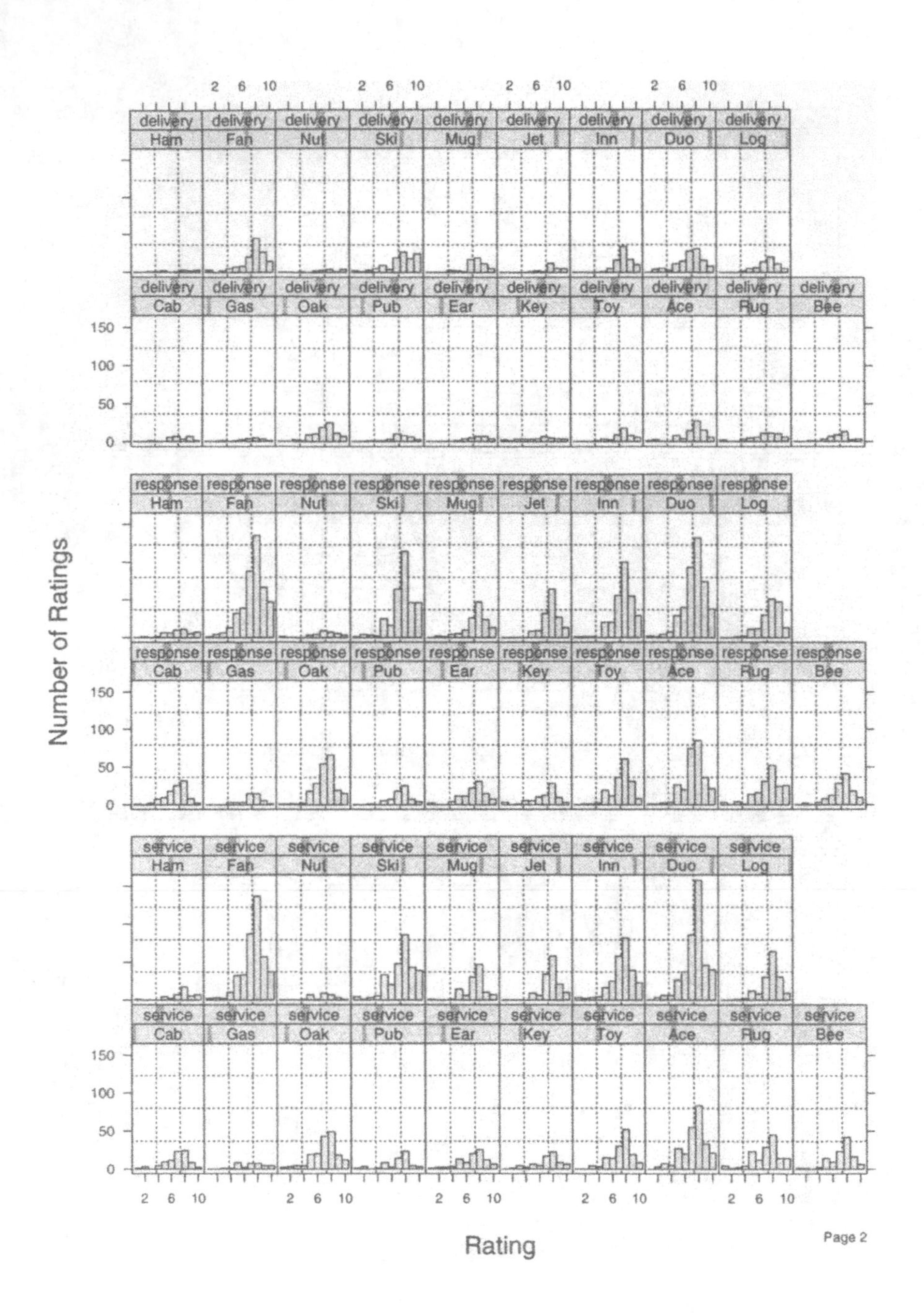

FIGURE 2. Histograms of Rating Given Company and Attribute

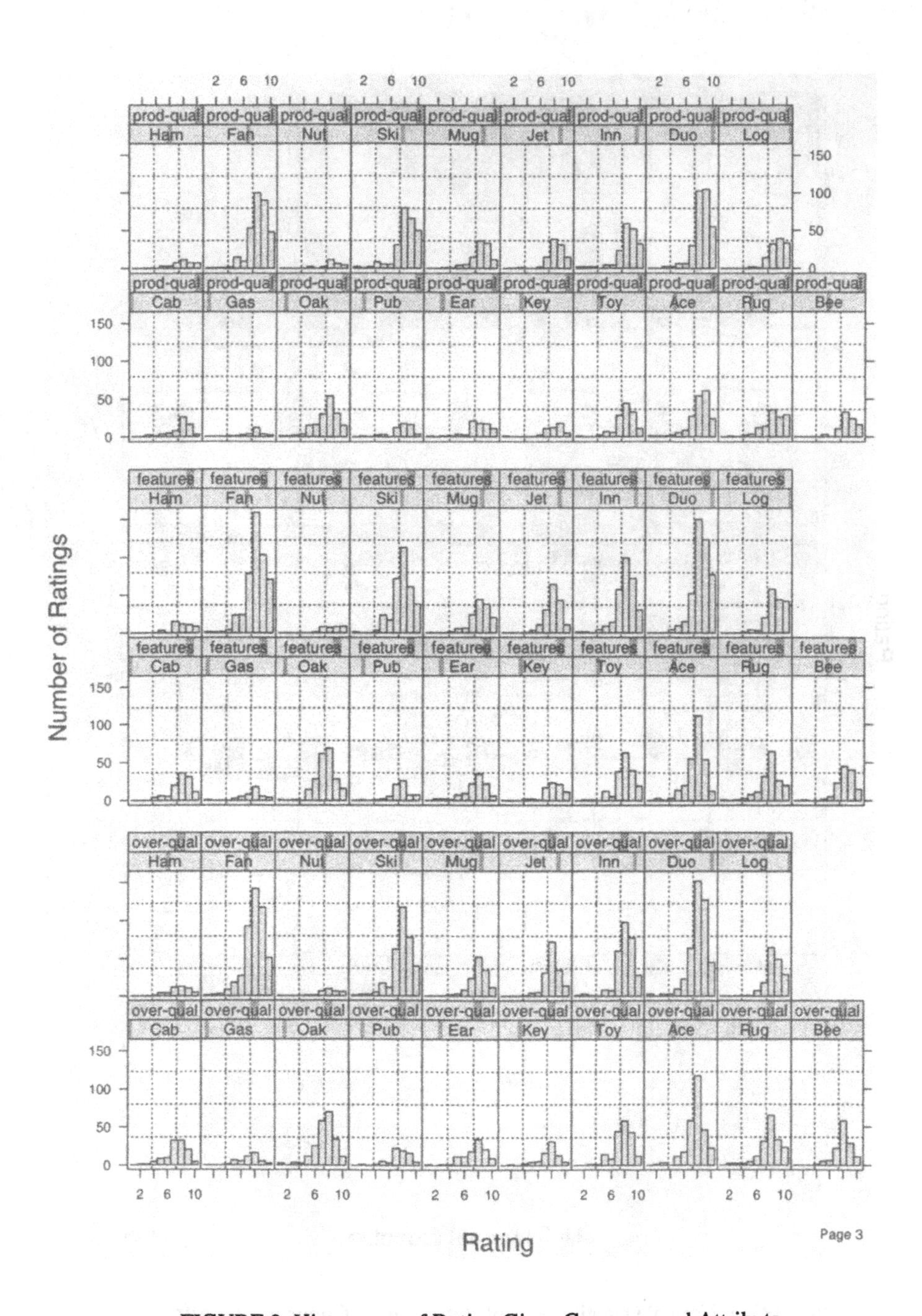

FIGURE 3. Histograms of Rating Given Company and Attribute

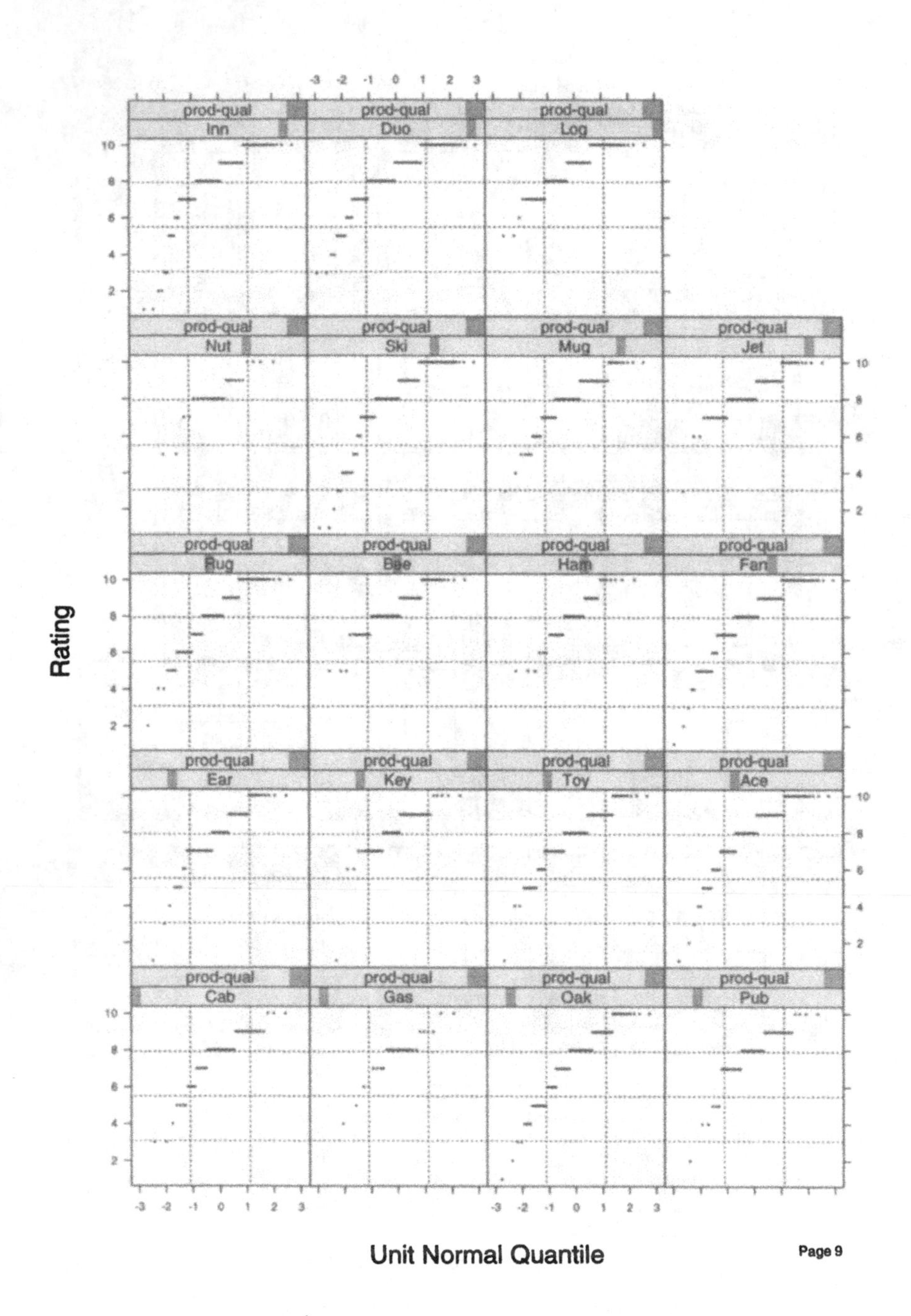

FIGURE 4. Normal Quantile Plots of Rating Given Company and Attribute

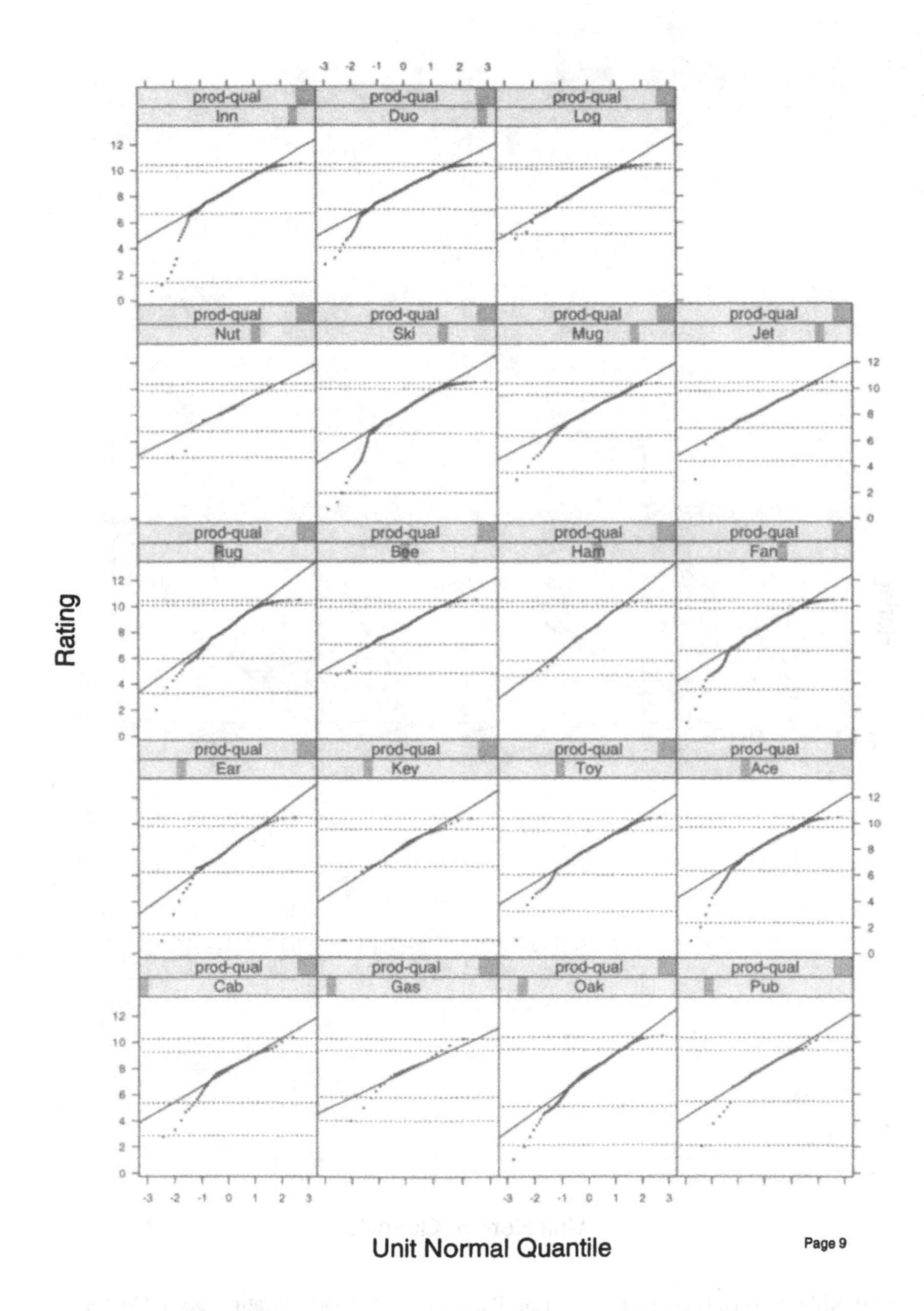

FIGURE 5. Normal Quantile Plots of Smoothed Rating Given Company and Attribute

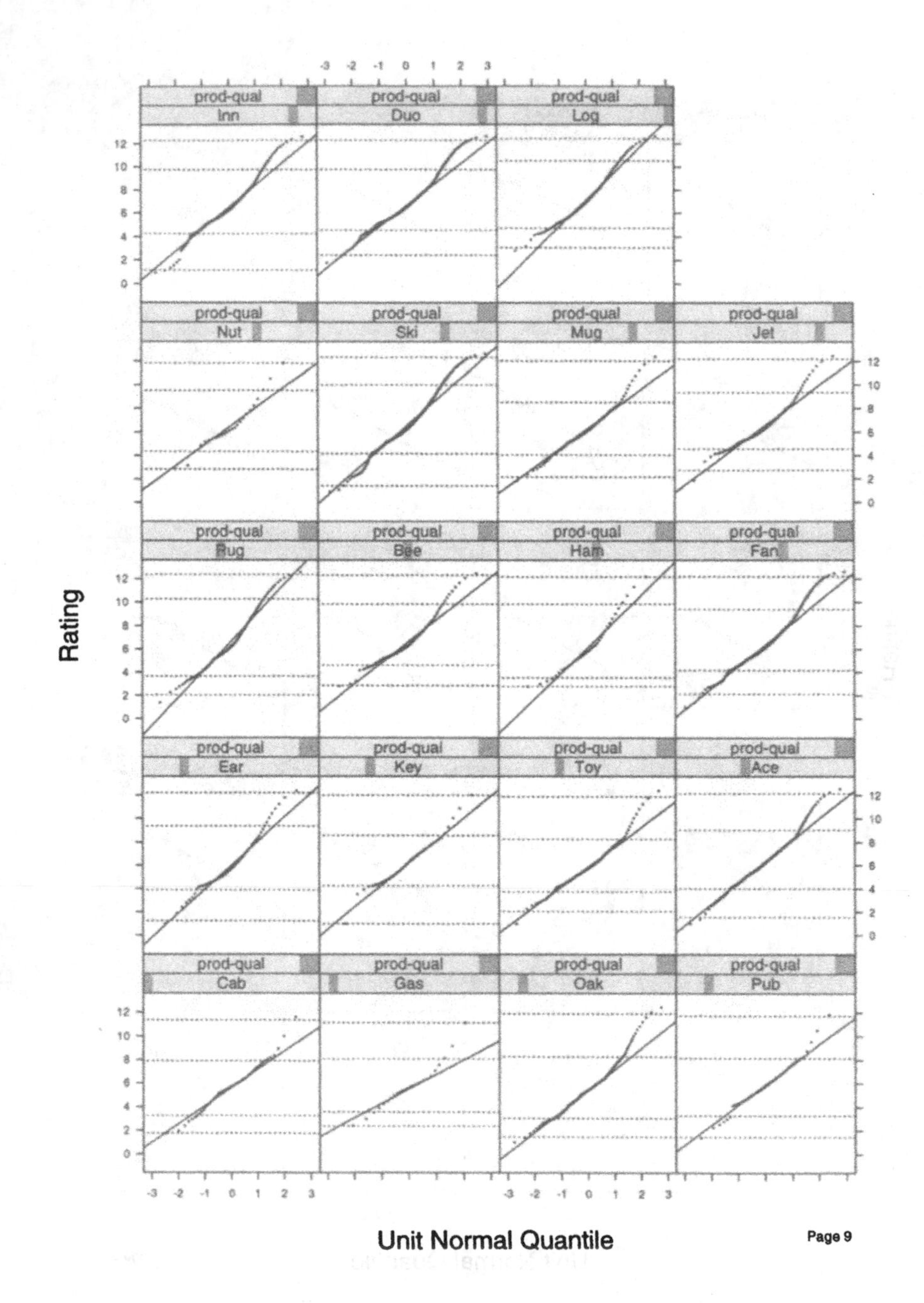

FIGURE 6. Normal Quantile Plots of Log Transformed Smoothed Rating Given Company and Attribute

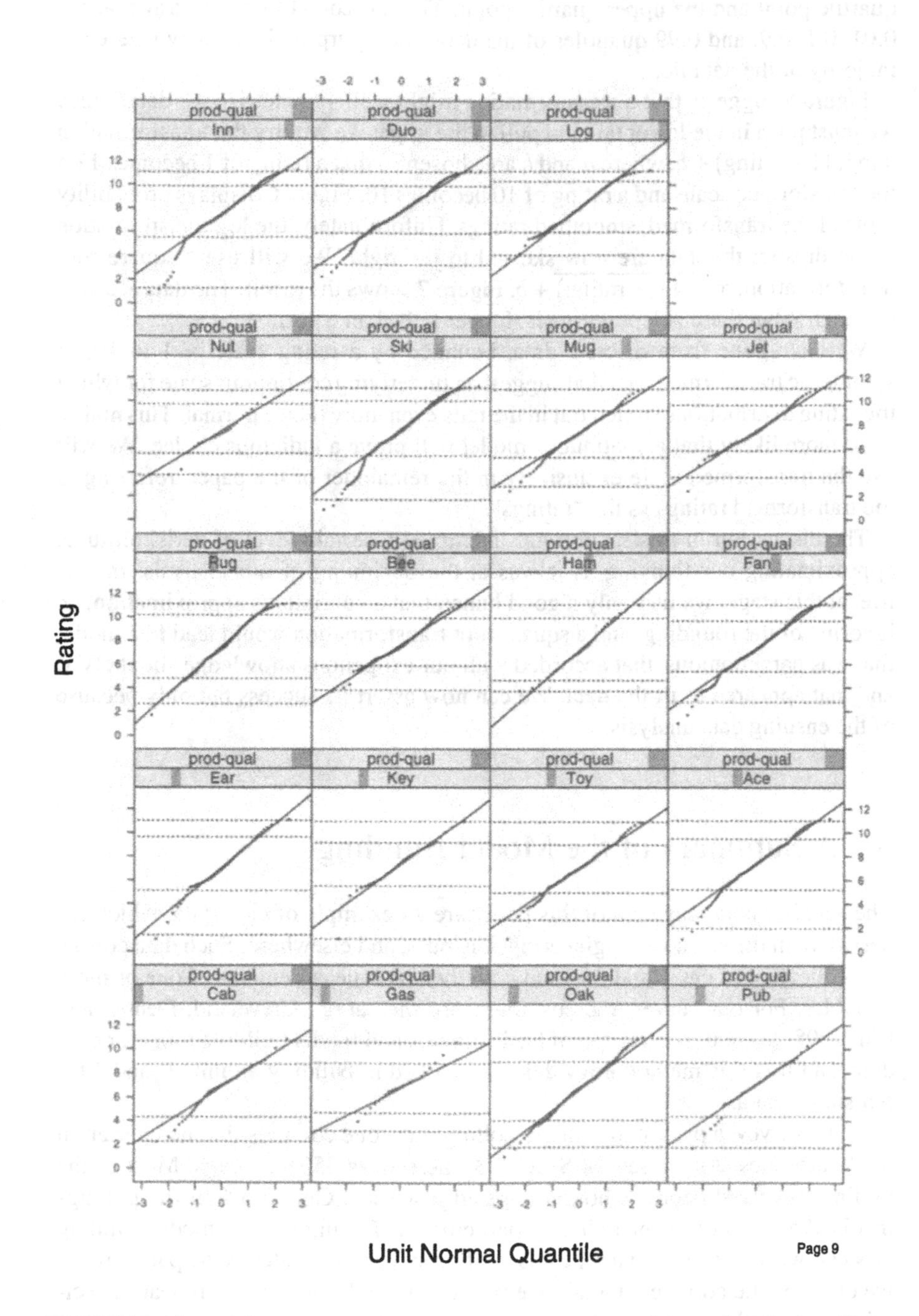

FIGURE 7. Normal Quantile Plots of Square-Root Transformed Smoothed Rating Given Company and Attribute

quartile point and the upper quartile point. The horizontal lines are drawn at the 0.01, 0.1, 0.9, and 0.99 quantiles of the data; their purpose is to show where the majority of the data lie.

Figure 5 suggests that a transformation might well symmetrize the data. Since we must push in the lower tail and pull in the upper we will try the transformation $a \log(11 - \text{rating}) + b$, where a and b are chosen so that a rating of 1 becomes 1 on the transformed scale and a rating of 10 becomes 10. Figure 6 displays probability plots of the transformed, smoothed ratings. Unfortunately, the log transformation is too drastic; the data are now skewed to the right. We will use a square root transformation, $a\sqrt{(11 - \text{rating})} + b$. Figure 7 shows the result. The data are now symmetric but there is leptokurtosis: longer tails than a normal.

We have gone from discrete data, bounded by a rating scale of 1 to 10, to symmetric transformed data that suggest an underlying continuous scale for which the rating distributions stretch out in the tails even more than a normal. This makes it far more likely that a continuous model will prove a judicious choice. We will use the transformed scale exclusively in the remainder of the paper, referring to the transformed ratings as the "ratings".

The discussion in this section has attempted a certain level of verisimilitude, approximating our thinking as it was at the beginning of our analysis. In real life, at this stage, we had only a good hunch that a continuous approximation, an ignoring of the rounding, and a square root transformation would lead to a model that was parsimonious, that accorded with our exogenous knowledge about CVA, and that appeared to fit the data. We can now assert its success but only because of the ensuing data analysis.

5 A Summary of the Model Building

The supplier company data of this paper are an example of case data, which are pervasive in the sciences, engineering, business, and elsewhere. Such data consist of a collection of cases and for each case there are measurements of one or more variables. For our survey, the customers are the cases. Cleveland, Denby, and Liu (1998) present a collection of basic models and model building tools for case data, and it is this methodology that we invoked in building an initial model for our survey data.

In the survey, a person in a single polling rates one company in one quarter on the 9 attributes. As we saw in Section 3, there were 3505 pollings. Most of the pollings involved people who participated just once. Only 3.3% of the pollings involved a person who had already participated. To simplify the model building process, we supposed that a repeat is a different person. (Later in the paper, when we describe the complete model, we will we will take account of repeats, specifying them in the model from exogenous information.)

Step 1: Overall Structure to Guide the Data Exploration

We entertained an initial structure on which to base our exploration of the data:

$$r_{ac} = \theta_{asq} + \mu_c + \alpha_c \epsilon_{ac}$$

where r_{ac} is the rating (on the transformed scale) by person (case) c on attribute a. θ_{asq} is the rating performance for supplier company s on attribute a in quarter q. (In our notation, r_{ac} does not explicitly show a dependence on s and q; because we have assumed that each polling is a separate person, s and q are determined by c.) μ_c is a person location effect, and α_c is a person scale effect. ϵ_{ac} is an error term. We allow for a dependence of the error variance on attribute by supposing $E\epsilon_{ac}^2 = \sigma_a^2$.

The equation provides a framework for data exploration that entertains as few specifications as possible. It allows us to study the few specifications that are made and to develop new ones.

Step 2: Modeling Company, Time, and Attribute Effects

We visualized two functions of the data to study specifications for θ_{asq}. The first was the sample means across c for each combination of a, s, and q. The second was the sample means across c and q for each combination of a and s. This led to certain tentative decisions about modeling the θ_{asq}.

First, for fixed a and s, the θ_{asq} were taken to be an ISD(1,1) process as a function of the time parameter q. This accords with our exogenous information that customer opinions have a substantial low-frequency component, but also allows for high-frequency quarter-to-quarter variation.

Second, while we have substantial exogenous information and information from the data about the relationships of the θ_{asq} for fixed c and q — that is, relationships among the attributes — we chose not to impose strong specifications to reflect this information to its fullest. The reason for this is discussed in Section 7.

Step 3: Adjusting for Company, Attribute, and Time Effects

To get at person effects, we adjusted the data for the company, attribute, and time effects by fitting loess curves to the ratings as a function of q for each combination of a and c, and then subtracting the fitted curve. In other words, we adjusted to remove the effect of the θ_{asq}. Because a large number of observations are pooled to estimate the θ_{asq}, we assume the estimates are the true values and, with a slight abuse of notation, will take r_{ac} to denote the adjusted data:

$$r_{ac} = \mu_c + \alpha_c \epsilon_{ac}. \tag{1}$$

Step 4: Estimating Attribute Variances

We used an iterative method based on Equation 1 to estimate the σ_a^2. Because a large number of observations are pooled to estimate the σ_a^2, we assume the estimates are the true values.

Step 5: Least Squares Fitting

For each c, we fitted Equation 1 by least squares to get estimates $\hat{\mu}_c$ and $\hat{\alpha}_c$, and residuals, $\hat{\epsilon}_{ac}$.

Step 6: Modeling the Error Distribution

Beta-quantile plots of studentized residuals, $\hat{\epsilon}_{ac}/\hat{\alpha}_c$, led to a specification of a normal distribution for ϵ_{ac}.

Step 7; Modeling the α_c Distribution

The assumption of a normal distribution for the ϵ_{ac} implies that the $\hat{\alpha}_c^2$ have a distribution that is a multiplicative convolution of the α_c^2 distribution with a $\chi^2\{d(c)\}/d(c)$ distribution where $d(c)$ is the number of ratings by person c minus 1. We were able to deconvolve using quantile plots, which led to a specification of the distribution of α_c^2 as a unit exponential; thus the distribution of α_c is a Weibull.

Step 8: Modeling the μ_c Distribution

The assumption of a normal distribution for the ϵ_{ac} and the assumption of a Weibull distribution for $\hat{\alpha}_c$ implies that the $\hat{\mu}_c$ have a distribution that is an additive convolution of the distribution of μ_c with a product of a Weibull and a normal. We were able to deconvolve through quantile plots, which led to the specification of the distribution of μ_c as a $t\{\nu\}$-distribution with $\nu = 6$ degrees of freedom.

6 A Bayesian Model

6.1 Missing-Data Mechanism and Repeat Polling

In Section 3 we discussed two issues about respondents' participation that are critical for the modeling.

First, some respondents were polled more than once, sometimes rating the same company and sometimes rating a different one. The fraction of such people here is quite small. We could simply ignore this, as we did in the model building, and treat a repeat as a new person. However, the data studied here is a subset of our currently available data, and in our complete data set we have a much greater fraction of repeats. For this reason we will incorporate repeats in our model here, but we will rely on our exploration with the larger data set and ask the reader to take on faith that our modeled repeat mechanism is in accord with the data and exogenous information.

Second, there is missing data in the sense that respondents do not answer all questions, either as a result of the survey design or because a respondent decided not to rate an attribute. In our modeling we will suppose that the absence of a rating is ignorable (Gelman, King, and Liu, 1997; Rubin, 1976, 1977). In fact, in data exploration not discussed here, we have reason to believe that the mechanism is not ignorable, but the effect appears to be quite small and no threat to the validity of our analysis.

Except for these considerations above, we follow the sharp specifications of Section 5.

6.2 Notation

Notation is somewhat more cumbersome than in Section 5 because of the small fraction of people who repeat. As before, c denotes a person (case), a respondent in the survey. k denotes the repeat; $k = 1$ is the first polling of a person, $k = 2$ is the second polling for a person who repeats, and $k = 3$ is the third polling of a person who repeats twice. (No one participated more than three times.) s denotes the supplier company and q, the quarter. But company and quarter depend on person and repeat in the sense that if we know c and k, then we know s and q. Finally, as before, a denotes the attribute.

6.3 The Overall Form of the Model

We suppose

$$r_{ack} = \theta_{asq} + \mu_c + \rho_{ck} + \alpha_{ck}\epsilon_{ack} \tag{2}$$

where r_{ack} is the rating of attribute a by person c for the kth time. θ_{asq} is the effect of company s for attribute a in quarter q. μ_c, ρ_{ck}, and α_{ck} describe the person effects as described below, and ϵ_{ack} is the error term. To make the variables well-defined we suppose

$$E(\mu_c) = E(\epsilon_{ack}) = E(\rho_{ck}) = 0,$$

and

$$E(\alpha_{ck}^2) = 1. \tag{3}$$

6.4 Person (Case) Effects and Errors

Following, largely, Section 5, we will make the following specifications for the distributions of the person effects and errors:

$$\mu_c \sim \sigma(\mu)t\{6\} \tag{4}$$

$$\rho_{ck} \sim \sigma(\rho)t\{6\} \tag{5}$$

$$\alpha_{ck}^2 \sim exp\{1\} \tag{6}$$

$$\epsilon_{ack} \sim \sigma_a(\epsilon)N\{0,1\}. \tag{7}$$

All of these random variables are independent of one another and of the θ_{asq}.

μ_c and ρ_{ck} are the person location effects. If $\sigma(\rho)^2 = 0$, then repeat locations for a person are the same. If $\sigma(\mu)^2 = 0$, then repeat locations for a person are distributed like those of different people. If both variances are positive then the location values for a person who repeats are not equal but tend to be nearer in value then those for different people. α_{ck} describe the person scale effects. Unlike for the location effects, repeat scales for a person are distributed like those of different people. ϵ_{ack} are the error variables.

6.5 Company, Attribute, Time Effects

For fixed a and c we suppose that for $q = 1$ to 9, the time series θ_{asq} is an ISD(1,1) process:

$$\theta_{asq} = \kappa_{asq} + \iota_{asq} \tag{8}$$

$$\kappa_{asq} - \kappa_{as(q-1)} = \lambda_{asq} + \lambda_{as(q-1)}, \tag{9}$$

where the λ_{asq} and ι_{asq} are independent of one another, and each is a gaussian white noise series with variances $\sigma(\kappa)$ and $\sigma(\iota)$. This ISD(1,1) process is also an IMA(1,1). Formulating the structure as an ISD leads to a more fathomable parameterization that allows better incorporation of exogenous information and more efficient computational methods (Cleveland and Liu, 1998b).

In Section 7 we give reasons for not wanting to impose strong restrictions on the θ_{asq} across a for fixed s and q even though there is a body of exogenous information that we might invoke to do so. However, we will place some mild impositions on the values. The reason is this: some companies have a very small number of observations compared with others. Without at least mild imposition the result is a posterior distribution with mass for company-attribute effects for these small-sample companies that appears unrealistic. When a posterior distribution appears counter-intuitive it is typically an indication that we did not properly assess our exogenous information. Thus we are prepared to impose conditions that rein in the errant values, a hierarchical distribution on the 19×9 values of κ_{as1} for all combinations of a and s. We take them to be independent with

$$\kappa_{as1} \sim N\{\tau_a, 1\} \tag{10}$$

where the 9 τ_a have improper uniform distributions.

6.6 Standard Deviations

The above model specifications have given rise to several variances: $\sigma_a^2(\epsilon)$, $\sigma^2(\mu)$, $\sigma^2(\rho)$, $\sigma^2(\kappa)$, and, $\sigma^2(\iota)$. We wish to place little restriction on the these parameters based on our exogenous information; we specify distributions for them by taking their inverses to be unit exponentials.

6.7 *Posterior Computation*

Advances in computational methods have made it possible to simulate posterior distributions of complex models such as ours using Markov chain Monte Carlo (MCMC). Among the large number of MCMC-related papers are Metropolis and Ulam (1949), Metropolis *et al.* (1953), Hastings (1970), Geman and Geman (1984), Tanner and Wong (1987), and Gelfand and Smith (1990), which are key references that stimulated much work in this research area. The Gibbs sampler and the Metropolis-Hastings algorithm are used to simulate the posterior distribution of the variables in our model.

To draw the person location parameters, μ_c and ρ_{ck}, we use the conventional hierarchical representation of the t-distribution using the *normal/independent* distribution with gamma distributed *weights*. Detailed implementation is straightforward and relevant references appear in various places (see Liu (1995, 1996) and Pinheiro, Liu, and Wu (1997) for recent examples).

To draw person scale parameters, α_{ck}, we use the Metropolis-Hastings algorithm. The conditional distribution of ratings given the θ_{asq}, μ_c, and ρ_{ck} is a mixture of normals with mean zero where the inverse variance is mixed by an exponential, that is, a gamma. Because it is computationally easy to simulate these variables when the gamma is replaced with an inverse gamma, we make use of this fact and then employ the Metropolis-Hastings algorithm to ensure correct MCMC. The jumping rate of the Metropolis-Hastings algorithm is greater than 80% when the degrees of freedom for the inverse gamma approximating $\chi^2_{\nu_\tau}/\nu_\tau$ is $\nu_\tau = 6$.

To draw the components in the time series structure, we use the methods proposed by Cleveland and Liu (1998b) for $\{\kappa_{asq}\}$. Taking draws of $\{\theta_{asq}\}$ requires taking draws from the normal distributions that are determined by the shrinkage between the (conditional) normal distributions for the polling ratings and the (conditional) normal distribution of $\{\kappa_{asq}\}$.

7 The Posterior Distribution

In Section 1 we set out two goals in the analysis of the survey data — (1) characterize the *relative performance* of the companies on the attributes; (2) study *CVA theory*, both to test old tenets and to develop new ones. In this section we briefly illustrate how we use information in the posterior distribution together with exogenous information not incorporated into the posterior to move toward these goals. First, we will discuss CVA theory, then, relative performance, and finally, CVA theory again.

7.1 *CVA Theory: Components of Variation*

The model equation that relates a rating to supplier company, attribute, quarter, person, and error effects, breaks the ratings into four additive components of variation:

$$r_{ack} = \theta_{asq} + \mu_c + \rho_{ck} + \alpha_{ck}\epsilon_{ack}.$$

The θ_{asq} are broken further into two additive components of variation:

$$\theta_{asq} = \kappa_{asq} + \iota_{asq}.$$

For fixed values of θ_{asq}, ratings vary because of person and error effects: (1) the person location effects, μ_c, with variance $\sigma^2(\mu)$; (2) the person location repeat effects, ρ_{ck}, with variance $\sigma^2(\rho)$; (3) the person scale effects, α_{ck}, with variance 1, times the error effects, ϵ_{ack}, with variance $\sigma_a^2(\epsilon)$. For each combination of a and s, ratings vary because of company-attribute time effects: (1) a smooth, or low-frequency, component κ_{acq} where

$$\kappa_{asq} - \kappa_{as(q-1)} = \lambda_{asq} + \lambda_{as(q-1)},$$

and λ_{asq} is a white noise series with variance $\sigma^2(\lambda)$; (2) a white noise component, ι_{asq}, with variance $\sigma^2(\iota)$.

Table 3 shows the posterior means of the variances of the components of variation. To our knowledge, this delineation and quantification of variation is the first in the area of CVA customer opinion polling, and it has important implications for CVA theory and the design of CVA surveys.

$\sigma^2(\lambda)$	0.012
$\sigma^2(\iota)$	0.024
$\sigma^2(\mu)$	0.71
$\sigma^2(\rho)$	0.71
$\sigma^2_{over-qual}(\epsilon)$	0.85
$\sigma^2_{response}(\epsilon)$	1.27
$\sigma^2_{prod-qual}(\epsilon)$	1.28
$\sigma^2_{value}(\epsilon)$	1.45
$\sigma^2_{pre-sup}(\epsilon)$	1.66
$\sigma^2_{features}(\epsilon)$	1.70
$\sigma^2_{service}(\epsilon)$	1.86
$\sigma^2_{delivery}(\epsilon)$	1.96
$\sigma^2_{cost}(\epsilon)$	2.07

TABLE 3. Posterior Means of Variances of Components of Variation

For fixed θ_{asq}, the posterior mean of the variance of a rating depends on the attribute and ranges from

$$E\{\sigma^2(\mu) + \sigma^2(\rho) + \sigma^2_{over-qual}(\epsilon)\} = 0.71 + 0.71 + 0.85 = 2.27$$

to

$$E\{\sigma^2(\mu) + \sigma^2(\rho) + \sigma^2_{cost}(\epsilon)\} = 0.71 + 0.71 + 2.07 = 3.49$$

The variation about θ_{asq} is substantial and has sobering implications about overall survey design. To illustrate, consider the `over-qual` attribute. In the information industry, scores appear to range from about 6.5 to 7.5, so worst to best has a range of 1. (We will see this shortly in our displays of company performance.) What sample size does it take to put the posterior probability of a company's performance in an interval of ± 0.3 to provide insight about which companies are at the top and which at the bottom? Instead of a full posterior predictive simulation we will be satisfied here with a back-of-the-envelope calculation to get a rough picture. Suppose the survey only asks about `over-qual` for one company and that the true variance is 2.27, whose square root is 1.51 Then with a diffuse exogenous distribution for the company performance, the posterior mass would be contained, roughly, within

$$\pm 1.96 \times 1.51/\sqrt{n} = 2.95/\sqrt{n}.$$

Thus we need a sample size of about 100 to achieve a ± 0.3 interval.

7.2 *Relative Performance*

To study relative performance we will focus on the κ_{acq}, the smooth component of κ_{acq}. The posterior distribution of the κ_{acq} conveys information about the relative performance of the companies on the attributes. Extensive study of this posterior has yielded important insight into performance. The following briefly illustrates some of the study.

The nine-quarter average performances are

$$\kappa_{as} = \frac{\sum_q \kappa_{asq}}{9}.$$

We use the normal distribution to approximate their posterior distribution. The means and 95% intervals of the approximating distribution are plotted in the trellis displays of Figures 8 and 9. Each plotted dot is the posterior mean for one attribute and one company. The line segment about the dot is the 95% posterior interval. The values on the two displays are the same but are organized differently. Figure 8 graphs the posterior values against attribute given company; each panel shows the 9 attribute means for one company. Figure 9 graphs the posterior values against company given attribute; each panel shows the 19 company means for one attribute. To enhance our perception of patterns in the data, both the attributes and the companies are ordered. The 19 companies are ordered by the 19 arithmetic averages of the posterior means across attributes. The 9 attributes are ordered by the 9 arithmetic averages of the posterior means across companies. In Figure 8, the attributes are ordered, going from bottom to top in the panels, so that the attribute averages increase. And, as we move from left to right and then bottom to top through the panels, the company averages increase. In Figure 9, the companies are ordered, going from bottom to top in the panels, so that the company averages

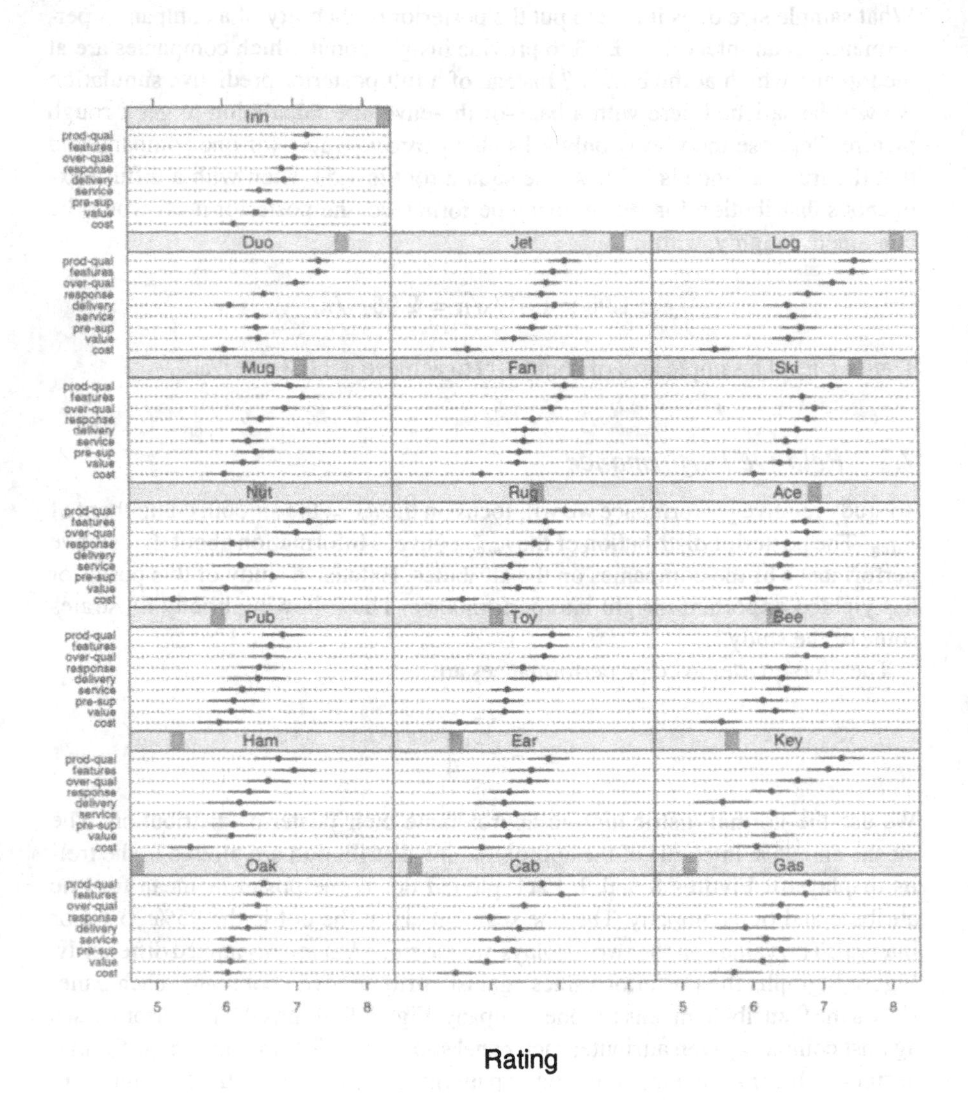

FIGURE 8. Posterior Distribution: Nine-Quarter Average Rating Given Company

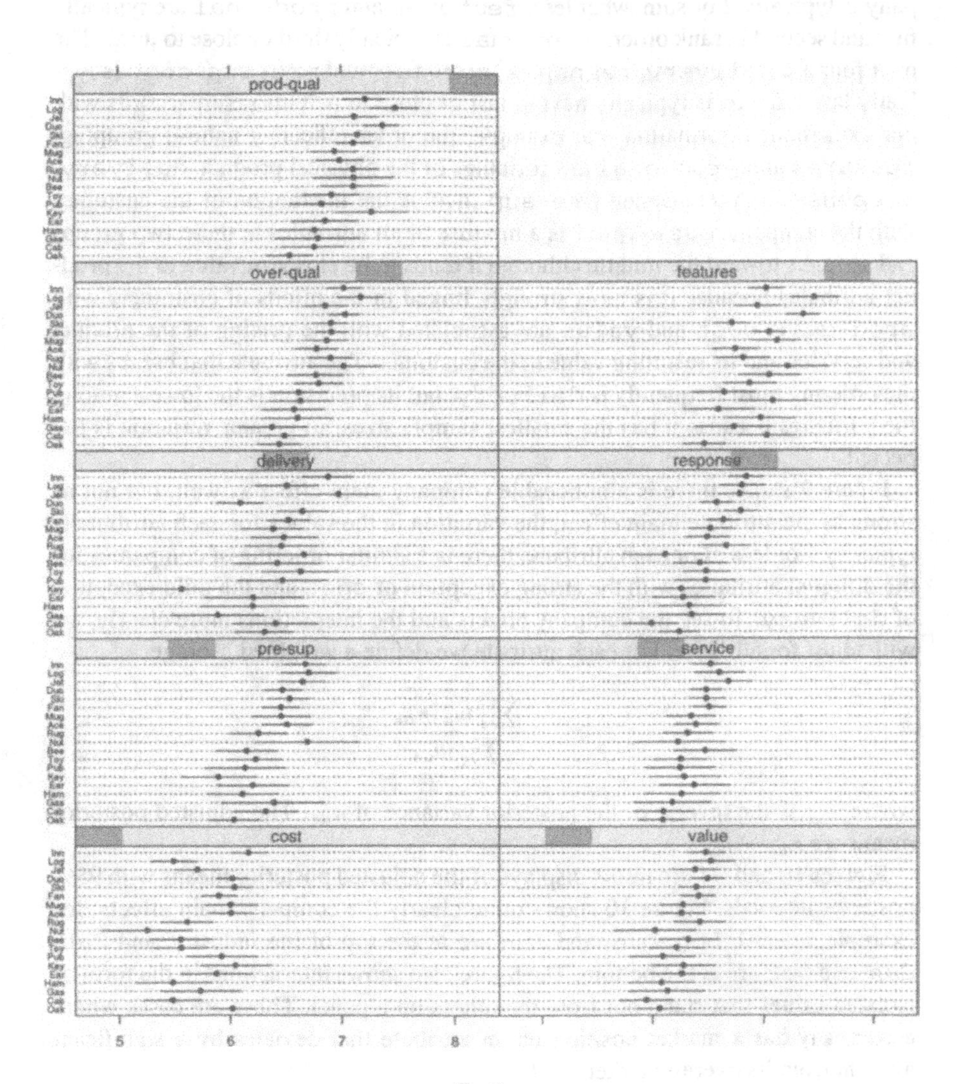

FIGURE 9. Posterior Distribution: Nine-Quarter Average Rating Given Attribute

increase. And as we move from left to right and then bottom to top through the panels, the attribute averages increase.

Figure 8 shows that a strong attribute main effect produces similar orderings of the posterior means for the companies. The variation in the values for each company is typically 2 or somewhat less. `features` and `prod-qual` are typically first and second in rank order. `over-qual` is typically third or close to third. The next four are `delivery`, `response`, `service`, and `pre-sup`. `cost` is typically last. `value` is typically next to last or close to it. This result accords with our exogenous information. For example, the order reflects a natural grouping. `features` and `prod-qual` are attributes of the physical product. `delivery`, `response`, `service`, and `pre-sup` involve the interaction of the customer with the company. `over-qual` is a mixture of all attributes in these two groups and so ranks toward the middle although it tends to be closer in value to the product attributes because it is most strongly linked in the minds of customers with `prod-qual`. `cost` and `value` are associated with the pricing of the product and services and its resulting value to the customer. The attribute that breaks with the ordering most frequently is `delivery`, but its precision is the lowest among the attributes because it has the smallest sample sizes, so greater variation is expected.

Figure 9 shows there is a noticeable company main effect as well, but not as strong as the attribute main effect; the variation in the values for each attribute is typically 1 or less . For each attribute, there is a similar ordering of companies for the different attributes with the strong exception of `cost` and the mild exception of `delivery`. To see the company effects and the interactions more clearly, we will adjust for attribute. For each attribute we define a weighted average

$$\kappa_a = \frac{\sum_s w_{as}\kappa_{as}}{\sum_s w_{as}}$$

where w_{as} is the inverse of the posterior variance of κ_{as}. The adjusted posterior means are $\kappa_{as} - \kappa_a$.

Figures 10 and 11 are trellis displays of the adjusted posterior means with 95% posterior intervals. Figure 10 shows quite clearly the company main effects; for example, overall, `Log`, `Duo`, and `Inn` are at the top of the industry, and `Cab`, `Gas`, and `Oak` are at the bottom. The figure also shows interactions in the form of isolated values that stand out from the others on a panel. These are cases where a company has a market position on an attribute that deviates by a significant amount from its overall market position.

Most of these isolated values occur for `delivery` and `cost`. A few isolated values involve other attributes. For example, `Ski` has a reduced market position on `features` while `Ham` has an increased one. It is displays such as these that are carefully studied to determine actions to improve performance. The tendency is to take action on an attribute for which a company is doing most poorly relative to competitors.

We turn now to the quarterly values κ_{asq} We will also adjust them by the

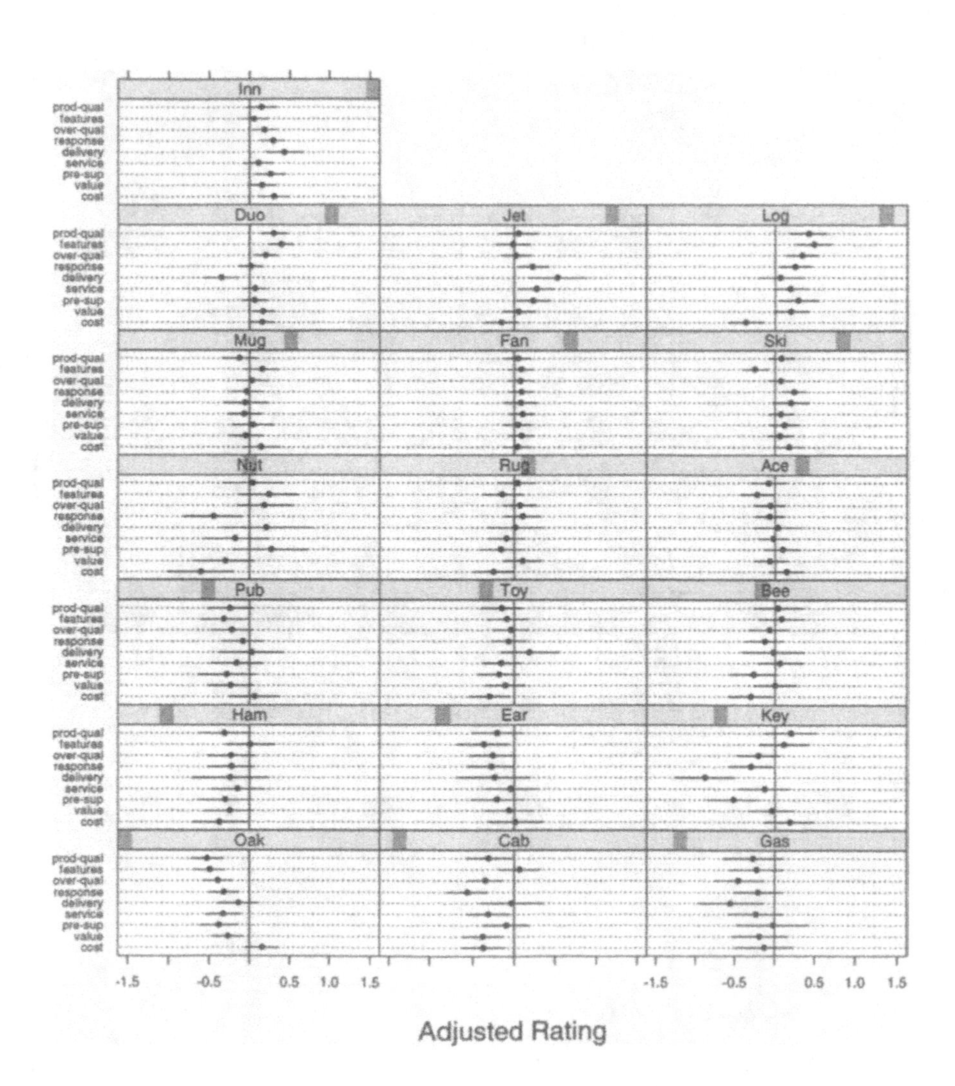

FIGURE 10. Posterior Distribution: Adjusted Nine-Quarter Average Rating Given Company

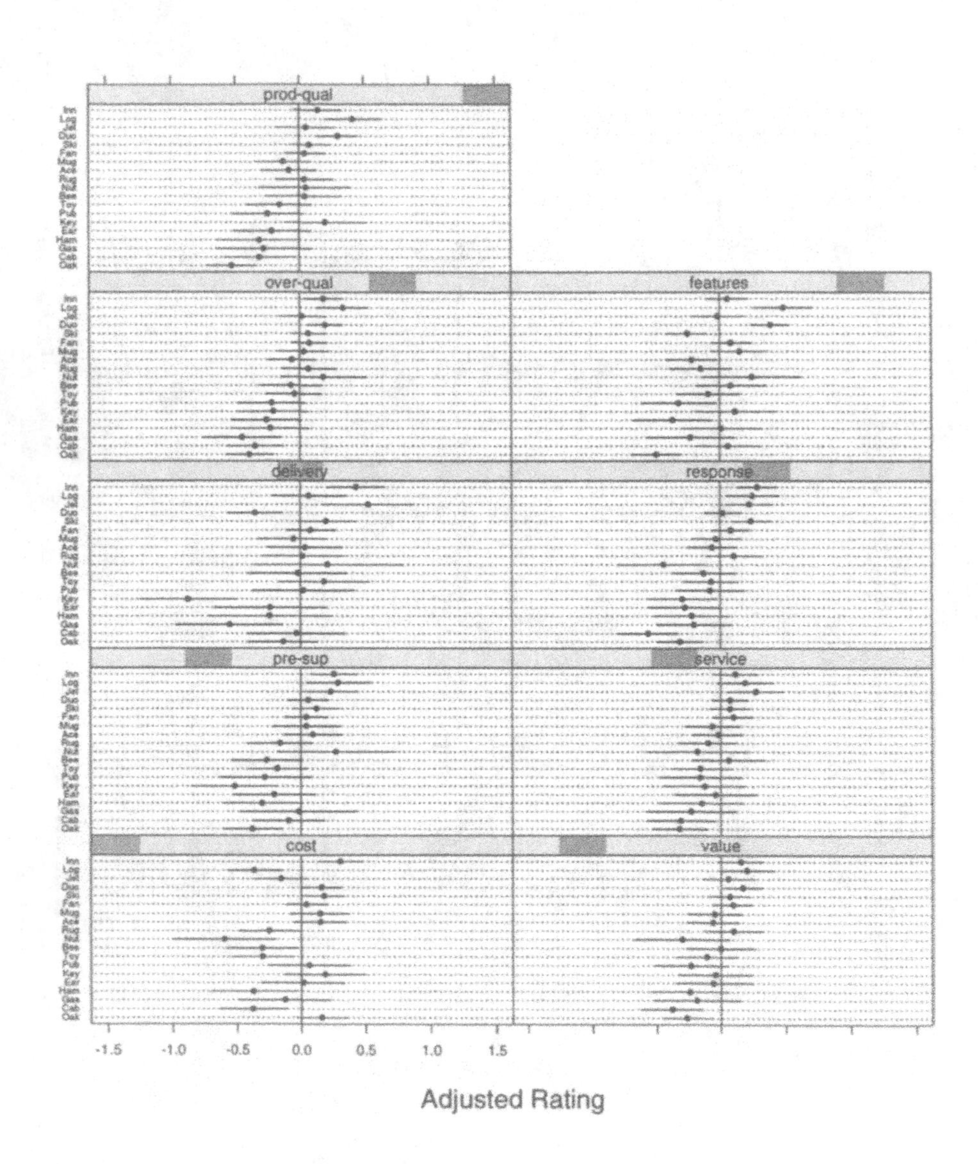

FIGURE 11. Posterior Distribution: Adjusted Nine-Quarter Average Rating Given Attribute

weighted attribute averages, studying

$$\kappa_{asq} - \kappa_{a}.$$

Figures 12 to 16 are a trellis display of the posterior means of the adjusted quarterly values and 95% posterior intervals. The companies are ordered by the number of observations for each; as we go from bottom to top and through the pages the number of observations decreases. The effect of the ordering is a general increase in the sizes of the 95% posterior intervals we go from bottom to top and through the pages. Only a few trends change by amounts that are not small compared with the precisions. For example, `Log`, at the top in `prod-qual` overall, has had a decrease that threatens its position.

7.3 CVA Theory: Relationships of Attributes

There is a substantial amount of exogenous information about relationships among the attributes in CVA surveys. The information comes from a knowledge of the nature of the attributes that are rated. It comes from an experience base of interacting with customers and carrying out *discovery*, a process of asking customers what is important and how attributes relate to one another. It comes from surveys other than ours.

The exogenous knowledge has evolved into a CVA theory. Here is one key piece of the theory. A customer's decision to purchase a product or service is assumed to depend strongly on the customer's perception of the attribute `value`, the worth of the product or service relative to its price. `value` is assumed to depend directly on `over-qual` and `cost`, and then `over-qual` is assumed to depend on the remaining attributes: `prod-qual`, `delivery`, `features`, `pre-sup`, `response`, and `service`.

As stated in earlier sections, we chose not to impose strong model specifications about the relationships of attributes. More specifically, while we made strong assumptions about the behavior of the θ_{asq} across q for fixed a and s, we made mild assumptions about the behavior across a and s for fixed q. For example, we did not build into our model any causal dependence of `value` on `over-qual` and `cost`. There were two reasons for this. First, we were reluctant to add to the complexity of an already complex model. Second, the exogenous information about the relationships are subject to substantial debate, and so, in our primary analysis, we wanted to withdraw from this debate by making no strong assumptions. Our reasons are nicely encapsulated by Gelman, Carlin, Stern, and Rubin (1995, pp. 56-57):

> In almost every real problem, the data analyst will have more information than can be conveniently included in the statistical model. This is an issue with the likelihood as well as the prior distribution. In practice, there is always compromise for a number of reasons: to describe the model more conveniently; because it may be difficult to

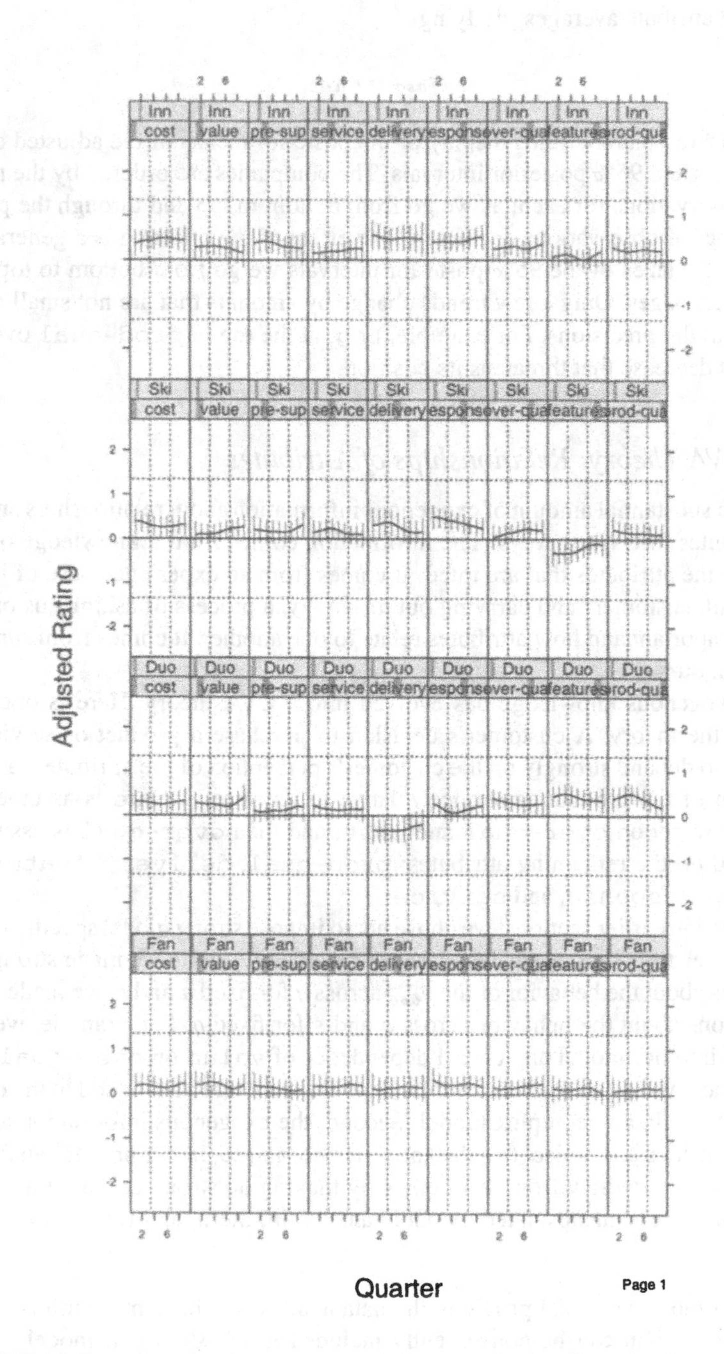

FIGURE 12. Posterior Distribution: Adjusted Quarterly Rating Given Attribute and Company

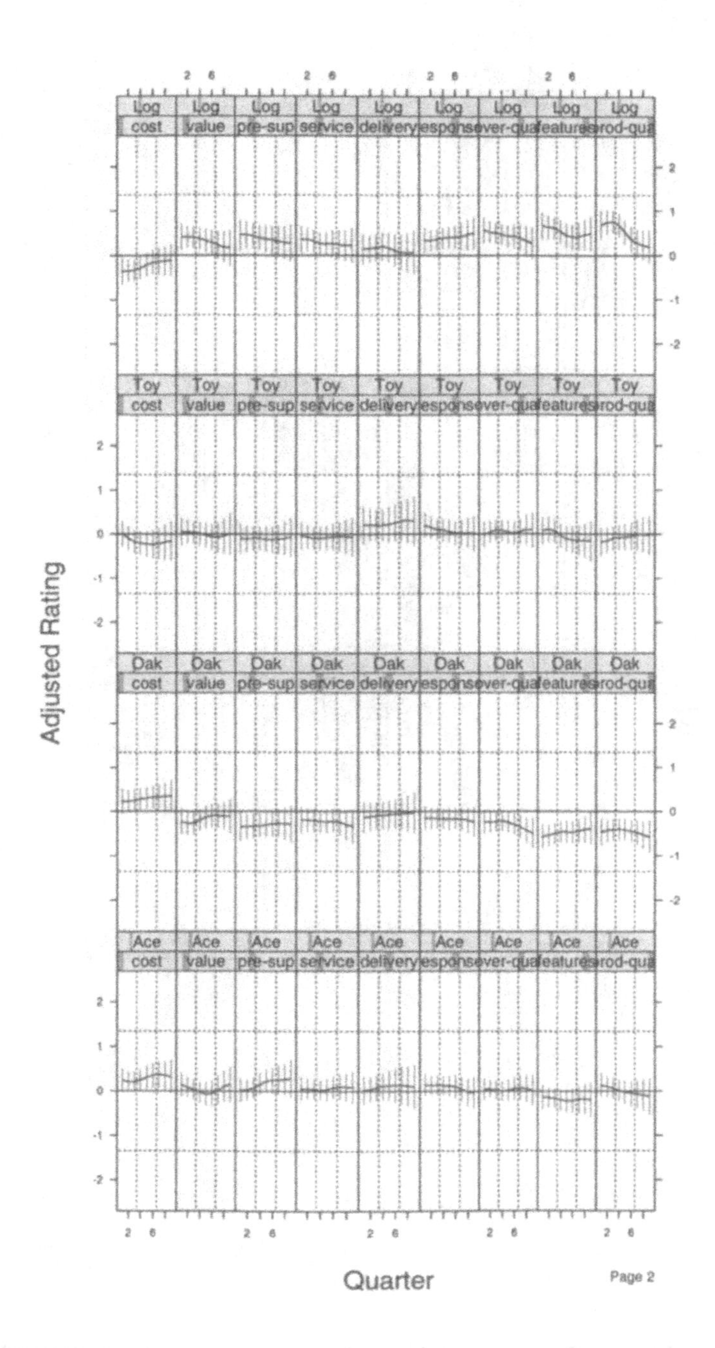

FIGURE 13. Posterior Distribution: Adjusted Quarterly Rating Given Attribute and Company

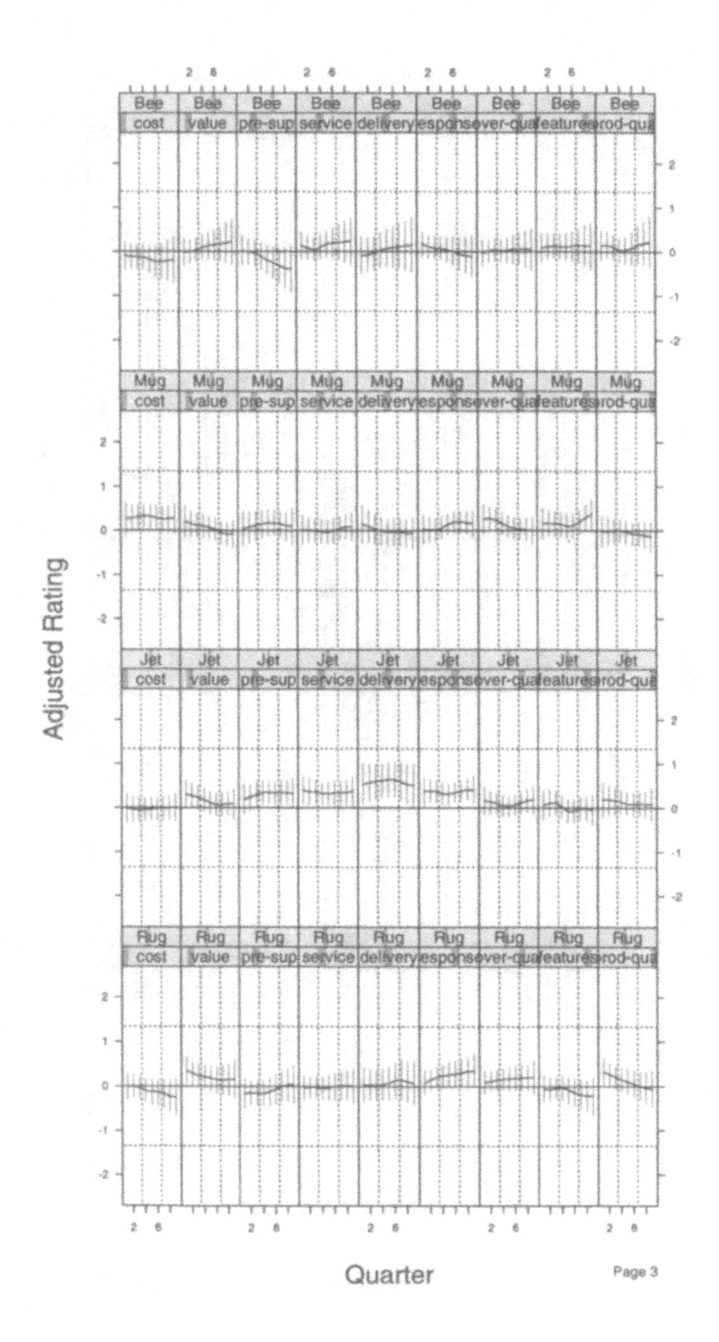

FIGURE 14. Posterior Distribution: Adjusted Quarterly Rating Given Attribute and Company

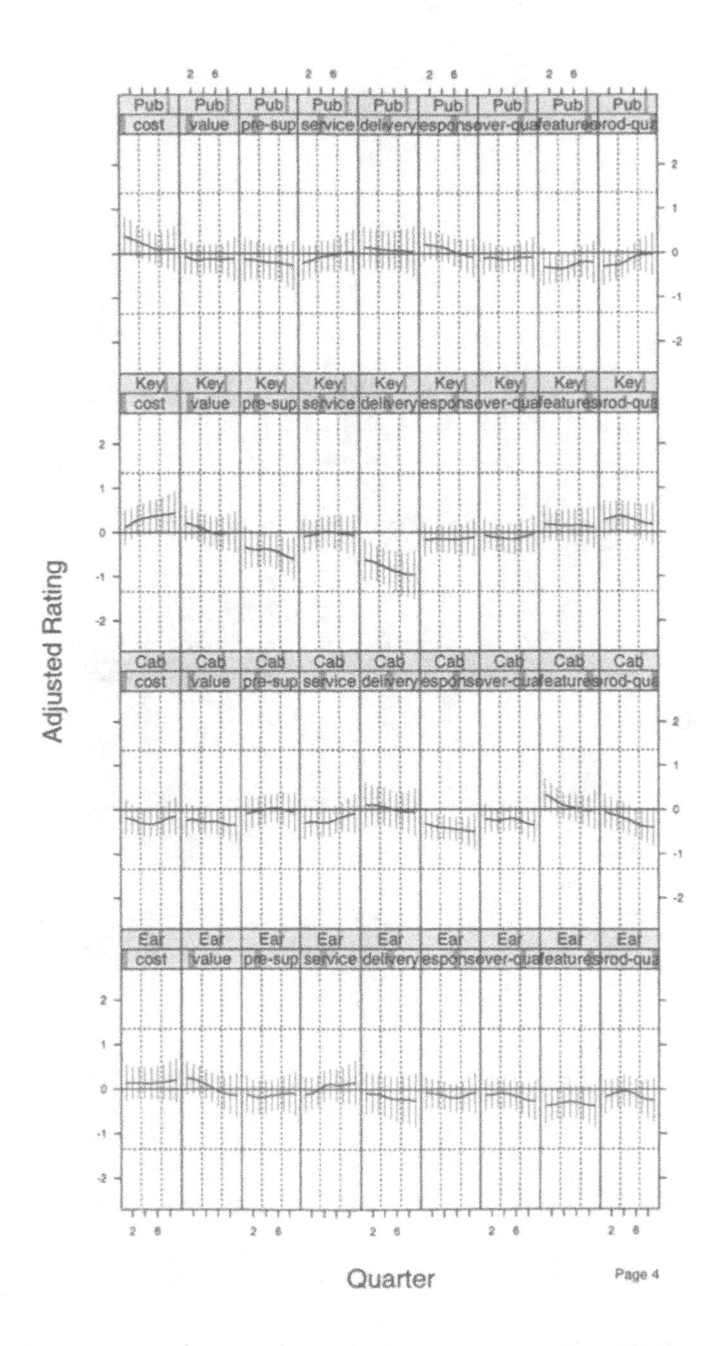

FIGURE 15. Posterior Distribution: Adjusted Quarterly Rating Given Attribute and Company

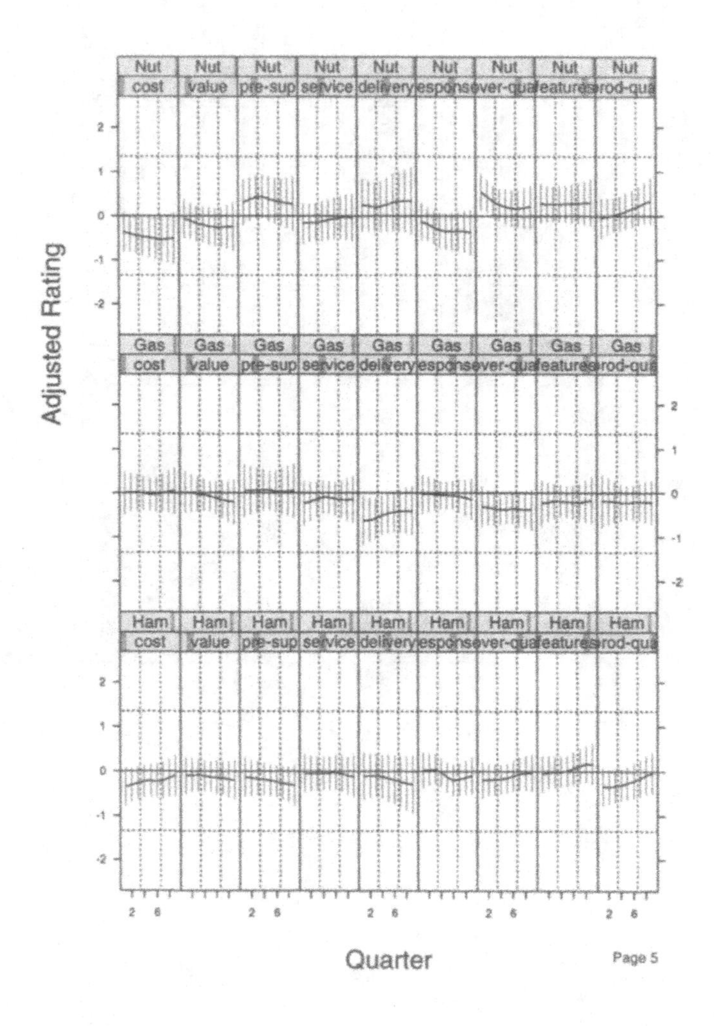

FIGURE 16. Posterior Distribution: Adjusted Quarterly Rating Given Attribute and Company

express knowledge accurately in probabilistic form; to simplify computations; or perhaps to avoid using a possibly unreliable source of information.

Instead of building strong assumptions about attribute relationships into our model at the outset, we decided to employ *post-posterior* analysis (Cleveland and Liu, 1998a). To make our analysis fathomable — specifically, to bring in our exogenous knowledge about relationships in a reliable way — we focused just on the κ_{as}. We studied their posterior distribution. We combined the posterior information with reasonable specifications about relationships based on our exogenous information to form a post-posterior distribution of the κ_{ac}. This amounted to a model building exercise using the posterior information and exogenous information.

Of course, another approach would have been to return to the full, original model, and add the exogenous information. We judge this to be a task that is too complex given our current state of knowledge, and therefore unreliable. Our more conservative approach of narrowing our analysis of attribute relationships appears to us more prudent. By dealing just with the κ_{as} we are able to more reliably specify exogenous information and to more clearly separate the influence of the data (in the form of the posterior distribution) and the influence of the exogenous information in drawing inferences about the attribute relationships. But we do hope as our experience base increases that a comprehensive model can be developed.

Acknowledgments

We are grateful to Eric Bradlow for a number of discussions about Bayesian models and customer survey data, and for a number of very helpful comments on the paper; to Don Rubin for discussions that at the beginning of our project helped us to chart our course; to Nick Fisher for astute comments about what material to include in the paper; to Brad Carlin for overseeing our participation in the Workshop and providing important comments on the paper; and to the other organizers of the Workshop for carrying on this fine tradition of a conference on applications.

References

Albert, J. H. and Chib, S. (1993). Bayesian Analysis of Binary and Polychotomous Response Data, *Journal of the American Statistical Association*, **88**, 669–680.

Anscombe, F. J. and Tukey, J. W. (1961). The examination and analysis of residuals, *Technometrics* **5**, 141–160.

Becker, R. A. and Cleveland, W. S. (1996). *Trellis Graphics User's Manual*, MathSoft, Seattle. Internet: b&w or color postscript (224 pages) available from site
cm.bell-labs.com/stat/project/trellis.

Becker, R. A., Cleveland, W. S., and Shyu, M. J. (1996). The Design and Control of Trellis Display, *Journal of Computational and Statistical Graphics* **5**, 123–155. Internet: b&w or color postscript (36 pages) available from site cm.bell-labs.com/stat/project/trellis.

Box, G. E. P. and Hunter, W. G. (1965). The Experimental Study of Physical Mechanisms, *Technometrics* **7**, 23–42.

Box, G. E. P. and Jenkins, G. M. (1970). *Time Series Analysis Forecasting and Control*, 2nd ed., Holden-Day, Oakland, CA.

Box, G. E. P. (1980). Sampling and Bayes' Inference in Scientific Modelling and Robustness, *Journal of the Royal Statistical Society A* **143**, 383–430.

Bradlow, E. T. and Zaslavsky (1997). A Hierarchical Latent Variable Model for Ordinal Data from a Customer Satisfaction Survey with "No Answer" Responses. Technical Report, Department of Marketing, Wharton School, University of Pennsylvania.

Cleveland, W. S. (1993). *Visualizing Data*, Hobart Press,
books@hobart.com.

Cleveland, W. S., Denby, L., and Liu, C. (1998). Random Location and Scale Models for Case Data, in preparation.

Cleveland, W. S. and and Liu, C. (1998a). A Theory of Model Building, in preparation.

Cleveland, W. S. and and Liu, C. (1998b). Integated Sum-Difference Time Series Models, in preparation.

Daniel, C. and Wood, F. S. (1971). *Fitting Equations to Data*, Wiley, New York.

Draper, D. (1995). Assessment and Propagation of Model Uncertainty, *Journal of the Royal Statistical Society, B* **57** 45–97.

Dempster, A. P. (1970). Foundation of Statistical Inference, *Proceedings of the Symposium of the Foundations of Statistical Inference, March 31 to April 9*, 56–81.

Draper, D., Hodges, J. S., Mallows, C. L., and Pregibon, D. (1993). Exchangeability and Data Analysis, *Journal of the Royal Statistical Society, A* **156**, Part 1, 9–37.

Edwards, W., Lindman, H., and Savage, L. J. (1963). Bayesian Statistical Inference for Psychological Research, *Psychological Review* **70**, 193–242.

Gale, B. T. (1994). *Managing Customer Value*, MacMillan, New York.

Gelfand, A. E. and Smith, A. F. M. (1990). Sampling Based Approaches to Calculating Marginal Densities, *Journal of the American Statistical Association* **85**, 398–409.

Gelman, A., King, G., and Liu, C. (1998). Multiple Imputation for Multiple Surveys (with discussion), *Journal of the American Statistical Association*, to appear.

Gelman, A., Meng, X-L., and Stern, H. (1996). Posterior Predictive Assessment of Model Fitness Via Realized Discrepancies, *Statistica Sinica* **6**, 733–807.

Geman, S. and Geman, D. (1984). Stochastic Relaxation, Gibbs Distributions, and The Bayesian Restoration of Images, *IEEE Transactions on Pattern Analysis and Machine Intelligence* **6**, 721–741.

Good, I. J. (1957). Mathematical Tools, *Uncertainty and Business Decisions*, edited by Carter, C. F., Meredith, G. P., and Shackle, G. L. S., 20–36, Liverpool University Press.

Hill, B. M. (1986). Some Subjective Bayesian Considerations in The Selection of Models, *Econometric Reviews* **4**, 191–251.

Hill, B. M. (1990). A Theory of Bayesian Data Analysis, *Bayesian and Likelihood Methods in Statistics and Econometrics*, S. Geiser, J. S. Hodges, S. J. Press and A. Zellner (Editors), 49–73, Elsevier Science Publishers B. V. (North-Holland).

Johnson, V. E. (1997). An Alternative to Traditional GPA for Evaluating Student Performance (with discussion), Statistical Science, **12**, 251–278.

Kass, R. E. and Raftery, A. E. (1995). Bayes Factors, *Journal of the American Statistical Association* **90**, 773–795.

Kass, R. E., Tierney, L., and Kadane, J. B. (1989). Approximate Methods for Assessing Influence and Sensitivity in Bayesian Analysis, *Biometrika* **76**, 663–674.

Liu, C. (1995). Missing Data Imputation Using the Multivariate t-distribution, *The Journal of Multivariate Analysis* **53**, 139–158.

Liu, C. (1996). Bayesian Robust Multivariate Linear Regression with Incomplete Data, *Journal of the American Statistical Association* **91**, 1219–1227.

Longford, N. T. (1995). *Models for Uncertainty in Educational Testing*, Springer, New York.

Metropolis, N., Rosenbluth, A. W., Rosenbluth, M. N., and Teller, A. H. (1953). Equations of State Calculations by Fast Computing Machines, *Journal of Chemical Physics* **21**, 1087–1091.

Metropolis, N. and Ulam, S. (1949). The Monte Carlo Methods, *Journal of the American Statistical Association* **44**, 335–341.

Naumann, E. and Kordupleski, R. (1995). *Customer Value Toolkit*, International Thomson Publishing, London.

Pinheiro, J., Liu, C., and Wu, Y. (1997). Robust Estimation in Linear Mixed-Effects Models Using the Multivariate t-distribution, Technical Report, Bell Labs.

Rubin, D. B. (1976). Inference and Missing Data, *Biometrika* **63**, 581–592.

Rubin, D. B. (1984). Bayesianly Justifiable and Relevant Frequency Calculations for The Applied Statistician, *The Annals of Statistics* **12** 1151–1172.

Rubin, D. B. (1987). *Multiple Imputation for Nonresponse in Surveys*, Wiley, New York.

Savage, L. J. (1961). *The Subjective Basis of Statistical Practice*, unpublished book manuscript.

Tanner, M. A. and Wong, W. H. (1987). The Calculation of Posterior Distributions by Data Augmentation (with discussion), *Journal of the American Statistical Association* **82**, 528–550.

Young, F. W. (1981). Quantitative Analysis of Qualitative Data, *Psychometrika* **46**, 357–388.

Discussion

N.G. Best
Imperial College, London

This paper deals with the problem of modelling survey data in which customers are asked to rate companies on various attributes. The data are typical of many surveys in economics, social science and medicine, in that repeated qualitative responses are elicited from a large number of individuals; there is considerable missing data which may or may not occur at random; and the response data are likely to have associated measurement error and individual scale and location effects due to the subjective nature of the survey questions. There is usually no well-defined biological or physical model by which such data are believed to arise. Hence considerable care is required to check the appropriateness and plausibility of any assumptions made when building a statistical model to describe and analyse this type of data.

Clark, Cleveland, Denby and Liu (CCDL) approach the above problem via a"relentless probing of the structure of the data... [using] visualization tools...to judge the likelihood of the data given a variety of specifications". They describe the framework for this strategy as a form of Bayesian inference in which information from the data visualizations is regarded as analogous to a likelihood function. A process of reasoning (rather than mathematical computation via Bayes theorem) is then used to combine this 'likelihood' with prior sources of information to derive an appropriate model. I fully commend the principles of this approach; my quibble relates to the somewhat artificial parallel drawn by the authors between this and the Bayesian paradigm. The methods described by CCDL represent a sophisticated and powerful exploratory analysis of the data which they use to inform the specification of a fully Bayesian model. However, many of the steps taken during the exploratory stage would appeal equally to committed non-Bayesians! This is not so much a criticism of the method, but rather a comment that there is nothing inherently Bayesian about what was done. The point to empahsize is the particular importance of carrying out detailed exploratory analysis before launching into Bayesian model estimation in view of the degree of model complexity which has become routinely entertained since the Markov chain Monte Carlo revolution. Nevertheless, I do not believe that such exploratory analysis can completely replace the need to examine posterior sensitivity of the model to the assumptions made. CCDL do not address this issue in their paper, and it is not clear to me what is the impact of some of the modelling assumptions they make. For example, how critically do the substantive conclusions depend on, say, the choice of transformation (logarithm versus square root) for the response data, the choice of distribution for the person location effects (Normal versus t), or the fact that the measurement rounding error was ignored?

Having arrived at a final model specification for their data, CCDL use Markov chain Monte Carlo (MCMC) methods to carry out fully Bayesian model estimation. Simulations are carried out using a mixture of Gibbs and Metropolis-within-Gibbs sampling algorithms. However, no mention is made of how convergence was assessed, nor how many samples were used to estimate the joint posterior distribution for their model. The model contains a number of parameters which are likely to exhibit high correlations (for example, the parameters of the time series model, or the person and attribute variances) which could lead to poor mixing of the chains.

As a postscript, CCDL continue their theme of sequential model building by carrying out a regression and factor analysis based on the posterior distributions estimated above. This *a posteriori* analysis was implemented via penalized maximum likelihood and was intended to further study the relationships amongst the company attributes. CCDL justify this approach by arguing that they prefered not to impose strong structure about the attribute inter-relationships in the original Bayesian model in order to avoid suppressing important features. However, this seems to contradict the central theme of the paper, namely to carry out detailed *a priori* exploration of the data in order to guide formal model specification. Furthermore, the data appear to be used twice — once as direct input into the *a posteriori* model, and secondly to estimate the posterior parameter distributions which are in turn used as 'data' in the *a posteriori* model. I would prefer to see an analysis in which the original Bayesian model is extended to incorporate additional hierarchical strucuture on the company attribute effects (including robustifying assumptions to allow for unexpected or outlying observations), and re-estimated using fully Bayesian methods.

The main emphasis of CCDL's work concerns the model building strategy via a process of probing and visualizing the structure of both data and parameters, and they have demonstrated an impressive range of graphical and exploratory

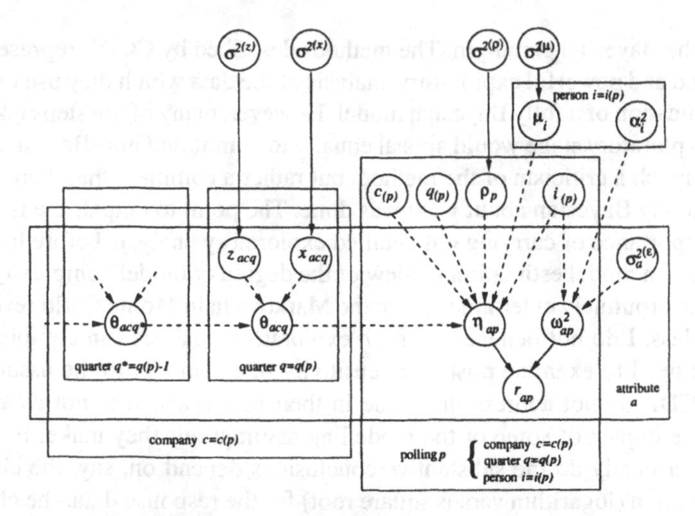

FIGURE 1. Graph of CCDL's customer survey model

techniques by which to achieve this. Below, I suggest an additional graphical tool — namely, Bayesian graphical models — to help visualize the structure of the *model* itself. In addition, I was disappointed that the authors did not attempt to more fully exploit the power and flexibility of modern Bayesian computational approaches to investigate features of their data such as measurement error, and to facilitate prediction and comparison of relative company performance. These issues are also discussed below.

Bayesian graphical models

Graphical models are a useful aid to the development, communication and computation of Bayesian analyses (Spiegelhalter et al., 1995a). A graphical representation of CCDL's customer survey model is shown in Figure 1. Nodes represent random variables, solid arrows indicate probabalistic dependencies between nodes, broken arrows indicate deterministic relationships, and missing links between nodes indicate conditional independence assumptions. The nodes $c_{(p)}$, $q_{(p)}$ and $i_{(p)}$ are indicator variables denoting the company, quarter and individual associated with polling p. The node r_{ap} denotes the rating scored on attribute a at polling p; this is assumed to arise from some underlying distribution with location parameter η_{ap} and scale parameter ω^2_{ap}. The location parameter is a function of θ_{acq} (the performance level on attribute a for company $c = c_{(p)}$ at time $q = q_{(p)}$) plus a person location effect μ_i (where $i = i_{(p)}$) and a polling-specific effect ρ_p. The scale parameter ω^2_{ap} is a function of σ^2_a (the variance of attribute a) and of a person scale effect α_i. The time-series nature of the data is indicated by the deterministic link between $\theta_{ac(q-1)}$ and θ_{acq}, the mean performance levels for company $c_{(p)}$ on attribute a in quarters $q_{(p)} - 1$ and $q_{(p)}$ respectively.

Measurement error

In section 3 of their paper, CCDL carry out exploratory analyses to determine whether a continuous distribution is appropriate for the response measurement scale. They chose to ignore the rounding error implicit in this assumption, which seems reasonable for the purposes of their subsequent data visualizations and exploratory analyses. However, ignoring such measurement error in the final Bayesian model may result in over-precise estimates of other model quantites. Incorporation of rounding error represents a straightforward extension of the basic model: the observed response r_{ap} is modelled as an imprecise measurement of the true unknown rating δ_{ap} rounded to the nearest integer; an appropriately truncated probability distribution is then assumed for δ_{ap} (see Spiegelhalter et al. (1995b) for further details of such models). An alternative model for the observed customer survey data is to treat the reponse as an ordered categorical variable, where each category corresponds to an interval with unknown endpoints on some underlying continuous latent scale. Best et al. (1996) describe the Bayesian analysis of such models.

Characterizing relative performance/presentation of results

The following are additional suggestions for posterior summary measures which may enhance the presentation and interpretation of the results of CCDL's analysis. These summary statistics take advantage of the fact that posterior distributions of arbitrary functions of model parameters may be readily calculated within the Bayesian MCMC approach.

Calculating rank order is a natural way of comparing performance between a set of institutions such as schools, hospitals or companies. However, ranks are a notoriously unreliable summary statistic, and Goldstein and Spiegelhalter (1996) discuss the computation of confidence or credible intervals on ranks to quantify uncertainty. Figure 2 shows the posterior median and 95% interval estimates for a set of ranks based on simulated data designed to mimic the structure of CCDL's customer survey ratings. Companies whose rank interval falls entirely within, say, the upper or lower quartile could be regardard as performing exceptionally well or poorly on that attribute respectively. Likewise, companies whose rank interval spans the inter-quartile range could be regarded as indistinguishable on that attribute.

Formal idenitification of a company with a 'single shot' attribute could be achieved by determining whether a company has a high posterior probability that the adjusted score for one of the attributes is more than, say, 50% greater than the median adjusted scores over all other attributes for that company (where adjustment is first made for the overall attribute means and variances). Figure 3 shows an example of the type of results which could be obtained for 2 different companies in CCDL's customer survey analysis; again, these results are based on simulated data.

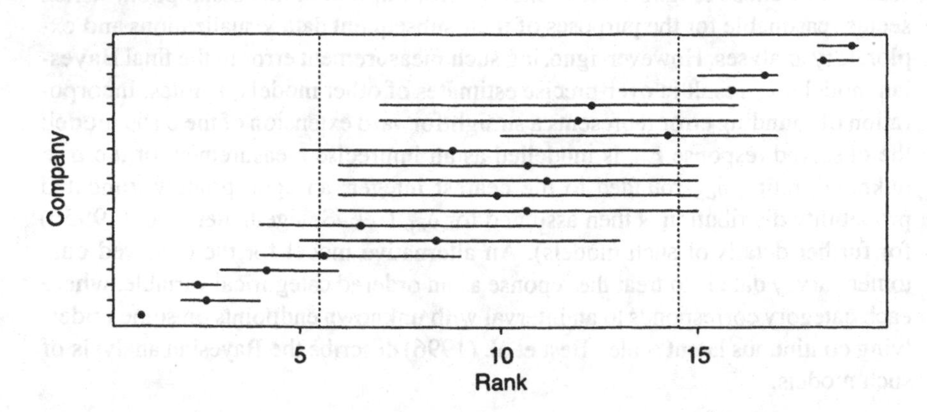

FIGURE 2. Posterior median (•) and 95% interval (——) for the rank of each company on attribute A. Vertical dashed lines indicate the inter-quartile range of the ranks

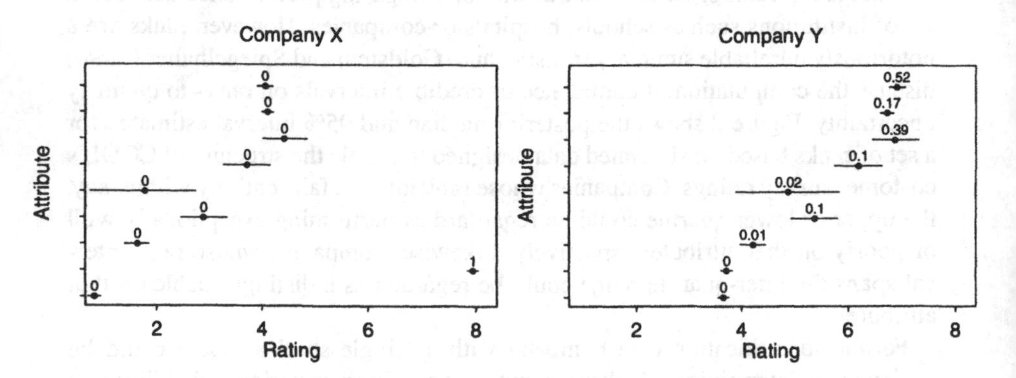

FIGURE 3. Posterior mean (•) and 95% interval (——) for the true rating on each attribute for companies X and Y. Numbers above the dots indicate the posterior probability that the attribute represents a 'single shot'

Prediction and Model Validation

An important omission from CCDL's analysis is the calculation of predicitive distributions. From a substantive point of view, it would seem natural to use the model developed thus far to predict a company's future performance on each attribute in order to target resources appropriately. Predictive distributions could also be used for model validation purposes, either by means of cross-validation techniques (see, for example, Gelfand et al. (1992)), or by predicting, say, the 9th quarter performance ratings using a model estimated from the data from quarters 1–8.

Conclusions

Until recently, acknowledging the full complexity and structure of real-life applications in a statistical model was difficult due to methodological and software limitations. The advent of MCMC methods has revolutionised this situation: Bayesians now have the scope to implement statistical models of almost arbitrary complexity. However, the temptation to fit highly complex models must be tempered by a realistic understanding of the limitations of the data and plausible substantive mechanisms. Furthermore, Bayesian methodology for model selection and model criticism has lagged behind the computational advances, leading to a situation where it is often difficult to carry out *a posteriori* checks on certain model assumptions. The model-building approach presented by CCDL offers a partial solution to this problem, by providing a framework for constructing realistic models based on reasoned decisions which are informed by detailed *a priori* exploration and visualisation of the data and parameters. This appraoch should not been seen as an alternative to posterior model criticism, but rather, as a means to eliminate inappropriate models and reduce the set of alternative specifications to a manageable size.

CCDL have tackled a challenging problem in this paper, and parallels may be drawn between this and large-scale survey studies in other application areas such as medicine and social science. I concur with the CCDL's policy of attempting to keep the final model as simple as possible. However, the exploratory analysis highlighted a number of potentially interesting features of the data which warrant further exploration, and which may be readily addressed within the Bayesian paradigm. I have focused on these aspects in my discussion, although I appreciate that it may be simply pressure of time that has prevented the authors from already carrying out many of the extensions I propose.

In conclusion, I would like to thank the authors for opening my eyes to the power and flexibility of the graphical techniques they employ to visualize their data and build a realistic model. In return, I hope my comments will encourage them to further exploit the power and flexibility of the Bayesian computational approach to probe their basic model and directly address the some of the goals of their analysis, namely comparison of company performance and attribute relationships, measurement error and prediction.

References

Best, N.G., Spiegelhalter, D.J., Thomas, A. and Brayne, C.E.G. (1996). Bayesian analysis of realistically complex models. *Journal of the Royal Statistical Society, Series A* **159**, 323-342

Dawid, A.P. (1979). Conditional independence in statistical theory (with discussion). *Journal of the Royal Statistical Society, Series B* **41**, 1-31.

Dellaportas, P. and Stephens, D.A. (1995). Bayesian analysis of errors-in-variables regression models. *Biometrics* **51**, 1085-1095.

Gelfand, A.E., Dey, D.K. and Chang, H. (1992). Model determination using predictive distributions with implementation via sampling-based methods. In Bayesian Statistics 4 (eds. J.M. Bernardo, J.O. Berger, A.P. Dawid and A.F.M. Smith), pp. 147-167. Oxford: Oxford University Press.

Goldstein, H. and Spiegelhalter, D.J. (1996). Statistical aspects of institutional performance: league tables and their limitations (with discussion). *Journal of the Royal Statistical Society, Series A* **159**,

Spiegelhalter, D.J., Thomas, A. and Best, N.G. (1995a). Computation on Bayesian graphical models. In *Bayesian Statistics 5* (eds. J.M. Bernardo, J.O. Berger, A.P. Dawid and A.F.M. Smith), pp. 407-425. Oxford: Clarendon Press.

Spiegelhalter, D.J., Thomas, A., Best, N.G. and Gilks, W.R. (1995b). *BUGS Bayesian inference Using Gibbs Sampling: Version 0.5*, MRC Biostatistics Unit, Cambridge.

Whittaker, J. (1990). *Graphical Models in Applied Multivariate Statistics*, pp. 71-77. Chichester: John Wiley & Sons.

Discussion

Eric T. Bradlow
Wharton School
Kirthi Kalyanam
Santa Clara University

Over the last decade marketing practitioners have been increasingly interested in the value of a product/service as perceived by a customer relative to competitive offerings. There has been a general buy in into the idea that customer perceived relative value is a leading indicator of market share. Consequently companies have been very interested in the tracking of this metric over time (Gale 1994; Slywotzky, 1996, Cleland and Bruno, 1997). This paper (henceforth CCDL) develops a statistical model for analyzing survey data regarding customer perceived value. While most of the published work in this area has focused on developing frameworks, relatively little attention has been paid to the issues related to the rigorous analysis of customer survey data. Hence this paper fills a very important gap in the literature. As much as we like the paper and appreciate the value it adds to the emerging field of customer value analysis, some critical comments are in order.

'Relative' Value or Absolute Value:

The value of a product or a service is almost always evaluated relative to some alternative. Hence it is relative value not absolute value that is of direct managerial interest. Consequently, most market share models in the literature relate observed share to relative utility or relative value. This has some implications for the design of the survey and the subsequent analysis of the data. With respect to the design of the survey a strong argument can be made that monadic tasks by respondents should be avoided. Respondents should be asked to consider different products simultaneously and then provide judgments about the value of each product, and the other associated measures. Simultaneous consideration could mitigate the left skewness and the tendency for the mode to be at 7 or 8 as observed in the data. These issues have been well appreciated in the context of Conjoint Analysis.

The other natural consideration is whether one should be analyzing absolute value or relative value? The literature suggests that managerial interest is in the latter measure. This latter measure can be obtained in the current data by simply dividing each respondents' ratings on a certain attribute by the average of the respondents' ratings across all the companies. If the relative ratings (a variable that is clearly continuous) have a more symmetric distribution, it may also simplify the statistical analysis.

Sharp specifications in the model

CCDL utilize (equations 6.4-6.7), a t(6), t(6), exp(1), and N(0,1) distribution respectively for various random effects in the model. While we recognize that any choice of model at a hierarchical stage is typically done out of convenience, and is mostly arbitrary, we must always avoid the desire to overfit the data. We wonder whether these results would be reproduced in subsequent analyses. If not, do we then make inferences based on this model? If we collect more data do we assume this model for all of the data or do we need to go through this entire refitting process again? Furthermore, we wonder how sensitive the inferences are to the exact specifications at these stages. Would a logistic (which is very close to a t(6)) yield different inferences?

Handling the Missing Data

CCDL report that some respondents chose not to rate all attributes about which they were asked. Hence, instead of obtaining the full set of 31,545 ratings only 25,309 ratings were obtained. In other words 20% of the data were missing, and given the large magnitude of this number, this issue should be considered further. Were most of these non-responses related to a certain attribute or a small subset of attributes? More specifically did respondents not choose to rate certain attributes because these attributes were not relevant to them or because they were too uncertain or did not consider them important?

It is important to distinguish between these three cases and handle them appropriately. In Bradlow (1994), and Bradlow and Zaslavsky (1997) they derive a non-ignorable model for the missing data process for customer satisfaction survey data. The reason we feel understanding this process is critical to this work especially is the process by which CCDL arrive at their model. Specifically, their process was to (1) assume a continuous model after transformation, (2) select a general additive model for the data (page 19), (3) fit distributions to random effects in that model, and (4) graphically check fit of the model. While we laud this process, suppose that the missing data process caused a lack of "building-up" at the extremes of the ordinal scale. Then at stage (1) of their process, CCDL would be utilizing a transformation and hence a continuous model that only holds because of the non-ignorable missing data mechanism.

What have I learned that I couldn't have learned through simple methods?

The complex modeling approach employed by CCDL leads to a nice summary of the data and an approach to derive inferences based on posterior distributions of model parameters. The underlying question that remains is: which of these inferences could I have learned through simple ordering of means, and looking at relative sizes of variances. We appreciate that CCDL present in the manuscript a very small percentage of their actual findings, however, of the ones presented many may be observed without the model.

Concluding Comments

In summary, we believe this manuscript: (a) contributes greatly to the current state of modeling of CVA data and is the first model-based approach to explicitly take into account the use of different scales by individuals, (b) brings into the forefront the use of graphical model building stemming from an assumed simple model and then building up, (c) describes how exogeneous information may be used to explain model findings and why it sometimes should not directly be included in the model, and (d) is a step in the right direction.

References

Bradlow, E.T. (1994), Analysis of Ordinal Survey Data with 'No Answer' Responses, Doctoral Dissertation, Department of Statistics, Harvard University.

Bradlow, E.T. and Zaslavsky, A.M., A Hierarchical Model for Ordinal Customer Satisfaction Data with "No Answer" Responses, unpublished manuscript.

Cleland Alan S. and Alebert V. Bruno (1997), The Market Value Process: Bridging Customer and Shareholder Value, San Francisco: Jossey-Bass Publishers.

Gale, Bradley T. (1994), Managing Customer Value: Creating Quality & Service that Customers Can See, New York: The Free Press.

Slywotzski, Adrian J. (1996), Value Migration: How to think Several Moves Ahead of the Competition, Boston: Harvard Business School Press.

Discussion

Peter E. Rossi
University of Chicago

In many industries, it has become routine to collect customer satisfaction data. Typically, customers are asked to rate a company/product/service on overall satisfaction and key dimensions of performance. Analysis of customer satisfaction data is a very important market research activity and this analysis influences many important allocation and product development decisions. While much to do is made up of this satisfaction data, it is rare that satisfaction data is matched up with equivalent usage data for the same set of customers. This would enable the researcher to test the key assumption in these analyses that there is a relationship between customer satisfaction and market performance. Since the authors are closer to the generation of these data than most people working in this area, I would encourage them to collect the information which would enable them to correlate the satisfaction data with market performance.

This study has two goals: 1) To describe the data with a reasonable formal inference machinery that would enable them to make probability statements about differences in satisfaction measures between companies and 2) to develop informal graphical methods to guide in the choice of model specification in Bayesian context. These are both challenging goals because of the high dimension of the data. The data used by the authors is four dimensional: company x variable x respondent x time. There are good reasons to believe that there will be considerable variation along all four dimensions. In particular, some respondents may only use the top portion of the ratings scale while others use the entire scale. Some attributes (such as cost) might have lower overall mean ratings than others. It is important to note that even if the only parameters of interest are mean ratings by company, it will be important to model the variation along the respondent and time dimensions in order to make correct inferences.

The authors assume that appropriately transformed ratings scores are continuously distributed and symmetric. Most of the modeling effort has gone into specifying the form of prior distributions of company-attribute means and the location/scale effects which characterize variation across individuals. The authors use a very sensible conditional analysis to choose the distribution of means and scale parameters across individuals. The distribution of means is chosen to be t(y) (a distribution hardly distinguishable from the normal) and an exponential(1) is chosen for the distribution of scale parameters across individuals. The exponential(1) is a somewhat unusual choice for a scale parameter since it puts mass near zero (the mode). This means that there are some households who use a lot of the scale while others use almost none of the scale (only one rating point). This might be worth more investigation. The authors find evidence against a more standard inverted gamma distribution; in my own work, I've found a log-normal distribution can be useful.

The posterior analysis confirms many of the findings which could be obtained by computing means and standard confidence intervals, except that the posterior intervals are somewhat tighter than the standard confidence intervals computed from an approximate likelihood analysis. It would be interesting to investigate whether or not the tighter posterior intervals are a result of the prior (more details on the prior hyperparameters would be useful) or as the result of more efficient use of the data in making inferences about company/attribute means.

Rejoinder

We are grateful to the discussants for their careful attention to the paper. They reviewed our work and made a number of excellent suggestions. To discuss their discussion we will invoke the email-style comment and response, with the discussant in italics.

Best

However, many of the steps taken during the exploratory stage would appeal equally to committed non-Bayesians. This is not so much a criticism of the method, but rather a comment that there is nothing inherently Bayesian about what was done. Yes, we fully agree that those who declare themselves non-Bayesians would be attracted to the methods. This does not make the methods non-Bayesian but rather says that non-Bayesians in word are typically Bayesian in deed when building models. A non-Bayesian who is a competent data analyst combines exogenous information and information from the data to build models.

I do not believe that such exploratory analysis can completely replace the need to examine posterior sensitivity of the model to the assumptions made. Nor do we. Our full analysis of the data included sensitivity analyses, model enlargements, and posterior predictive checking. Our paper did not report on the full scope of the analysis and iterative model building. If we had a record of the complete path it would fill a book. For example, we incorporated explanatory variables not discussed in the paper, but found them to have little influence. There were several complete Bayesian models that were tested and ultimately altered. The model building summarized in the paper and described in full by Cleveland, Denby, and Liu (1998) constitutes a partial demonstration of validity for part of the model.

... no mention is made of how convergence was assessed, nor how many samples were used to estimate the joint posterior distribution. Yes, assessing the convergence of the iterative simulation is very important in Bayesian computation using MCMC methods. Many convergence-diagnostic methods have been proposed in the literature (Cowles and Carlin 1996). Our numerical model building tools such as the many little case regressions (Cleveland, Denby, and Liu 1998) and intermediate computational results for the model such as maximum likelihood estimates gave us guides that could be used to assess convergence of posterior distributions. Of course, we also monitored the posterior means and variances of the parameters of interest by sequences of various lengths. The final results were calculated based on the last 10,000 draws of a single sequence of 12,000 iterations. Running multiple sequences (Gelman and Rubin, 1992) is currently under consideration. The major difficulties include creating over-dispersed starting distributions; we are taking the approach of Liu and Rubin (1996, 1998).

CCDL continue their theme of sequential model building by carrying out a regression and factor analysis based on the posterior distributions estimated above... the data appear to be used twice — once as direct input into the posterior model, and secondly to estimate the posterior parameter distributions which are in turn used as 'data' in the a posterior model. It is best to think of what we did as combining information. Let us break the exogenous information into two sets: A and B. First, we combined A with the data to form the posterior information, more precisely, the combined information in A and the data. Then we combined the information in the posterior with B. In effect, we added more specifications but worked through the posterior to keep it simple. The results need to thought of as very tentative, awaiting a more thorough analysis after we gain more experience from other data.

A graphical representation of CCDL's customer survey model is shown. This is an appealing diagram for conveying the structure of the model.

... ignoring such measurement error in the final Bayesian model may result in over-precise estimates of other model quantities. We recant. Our descriptor of the data as a result of rounding seems less sensible now than simply viewing our model as a continuous approximation of discrete data. The recantation occurred when we realized that the concept of rounding implies that the continuous rating must have a smaller variance than the discrete rating. It would not hurt, of course, to run some simulations to check the effects on our quantities of interest, but various back-of-the-envelope calculations suggest the approximation is excellent for the estimation of company-attribute effects.

An alternative model for the observed customer survey data is to treat the response as an ordered categorical variable, where each category corresponds to an interval with unknown endpoints on some underlying continuous latent scale. This is a good point. In our paper, we transform and then model distributions. If one uses the categorical-interval approach, a transformation of the data is estimated. But we must make a distributional specification of some sort because any monotone transformation of the data obeys such a model as well. To further check our model, we could use our square root transformation, make the distributional assumptions from our model building process, fit the endpoints, and see if they are consistent with our model.

Bradlow and Kalyanam

The other natural consideration is whether one should be analyzing absolute value or relative value. The literature suggests that managerial interest is in the latter measure. It is correct that we need a performance measure that takes the companies' performance measures on each attribute and corrects for an industry average. We do this is Section 7. Figures 10 and 11 are trellis displays of posterior means minus an industry average for each attribute. We could have divided rather than subtracting but we question the meaning of a ratio for a subjective rating scale. The discussants add: *This latter measure can be obtained in the current data by simply dividing each respondent's ratings by the average of the respondents' ratings across all companies. If the relative ratings have a more symmetric distribution, it may also simplify the statistical analysis.* The discussants' divisor would have a quite small variance compared with the variance of a rating; thus the skewness would be altered very little. We believe it makes more sense to model uncorrected data and then compute, at the end, whatever measures might seem sensible.

While we recognize that any choice of model at a hierarchical stage is typically done out of convenience, and is mostly arbitrary, we must always avoid the desire to overfit the data. There are two dangers. One is to overfit. The second is to choose specifications that do not fit the data. If we carefully study the data, mixing the study with reason and exogenous information, we can discipline the inclination to overfit. But if in the name of protecting ourselves from overfitting we avoid model building, we run great risks of choosing specifications that do not fit, that is, specifications that are manifestly disproved by the data. Our observation is that the latter is the serious problem in the practice of statistics, both among declared frequentists and declared Bayesians.

The complex modeling approach employed by CCDL leads to a nice summary of the data and an approach to derive inferences based on posterior distributions of model parameters. The underlying question that remains is: which of these inferences could I have learned though simple ordering of means, and looking at relative sizes of variances. The model provides a full description of the structure and variability in the data. It becomes a fundamental tool that can then be used for many purposes, not just those that can be satisfied by looking at means and rough estimates of variability. For example, we can use it to redesign our survey instrument or to design a new one; in this case, the description of the person effects is critical.

In Bradlow (1994) and Bradlow and Zaslavsky (1997) they derive a non-ignorable model for the missing data process for customer satisfaction survey data. This work is quite interesting. While 20% of our data is missing, only 9% is missing not by design (randomly). Some of our checking of the Bayesian model (not reported in the paper) showed that this random missingness was informative, but the magnitude of the effect was quite small.

Rossi

I would encourage them to correlate the satisfaction data with market performance. Within corporations there is enormous pressure already to link company process data with customer ratings of attributes from surveys such as ours. The goal is to find important drivers: company actions that can have major influences on customer opinion. The company processes and customer perceptions are governed by complex causal links. It is not possible to undergo extensive experimentation, so one must rely for the most part on the natural variation that occurs across companies and across time. But this natural variation cannot necessarily support the isolation of the causal links. There is a tendency to take at face value results from regression analyses and factor analyses. In some cases these analyses are carried out without accounting for case (person) effects, which invalidates the results. The case effects, when ignored, distort the estimation, quite substantially if the variances of the case distributions are large. Often, an outcome of the distortion, one that results from the case location effects, is positive, significant estimates of parameters that are in theory positive. This creates a false sense of well being (Cleveland, Denby, and Liu 1998) even though the isolation of causal links can be utter nonsense.

It would be interesting to investigate whether or not the tighter posterior intervals [tighter than intervals for sample means] are a result of the prior or as the result of more efficient use of the data in making inferences about company-attribute means. In Section 7 the nine-quarter average performance for supplier company s on attribute a,

$$\kappa_{as} = \frac{\sum_q \kappa_{asq}}{9},$$

is discussed and posterior means and 95% posterior intervals are plotted. Let $\gamma(\kappa_{as})$ be the posterior standard deviation of κ_{as} and let $\gamma(\bar{r}_{as})$ be the standard deviation of the sample mean of r_{ap} across all p for supplier company s

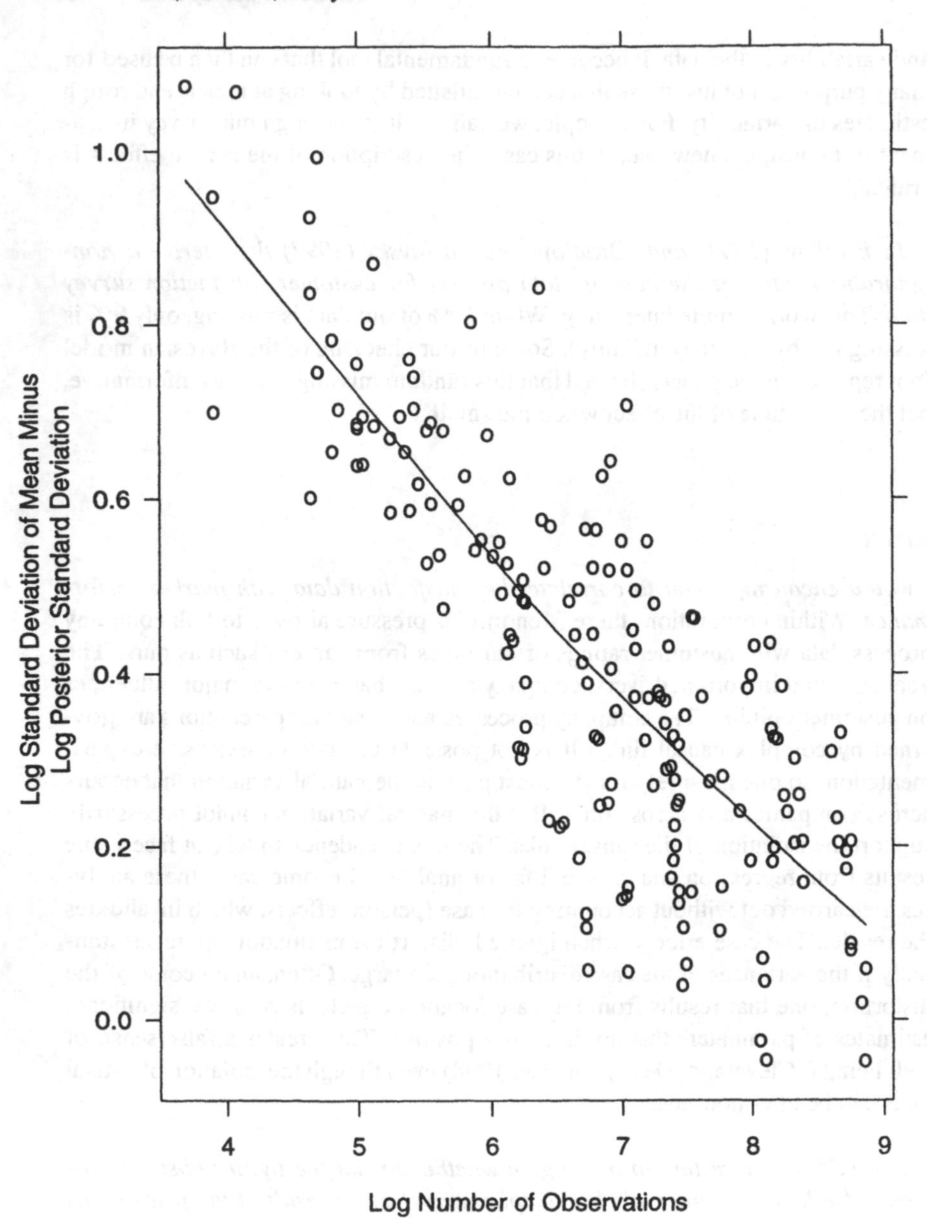

FIGURE 1. Log Base 2 Standard Deviation of Posterior Means Minus Log Base 2 Standard Deviation of Sample Means vs. Log Base 2 Sample Size

and attribute a, that is, the square root of the posterior variance of $\bar{r}_{as}$. Figure 1 graphs $\log_2(\gamma(\bar{r}_{as})) - \log_2(\gamma(\kappa_{as}))$ against $\log_2(n_{as})$ where n_{as} is the number of ratings of attribute a for company s. We see that the differences decrease substantially with $\log_2(n_{as})$. This suggests to us that the shrinkage resulting from the prior distributions accounts for the reduction. Were we to study the quarterly means in a similar way we would find a considerably bigger difference due to both shrinkage and the smoothness imposed by the time series specification for the quarterly effects.

Additional References

Cowles, M. K. and Carlin, B. P. (1996). Markov chain Monte Carlo convergence diagnostics: a comparative review, *J. Amer. Statist. Assoc.*, **91**, 883-904.

Gelman, A. and Rubin, D. B. (1992). Inference from iterative simulation using multiple sequences, *Statist. Sci.*, **7**, 457-511.

Liu, C. and Rubin, D. B. (1996). Markov-normal analysis of iterative simulations before their convergence, *J. Econometric*, **75**, 69-78.

Liu, C. and Rubin, D. B. (1998). Markov-Normal analysis of iterative simulations before their convergence: reconsideration and application, Technical Report, Bell-Labs, Lucent Technologies and Department of Statistics, Harvard Univ.

Functional Connectivity in the Cortical Circuits Subserving Eye Movements

Christopher R. Genovese
John A. Sweeney

ABSTRACT The eyes move continually during visual processing, usually without explicit control or awareness, to scan the key features of a scene. This is essential to our ability to attend to multiple features of our environment and to extract useful information from complex visual stimuli. Eye movement abnormalities are a significant and reliable neurobehavioral marker for a number of major neurological, developmental, and psychiatric disorders, so enhanced understanding of how the brain controls eye movements can yield insights into the brain abnormalities at the root of these conditions.

Eye movements have been studied extensively in both humans and monkeys. A general picture has emerged from these data regarding what areas of the brain subserve eye-movement processes. Although there is a strong homology between humans and monkeys, there are also notable differences, and a detailed delineation of the system in the human brain is still needed. To develop and test a complete theory of how eye movements are implemented requires study of the component sub-processes of the human system, of the interactions among these sub-processes, and of their functional connectivity. The advent of neuroimaging has opened the door to exciting new advances in this area.

We use functional Magnetic Resonance Imaging (fMRI) to study the eye-movement system in humans. Among neuroimaging techniques, fMRI offers a superior combination of spatial and temporal resolution. Each experiment yields as data the realization of a complicated spatio-temporal process that contains information about the dynamics of neural processing during eye movements. We apply a Bayesian hierarchical model for these data to make inferences about the system. In particular, we address an open question: What is the functional relationship between the neural systems subserving saccadic eye movements (rapid repositionings) and smooth visual pursuit of a target? We also illustrate several computational and statistical issues that arise in making inferences from these data.

1 Introduction

The human brain is roughly three pounds of protoplasm, blood, and salt water that could be held in one hand, but it is arguably the most complex object known. The brain contains approximately 100 billion neurons, each of which typically has thousands of synaptic connections to other neurons. The operation of an individual neuron is conceptually simple: it maintains a resting electrical potential across its membrane and, when suitably stimulated, "fires" to send chemical messages (neurotransmitters) to other neurons. It is the richness and plasticity of these connections that enable the brain to store information and build associations at many levels. The challenge of understanding the brain is to understand how complex behaviors, from sensory and motor functions to intelligence and problem solving, can emerge from a combination of such relatively simple units.

In recent years, this challenge has been addressed from several scientific directions. Neurophysiologists have used animal models to learn how neuronal activity changes in response to various tasks and stimuli, and from this they have uncovered the workings of many specialized brain regions and neural circuits. Neuropsychologists study behavioral processes in both humans and animals to discover how specific brain areas subserve different sensory, motor and cognitive functions. The field adopts a primarily "bottom up" methodology in which the behavioral effects of circumscribed brain lesions are measured in a controlled fashion. In contrast, cognitive psychologists take a "top down" approach, building abstract models of memory, cognition, and problem solving that characterize the structure of the mind and that make testable predictions for empirical studies. Psychiatrists and neurologists gain insight into the brain by studying the effects of various clinical disorders in patients. Neurosurgeons learn about brain-behavior relationships by assessing the impact of brain surgery on their patients' functioning and from electrically stimulating the brain during neurosurgical procedures. The combined knowledge derived from these fields forms the basis of our current understanding of brain organization and function.

Great strides in understanding brain function have come in recent years from the application of new technologies that allow non-invasive physiological monitoring of the human brain in action. This field—called functional neuroimaging—enables characterization of regional changes in brain activity in response to varying task demands. Although it began with Positron Emission Tomography (PET), the field now has shifted largely to functional Magnetic Resonance Imaging (fMRI) as the technique of choice because of its superior temporal and spatial resolution. Functional neuroimaging has advanced each neuroscientific discipline separately and has spurred multiple integrative efforts across disciplinary boundaries. Animal physiologists can learn about the similarities and differences between humans and non-human primates and can study physiologic changes across the entire brain simultaneously rather than neuron by neuron. Clinical investigators can directly localize brain pathology non-invasively and *in vivo*. Neuropsychologists and cognitive psychologists can directly test their models. This synergy has greatly advanced one of the great frontiers of science—understanding how the brain works.

In this paper, we concentrate on the oculomotor system, the brain system that controls eye movements (Leigh and Zee, 1991 and Kennard and Rose, 1988). From neuronal recordings in non-human primates and from clinical studies of humans, this system is known to consist of multiple components that are widely distributed across the brain. It involves both low-level functions for motor control and high-level processes that integrate sensory input and cognitive functions in more sophisticated ways. Neuroimaging is especially informative regarding this system for several reasons: (i) a great deal is already known about the neurophysiology of the oculomotor system from extensive monkey studies, and neuroimaging allows for the study of this system in humans, with guidance from the monkey studies and the well established parallels between human and non-human primates; (ii) the oculomotor system is known to involve the interaction of multiple brain areas that are widely distributed, and neuroimaging allows for the simultaneous study of distributed areas where other techniques cannot; (iii) critical distinctions in oculomotor processing occur at a fine-scale across both space and time; neuroimaging offers a excellent spatial and temporal resolution with the capability of choosing a balance between them to meet specific needs; and (iv) previously established associations between eye-movement impairment and various neurophysiological abnormalities yield specific predictions that can be tested with a neuroimaging experiment.

We study the oculomotor system using fMRI. During an fMRI experiment, a participant performs a sequence of behavioral tasks while Magnetic Resonance (MR) images of the subject's brain are acquired at regular intervals (on the order of a second). The brain's response to task performance gives rise to small changes in the images over time. A central challenge in the statistical analysis of fMRI data is to identify and characterize these task-related signal changes. The challenge of interpretation is even greater: to determine from the task-related signal changes (if any) how the brain subserves the processes under study. We apply a modeling framework described in Genovese (1997) to address both these challenges by answering specific scientific questions about the workings of the oculomotor system.

We begin in the next section with a detailed description of the oculomotor system and the potential scientific and clinical gains to be achieved by studying this system. Within this context, we discuss the questions and issues that motivate our experiments, which in turn will be discussed in later sections. In Section 3, we present an overview of fMRI, from the fundamentals of signal acquisition to the statistical methods for analysis of the data. Section 4 describes the model for fMRI data on which we base our analysis. This is a hierarchical model that attempts to capture the critical sources of variation in fMRI data, originally proposed in Genovese (1997). We give the details of our experimental methods in Section 5 and describe the data and results of our analyses in Section 6. These data and analyses are only the first step in a long term study of the oculomotor system, and they give rise to a variety of more detailed questions. Finally, in Section 7, we discuss the impact of these results from a wider perspective and highlight some key issues and some future directions for our research.

To set these ideas in a proper context and to guide the reader through the details of later sections, it will be helpful to describe first the basic anatomical organization of the brain. To put the strengths and weaknesses of fMRI in perspective, we also briefly summarize the range of techniques that can be used to study the brain. Before proceeding, the reader should be warned that the words "function" and "functional" are used quite frequently in this paper but in several different senses. In a neuroscience context, "function" refers to the workings of the brain as a control and information processing system as opposed to "structure" which relates to the anatomical and physical layout of the brain. Thus, *structural* MRI attempts to infer anatomical features of the brain from the images; this is what one would use after being bumped on the head. *Functional* MRI, on the other hand, focuses not on the brain images themselves but on fluctuations in the measurements that relate to the firing of neurons in the active brain. In a mathematical context, function and functional take on their more familiar meanings. We hope that the intended meanings are clear from the text.

1.1 The Gross Anatomy of the Brain

The brain is subdivided into three basic parts: the cerebral hemispheres, the cerebellum, and subcortical areas including the brain stem. Refer to Figures 1, 2, and 3 throughout this discussion. It will be helpful, here and below, to have some terminology for discussing locations in the brain. Suppose we choose an arbitrary reference point in the brain. Any structure closer to the front (back) of the brain than the reference point is *anterior* (*posterior*) to that point. Any structure that is higher (lower) than the reference point is *superior* (*inferior*) or, equivalently, *dorsal* (*ventral*) to it. The plane between the left and right hemispheres demarcates the midline of the brain. Any structure closer (farther) to the midline than the reference point is *medial* (*lateral*) to it. Left and right are used to indicate lateral directions taken from the back of the brain looking forward. All of these directions are indicated on Figure 1. MR images whose slices are oriented perpendicular to the body axis are called *axial* images. By radiological convention, these images are arranged as though looking at brain from the perspective of a doctor facing a patient; thus, left in the image is right in the body and vice versa.

The two cerebral hemispheres are approximate mirror images of each other, separated by the deep longitudinal or inter-hemispheric fissure and connected by a mass of fibers called the corpus callosum. The surface of the hemispheres consists of convoluted folds of "grey matter", a sheet of neurons varying in thickness from two to five millimeters; this is the *cerebral cortex* and contains the neurons responsible for most higher brain functions. The neurons in the cerebral cortex are arranged in six distinct and identifiable layers containing different types of neurons and connections. The thickness of these layers varies across the brain. The cortex can be thought of as a continuous sheet that is folded in a complicated but apparently systematic way. The crest of the convolutions are called *gyri* (singular: gyrus), and the canyons between the gyri are called *sulci* (singular:

sulcus). Beneath the cortex is a mass of fibers ("white matter") that carry information between nearby cells in the grey matter and other parts of the brain. Deep in the brain, above the brain stem and under the cortex, is a collection of large, discrete clusters of cells called the *basal ganglia* that manage a variety of motor functions. Another large collection of nuclei (neuronal clusters) comprise the thalamus, which is a major relay station that transmits sensory information to the cerebral cortex and connects cerebellum, basal ganglia, cortex, and brain stem.

The cerebral cortex is divided into four main lobes: frontal, parietal, temporal, and occipital. (See Figure 1.) These lobes appear somewhat arbitrary anatomically but are distinct both histologically and functionally. The *occipital lobe*, in the most posterior section of cortex, is exclusively devoted to processing visual input. The *parietal lobe* controls higher visual and somato-sensory processing, navigation, and management of internal representations of the environment. The *temporal lobe* is primarily responsible for auditory processing, the integration of diverse sensory information, language comprehension, and object recognition. The *frontal lobe* controls motor function (including eye movements), speech, and many higher cognitive functions. The most anterior part of the frontal lobe, called the *prefrontal cortex*, appears to be quite special. Relative to other mammals, and even other primates, the prefrontal cortex is substantially enlarged in humans. The role of the prefrontal cortex in high-level cognition is still being investigated, but it currently appears to be involved in the context-sensitive modulation of attention and the voluntary control of behavior. The prefrontal cortex thus interacts with most other systems and is, in a sense, the primary player in high level functions such as planning and problem solving.

Knowledge of a few key neuro-anatomical landmarks is helpful for navigating around the brain. The *central sulcus* separates the parietal and frontal lobes. On the anterior side of this sulcus and coming up onto the gyrus there (called the precentral gyrus) is the primary motor area. On the posterior side of this sulcus and coming up onto the gyrus there (called the postcentral gyrus) is the primary somatosensory area governing perception of such sensations as touch and vibration. The lateral sulcus separates the temporal lobe from the frontal and parietal lobes. On the lateral surface of the temporal lobe itself, there are three identifiable gyri, one above another: the superior, middle, and inferior temporal gyri. Within the superior temporal gyrus near the base of the primary somatosensory area is the area involved in basic auditory perception. At the posterior aspect of this same gyrus in the left hemisphere is Wernicke's area which plays an important role in language comprehension. The lateral surface of the frontal lobe is similarly divided into gyri: superior, middle, and inferior. Roughly on the posterior aspect of the inferior frontal gyrus in the left hemisphere, anterior to the precentral gyrus, is Broca's area which plays the primary role in language expression.

Such a manifest of function-location pairs may give the mistaken impression that the brain is just a collection of specialized units. While true in terms of elementary processes, higher cognitive processing emerges from a complex interaction among these different systems. These functions tend to be organized into integrated pathways that allow hierarchically elaborative processing of stimuli.

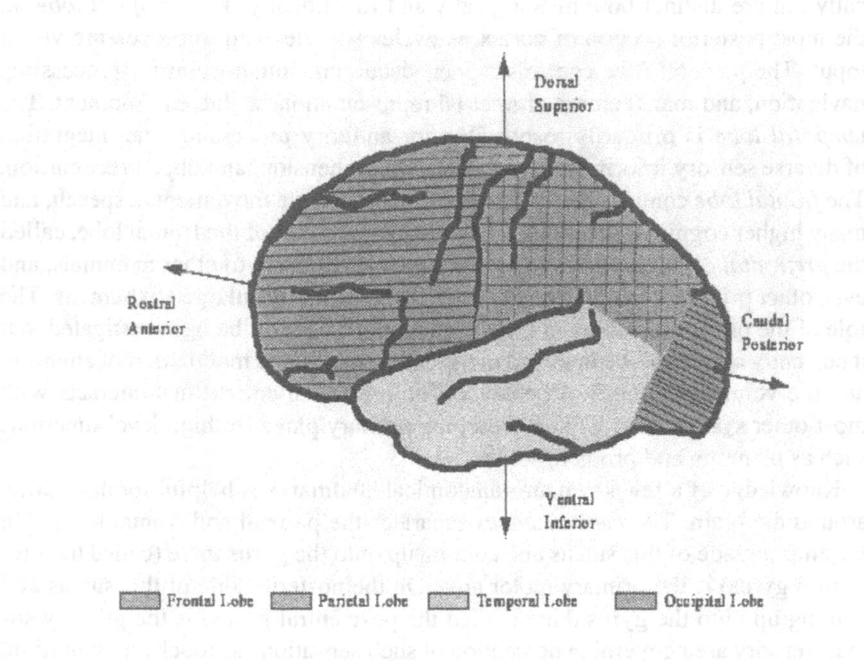

FIGURE 1. A schematic of the brain in external side view, showing a single hemisphere divided into lobes. The arrows indicate the canonical directions, each of which can be referred to by two synonymous terms (e.g., Anterior or Rostral, Posterior or Caudal).

As we will see below, the oculomotor system is an exemplar of the hierarchical organization of the brain. Other examples include the so-called "what" and "where" pathways for object recognition (Haxby, et al. 1991). The recognition of a visual pattern as a particular object starts in the occipital cortex with a decomposition of the input into various visual features, such as angles, edges, colors, shading, perspective, motion. This information is integrated along a pathway—the "what" pathway—that proceeds (ventrally) from primary visual cortex into the temporal lobe. The "where" pathway, which selectively localizes objects in space rather than identifying them, also begins with a collection of visual features but its outputs from visual cortex are different. They proceed (dorsally) up to the parietal lobe, which processes information to place an object within an internal representation of the environment.

The *cerebellum* lies beneath the occipital and posterior temporal lobes and directly posterior to the brain stem; it is clearly identifiable by its very finely folded surface. The cerebellum consists of two lateral cerebellar hemispheres and the midline vermis. The vermis is phylogenetically the oldest area of the cerebellum. It is responsible for the fine regulation of movement, such as enabling one to reach accurately for an object on a direct line without meandering. An individual with damage to the vermis shows motor function similar to a very intoxicated person in that reaching efforts are grossly imprecise, e.g., as in repeated finger to nose pointing. The cerebellar hemispheres are less well understood. They are each arranged in a series of lobules which have a variety of functions. Portions of this region appear to govern fine regulation of timing for movement, such as when to reach to catch a moving object. The lateral regions of the cerebellar hemispheres are phylogenetically newer structures, showing a pattern of development in size similar to the frontal lobe of the cerebral hemispheres.

The *brain stem* is the phylogenetically oldest part of the brain. The brain stem consists of small clusters of neurons (called nuclei) that serve very specific functions. It is the gateway for motor commands from the brain to the rest of the body; it regulates respiration and heart rate; and it has an important role in primitive functions such as sleep. The brain stem also contains the vast majority of the cell bodies for neurons that release modulatory neurotransmitters—the complex chemical messages that modulate the activity of excitatory and inhibitory pathways. These cell bodies project to points all over the cerebral cortex in order to deliver these vital chemicals.

1.2 Studying the Brain

Historically, the study of the brain has progressed at two levels, the macro-level "systems" approach and the micro-level "cellular" approach. The systems approach focuses on the workings of entire brain systems and their relation to observed behavior. Detailed analysis of postmortem brains provides the opportunity to establish clinico-pathological associations by identifying observational correlations between particular clinical/behavioral dysfunctions and the location of lesions or other brain damage. A more advanced method is electrical stimulation of the brain, where an electrical impulse is applied at a particular location either to the surface of the brain or deeper and the corresponding behavioral response recorded. Stimulation has been performed on humans prior to brain surgery and more commonly using animal subjects. Empirical observations of behavior (esp. perception, memory, and problem-solving) provide another tool. This is the basis of cognitive psychology, which attempts to model the function of the brain at an abstract level and understand the capabilities required for high-level cognition to emerge. Results from cognitive psychology set constraints on how the brain can operate. Studies of human development offer further constraints.

The cellular approach focuses on the workings of individual neurons and specialized neural circuits. It involves invasive techniques that continue to be very important for understanding brain mechanisms. The most basic of these techniques is the evaluation of postmortem brain tissue using a variety of dyes and stains that differentially mark various types of cells or neuronal parts. The detailed cellular organization of the brain, particularly the cerebral cortex, has been revealed in this way. Direct cellular (or single unit) recording allows much greater precision but is so invasive that it cannot be used on human subjects. In this technique, a probe electrode is inserted into the brain in close proximity to a single neuron. The electrode records the electrical activity of that neuron over time, and its degree of task-related activity at any time is indicated by its frequency of firing. This method gives extraordinarily detailed information about single neurons, and by painstaking work, the function of many specific circuits has been revealed. Multiple neurons can be recorded simultaneously, but it is currently infeasible to do so if the neurons are widely separated. Hence, this method cannot measure interactions among the distributed neural circuits that are responsible for high-level functions.

In recent years, a synthesis has been forming between the systems and cellular approaches. Structural neuroimaging technologies offered the first non-invasive way to identify, diagnose, and track the development of lesions, tumors, and other brain pathologies. The first of these methods to be widely used is Computed Tomography (CT, the "cat" scan) in which focused beams of x-rays are transmitted through the subjects head from many angles. Because the rate of x-ray transmission through tissue depends on the tissue density, it is possible (with sufficiently many views) to reconstruct an image of tissue density within the brain. CT imaging revolutionized diagnostic radiology, and it was the first radiological technique that required non-trivial statistical methodology to construct the images. Magnetic Resonance Imaging (MRI) measures the density of particular atomic nuclei in tissue by careful manipulation of magnetic fields (see Section 3). Structural MR images can be acquired at very high spatial resolution and provide excellent contrast for clinical radiology.

The advent of functional neuroimaging took this a step further with the non-invasive study of the *active* brain, allowing direct connections to be made between the systems and cellular levels of inquiry. Functional neuroimaging differs from structural neuroimaging in that the technique must be sensitive to physiological changes in the brain that are correlated with varying task demands. Event-

Related Potentials (ERP) measure functional changes directly from the electrical activity of the brain. This can be based on variation in electric fields, Electro-Encephalography (EEG), or magnetic fields, Magneto-Encephalography (MEG). The electrical activity of neurons firing during the task causes measurable changes in these fields. In ERP research, field variations are measured on several "channels" as the subject repeatedly performs some behavioral task (e.g., viewing blinking lights), and under some assumptions, features of the neural activity can be reconstructed from the measured signal. In practice, ERP's provide excellent temporal resolution, but because the underlying reconstruction problems are terribly ill-posed, it offers very poor spatial resolution.

Positron Emission Tomography (PET) images are acquired by injecting a radioactive tracer into the subject's bloodstream Vardi, Shepp and Kaufmann (1985). As the tracer decays, it emits positrons that collide with electrons in the tissue after moving a short distance; the collision gives off a pair of photons (gamma rays) that move in opposite directions. A detector ring around the subject's head can count these photon pairs, and the tracer density in the tissue is estimated from these counts. PET can be used to obtain images that describe a variety of brain processes, including a measure of cerebral blood flow that relates to brain activity(Mintun, et al., 1984). PET images have moderate spatial resolution, but because of limits on the amount of radioactive material to which a subject can be exposed, it is common to average PET images across subjects, further blurring the results. Moreover, a dose of the tracer must decay sufficiently (often over several minutes) before the next image can be obtained, and each image represents an average over this decay period. Subject movement in the scanner thus becomes a critical concern. PET has tremendous potential as a way of tracing other metabolic processes, such as regional glucose metabolism, that are inaccessible to other methods, but its role in functional neuroimaging is being largely subsumed by fMRI.

Among current neuroimaging methods, fMRI offers a superior combination of spatial and temporal resolution, and because the technique involves no known toxicity, subjects can be imaged as often as needed. Functional MRI is having a powerful impact on scientific exploration in many fields and promises to play a vital role in our efforts to understand the brain. We discuss the details of fMRI acquisition in Section 3.

Figure 4 compares the spatial and temporal resolution among all these techniques. The potential for functional neuroimaging, and fMRI in particular, lies in its ability to study higher cognitive processes in humans and to reveal the processing throughout the brain simultaneously.

2 The Oculomotor System: Importance and Function

Of all the senses, vision plays the largest role in helping us interact with our environment. As any computer vision specialist can testify, the challenge of making sense of the myriad visual stimuli we receive is immense (Marr, 1982). We perceive objects, but the eye sees only gradations of intensity and color in the incident light. We perceive depth and distance, but the eye sees only a projection. We perceive a stable world around us even as we move, but the eye sees only a set of disparate images from different angles and perspectives. Vision requires more than a passive recording of information; it requires a coordinated, integrated decomposition of sensory input at many levels. This is the task of the visual system.

The windows to this system are the eyes. The retina, the light-sensitive region on the back of the eye, does substantial processing of visual input and then passes the results to the brain's visual system via the optic nerve. The center of the retina, called the fovea, is the part of the retina that is most sensitive to color and fine-scale features. Because the fovea is relatively small, eye movements are a critical part of visual processing; in primates, eye movement is the primary mechanism by which the fovea can be aimed at objects of interest. Through eye movements, the brain controls what visual input is processed, and thus the oculomotor system is very much involved in the processes of perception. The development of the fovea and the primary use of eye movements (rather than head movements) to focus the eyes on different targets are phylogenetically late developments present only in higher mammals. It is an advance that supports many high-level perceptual activities in humans such as reading text (Just and Carpenter, 1980).

There are two main systems—saccades and pursuit—by which the brain positions the fovea on objects of interest. Saccades are the gross re-positionings of eye fixations by rapid shifting of gaze and attention from one point to another. Saccadic eye movements are very fast (lasting 10's of milliseconds depending on their size) and precise (capable of shifting the eyes a large distance to within a degree of angle of the intended location) (Becker, 1989) and they typically occur without explicit control or awareness. Pursuit eye movements keep the eyes focused on slowly moving targets by matching the angular velocity of the eyes to that of a target of interest (Carl and Gellman, 1987).

The adaptive value of these eye movements is clear: if you want to find and catch dinner for yourself and not be dinner for someone else, it is helpful to be able to scan your surroundings quickly and precisely. In daily life, the eyes scan the key features of the environment approximately 3-4 times per second with some combination of involuntary saccades and pursuit. If there is potentially important information at a location away from the current focus of the eyes, the oculomotor system is engaged to shift the eyes to that new location. Hence, by determining what visual information is processed, the oculomotor system controls the perceptual evaluation of the visual environment.

The process underlying control of eye movements is more complicated than it may initially appear. Maintaining a stable visual perception through the blur of information that occurs during a saccade or pursuit requires holding internal representation of the environment that is separate from the immediate visual input. In addition, whenever the eyes move, the brain must transform its internal coordinate systems, the spatial representations by which it integrates different views into a coherent sense of what and where (Colby, Duhamel and Goldberg, 1996).

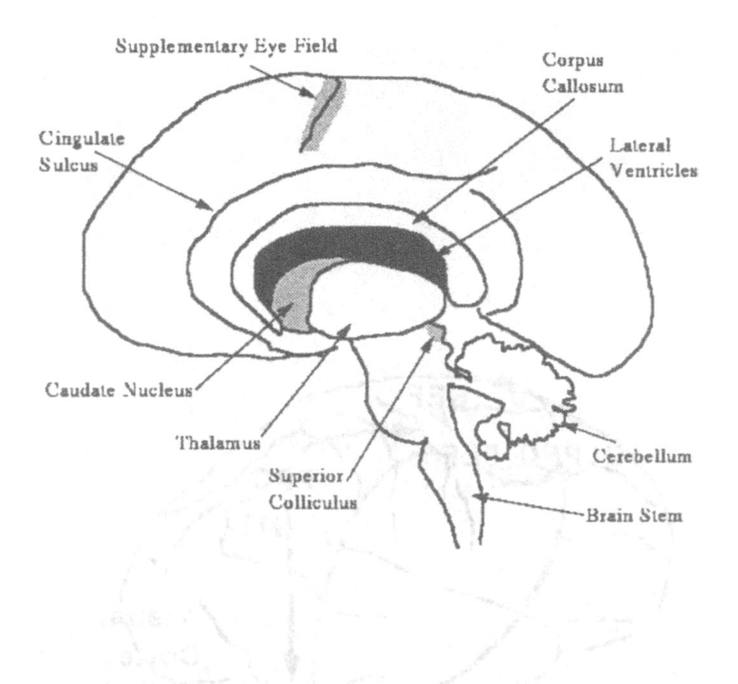

FIGURE 2. A cut-away side view of the brain at the midline plane between the left and right hemispheres. Various anatomical structures that are relevant to eye movements are labeled.

The transformation of these spatial representations are intimately involved in the planning and coordination of motor actions. Eye movements are also modulated by the higher-order processes controlling visual attention and working memory (Sweeney, et al., 1996b; Funahasi, Bruce and Goldman-Rakic, 1989) These interactions indicate that the oculomotor system is embedded into many levels of cognitive processing. Consequently, the study of eye movements offers insights into a variety of brain systems, with implications for both basic and clinical neuroscience. In this section, we describe in some detail the central questions and potential gains that motivate our study of the oculomotor system.

2.1 *Layout of the System*

The oculomotor system is controlled by the complex interaction of several functional components that are distributed throughout the brain. Before examining the underlying scientific issues, it will be helpful to paint a schematic of the system as it is currently understood. The main areas involved in the oculomotor network are depicted in Figures 2 and 3. The basic functions of these areas are as follows.

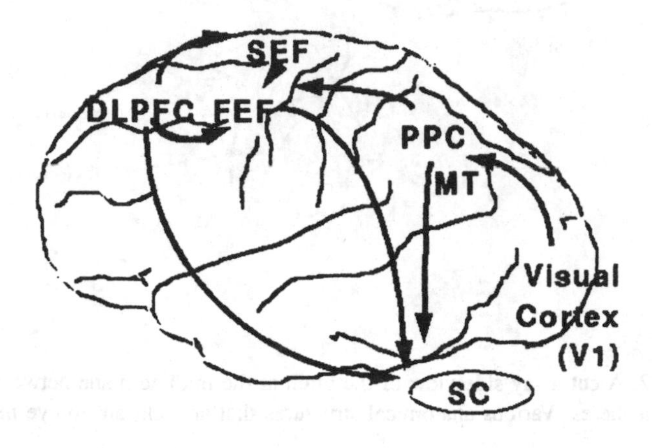

FIGURE 3. A schematic of the oculomotor system, including the principal brain regions involved in the system and the principal pathways among them. Here, SEF=Supplementary Eye Fields, FEF=Frontal Eye Fields, PPC=Posterior Parietal Cortex (particularly the Intraparietal sulcus), DLPFC=Dorsilateral Prefrontal Cortex, and SC=Superior Colliculus. MT (a.k.a. V5) is a motion sensitive component of the visual system. The Visual Cortex contains the early visual areas V1, V2, and so forth. V1 is the largest of these.

The Eyes. As described above, the eyes control and process the input into the system. Six muscles surrounding the eye precisely control its positioning and movement via three cranial nerves.

Thalamus. The thalamus is a complex relay station located superiorally to the brain stem; it serves as the gateway for almost all sensory input to the cortex. The lateral geniculate nucleus in the thalamus receives visual input from the eyes and sends it on to the visual areas in the occipital lobe of the brain. The lateral geniculate is where the initial sorting of visual input takes place, including early differentiation of signals related to color and form (to be processed by the "what" pathway) from those related to location and motion (to be processed by the "where" pathway).

Superior Colliculus. The superior colliculus (SC) lies at the superior and posterior aspects of the brain stem. It is divided into two layers, superior and inferior. The superior layer of the SC receives and processes sensory input and passes information to the inferior layer. This system provides a mechanism for quick-and-dirty sensory processing for tasks or situations where eye movements need to be initiated very quickly. The inferior layer of the SC sends eye movement commands down to oculomotor nuclei in the brain stem. It is arranged in a coarse representation of the visual field that is used to control the path of eye movements (Wurtz and Goldberg, 1972). It integrates input from the superior layer and from several cortical regions. When an eye movement is initiated, a burst of activity occurs in the inferior SC at the represented vector difference between the *desired* and current positions, and as the eyes are moved, a wave of activity proceeds back along that vector to the "fixation zone" representing the origin of the coordinate system (Munoz and Wurtz, 1992). In the fixation zone, sustained activity is resumed to help hold the eyes fixed in place between saccades.

V1, V3, V4 and MT/V5. The visual system in the occipital lobe is partitioned into a number of distinct areas, typically labeled with a V followed by a number. V1 represents the most basic and initial area for cortical processing of visual input. It contains a detailed representation of what is currently in the visual field. This information is then passed through more specialized areas that decompose and recombine the input to isolate particular features. For instance, V3 and V4 appear to process color and form, while V5 (a.k.a. Area MT) is primarily sensitive to motion. V5 is particularly involved in the visual pursuit eye movements, providing information about the direction and velocity of moving targets (Tootell et al., 1995).

Frontal and Supplementary Eye Fields. The frontal eye fields (FEF) and supplementary eye fields (SEF) are the output regions in the cerebral cortex for the generation of eye movements (Bruce and Goldberg, 1985; Schlag and Schlag-Rey, 1987). The frontal eye fields are anterior to the primary motor area, while the supplementary eye fields are anterior to the supplementary motor area (Sweeney et al., 1996b). (The primary and supplementary motor areas control other aspects of voluntary body movement, such as the hands and legs.) Currently, there is little evidence to distinguish the function of the SEF from the FEF. There has been some suggestion in the recent literature that the SEF is more heavily involved with handling *sequences* of eye movements, which involve different logistics and planning (Gaymard and Pierrot-Deseilligny, 1990). Differentiation between the two areas nevertheless still remains an open question.

Posterior Parietal Cortex. The Posterior Parietal Cortex (PPC), particularly what is called the Intraparietal Sulcus, lies along the "where" pathway and is responsible for re-mapping the internal coordinate system as the eyes move. The computational burden here is rather stunning: whenever the eyes move, all of the neurons sensitive to particular locations in the visual field must re-map their sensitivity to the new part of the image. The brain appears to maintain multiple coordinate atlases simultaneously and the re-mapping may be made feasible by encoding coordinate transformations that are more easily changed (Colby, Duhamel and Goldberg, 1996).

Dorsilateral Prefrontal Cortex. This area (DLPFC) on the middle frontal gyrus is associated with maintaining spatial information over time to guide movements that are expected to be initiated in the near future (on the order of seconds) (Sweeney et al., 1996a). It is also believed to provide contextual information that modulates the commands of the eye movement system.

These are the main players in the oculomotor system, but understanding the functioning of these individual areas is less than half the story. Figure 3 also labels the principal *pathways* along which information flows during the preparation for eye movements. The system consists of two main pathways that process the input at different levels. The first and fastest—from the eyes to the lateral geniculate nucleus to the SC—involves the most basic visual processing and motor operations. This pathway is fast and approximate, using only the most salient features of the input and ignoring many aspects of the environment and their contextual relevance in order to get the fastest possible response to unpredictable visual targets. Think of this as generating the command that shifts attention to the nearby small buzzing object (it's a wasp but we don't know that yet).

The second pathway allows for the cognitive modulation of eye movements. Here, some information from the lateral geniculate is passed to V1; from there it is distributed to V3, V4, MT and then to the intraparietal sulcus and temporal lobe. This information is combined and decomposed in a complex way, enabling recognition of objects and other important associative features of the scene. The output is passed to the frontal eye fields and to the prefrontal cortex. The prefrontal cortex integrates sensory input (e.g., what and where) with information about the current context and feeds higher-order commands to the frontal eye fields to modulate the incipient eye movement command. Both the eye fields and the intraparietal sulcus feed into the inferior SC and directly to the lower brain stem. This pathway integrates a wider variety of information about the environment and current context than subcortical pathways in order to guide adaptive decisions regarding the generation or suppression of eye movements. Think of this as generating the command that, realizing that we are falling from a branch of a tree, decides to ignore the wasp and make an eye movement to the object in the visual field that may be (and is, phew!) a nearby branch with which we can save ourself.

These pathways are quite distinct, even though they involve several common areas. The inferior layer of the SC and other parts of the brain stem integrate the output of these pathways, generating or suppressing the eye movement using the most accurate coordinates available under the time constraints on action. The intraparietal sulcus, prefrontal cortex, and frontal and supplementary eye fields relay information both to the SC and directly into the lower brain stem.

While this description of the oculomotor pathways is necessarily simplified, it does reveal some interesting principles underlying the design of the brain. First, there are multiple partially redundant pathways and connections arranged hierarchically. The redundancy provides robustness against gross loss of vital function in the event of focal trauma or disease. At the same time, the incorporation of distinct functions subserved by similar regions allows the system to function more efficiently than if the components were purely redundant. Thus the first pathway provides a way to make a crude but fast eye movement in the absence of other information or if there is need for quickness, while the other pathways do the same job but in a broader environmental context.

Second, as processing proceeds to higher levels, the neural circuits become more general. For example, the part of the visual field to which a given neuron is sensitive (called its receptive field) is small for low-level processing (e.g., the first node in the cortical pathway) but gets larger as the processing becomes more abstract (e.g., later nodes in the pathway). Similarly, the system integrates more and more disparate kinds of input in higher cortical functions. Only basic motor operations and crude sensory information about target location are involved in the first pathway, but later, motion, form, and color information are extracted, then object recognition occurs, then input from other senses is integrated so we can look to where we touched or heard a sound, then memory and other context information are brought to bear progressively to prioritize and plan possible responses.

Third, attention modulates the functioning of the pathways. In our example, if we are falling off the tree branch in the wind and attending to the need to regain control, the fast moving, loud buzzing, possibly pain inducing wasp does not induce an eye movement. The command is generated by the first pathway, but the context of the situation, the relative costs and benefits of the possible actions (having been learned by association with long experience) in some sense determine the focus of our attention. If we are attending to averting a life-threatening fall, our system tends to down-weight the importance of the wasp, and the eye movement to look toward it, as it settles on an arm, may not be made.

Finally, neural computation appears to proceed by successive and repeated splitting and merging of the input in a highly nonlinear way. Each step in the process sorts the data on different features and integrates other features; the result is passed on to other areas where it will be further sorted and integrated.

2.2 *Why Study Eye Movements?*

Basic Neuroscience

One of the basic challenges of neuroscience is to determine how the various functions of the brain are implemented. This requires learning not only about the conditions under which individual neurons fire but also about the interactions among different neural circuits. The complexity of the brain makes it necessary, however, to focus this effort as much as possible, for instance on a particular functional system. The more specialized a system, the easier it is to understand its implementation, but the more integrated a system is with other functions, the greater the insight to be gained. The oculomotor system offers a good balance of these features; it has a specific and measurable purpose and numerous distributed connections into other major systems in the brain.

The study of such a neural system can proceed at four different levels:

1. Localization, in which the anatomical location of the system components is identified,

2. Characterization, in which the basic function of each component is determined,

3. Differentiation, in which related but distinct functions within the system are distinguished, and

4. Integration, in which the interactions among the system components that produce the final behavioral responses are understood.

Although questions at all four levels can be addressed concurrently, this ordering tends to represent the path (from the first to the last) along which the science progresses. Note that the second and third levels here are strongly related, but we distinguish them because they represent the underlying steps in an iterative process of identifying functionality. It is also worth noting that all the various tools used to study brain systems—from single-unit recording to fMRI—also tend to develop along the same path. Part of the goal of this paper is to show how questions at the higher levels can begin to be addressed. For the oculomotor system, interesting scientific questions arise at all four levels.

Localization. Most of what is known about the organization of the oculomotor system was obtained using non-human primate models (Wurtz and Goldberg, 1989). With single-unit recording, the behavior of neurons at a very specific location can be monitored while a monkey performs a variety of eye movement tasks. In this way, the functions of the areas in Figures 2 and 3 have been slowly revealed by showing that a high percentage of cells in a given region exhibit significant alteration in their firing rates in response to specific eye movement tasks. An important question remains, however: How closely related are the organizations of the oculomotor system in humans and non-human primates? Current evidence suggests that the homology is quite good, but even beyond gross morphological differences, there are several examples in which the correspondence breaks down (Luna et al., 1998)(see Section 2.3). Another issue for mapping out the locations of components in the system is variability across individuals in the anatomical geometry of the cerebral cortex. An important task is to assess the magnitude and nature of these variations.

Characterization. Consider what is called the visually-guided saccade task: a subject holds her gaze fixed on a marked point in the center of the visual field, and when a point of light appears at a random location in the visual field, she makes a saccade to the location of the light. This is a simple example of an eye movement task that is used to map out the oculomotor system. Increased neuronal activity close to the time of the saccadic response identifies neurons that play a role in generating the saccade. Different neurons will show sensitivity to the task for saccades at specific direction or distance. Although this is potent information, the story becomes more complicated when other tasks are included. It was long thought that brain areas showed tight specialization in their function. For example, the frontal eye fields are involved in voluntary motion of the eyes, an explicitly motor operation. But more recently, it has become clear that there is substantial overlap in the functions carried out by different components in a system. For example, the frontal eye fields also contain cells involved explicitly in cognitive and sensory operations, even though other parts of the oculomotor network seem to carry out these functions as well (Sweeney et al., 1996b; Sommer and Tehovnik (in press)). How and why this sharing takes place remains to be understood, but it does imply that great care is needed in characterizing functionality, even in low-level sensory-motor systems.

Differentiation. There appears to be a hierarchical organization to functionality in brain systems: at one level, an area will appear to be quite specialized to a particular type of processing, but as we look deeper, we see systematic subdivisions into quite distinct functionality. For example, in the macaque monkey, pursuit and saccade areas within the frontal eye fields are clearly separated (Gottlieb, MacAvoy and Bruce, 1994; Tian and Lynch, 1996). In humans, the separation of these systems in the frontal eye fields is not yet clear (Berman et al., 1996; Petit et al., 1997).

Integration. Having identified the nodes of a functional network like the oculomotor system, the next critical step will be to assess how those nodes interact. How does information flow through the system? How are the functions of the different areas coordinated? How do they share seemingly redundant functionality? What is the temporal sequence of processing? What is the impact on the system if parts of the network are impaired? These questions relate to the functioning of the system as a whole, to the emergence of observable behaviors from distinct functional units. For example, there are two competing views regarding the cognitive control by the prefrontal cortex of the oculomotor and other systems. The first is the "executive" model in which the prefrontal cortex is the central regulator of high-level processing, a sophisticated control system sending commands that coordinate the activity of other systems throughout the brain. The second is the "distributed processing" model in which the coordination is automatic, built into the system through complex pathways and interconnections with higher functions arising as an emergent property of the system integration. In the distributed processing model, the prefrontal cortex is important as a supply of cognitive resources that improves the achievable complexity of the operations but does not control other areas. Both of these views seem reasonable, but they make very different predictions about high-level cognitive processing. This question is concerned completely with the integration of distinct components and with the basic issue of how high-order cognition arises from the brain's architecture. The oculomotor system is an ideal testing ground for studying the various theories, and effective statistical methods are crucial for being able to distinguish fine structure in the brain processes.

Clinical Issues

Eye movement abnormalities are a common neurobehavioral associate for a variety of psychiatric and neurological disorders (Leigh and Zee, 1991). For some disorders, the affected patient exhibits such abnormalities, allowing identification of the disturbed neural pathways and enabling physicians to monitor treatment response. In other cases, biological relatives of an individual affected with an illness exhibit the eye-movement abnormalities without expressing the disorder, indicating that the abnormality may be a familial or genetic marker for a predisposition to develop the illness later in life (Holzman et al., 1974). An improved understanding of the oculomotor system thus promises to yield insights into the changes at the root of these conditions, to help identify risk factors for these disorders, and perhaps to facilitate new diagnostic techniques. Most importantly, oculomotor performance can reflect cortical integrity, enabling the identification of abnormalities in specific cortical systems.

A neurological disorder involves a disturbance in nervous system function. A critical aspect of understanding disorders of the central nervous system is to identify the nature of these disturbances and their precise effect on the brain. While many insights can be gained using animal models or examining postmortem tissue, it would be most desirable to study the functional effects of disturbances in brain tissue as they occur in patients afflicted with a given illness. Functional neuroimaging has made this possible. One basic approach is to study the impact of the disorder on a selected system, and the greatest gains are likely to be realized when the chosen system satisfies the following:

1. Allows accurate measurement of the system's inputs and outputs,
2. Operates at all levels of processing, from basic sensory-motor to abstract cognitive domains,
3. Involves regions that are distributed throughout the brain,
4. Can be studied with relatively simple tasks that can be performed by both patients and healthy individuals over a wide age range.

The oculomotor system has all of these properties. The timing of input stimuli can be precisely controlled and the system output (the eye movement itself) can be precisely measured, both in time and location. Because the visual input is controlled and the motor response readily observable, the dynamics of the task can be monitored. As we have described, the eye movement system integrates all levels of processing in a widely distributed network. Moreover, there is a range of

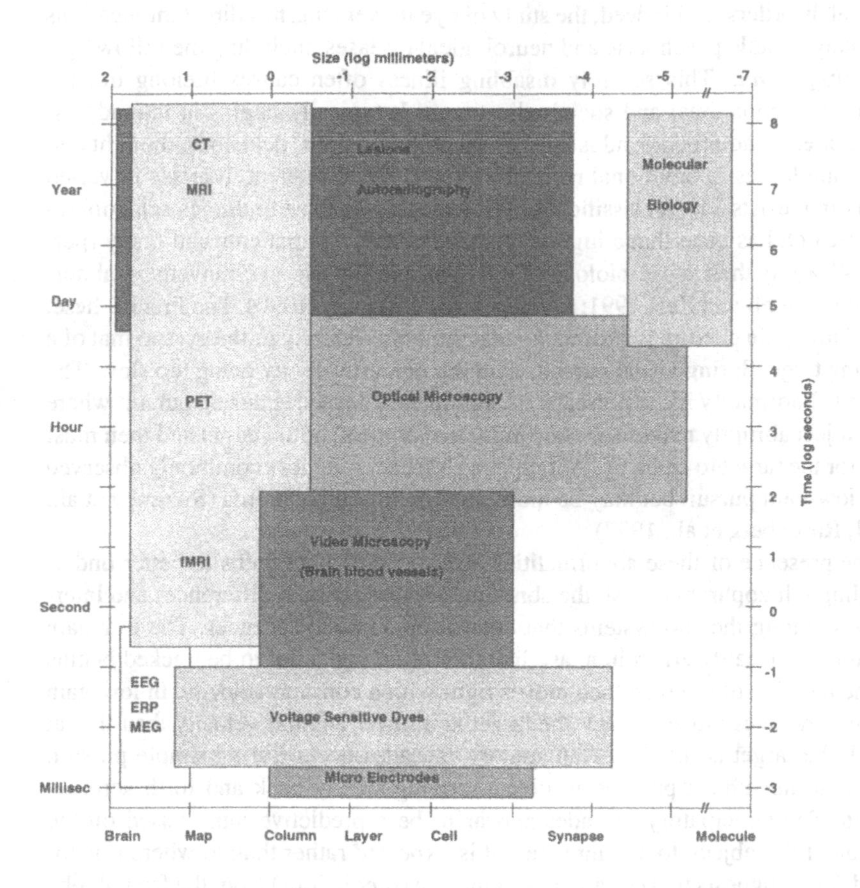

FIGURE 4. A heuristic comparison of the spatial and temporal resolutions of various techniques for studying the brain and brain function. The boxes for each technique show the range of resolutions that can be explored with that technique. This display is based on a similar diagram in Posner and Raichle (1994).

eye movement tasks that exercise the system in different ways (see Section 5), tasks that can be learned and performed by non-human primates, children, and elderly patients. In contrast, complex cognitive tasks can be difficult for patients to grasp and perform and can induce subject movement in the imaging scanner. Eye movement studies also often provide direct linkages to animal models.

The oculomotor system is therefore a powerful tool for studying a number of clinical disorders, and indeed, the study of eye movements has direct implications for many notable psychiatric and neurological diseases, including the following.

Schizophrenia. This severely disabling illness often causes lifelong impairments in occupational and social adjustment. It typically begins in late adolescence or early adulthood and is marked by hallucinations, delusions, thought disorder and losses in emotional responsivity and social interest. It arises in varied forms and resists a tight classification. However, one robust finding in schizophrenia research has been that a high percentage of affected patients and many (perhaps 40%) of their close biological relatives exhibit two eye-movement abnormalities (Leigh and Zee, 1991; Clementz and Sweeney, 1990). The first of these, called low-gain pursuit, is a difficulty matching the velocity of the eyes to that of a moving target during visual pursuit, with the pursuit velocity being too slow. The second abnormality is the presence of anticipatory saccades during pursuit where the subject abruptly makes saccadic movements ahead of the target and then must wait for the target to catch up. Anticipatory saccades are less commonly observed than low-gain pursuit but may be more specific to schizophrenia (Sweeney et al., 1994; Rosenberg et al., 1997).

The presence of these abnormalities offers two distinct paths to better understanding schizophrenia. First, the abnormalities reflect both differences and interactions among the sub-systems that control pursuit and saccades. The low-gain pursuit abnormality arises in a task like the following: a dot to be tracked begins on the left side of a screen then moves right with a constant velocity. In low-gain pursuit, the eyes move to track the target at a lower angular velocity than that at which the target is moving. Anticipatory saccades occur during simple pursuit, in which the subject pursues an object moving slowly back and forth across a screen. The anticipatory saccades appear to be a predictive jump based on the motion of the object, to a point where it is expected rather than to where it is located. This suggests that cognitive information (predictions) from the frontal lobe may be drowning out the motion information from MT. Both of these effects are consistent with the idea that certain pathways to the pursuit system are being disturbed by the disorder. If we can characterize these sub-systems and distinguish their implementation, we may be able to identify some of the specific focal brain abnormalities associated with the disorder.

The second path to understanding schizophrenia is that these eye-movement abnormalities serve to identify families prone to schizophrenia and thus provide more generally homogeneous populations to study. Comparisons within these families can yield insight into how inherited factors influence brain physiology and also identify risk factors that suggest who in the family needs preventative intervention. The genetic basis of schizophrenia is not nearly understood, but the transmission of eye-movement abnormalities to offspring gives information about the genetic basis of the disorder.

Parkinson's Disease. This disease, which typically appears in patients older than 50 years, disrupts the brain's motor systems, especially the voluntary initiation of movement. Reflexive movements are generally unaffected. In the oculomotor system, the disease interferes with the pathway from the frontal eye fields to the basal ganglia to the superior colliculus (see Figure 3). This results in inaccurate voluntary saccades (Bronstein and Kennard, 1985; Lueck et al., 1990). Saccadic accuracy in a variety of eye-movement tasks provides an effective measure for monitoring treatment response (e.g., to L-DOPA) of Parkinson's patients.

Childhood Autism. Autism, or Pervasive Developmental Disorder, is a complex neurologic disorder associated with abnormalities in language skills, social interaction, and higher order cognitive processes. Its onset occurs in the earliest years of life. An ongoing debate about the pathophysiology of this disorder surrounds the question of whether autism arises from pathology in the cerebral cortex, primarily in frontal lobe systems, or in the cerebellum, particularly the posterior lobules of the cerebellar vermis (Lobules VI and VII) (Minshew, Sweeney and Bauman, 1998). Structural neuroimaging studies are not conclusive on this point, in part because they have been confounded by including patients with a number of complicating factors such as fetal rubella and mental retardation that on their own can cause neurological deficits. Little data exist at present directly linking functional deficits with focal anatomic disturbances in this illness.

Studies of eye movements provide a valuable opportunity to test the competing models of frontal and cerebellar origins of autism, because these brain regions are known to play different specific roles in the regulation of saccadic eye movement activity. The control of dynamic aspects of saccades, particularly their accuracy, is known to be regulated by the specific regions in cerebellar vermis (posterior lobules VI and VII) which are believed to be abnormal in autism. The frontal lobe regions believed to be compromised in autism would be expected to be associated not with a disturbance in basic saccade dynamics, but rather with abnormalities in the higher-order voluntarily control of saccades, such as is required to suppress saccades to compelling visual targets and for making saccades to remembered locations without sensory guidance. While our behavioral studies suggest that disturbances in autism are specific to the higher order control of saccades (Minshew, Sweeney and Furman, 1995), fMRI studies are needed to directly document and examine the nature of information processing disturbances in cortical areas of interest, and to link functional and anatomic disturbances to establish clinicopathologic relationships.

Alzheimer's Disease. Alzheimer's disease is the most common dementing condition of late life. It typically begins with memory disturbance, but all cognitive functions become compromised later in the course of the illness. The degeneration of cortical neurons appears to lead to progressively severe inefficiencies in computational activity that are the likely cause of the cognitive manifestations of the disorder. One way to conceptualize such a progression is through a "left-shift" in the relationship between brain activation and task difficulty. In this view, the peak sustained activation would occur during less difficult versions of tasks in Alzheimer's patients than in normal individuals because they need a maximal response from affected regions to perform even simpler cognitive tasks. A second testable prediction is that brain regions (such as the anterior cingulate cortex) in which activation generally parallels increases in task difficulty especially those imposed on prefrontal cortex, would exhibit activation during processing of even simpler tasks under which activation would not be detectable in healthy control subjects. Our preliminary data indicate that activation in the anterior cingulate is robust in Alzheimer's patients performing even a simple visually guided saccade task, while this is not seen in healthy individuals.

Stroke. A stroke, or cerebral vascular accident, is a clinical event in which a reduction of blood flow (e.g., a clot in an artery) disrupts and potentially kills brain neurons because of several factors, including a loss of oxygen needed for cellular metabolism. As the brain recovers from a stroke that permanently destroyed a specific area of the brain, it demonstrates a stunning level of plasticity in its reorganization. This is believed to occur both as areas in the affected hemisphere take over functions that they did not previously serve, and as the unaffected hemisphere takes over functions previously served by the other half of the brain. Understanding the recovery process is important for learning about the plasticity of brain reorganization: to what degree it is reorganizing function and how this reorganization varies with demographic covariates such as gender and age. On the clinical side, this understanding may lead to the development of new treatments to improve recovery from stroke and other localized lesions. Neuroimaging studies of brain reorganization could prove invaluable in developing and validating such novel treatments since they provide a window for directly monitoring physiologic reorganization. Eye movement activation tasks performed in an fMRI environment provide an effective way to track recovery and characterize functional re-mapping because they elicit multiple robust and widely-distributed patterns of activation throughout the brain. Our preliminary data provide striking examples of brain reorganization in the oculomotor system during the months after stroke.

Brain Development and Developmental Disorders. With respect to brain anatomy, one of the most dramatic evolutionary developments in higher primates is the growth of the frontal lobes. The frontal lobes, thought to be critically involved in higher cognitive functions, are especially pronounced in humans, comprising a much higher percentage of brain tissue than in other primates. Recent data indicate that neurobiological development in the frontal cortex continues well into adolescence (Rakic, 1995), a pattern that closely parallels the age-related maturation of higher executive cognitive processes thought to be subserved by this brain region. Single-cell recording studies of non-human primates yield sharper results about how neurons in prefrontal cortex control higher-order eye movements than perhaps any other function. Hence, the study of normal developmental changes in prefrontal activation as individuals perform voluntary saccade tasks provides an especially promising tool to discern the interrelationship between brain maturation and cognitive development. Functional MRI is particularly well-suited to such a study because the injection of radioactive tracers may pose an unacceptable health risk for studies of healthy children and adolescents. This approach will address several fundamental questions in cognitive neuroscience and clinical neuropsychiatry. For example, a number of neuropsychiatric illnesses, such as schizophrenia and obsessive compulsive disorder, are believed to result from abnormalities in later development. Functional MRI studies on children and adolescents whose parents are affected with such disorders provide a novel way to find associations between abnormal developmental processes and the various clinical conditions.

2.3 *Specific Objectives*

Given the importance of pursuit and saccades as the two primary eye-movement mechanisms, a basic question of interest about the oculomotor system is as follows: How do the systems that manage saccadic eye movements and visual pursuit of a target relate to each other functionally? In this paper, we begin to address this question by attempting to differentiate pursuit and saccades, with an eye towards differences across primate species:

> **Target Question.** Are saccades and pursuit implemented by separate sub-systems in both humans and non-human primates?

One of the key distinctions between humans and macaque monkeys is the anatomical classification of the brain areas that subserve pursuit and saccades. In the monkey, the pursuit and saccade areas span the boundary of the prefrontal cortex (one lies in what is called Broadman's area 6 and the other in area 8). On the other hand, both pursuit and saccades appear to be implemented outside the prefrontal cortex in humans (both in Broadman's area 6) (Berman et al., 1996; Petit et al., 1997). Differentiating the pursuit and saccade areas would provide important information regarding the neurophysiology of the cerebral cortex in humans with some clues to the developmental differences between the two species. Specifically, some cellular properties of prefrontal cortex are different than the rest of the cortex (in cortical layer four, the prefrontal cortex is called "granular" because of the types of synaptic and neural elements it contains). Prefrontal cortex also receives inputs from different nuclei in the thalamus than does the rest of the frontal cortex. If pursuit and saccades are differentiated in humans as they are in the monkey, it would suggest that there was an evolutionary "reallocation" of the frontal eye fields in the human, in which eye movement systems were moved. If the two functions are not differentiated, it will provide a significant example in which functional homology between monkeys and humans fails.

3 Overview of fMRI

An MR scanner is a tube containing a very strong, uniform magnetic field along with some mechanism for manipulating the magnetic field within the tube. At its most basic level, the scanner serves as a tool for measuring the density of a single nuclear species (e.g., hydrogen) within a volume of tissue. It is possible to isolate a single nuclear species because of the phenomenon called Nuclear Magnetic Resonance (NMR). By carefully modulating the magnetic field, it is possible to encode spatial information in the measured NMR signal so that an MR image can be constructed, showing the nuclear density in an array of smaller volumes. Functional MRI arises from the sensitivity of these measurements to a blood-flow response in the brain that is related to brain activity. In this section, we describe all of these phenomena to clarify the basic aspects of fMRI data acquisition and to indicate what fMRI can say about brain function. We start at the sub-atomic level.

3.1 Principles of Nuclear Magnetic Resonance

Both protons and neutrons, the basic constituents of atomic nuclei, act and interact like small bar magnets. The strength of their magnetic field is described by a vector quantity called a *magnetic moment*. When particles are grouped together, as in an atomic nucleus, their magnetic moments combine by superposition, and the resulting aggregate acts like a larger bar magnet. Protons and neutrons in an atomic nucleus tend to pair with like particles in such a way that the net magnetic moment of the pair is effectively zero. Thus, only nuclei with an odd number of protons or neutrons will have a non-trivial magnetic moment. Such nuclei can thus be influenced by external magnetic fields.

When a nucleus with a non-zero magnetic moment is placed in a strong, *uniform* magnetic field, the nucleus—acting like a small magnet in a much stronger field—tries to align its magnetic moment either parallel or anti-parallel to the magnetic field. It does so because these are the two orientations of lowest energy among all possibilities, with the parallel orientation slightly lower in energy than the anti-parallel orientation. However, the magnetic moment cannot align itself exactly in either direction because the nucleus has a non-zero angular momentum which must be conserved. Instead, the nuclear magnetic moment *precesses* about the field at some fixed frequency. This angular momentum and precession are the result of quantum mechanical effects rather than any physical rotation, but the conceptual model of a spinning top precessing about the vertical is still a useful way to visualize what is happening here. The precession frequency of the nucleus is determined by two factors: (i) the type of nucleus and (ii) the strength of the uniform field. Specifically, the precession frequency can be expressed as γB_0 where B_0 is the strength of the magnetic field and γ, the gyromagnetic ratio, is a constant characteristic of the nuclear species (e.g., hydrogen, sodium, etc.).

The superconducting magnets typically used for imaging humans have fields in the range of 1.5 to 4 Tesla (T), on the order of one million times the strength of the Earth's magnetic field at the surface. Among the elements whose nuclei contain an odd number of protons or neutrons, there are several that occur naturally in living tissue in sufficient quantity to produce a non-negligible signal: Hydrogen-1, Carbon-13, Sodium-23, and Phosophorus-31. Hydrogen, with a single proton and no neutrons, has the strongest magnetic moment among these and, being a key component of both water and lipids, is far more abundant in the human body than any of the other elements. Consequently, the NMR signal obtainable from hydrogen is about 1000 times stronger than that from any other element, and hydrogen is thus the species to which images are most commonly calibrated. Hydrogen (^{1}H) has a gyromagnetic ratio of approximately 43 MHz/T, so that its precession frequency lies in the radio band of the electromagnetic spectrum.

Now we step back to consider a mass of material containing such nuclei. In many materials, particularly living tissue, there is no preferred orientation, so the orientations of the nuclear magnetic moments are distributed approximately uniformly on the sphere. The superposition of all these moments consequently sums to essentially zero, and the material shows no magnetic properties. This changes when the material is placed in a strong, uniform magnetic field. The magnetic moments of the nuclei within the material align themselves and precess as described above, clustering about the field lines. Since the parallel orientation is a state of slightly lower energy than the anti-parallel orientation, just over half of these nuclei align with the field, and the net magnetization of the material becomes non-zero, i.e., the material acts like a magnet.

At this point, the nuclear magnetic moments (of a given species such as hydrogen) within the field are all precessing about the uniform field with the *same frequency*, but the phases of this precession are distributed uniformly about the circle. The mass of material thus has a constant magnetic moment. If we could get the nuclear magnetic moments to precess *in phase* with each other, the net magnetic moment of the material itself would then also precess with that frequency. Why would we want that? A varying magnetic field induces a current through any nearby loop of wire. If the mass of material had a precessing magnetic moment, the amplitude of the current induced in a nearby receiver coil—the NMR signal—would be proportional to the density of nuclei (of a given species) within the material.

This is where resonance comes into play. To make the nuclei precess in phase, we expose the material to a radio-frequency (RF) pulse of the exact frequency at which the nuclei are precessing. This pulse *excites* the nuclei by "tipping" their magnetic moments away from the main field; whereas they had been precessing in a small circle about the field, they are now precessing in a large circle about the field and *in phase*. Only those nuclei precessing at the right frequency are affected by the tipping pulse; this is resonance. Moreover, the effect of the tipping decays exponentially with time, in two ways. The first results from the gradual realignment of the tipped magnetic moments with the strong, uniform magnetic field. The second results from a gradual loss of phase coherence, where local magnetic irregularities move the precessing magnetic moments out of phase with each other and reduce the measurable NMR signal. This second source of decay tends to be faster than the first and plays an important role in fMRI, as described below.

3.2 *Magnetic Resonance Imaging*

The main magnet in the MR scanner is designed and adjusted to generate a magnetic field that is as uniform as possible within the imaging volume. As such, the NMR phenomenon allows an MR scanner to measure the total density of a nuclear species such as hydrogen within a volume of tissue. In order to acquire detailed spatial images of how this density varies in the tissue, however, it is necessary to carefully encode spatial information into the field of precessing nuclei. There are countless variants and details of acquisition that apply in practice, but the following is a simplified conceptual description of MR images acquisition, using hydrogen imaging as an example. (See also Brown and Semelka (1995).)

Introduce a Cartesian coordinate system (x, y, z) where, for definiteness, we take the z-axis to be aligned with the axis of the subject's body. We will consider acquiring an M-slice image, where each slice is an $N \times N$ array of contiguous volume elements, or *voxels*. For the purposes of this discussion, we will take $N = 4$ and $M = 1$. (In typical fMRI studies, N is either 64, 128, or 256; and M ranges from 1 to 14. Typical voxels have sides from 1mm to 3mm in plane with slice thickness ranging from 3mm to 5mm.) Let $\rho_{jk} \geq 0$ for $j = 0, \ldots, N-1$ and $k = 0, \ldots, N-1$ denote the densities of hydrogen nuclei in our 16 voxels, with j varying in the x direction and k varying in the y direction. The task is to use the NMR phenomenon to extract the ρ_{jk}'s. Spatial information is encoded into the magnetic field by introducing gradients in the magnitude, ***but not the direction***, of the main magnetic field. There are three basic parts of the acquisition sequence: (a) slice selection, (b) phase encoding, and (c) frequency encoding.

Although it is possible to obtain fully 3-dimensional images, most current acquisition methods acquire each image as a sequence of distinct, and possibly separated, slices. Let ν_0 be the known precession frequency in the main magnetic field for the target nuclear species (e.g., hydrogen). Because the RF pulse excites only the nuclei that are precessing at frequency ν_0 and because the precession frequency of a nucleus depends linearly on magnetic field strength, we can control which nuclei get excited by adjusting the gradients. Slice selection is accomplished by applying a constant gradient to the magnitude of the main magnetic field in the z-direction, which makes the magnetic field strength vary linearly as a function of z coordinate but remain constant in x and y for fixed z. The gradient causes only those hydrogen nuclei in a target slice of tissue to be precessing at frequency ν_0; those nuclei outside the slice are precessing faster or slower because the magnetic field is larger or smaller there. The selected slice lies in the x-y plane. By changing the gradient, we can select which slice of tissue gets excited by the pulse and how thick that slice is. After the slice selection gradient is applied, the RF excitation pulse is delivered, and because of the NMR phenomenon, only the hydrogen nuclei in the target slice are excited. The slice-selection gradient is then turned off, but the excited nuclei continue to precess in phase at frequency ν_0, generating a time-varying magnetic field. (In practice, there is decay and de-phasing as described earlier, but we ignore that here.)

Figure 5a shows a snapshot of our 16 voxel image just after the slice selection

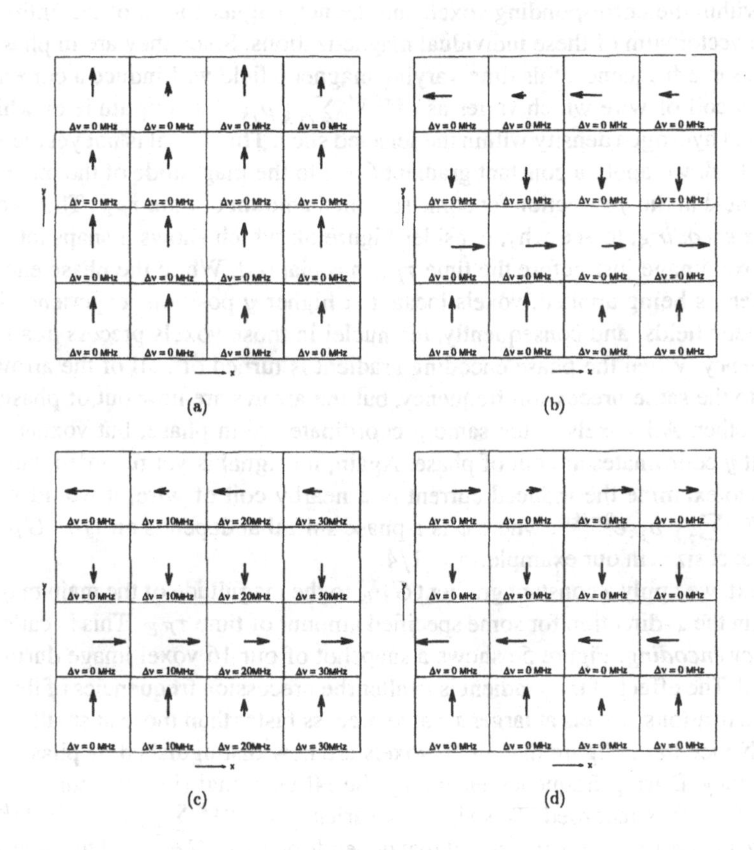

FIGURE 5. Four snapshots of voxel magnetizations at different points during the acquisition of a 4 × 4 × 1 image. The arrows denote the in-slice component of the net voxel magnetization; think of them as rotating clockwise (and clock like) around their flat end. The lengths of the arrows are proportional to the hydrogen densities within the voxel. The directions of the arrows show the relative phase of precession. The value of $\Delta\nu$ for the voxel indicates the difference between the precession frequency for the voxel magnetization (the frequency with which the arrow is rotating) and the base precession frequency ν_0 that would hold in the absence of gradients. The four panels show the configuration in the following cases: (a) just after slice selection, (b) at the end of phase encoding, (c) during frequency encoding, and (d) at the end of phase encoding (using a different phase shift than in (b)).

gradient is turned off. The arrows show the in-slice component of net magnetization for each voxel. These are all precessing in phase (indicated by the arrows pointing in the same direction) and at the same frequency (indicated by $\Delta\nu = 0$ within the cells). The length of each arrow is proportional to the hydrogen density ρ_{jk} within the corresponding voxel, and the net magnetization of the entire slice is the vector sum of these individual magnetizations. Since they are in phase and at the same frequency, this time-varying magnetic field will induce a current in a nearby coil of wire which varies as $e^{2\pi i\nu_0 t}\sum_{j,k}\rho_{jk}$, the amplitude of which is the total hydrogen density within the selected slice. This signal is not yet recorded.

Instead, we apply a constant gradient G_{PE} to the magnitude of the main magnetic field in the y-direction for some specified amount of time τ_{PE}. This is called *phase encoding*; to see why, consider Figure 5b which shows a snapshot of our 16 voxel image just before the time τ_{PE} has elapsed. While the phase encoding gradient is being applied, voxels located at higher y positions experience larger magnetic fields, and consequently, the nuclei in those voxels precess at a higher frequency. When the phase encoding gradient is turned off, all of the arrows return to the same precession frequency, but the arrows are now out of phase with each other. All voxels at the same y-coordinate are in phase, but voxels at different y coordinates are out of phase. Again, no signal is yet recorded, but if we were to examine the induced current in a nearby coil of wire, it would vary as $e^{2\pi i\nu_0 t}\sum_{j,k}\rho_{jk}e^{2\pi ik\phi}$, where ϕ is a phase shift that depends on τ_{PE}, G_{PE} and the voxel size. In our example, $\phi = 1/4$.

Next, we apply a constant gradient G_{FE} to the magnitude of the main magnetic field in the x-direction for some specified amount of time τ_{FE}. This is called *frequency encoding*. Figure 5c shows a snapshot of our 16 voxel image during this period. The effect of the gradient is to alter the precession frequencies of the voxel magnetizations; voxels at larger x value precess faster than those at smaller x values. Notice in the figure that all 16 voxels are now distinguished by phase and/or frequency. During frequency encoding, the NMR signal (i.e., the current in the coil of wire) is recorded. This signal is varies as $e^{2\pi i\nu_0 t}\sum_{j,k}\rho_{jk}e^{2\pi i(j\nu_1 t+k\phi)}$, where ν_1 is a frequency increment that depends on τ_{FE}, G_{FE}, and the voxel size. In our example, $\nu_1 = 10$MHz.

Thus after one pass of slice selection, phase encoding, and frequency encoding, we have recorded a complex linear combination of the quantities ρ_{jk} which we wish to recover. Note that we can completely describe this recorded signal by saving the values at times $t = 0, \ldots, N-1$. To completely reconstruct the ρ_{jk}'s, we need to execute the above process a total of N times, using phase shifts $s\phi$ for $s = 0, \ldots, N-1$. Figure 5d shows a snapshot of the image using a phase shift value of $3/4$.

The resulting measurements are a matrix R_{st} of the form

$$R_{st} = e^{2\pi i\nu_0 t}\sum_{j=0}^{N-1}\sum_{k=0}^{N-1}\rho_{jk}\, e^{2\pi i(j\nu_1 t+k\phi s)},$$

for $s = 0, \ldots, N-1$ and $t = 0, \ldots, N-1$. In words, after a demodulation at known frequency ν_0, the matrix R is just the Fourier Transform of the matrix ρ. To reconstruct ρ, we just apply the inverse transform to the demodulated measurements.

3.3 Functional Magnetic Resonance Imaging

The BOLD Effect

Among the first studies to use MRI to assess functional neural activity in humans are (Belliveau et al., 1992; Kwong et al., 1992 and Ogawa et al., 1992). The latter two introduced the Blood Oxygenation Level Dependent (BOLD) effect for characterizing activation in the brain. Most fMRI experiments are based on the BOLD effect because it gives roughly an order of magnitude better sensitivity to task-related changes than do the available alternatives.

The BOLD effect owes its existence to two distinct phenomena, one chemical, one biological. First, the presence of atoms like iron with magnetic properties can speed the decay of the MR signal by de-phasing the precession of nearby excited nuclei. This is an irregularity, an artifact, where randomly scattered iron atoms exert a differential magnetic force on the excited nuclei, pulling them out of phase more quickly than in homogeneous tissue. The net result is that the local strength of the MR signal is reduced. Iron is an important component of hemoglobin, the molecule in blood responsible for carrying oxygen. When hemoglobin is oxygenated, the attached oxygen molecule shields the magnetic properties of the iron, eliminating its impact on the MR signal. In de-oxygenated hemoglobin, the iron is unshielded. Consequently, as predicted by Thulborn et al. (1982), oxygenated blood gives a larger MR signal than de-oxygenated blood with the same hydrogen density. The second phenomenon underlying the BOLD effect is that the brain responds to concentrated activity with an in-flow of oxygenated blood to the active region. This *hemodynamic response* leads to an increase in the relative volume of oxygenated to de-oxygenated blood in the active region. The result is that a voxel tends to show a slightly (about 1%-5%) larger signal during periods of activity than during periods of inactivity. This is the BOLD effect.

fMRI Experiments

During an fMRI experiment, the subject performs a specific sequence of behavioral tasks in the MR scanner while images of the subject's brain are acquired at regular intervals. These tasks are designed to exercise specific behavioral processes and can range from active reasoning and cognition through passive attention to sensory stimuli. The observed pattern of responses in the brain offers insight into how the processes of interest are implemented. Each distinct task defines a single experimental *condition*. The experimental run is typically divided into blocks of time in which a single condition applies, with the corresponding task repeated as necessary to fill the block. These blocks of time are called task *epochs*, and within the experiment, there are multiple epochs nested within each condition distributed throughout the experimental run.

Figure 6 illustrates the design of a simple fMRI experiment, in which two con-

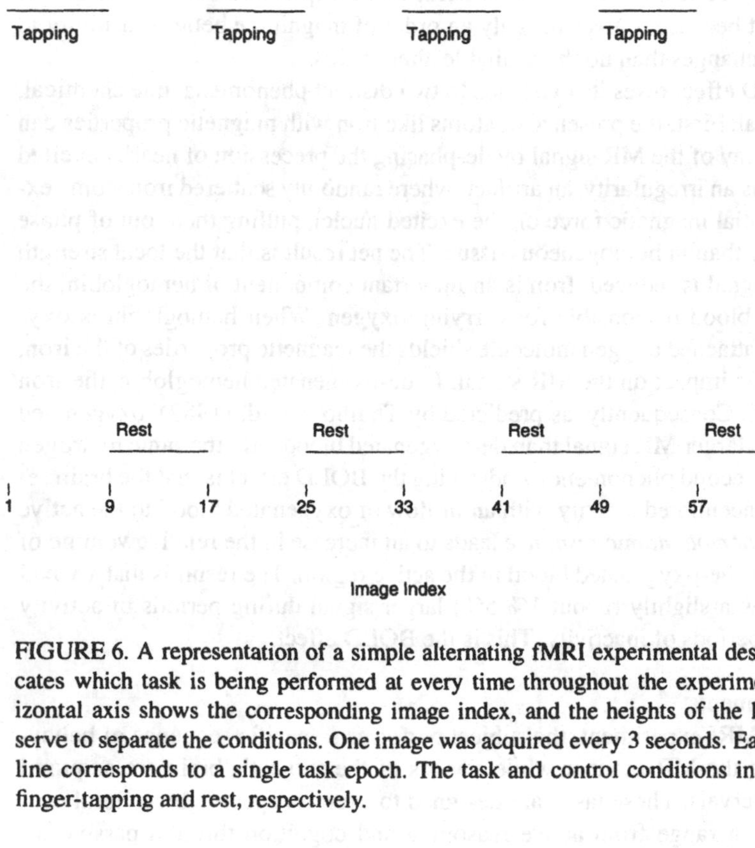

Tapping Tapping Tapping Tapping

Rest Rest Rest Rest

1 9 17 25 33 41 49 57

Image Index

FIGURE 6. A representation of a simple alternating fMRI experimental design that indicates which task is being performed at every time throughout the experiment. The horizontal axis shows the corresponding image index, and the heights of the line segments serve to separate the conditions. One image was acquired every 3 seconds. Each horizontal line corresponds to a single task epoch. The task and control conditions in this case are finger-tapping and rest, respectively.

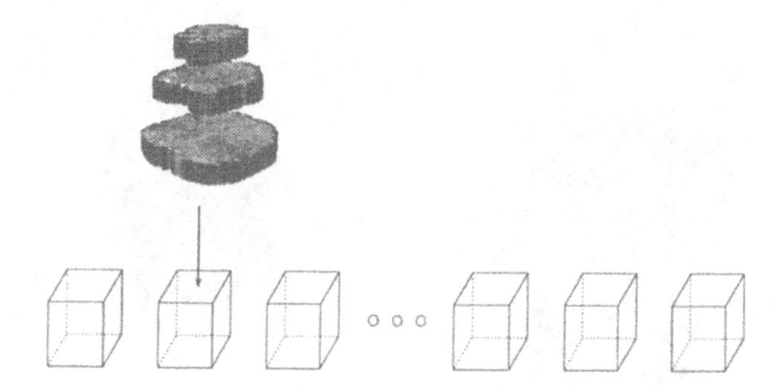

FIGURE 7. A representation of an fMRI data set. A series of three-dimensional images is acquired at regular intervals. Each of these images (the cubes) consist of an array of values, one value per volume element (voxel). The brain is typically imaged in a series of two-dimensional slices as depicted.

ditions alternate in epochs of equal length. In one condition, labeled "Task", the subject performs a task that involves a specific behavioral process of interest. It also necessarily involves other process of lesser interest. Consider for example a simple task for studying eye movements: moving one's eyes from a mark at the center of the presentation screen to the location of a light flash. This requires attending to the mark at the center, recognizing the light flash, and moving the eyes to the desired location, only the last two of which are of direct interest. In order to isolate these processes of interest, our simple design also includes a condition, labeled "Control", that (ideally) involves all of the processes in the "Task" condition except those explicitly being studied. Hence, in the example just described, the control condition may involve fixating one's gaze at the mark in the center of the screen. Any observed differences in the brain's response to the two conditions may be attributable to the isolated processes of interest. This approach to fMRI design requires that some strong assumptions be met regarding the set of processes involved in the two conditions; nevertheless, it is pervasive in the field and generalizes to much more complicated designs.

The data from an fMRI experiment is a time-series of three-dimensional images, where each image is composed of an array of volume elements, or voxels, ranging in volume from roughly 3 to 30 mm^3. Figure 7 shows a schematic of an fMRI data set. There is a trade-off in image acquisition between spatial and temporal resolution, where the cost of finer spatial resolution is longer acquisition time. Figure 8 shows two image slices, one with very fine resolution ($\sim 4mm^3$) voxels but acquired slowly and the other with coarser ($\sim 27mm^3$) voxels but acquired rapidly.

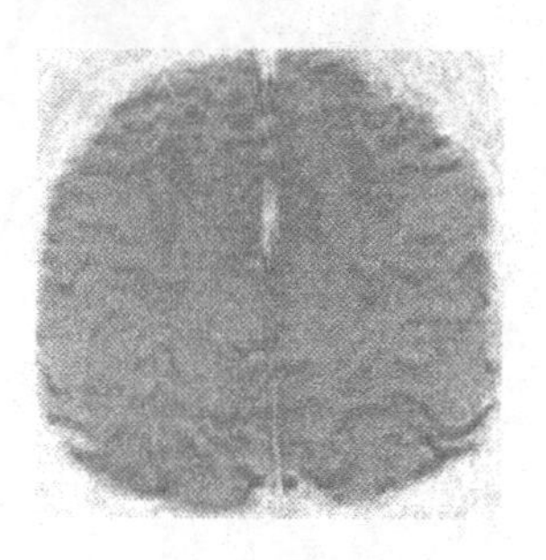

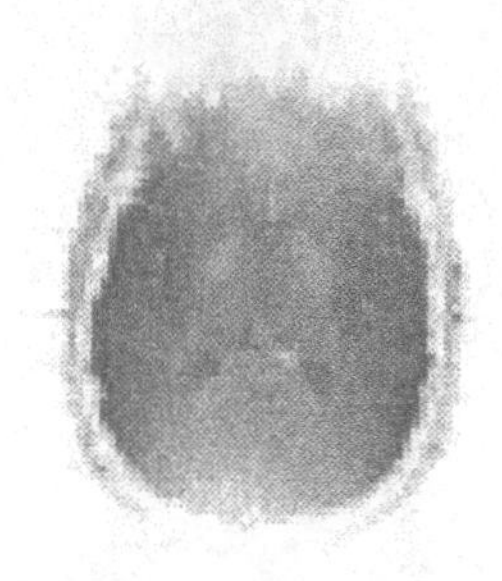

FIGURE 8. Two slices of functional MR images showing the capability of trading-off temporal and spatial resolution.

What makes *functional* MRI distinct from structural MRI is that the primary interest centers not on the images themselves, but on subtle changes in the measured signal across time, changes caused by physiological effects (i.e., the hemodynamic response) related to neural activity. If we consider the measurements associated with a particular voxel, we obtain a time-series describing the density of hydrogen (or some other nuclear species) within the volume of the voxel. See Figures 9 and 10. However, these density measurements are distorted by a variety of magnetic effects, including the BOLD effect discussed above. Consequently, the hemodynamic response to a period of neural activity perturbs the measured hydrogen density in a systematic way. Figure 11 shows a voxel time series exhibiting a large BOLD perturbation that is correlated with performance of the "Task" condition. The scientific challenge in fMRI is to determine from such task-related changes how the brain subserves the processes under study. The statistical challenge in fMRI is to infer the pattern of task-related changes, if any, from the voxel time-series.

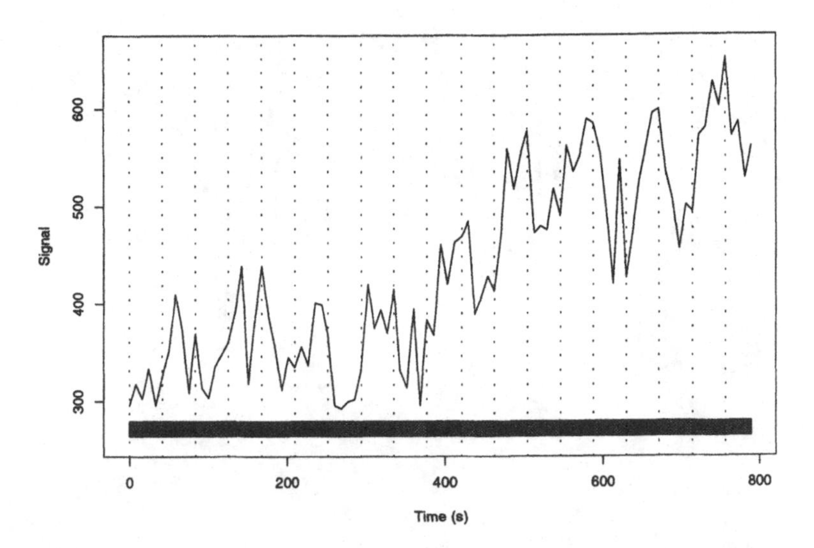

FIGURE 9. A single voxel time series, with time on the horizontal axis (seconds) and MR signal strength on the vertical axis (arbitrary units). The hashed band at the bottom of the figure shows the experimental design, with the task of interest being performed during \\\-hashed epochs and the control task during ///-hashed epochs.

Statistical Techniques

If the neurons within a voxel respond strongly to one condition but not another, then the resulting hemodynamic response in the brain will lead to a detectable perturbation in that voxel's time series through the BOLD effect, cf. Figure 11. This suggests a statistical approach for identifying such "active" voxels: Compare the average signal during each condition, and declare a voxel "active" if the difference between these averages is sufficiently large. Most current methods for the statistical analysis of fMRI data generalize this idea and are based on classification of voxels, typically via statistical hypothesis tests. For example, the two-sample t-test is perhaps the most commonly used method to compare the average signal in two conditions. Others in current use include correlations with a fixed reference function (e.g., a lagged sinusoid) (Bandettini et al., 1993), non-parametric tests (e.g., Kolmogorov-Smirnov) (Baker et al., 1994; Lehmann, 1975), spectral analysis of periodic designs (Weisskoff et al., 1993; Friston, Jezzard and Turner, 1994), split-sample t-tests (Friston, Frith and Frackowiak, 1994), cluster-size modulated t-tests (Forman et al., 1995), and F-tests based on the general linear model (Cohen et al., 1994; Cohen, Noll and Schneider, 1993; Friston et al., 1995; Worsley and Friston, 1995). More general classification methods have also been brought to bear, including principal components, cluster analysis, and neural networks (Weaver et al., 1994; Poline and Mazoyer, 1994). The implicit goal underlying all of these methods is to find the location of the activity associated with the processes of interest.

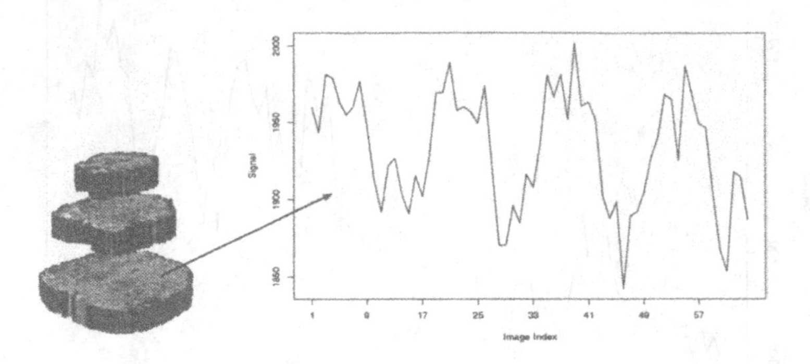

FIGURE 10. The relationship between the voxel time series and the voxel in the image: each measurement in the time series corresponds to the measured nuclear density in the labeled voxel over time.

While these classification approaches can provide reasonable identifications of active voxels, they suffer from two basic limitations: statistical limitations in the underlying models and inferential limitations in the scope of questions that can be addressed. First, the assumptions underlying many of the testing and classification methods are typically simplistic regarding both the noise process and the structure of the activity-induced signal. With respect to the latter, the hemodynamic-response that gives rise to the BOLD perturbation does not have a simple, uniform structure. Variations in the shape of this response can be highly informative and can significantly affect the quality of fit. Nonetheless, most current methods based on hypothesis tests assume a fixed, and often simplistic, response shape, while the available nonparametric methods offer no way to systematically constrain the response shape. See Genovese (1997), Lange and Zeger (1997), and Eddy (1997) for further discussion of these points. With respect to modeling the noise process, fMRI data is subject to diverse sources of variation, but it can be very difficult to account for these complex features of the noise. For example, recent work (Weisskoff, 1997) suggests that non-local, anisotropic correlations between voxels can occur. Typically, fMRI data is passed through several stages of pre-processing before analysis to "correct" for nuisance sources of variation one at a time. The distortion to the data is estimated and residuals are passed to the next stage. This serial estimation detracts from the quality of fit, and because the uncertainties at each stage are not propagated, it makes the final assessment of uncertainty in the inferences optimistic.

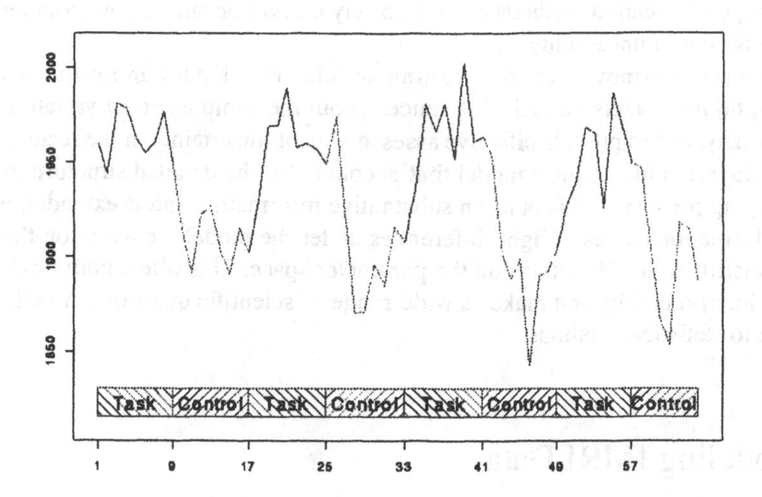

FIGURE 11. This voxel time series shows the correspondence between task performance and the BOLD perturbation. The horizontal axis gives the image index; one image was acquired every 3 seconds. The vertical axis gives the MR signal strength in arbitrary units. The bar along the horizontal axis indicates the experimental design as follows. The \\\ -hashed blocks (red) correspond to the "Task" condition where the subject performs the task of interest. The ///-hashed blocks (blue) correspond to the "Control" condition where the subject performs the control task. The lines on the time series curve are styled to indicate the timing of each data point: solid (red) lines correspond to "Task" and dotted (blue) lines correspond to "Control". Note the slight lag in the perturbation relative to task performance.

The second and much more important limitation of classification-based approaches is that they can primarily address a single basic question: where did the activity occur? While finding the location of activity is an important part of an analysis, it is for many scientific applications of fMRI only a first step. For cognitive psychologists, the primary interest centers on theories that describe the integrated function of the brain, and their goal is to test the predictions of these theories. This requires the ability to use fMRI data to answer questions about relationships among the task responses both within voxel and across regions of the brain. Examples of such questions are described in detail in Genovese (1997). Many of these cannot be addressed directly with classification methods. Investigators using classification methods must then rely on *ad hoc* analysis without an effect assessment of uncertainty.

Our approach is to move away from testing and classification towards estimates of meaningful parameters, to make inferences about the components of variation simultaneously, and to provide effective assessments of uncertainty in the results. We use a Bayesian hierarchical model that accounts for the detailed structure of the underlying processes; it is built on substantive information and is extendable as new information comes to light. Inferences under the model are based on the posterior distribution of functions on the parameter space. This offers both flexibility and interpretability and makes a wide range of scientific questions directly accessible to statistical methods.

4 Modeling fMRI Data

The data from an fMRI experiment result from a sequence of complex processes with biological and technological sources of variation intertwined. The goal of the analysis is to make inferences about the slight, activation-induced perturbations to the signal that enable scientists to gain insight into the intricacies of the brain. The statistical methods for analyzing fMRI data must enable a wide range of inferences and give appropriate assessments of uncertainty. To support inference with such complicated spatio-temporal data, a detailed and flexible model is needed. Three principal challenges arise in modeling these data. First, the underlying noise process is composed of several distinct sources of variation, many of which have an intricate structure. Subject movement during the experiment can cause drastic and irregular signal changes. Much of this movement is non-rigid and three-dimensional and thus is difficult to capture. The subject's physiological cycles—respiration and heart beat—cause large fluctuations in the voxel time series that may be nearly confounded with the task design. These fluctuations are enhanced near tissue boundaries, where much of the interesting activation takes place. Outliers are a persistent problem, and a variety of scanner instabilities and calibration errors can give rise to localized anomalies. The measured MR signal exhibits nonlinear and inhomogeneous drifts that can be large in magnitude relative to the background noise. These drifts can take diverse forms and are spatially

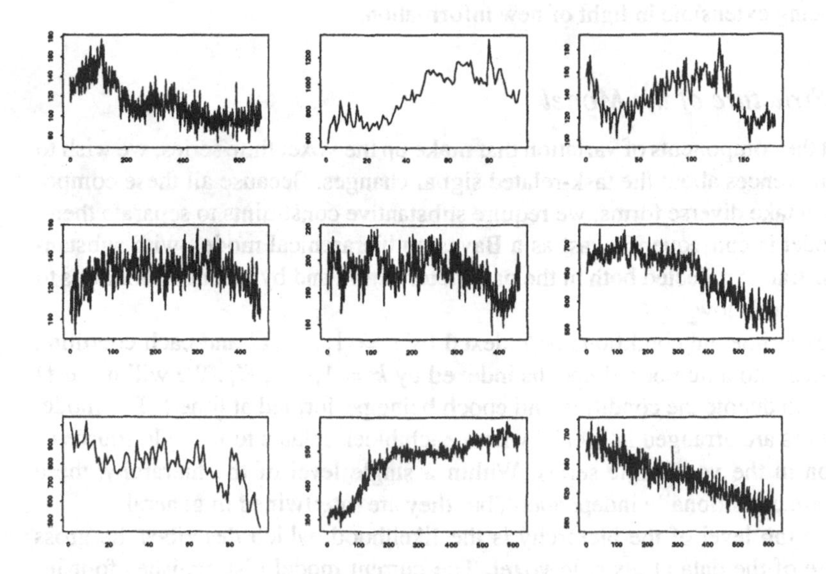

FIGURE 12. A collection of voxel time series highlighting some of theheterogeneity of the signal drift. The horizontal axes representthe image index and the vertical axes the strength ofthe MR signal.

irregular; for example, Figure 12 shows a sample of time courses with a variety of drift profiles.

The second principal challenge that arises in modeling fMRI data lies in capturing the structure of the hemodynamic response. The signal generated by the hemodynamic response in the brain arises from a complex mechanism that is only incompletely understood. The shape of these signal changes varies across the tissue and depends on the distribution of vasculature within a given voxel. Moreover, the magnitudes of the task-related changes are small relative to the noise level and vary over time even in response to the same task. The importance in capturing the shape of the hemodynamic response accurately is that it provides a rich source of information regarding the timing and structure of the underlying activation.

A third and much more difficult challenge is the complicated spatial structure across the brain in both the noise components and the task-related response. Taking advantage of these spatial relationships poses a number of statistical problems but promises to improve the precision of inferences markedly. This is a direction of development for our research, but for now we focus on the temporal aspect alone.

The model we use for fMRI data is described in detail in Genovese (1997), but we review its structure here. Currently, the model takes each voxel to be independent both structurally and stochastically. We are consequently treating each voxel time series as an independent data set with its own parameter values. The model is designed to account for the different sources of variation that affect the voxel time series; the goal is to depict the underlying structure as accurately as possible while being extensible in light of new information.

4.1 *Structure of the Model*

Amidst the components of variation that make up the voxel time series, we wish to make inferences about the task-related signal changes. Because all these components can take diverse forms, we require substantive constraints to separate them. The model is consequently cast as a Bayesian hierarchical model, with substantive constraints reflected both in the parameterization and by prior distributions to be described below.

Let experimental conditions be indexed by $c = 1, \ldots, C$ and each condition be divided into a number of epochs indexed by $k = 1, \ldots, K_c$. We will use $c(t)$ and $k(t)$ to denote the condition and epoch being performed at time t. The model parameters are arranged in blocks where each block relates to a single source of variation in the voxel time series. Within a single level of the hierarchy, these blocks are conditionally independent, but they are intertwined in general.

At the top level of the hierarchy is the likelihood, which describes the gross structure of the data at a single voxel. The current model distinguishes four interpretable components: the baseline signal, the signal drift, the activation profile, and the noise, as follows:

$$Y(t) = \mu + d(t; \boldsymbol{\delta}) + a(t; \tilde{\gamma}_{c(t),k(t)}, \boldsymbol{\theta}, \mu) + \epsilon(t; \boldsymbol{\phi}) \qquad (1)$$

The baseline signal, denoted by the real-valued parameter μ, is the mean level of the signal in the absence of activation and drift, the underlying measure of hydrogen density in the voxel. The drift profile $d(t)$ is modeled as the equivalent of a smoothing spline with respect to a fixed B-spline (de Boor, 1978) or orthogonalized spline basis (where the latter is derived by orthogonalizing the power basis). For both bases, $d(t)$ is explicitly constructed to be orthogonal to constants with respect to the empirical inner product, i.e., $\sum_i d(t_i) = 0$ where the sum is over image acquisition times. The parameters in $\boldsymbol{\delta}$ represent the coefficients of the profile in the corresponding basis. The activation profile $a(t)$ represents the task-related signal change as a function of time; it is a parameterized family of smooth curves, with the parameters describing the magnitude of the response for each condition and the shape with which the response manifests itself in the signal. The precise structure of this curve depends explicitly on the known experimental design through the times at which task performance begins and ends. The parameter $\gamma_{c,k}$ denotes the response amplitude for epoch k within condition c as a proportion of baseline signal; the amplitudes for epochs within a condition c are centered on the average response amplitude γ_c, as described below. The parameter vector $\boldsymbol{\theta}$ describes the shape of the response function for a single epoch of task performance, where each component specifies one feature of the response function. Finally, $\epsilon(t)$ denotes the temporal noise process parameterized by the vector $\boldsymbol{\phi}$. In the model fits presented in this paper, $\boldsymbol{\phi} = (\sigma^2)$ for $\sigma^2 > 0$ and the ϵ's are taken as independent $\text{Normal}(0, \sigma^2)$. In more general versions of the model, this term can include autocorrelations, physiological fluctuations, and heavy-tailed components.

The additive combination of components in the likelihood would not be identified without strong constraints on the components. These constraints are not arbitrary but reflect the underlying structure as accurately as possible. For example, a great deal is known about the basic shape of the hemodynamic response function, and the signal drifts tend to be very smooth. The priors used in the model are based on substantive information—theory, experiment, and observation about the basic processes generating fMRI data—as described in Genovese (1997).

The next level of the hierarchy describes variation in the response magnitude across epochs but within conditions. Specifically,

$$\tilde{\gamma}_{c,k} \mid \tilde{\gamma}_{-(c,k)}, \gamma, \cdots \sim \begin{cases} N_{+}(\gamma_c, \tau_0^2) & \text{if } \gamma_c > 0 \\ \text{point mass at } 0 & \text{o.w.,} \end{cases} \tag{2}$$

where N_{+} is a normal distribution truncated at 0, γ is the vector of average response magnitudes for all conditions and $\tilde{\gamma}_{-(c,k)}$ refers to all of the response magnitude parameters except that associated with the specified epoch. All of these parameters are non-negative and can take on the value 0, and the variance hyperparameter τ_0^2 is fixed (e.g., 2.5×10^{-5}). This level of the hierarchy can be excluded by setting $\tilde{\gamma}_{c,k} \equiv \gamma_c$ for all conditions and epochs; in this case, $\gamma_{c(t)}$ is used in the likelihood described above.

The remaining levels of the hierarchy enforce constraints on the variations of the components within a voxel. The drift is represented by a high-dimensional spline basis, but without constraint it would be grossly over-parameterized. The prior for the drift regularizes the profile $d(t)$ by penalizing both its magnitude and curvature, as follows:

$$\delta \mid \sigma^2, \lambda, \theta\gamma \sim A\exp(-Q(\delta)/2\lambda)), \tag{3}$$

where A is a normalizing constant. Here, Q is a quadratic form defined by

$$Q(\delta) \propto \left[a_n \parallel d \parallel_2^2 + a_c \parallel d'' \parallel_2^2\right],$$

where $\parallel f_2 \parallel_2^2 = \int |f(t)|^2$ is the $\mathcal{L}^2$ norm. The constants $a_n \geq 0, a_c > 0$ are fixed, and specify the trade-off between the norm and curvature penalties; typical values are 0.01 and 1 respectively. Since d is explicitly orthogonal to constants, a very large a_n prevents any drift, whereas a very large a_c forces the drift to be linear. The smoothing parameter λ controls the overall degree of smoothing. This is is a hyper-parameter of the model with prior given by

$$\lambda \mid \sigma^2, \gamma, \theta \sim \text{Exponential}(1/\sigma^2\lambda_0). \tag{4}$$

The mean of this distribution depends on the fixed hyper-parameter λ_0 and is targeted to achieve a desired "degrees of freedom" for the drift profile, as measured by the trace of the corresponding smoothing matrix (Hastie and Tibshirani, 1990). The dependence of this distribution on the noise level is a technical rather than substantive feature, since the smoothness of the drift is not known to depend explicitly on the noise level. However, this noise dependence makes the drift smoothing more adaptive and provides an interpretable way to control the impact of the prior. (It is easier to think in terms of degrees of freedom than in terms of values of the weights.) The other important feature of this distribution is that it shrinks strongly towards zero. When $\lambda = 0$, there is no drift (or only linear drift if $a_n = 0$), so the prior encourages simple drift structure unless sufficient likelihood gain can be achieved with a more complex profile.

The prior for the average response magnitude parameters takes the form

$$\gamma_c \mid \gamma_{-c}, \boldsymbol{\theta}, \sigma^2 \sim \eta \times \text{point mass at } 0 + (1-\eta)\,\text{Gamma}(\rho_{1c}, \rho_{2c}) \qquad (5)$$

where $0 < \eta < 1$, ρ_{1c}, and ρ_{2c} are fixed hyperparameters. Typical values are $\eta = 0.99$, $\rho_{1c} = 1$, and $\rho_{2c} = 50$. The Gamma component of the prior is chosen to cover the interval from 1%–5% of the baseline signal which is roughly the expected range of task-related signal changes for an "active" voxel. The vast majority of voxels will show no response at all to a particular task condition, and the point-mass at zero reflects this possibility. Including such atomic components in the prior is equivalent to specifying a number of distinct sub-models in which various combinations of conditions are excluded. All inferences are generated by averaging over these sub-models. See Genovese (1997) for a further discussion.

The shape of the response is more complicated. Observations have shown that the response manifests itself in the signal roughly according to the following basic template: a lag between beginning of task performance and signal change, a signal rise over approximately 6 to 8 seconds, signal plateau while the task continues, another lag after task performance ends, followed by a slow decay over 8 to 20 seconds with a possible dip below baseline before returning. The shape parameters $\boldsymbol{\theta}$ each capture one of these features, and the model parameterizes the response by a family of piecewise polynomials adapted to this template (Genovese, 1997). The priors for the shape parameters take the components as independent and are thus of the form

$$\boldsymbol{\theta} \mid \sigma^2 \sim \prod_s \text{Gamma}(\nu_{1s}, \nu_{2s}) \qquad (6)$$

where the product is over the component indices of $\boldsymbol{\theta}$ and the parameters ν_{1s} and ν_{2s} are chosen based on current understanding of the blood-flow response. Typical values are $\boldsymbol{\nu}_1 = (2, 4, 2, 4)$ and $\boldsymbol{\nu}_2 = (4.3, 1, 4.3, 0.43)$ for the four parameter response model. The choice of hyperparameters depends in part on the experimental design, as the responses to very short periods of task performance may differ qualitatively from the responses to longer (and more common) periods.

Finally, the noise model used in this paper requires only a parameter to describe the noise variance. The noise level depends on the acquisition parameters, in particular the voxel size and the rate of image acquisition. This dependence can be expressed theoretically for a particular acquisition scheme, and it can also be measured empirically using recorded measurements of "signal-to-noise" tracked by the scanner technicians for quality control. Our prior for the noise level is a proper and diffuse distribution of the form

$$\sigma^2 \sim \text{Inverse Gamma}(\xi_1, \xi_2) \tag{7}$$

where the distribution is centered on the predicted noise level for the given scanner and acquisition scheme. Typical values for the hyperparameters are $\xi_1 = 1.6$ and $\xi_2 = 100$. The spread of the distribution is kept rather diffuse to account for the inevitable variations across voxels in the tissue content and physiological effects. Even more precise information can be brought to bear voxel by voxel if pilot data are acquired prior to the experiment, using the same subject, scanner, and voxel coordinates.

4.2 Fitting the Model

We use two different approaches for fitting the model to fMRI data: direct posterior maximization and posterior sampling using Markov Chain Monte Carlo (MCMC). In the first approach, the location of the posterior mode is used to derive parameter estimates, and standard errors are derived from the normal approximation about the mode with the inverse observed Fisher information matrix. This approach is computationally efficient even for large data sets and is sufficient for most common inferential objectives. Comparisons with the full posterior sampling suggest that it gives good approximations in practice. For numerical optimization to be efficient, it is necessary that the posterior be smooth, so the response amplitude priors with atomic components must be represented as a set of distinct sub-models. Each sub-model corresponds to a constraint that a subset of γ_c parameters are zero. The posterior is maximized separately for each sub-model, and the Bayes Factors are estimated with a form of the Laplace approximation (DiCiccio et al., 1997) or with the Schwarz criterion (Schwarz, 1978; Pauler, 1996). From this, we derive the posterior probabilities of each sub-model, and average results across models using these probabilities as weights.

The second approach, MCMC, is used when more refined inferences are required, for example to study the distribution of complex functions of the parameters or to elucidate the detailed structure of the posterior. The MCMC approach is essential for making inferences about detailed spatial or temporal relationships in the responses across conditions. The sampling strategy is a mix of Metropolis and Gibbs steps (Tierney, 1994), depending on the priors used. Sub-models are handled using the reversible jumps of Green (1995), which yields direct estimates of the posterior probabilities across sub-models. Sampling occurs in three stages: a pre-scan where the initial Metropolis jumping distributions are adjusted, a period of burn-in where no output is recorded and the chain equilibrates, and the final sampling. The MCMC approach is computationally intensive and quite time consuming for large data sets, but it allows variations in the qualitative structure of the model and provides a way to tackle questions that cannot easily be addressed otherwise.

The data analysis process begins with image reconstruction and pre-processing using the FIASCO (Functional Image Analysis Software, Computational *Olio*) software package (Eddy et al., 1996). This includes an algorithm (Eddy, Fitzgerald and Noll, 1996) to align successive images that reduces movement artifacts. Maximum posterior estimates are then computed, and if MCMC sampling is to be done, these estimates are used to generate initial values and Metropolis jumping distributions.

4.3 Using the Results

The result of the model fit is an estimate of the posterior distribution of the parameters given the data at *every* voxel. Of primary interest in most cases are the response amplitude parameters γ which relate to the level of neural activity involved in processing the corresponding tasks. However, other model parameters, particularly those describing the response shape, can be highly informative regarding particular questions. There are two challenges in using these results effectively: we must construct comparisons that address the scientific questions of interest, and we must present these comparisons in an interpretable way. The latter issue is complicated by the need to communicate the results within the fMRI and scientific communities, many of which are unfamiliar with how to interpret a posterior distribution.

In the fMRI literature, statistical results are often reported as maps showing the value of some estimate or test statistic for every voxel, perhaps averaged over subjects. These maps are usually thresholded to classify which voxels are "active", and the thresholded maps are usually overlayed on an anatomical image to associate the observed pattern activity with anatomical structures.

To the degree that familiarity in form is a benefit to communication, we can use the posterior from our model fit to produce similar but more sensitive maps. However, we eschew direct subject averaging as it is currently implemented: rescaling the brains relative to a select few gross measurements (Talairach and Tounoux, 1988) and then mapping classifications onto a common coordinate system. Such across-subject averaging can significantly blur results. We present single subject results here. The problem of integrating results from multiple subjects is an ongoing area of research. See Genovese (1997) for a discussion of other options.

The simplest way to summarize the fit is to map of the marginal probability that $\gamma_c > 0$ for each condition c. These probabilities reflect the strength of evidence in the data that there is some response in the voxel to the corresponding tasks. These probabilities tend to be very small over most of the voxels yielding an unthresholded map with only a few prominent features. If it is desired to overlay this map on anatomical images or if pure classification is of interest, then these probabilities can be thresholded. (Bayes' rule for thresholding with a binary loss function is a reasonable way to do this (Schervish, 1995), although it demands some attention to the relative costs.) Maps of the estimated amplitudes $\hat{\gamma}_c$ are also useful, but unlike the probability maps, these provide no graphical indication of uncertainty.

The unfamiliarity of posterior probabilities raises some concerns from scientists interpreting these images. (It is worth noting however, that scientists often improperly interpret the p-values as posterior probabilities when using test statistics such as the t-test.) A more familiar alternative is to use what we call normalized contrast maps. These are maps whose pixels give the values of a t-like statistic measuring the magnitude of a contrast among conditions relative to the uncertainty:

$$NC_{c,c'} = \frac{\hat{\gamma}_c - \hat{\gamma}_{c'}}{\text{Standard Error}(\hat{\gamma}_c - \hat{\gamma}_{c'})}. \tag{8}$$

Here, both the estimates and standard errors include the variation over sub-models. Under the maximization regime, these standard errors are computed from the estimated covariance matrices. These maps allow us to consider particular comparisons among the conditions; there is of course no need to restrict to pairwise contrasts, any linear contrasts of the amplitude parameters are acceptable.

All of the maps thus far discussed are useful for identifying and localizing active voxels, but this formalism can address many more complex questions expressed as functions on the global parameter space. Many more complex questions can be addressed using this formalism, which is a big advantage of posterior inference in this context. Questions of relationships among the conditions, of variation in the response shape, of spatial relationships in specified regions can all be addressed with a measure of uncertainty and without selection biases. Examples and discussion of this point is a main theme of Genovese (1997). In the study described here, the inferential goals are more modest, and we focus on the response amplitudes at each location.

5 Experimental Methods

In this section, we describe three fMRI experiments designed to help address our target question: are the pursuit and saccade systems differentiated. Our analysis of the data obtained from the first two experiments is presented in the next section. These two experiments were run relatively early in our work when the experimental designs were relatively simple, as largely dictated by the limitations of the statistical tools then available. In this paper, we re-analyze these data using the model and methods discussed above. As an indication of how methods in the field have developed, we also describe a third experiment with a more sophisticated design that makes more subtle comparisons possible. The analyses of this later experiments will be presented in a future paper. All of the data discussed herein were acquired on the 3T GE Signa Scanner at the University of Pittsburgh Medical Center Magnetic Resonance Research Center.

There are several ways to differentiate the brain systems that subserve pursuit and saccadic eye movements with fMRI data. The most direct is to determine if the observed location of activity ascribed to these two processes are physically separated. Specifically, we ask if the voxels classified as active (relative to a suitable control) under the pursuit and saccade tasks are different, and if these differences in location correspond to reasonable anatomic distinctions. Experiments 1 and 2, described below, are motivated by this approach. Such "location dissociations" require careful interpretation because the changes detected by fMRI involve blood-flow responses which act only as a proxy for neural activity. Nevertheless, given what is currently known about anatomical and functional structure of the oculomotor system, we believe that location dissociations can provide strong evidence for addressing our target question.

A second way to differentiate the neural controls for pursuit and saccades is to design an experiment that manifests different temporal patterns of activation from the two types of tasks. For example, both pursuit and saccade are known to activate the opposite hemispheres for a given direction of eye movement (Leigh and Zee, 1991). Pursuit to the left is primarily controlled by the left hemisphere (ipsilateral), and saccades to the left are controlled by the right hemisphere (contralateral). Similarly, the size of a saccade (and possibly the speed of pursuit) change the location of activity within the associated brain area in a very specific way. Experiment 3 is motivated by these considerations.

Experiments 1 and 2 both involve the same simple, two-condition design: the subject alternates between 42 second epochs of a control task and an eye-movement task. In both experiments, the control task, called fixation, requires the subject to hold her gaze on a marked point projected onto the center of her visual field during the entire 42 second epoch. The eye-movement task differs between the two experiments. Experiment 1 uses the Visually-Guided Saccade (VGS) task, and Experiment 2 uses the Pursuit task, both described below. Each of these tasks is repeated continually throughout each 42 second eye-movement epoch.

In the VGS task, the subject is instructed to make saccades to targets that step three degrees of visual angle either to the left or right of the current fixation point. The target starts in the center of the screen at the beginning of each VGS epoch, and the direction of each step is chosen randomly. For example, from the center of the screen, the target would jump randomly either 3 degrees to the left or right, and the subject would shift her gaze to that new target. The target would then jump randomly 3 degrees to the left or right of that new position for the next saccade and so forth. Target jumps occurred every 750 milliseconds. When the target reaches 9 degrees to the left or right of center, it moves back toward center to remain in the subject's field of view from within the scanner.

In the Pursuit task, the subject is instructed to visually follow a target that moves horizontally across the projection screen at a slow speed. When the target reaches 9 degrees to the left or right of center, it returns in the other direction. The target begins the epoch in the center at the fixation point. The target velocity is cosinusoidal as a function of angular distance from the center—it speeds up toward the center of the screen and slows down to zero velocity at the edges—to help the subject reverse pursuit direction easily. This appears to the subject as a pendulum with motion only in the horizontal plane. The speed of pursuit is timed so the target returns to the central fixation point at the end of the epoch.

The data for Experiments 1 and 2 are acquired using two-shot Echo-Planar Imaging (TR=4.2s, TE=25) to obtain very high-resolution voxels of size 1.6mm × 0.8mm × 3mm. The images consist of 6 slices separated by 1mm; these were prescribed by hand using scout images obtained before the experiment and selected to include the region of the FEF within the field of view. Images were acquired every 8.4 s.

The experiments just described involve pursuit and saccades in both directions and thus activity in both hemispheres, making it harder to distinguish the two processes. In Experiment 3, we integrate both the VGS and Pursuit tasks and fixation into the same design with careful control of the timing and direction of eye movements separate the pursuit and saccade activation by hemisphere. Specifically, the experiment consists of a single 30 second fixation period followed by a sequence of identical 24.17 second blocks each of which consists of four epochs, followed by a single 30 second fixation at the end of the experiment.

An important issue is that at the behavioral level, pursuit and saccades are not completely separable. Failure to keep up with the target during pursuit initiates what are called "catch-up" saccades, small saccades that rapidly bring the eyes back to the target during pursuit. The propensity for catch-up saccades is directly related to the speed of pursuit, with a fast moving target requiring a greater number of saccades. However, catch-up saccades do occur during pursuit at any target velocity. Note that the activation from pursuit and the catch-up saccades is expected to lie in different hemispheres since the eye movements are in the same direction. Throughout the task epochs in Experiment 3, pursuit and saccadic eye movements in *opposite* directions are always separated in time enough so that their temporal patterns of activity will be completely out of phase with each other.

Each 24.17 second block in Experiment 3 is divided into four epochs: Pursuit-R, Fixation-R, Pursuit-L, Fixation-L. The block begins with the subjects gaze focused on a fixation point at the left side of the visual field. A target appears there and moves across the screen to the right for 5.8 seconds. This is Pursuit-R, and the subject is instructed to pursue the target as it moves. Catch-up saccades occur throughout this pursuit period, but the activation of saccade and pursuit show up in opposite hemispheres. When the target reaches the other edge of the screen it stops, and the subject is instructed to fixate on the target. This is Fixation-R, and it lasts for 5.8 seconds. The target then moves horizontally left across the screen for 5.8 seconds. This is Pursuit-L. Finally, the target stops at the left side of the screen, and the subject holds fixation for 6.77 seconds. This is Fixation-L.

The timing of the four epochs is important. During 5.8 seconds, a single 6 slice image is acquired. Within a block, the signal measured within one slice occurs at the same point in the task for both the L and R sections. The extra 0.97 seconds of fixation at the end of the block rotates the acquisition by exactly 1 slice, so that each slice is acquired in the same time interval in the Pursuit-L and Pursuit-R epochs within a block. Thus, in the first block of the experiment, slice 1 records what happens at the beginning of pursuit for both L and R, but in the second block, slice 2 records the beginning pursuit and slice 1 records the end. This aspect of the design controls for variation in activity during task performance when making comparisons across hemispheres. The design ensures that simultaneous saccade and pursuit activation occurs in opposite hemispheres. The combined pattern of activity provides an uncorrupted picture in each hemisphere of where pursuit and saccade activation take place.

The data for this experiment are again acquired with two-shot Echo-Planar Imaging (TR=5.8,TE=25) with the high resolution (1.6mm × 0.8mm × 3mm) voxels and 6 slices chosen anatomically. One practical issue is that the speed of pursuit must be carefully selected. Pursuit must be fast enough to engage the saccade system (i.e., manifest an adequately high catch-up saccade rate), which we can test outside the scanner using behavioral data measured in the laboratory (Friedman, Jesberger and Meltzer, 1991). On the other hand, pursuit must be slow enough so that target movement across the screen allows enough time to build up a detectable fMRI signal change. This movement is limited by the angular view of the subject, for once the target reaches the edge of the presentation screen, the only options are (i) to switch to fixation, (ii) move in the other direction, or (iii) jump the target across the screen and continue motion in the same direction. The second option here makes anti-directional pursuit epochs adjacent, and the latter introduces a large saccade in the direction opposite the most recent pursuit. Note that the eye-movement task conditions here are lasting for a much shorter time than in Experiments 1 and 2. This is required if we are to separate activation associated with movement in the two directions, but it reduces the duration and probably the peak magnitude of the BOLD response. We attempt to compensate for this reduction by running a large number of blocks. To avoid tiring the subject, we include several periods of rest during which images are acquired while the subject holds fixation. We also take pains to reduce head motion during acquisition by using a carefully trained and thoroughly instructed subject, as well as dense padding around the head. The slow pursuit speed helps reduce movement as well.

6 Results and Discussion

The basic question in this study is the degree to which the brain's systems for implementing saccadic and pursuit eye-movements differ. This question is interesting methodologically because we know from monkey studies that the regions involved are relatively small and spatially contiguous, so that discriminating these functionally distinct regions represents a significant challenge to the spatial resolution of fMRI. We replicated Experiments 1 and 2 described above on eight subjects and here present the results of our analyses. Because these experiments manipulate the two eye-movement tasks independently and compare them separately to the fixation condition, subtle task-related differences in the temporal patterns of response are washed out by variations across scans. Hence, we use the location of activity as our primary tool for separating these systems with these experiments. Fortunately, the methods described in this paper provide better precision and resolution than standard data analytic methods used in fMRI, and we are able to detect systematic (even striking) differences that were obscured in previous analyses.

In this paper, we focus on two specific inferential measures: posterior probabilities that a voxel is responsive to a particular task and normalized contrasts between the task and control conditions. In general, the modeling approach presented above enables more refined comparisons; indeed, the studies in this paper represent the initial stage of work in progress. All of the figures in this section show the results for the same subject. This subject was chosen, quite literally, at random from among the eight. Our results appear to be consistent across all our subjects, and we highlight any notable differences in the discussion.

Figures 13a and 13b shows thresholded posterior probability maps of activation over six brain slices for the VGS (Experiment 1) and Pursuit (Experiment 2) tasks respectively. The probabilities shown in the maps are the posterior probabilities that the response amplitude γ_c is non-zero for the corresponding task condition. The threshold of 0.2 was chosen for visual clarity, not for classification. The probabilities across all voxels divide into two groups, one concentrated near zero and a smaller group with non-trivial values. The value 0.2 lies in the lower tail of the latter cluster, and the vast majority of voxels have a much smaller probability of non-zero response. The patterns of activity in these maps reveal structure where we would expect it for both of these tasks on the basis of previous studies and the animal literature. The activation follows the sulci quite closely, and the Frontal Eye Fields, Supplementary Eye Fields, and the parietal eye fields along the Intraparietal Sulcus all show activation. Even at this level we can see some gross differences between activations present during the pursuit and saccade tasks. One striking feature of these maps is the minimal structure outside the brain. Voxels outside the brain can show no true activity, but naturally, chance fluctuations can lead to significant test statistics. The t-maps we had used for our earlier analyses of these data showed a substantial number of active voxels outside the brain. In contrast, using the methods in this paper, the likelihood gains from fitting these fluctuations are small because the noise fluctuations outside the brain are less likely to conform to the constraints on the activation profile. Consequently, the shrinkage implicit in averaging over sub-models eliminates the vast majority of these spurious structures. This is one example of how the constraints embedded in our model can improve the sensitivity of the inferences.

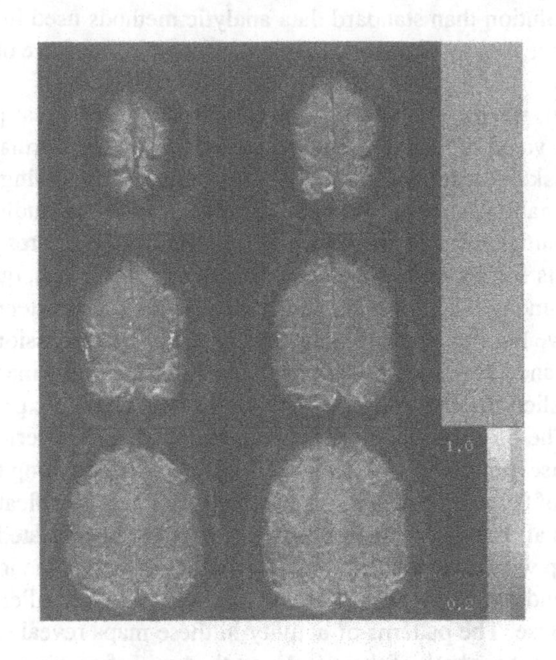

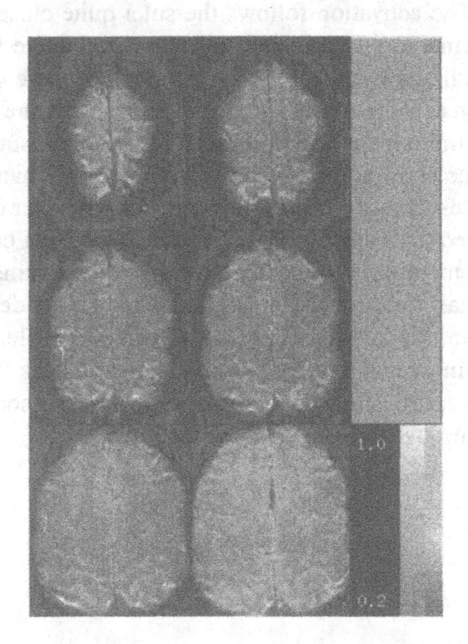

FIGURE 13. Posterior probabilities of non-zero response amplitude for (a) VGS (Experiment 1) and (b) Pursuit (Experiment 2) overlaid on anatomical images for six slices of the brain.

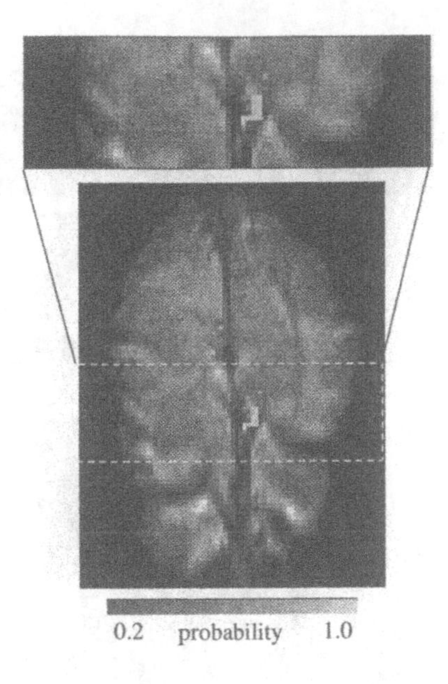

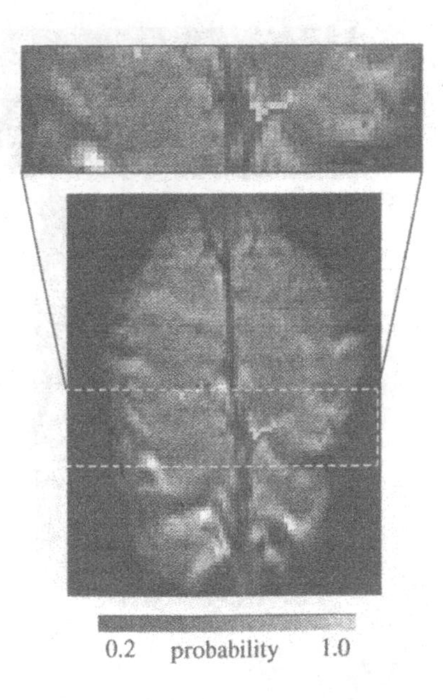

FIGURE 14. Close-up view of posterior probability maps in a single slice for (a) VGS (Experiment 1) and (b) Pursuit (Experiment 2). This image highlights activation differences along the midline: saccade and pursuit activation appear to diverge along distinct sulci; note the distinct directions of the central activation clusters.

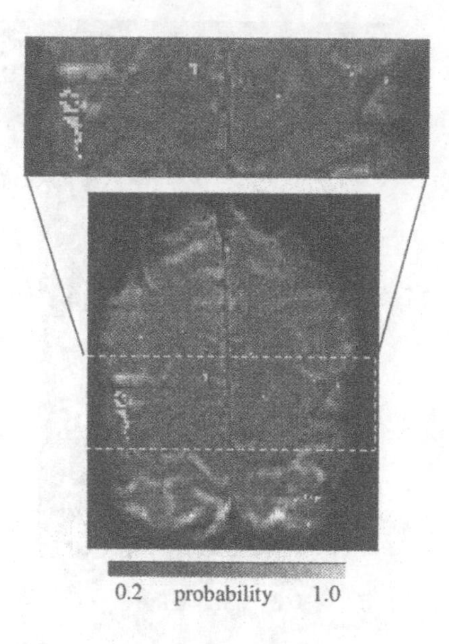

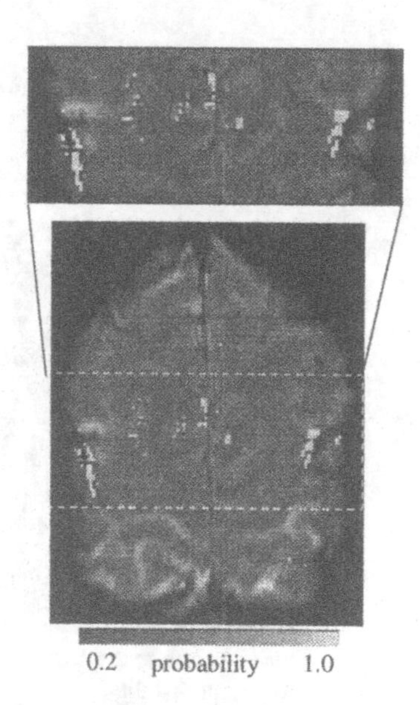

FIGURE 15. Close-up view of posterior probability maps in a single slice for (a) VGS (Experiment 1) and (b) Pursuit (Experiment 2). This image highlights activation differences in the Frontal and Supplementary Eye Fields (activity along the precentral sulcus and near the midline respectively).

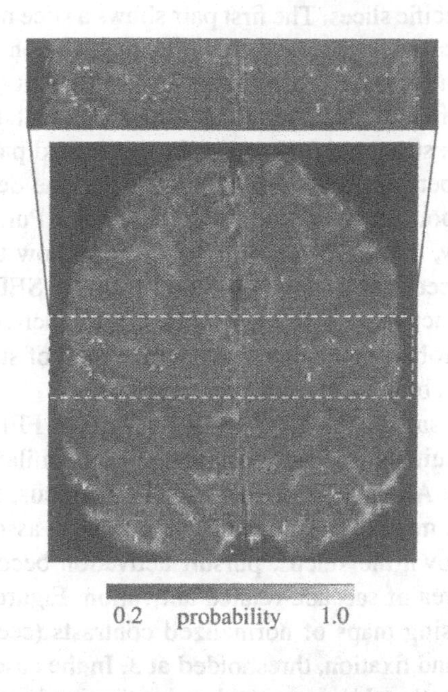

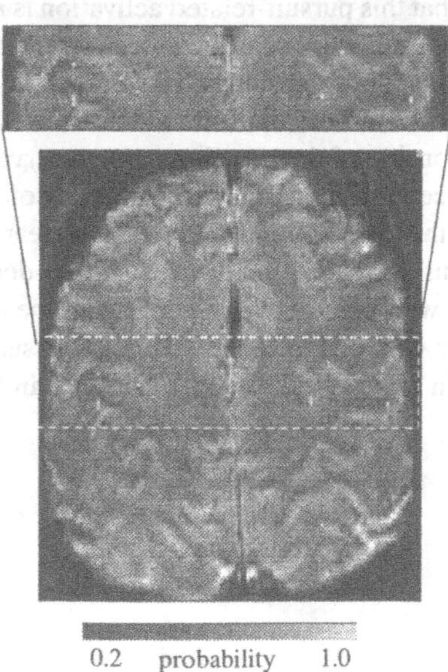

FIGURE 16. Close-up view of posterior probability maps in a single slice for (a) VGS (Experiment 1) and (b) Pursuit (Experiment 2) This image highlights activation differences in the Frontal and Supplementary Eye Fields (lateral and medial clusters of activity respectively).

Three pairs of figures Figures 14ab, 15ab, and 16ab compare activation in FEF and SEF for two specific slices. The first pair shows a slice near the top of the brain for which the activation is concentrated in a small region near the midline. The pattern of activation for VGS (Experiment 1) and Pursuit (Experiment 2) appear to diverge along different sulci (note the different orientations of the activation clusters), showing a structural difference. In the second pair of figures, the most notable difference between VGS and Pursuit is that the dense lateral clusters of activity appear on both sides of the brain whereas for Pursuit they are primarily on one side. Finally, the third pair of figures shows how the activation for both tasks follow the precentral sulcus (see below). In the SEF (the medial areas of activation), Pursuit activation was very modest. VGS activation, when present in the SEF, was most robust at the extreme lateral extent of small sulcal involutions of the midline of the brain.

In the precentral sulcus, the location of the human FEF, the functional organization of the pursuit and saccade regions appears similar to that present in the non-human primate. Along the anterior wall of this sulcus, as it comes down from the cortical surface, most of the activation is primarily associated with saccades. However, deeper down the sulcus, pursuit activation becomes evident immediately beneath the area of saccade-related activation. Figures 17 and 18 illustrate these differences using maps of normalized contrasts (see equation 8) between saccade or pursuit and fixation, thresholded at 3. In the case presented in the Figure 18, it can be seen that this pursuit-related activation is restricted to the anterior wall of the sulcus, and is not evident as the sulcus wraps back toward the cortical surface on the posterior wall of the sulcus. In contrast, in Figure 17, the saccade-related activity is restricted to the opposite side of the sulcus. In our other subjects, the spatial relationship of these regions are preserved, including the spatial ordering along the precentral sulcus and the adjacency of the saccade and pursuit activations. However, the saccade-related area can appear deeper toward the base of the precentral sulcus with the pursuit area shifted deeper to the base of the sulcus and sometimes wrapping between 1 to 2 centimeters up the deepest-most aspect of the posterior wall of the sulcus. This is consistent with the results of single-unit recording in monkeys (Gottlieb, MacAvoy and Bruce, 1994).

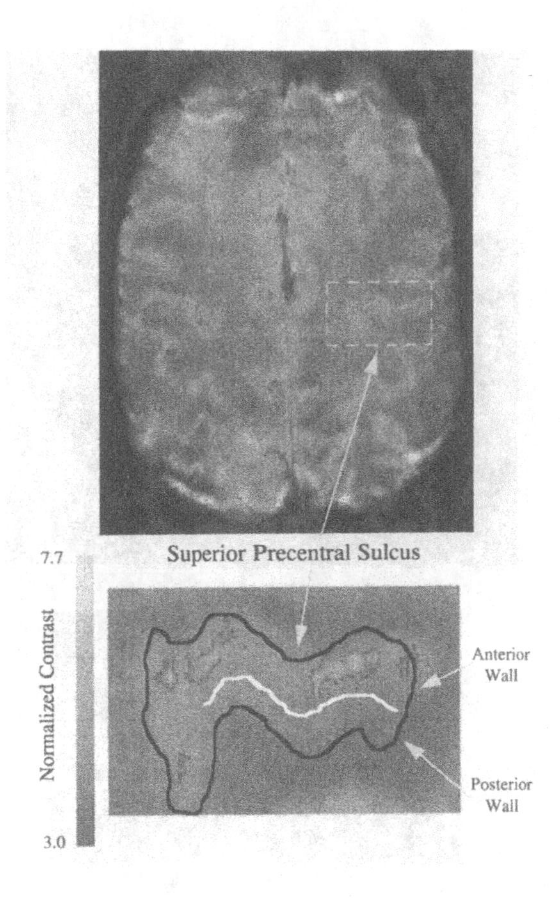

FIGURE 17. Normalized contrast map for VGS activation (Experiment 1) in a single slice, thresholded at 3 and overlaid on an anatomical image. The area surrounding the precentral sulcus is highlighted and outlined. See equation (8) for definition of normalized contrasts.

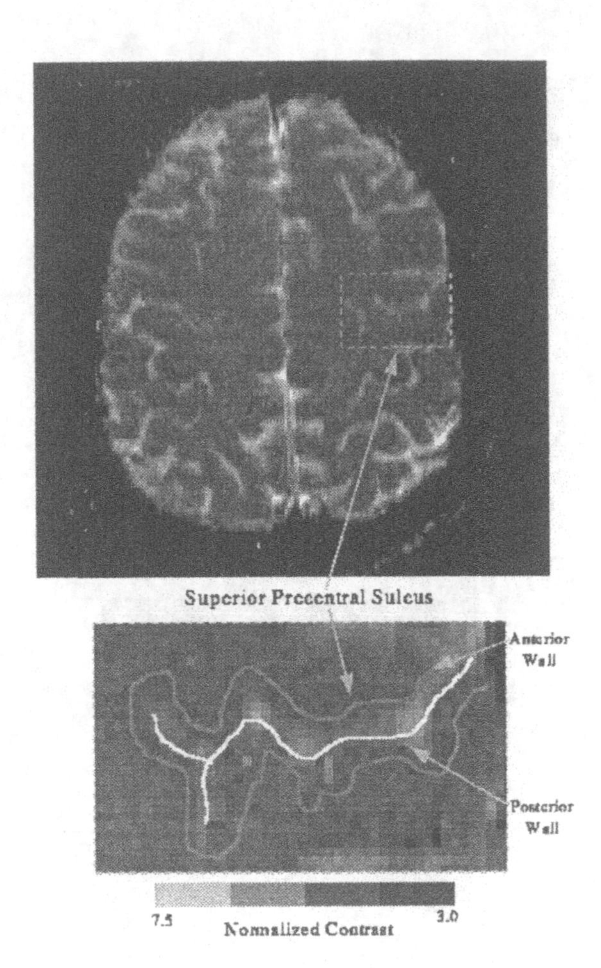

FIGURE 18. Normalized contrast map for Pursuit activation (Experiment 2) in a single slice, thresholded at 3 and overlaid on an anatomical image. The area surrounding the precentral sulcus is highlighted.

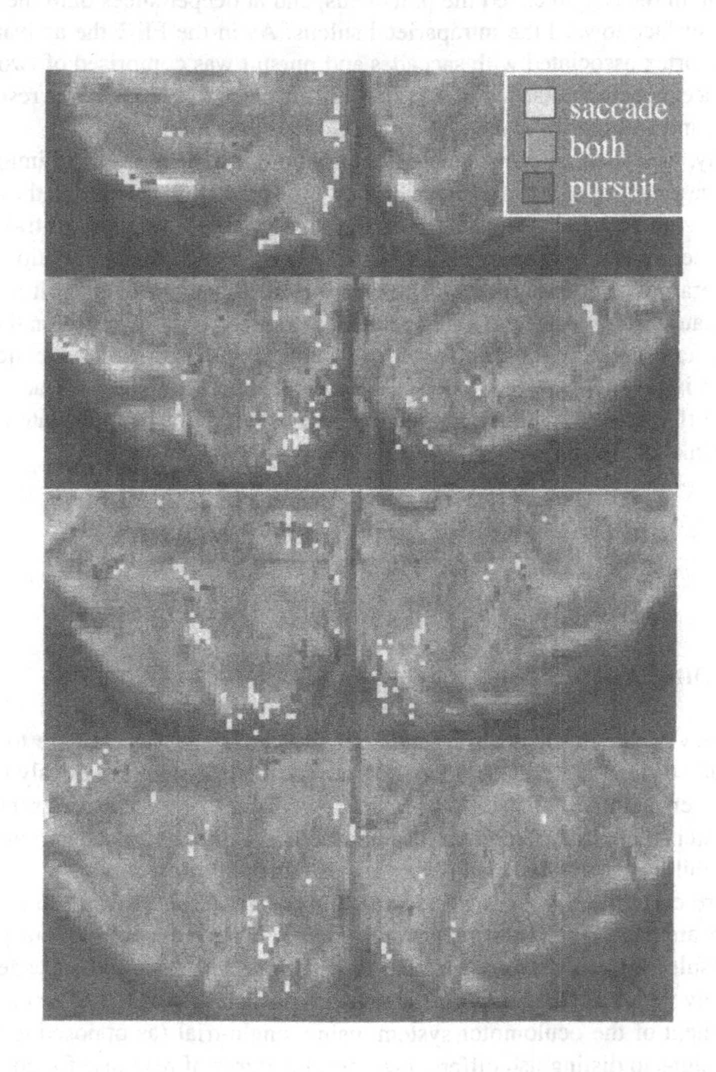

FIGURE 19. Sections of the parietal cortex for several slices arranged from the top to the bottom of the brain. The colored voxels indicate which areas activate in response to each task. Classification based on normalized contrast map thresholded at 3.

Figure 19 shows classification maps based on thresholded normalized contrasts for part of six slices from the top of the brain to the bottom. In this figure, the Pursuit and VGS areas show systematic and three-dimensional differences, with minimal overlap. Moving from the top of the brain to the bottom, we can see the movement of the pursuit and saccade areas along the posterior midline of the brain, in the region called the precuneus, and at deeper slices onto the lateral cortical surface toward the intraparietal sulcus. As in the FEF, the activation in parietal cortex associated with saccades and pursuit was comprised of two small and adjacent regions that we had found indistinguishable with lower resolution imaging and without the statistical advances described above.

Finally, in several of our subjects, examination of their anatomical images revealed the presence of a small sulcus bridging the precentral sulcus and the central sulcus in some subjects, The central sulcus runs parallel to the precentral sulcus and 1-2 centimeters posterior to it. This is a known anatomical variation in the human brain, and it is particularly interesting for the study of pursuit and saccade because in subjects who did have this anatomic variant it was often the most robustly activated area of the frontal eye fields. Figure 20 shows a pattern we observed in all such subjects. Along this special sulcus, pursuit and saccade are clearly differentiated, with the pursuit and saccadic regions being isolated to different banks of the sulcus.

7 Conclusions

Using new methods for imaging and statistical analysis, we are able to detect systematic differences between pursuit and saccade that were previously unclear with earlier methods. The statistical model we used offers improved sensitivity and greater inferential scope over current methods. This approach demonstrates the scalability of Bayesian methods to large and complex problems.

We are currently developing this work in several directions. On the scientific side, we are analyzing Experiment 3 for data from several subjects; our preliminary results suggest that we are able to distinguish pursuit and saccade more effectively with this design. Our next steps involve studying the higher cognitive involvement of the oculomotor system, using single-trial (as opposed to block-trial) designs to distinguish different temporal patterns of response for cognitive, motor, and sensory sub-processes. On the statistical side, we are attacking three critical problems: incorporating spatial structure into the model, combining inferences across subjects, and using the estimated temporal patterns of activation in the different oculomotor areas to characterize the functional connectivity among the areas. We are also extending the model to handle more complex features of the noise.

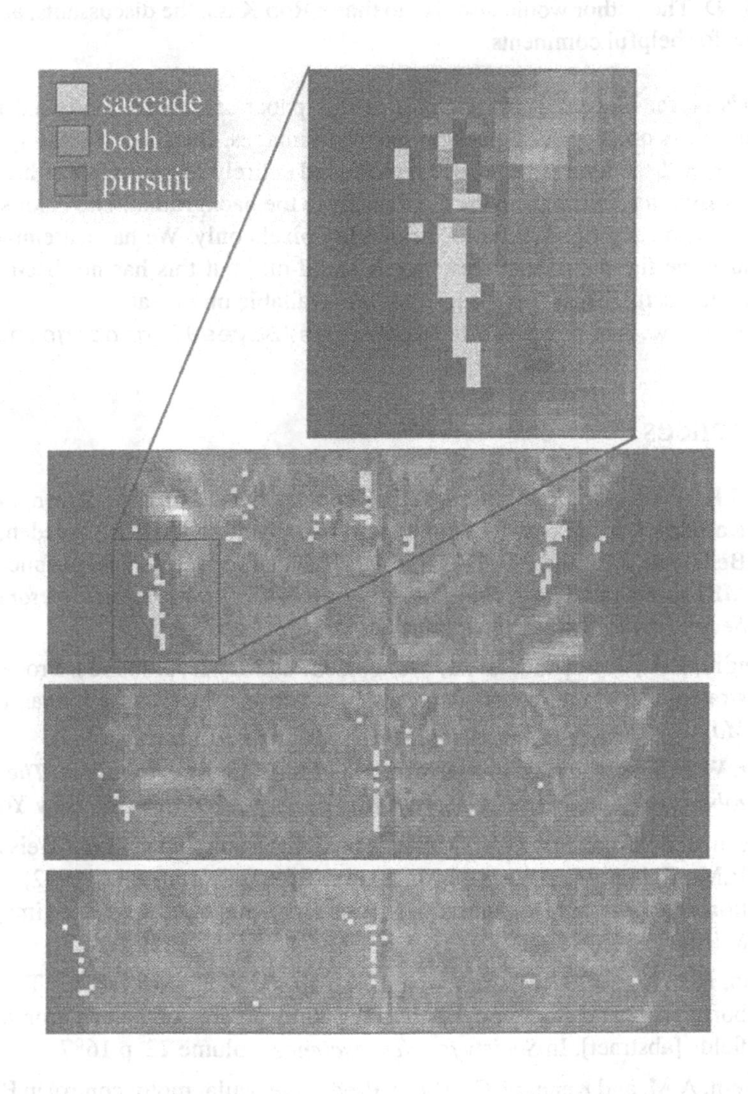

FIGURE 20. Area near the Frontal Eye Fields in three adjacent slices for a single subject. Inset highlights the special sulcus detected in several subjects. The colored voxels indicate which areas activate in response to each task. Classification based on normalized contrast map thresholded at 3.

Acknowledgments

This work is supported by NSF Grants DMS 9505007 and DMS 9705034 and NIH Grants MH 422969, MH 45156, MH 01423, NS 35949, HD 35469, and NARSAD. The author would also like to thank Rob Kass, the discussants, and the referees for helpful comments.

Editor's Note: Figures 13 through 20 in this paper are, in their original form, color overlays on grey-scale background brain images. Because of printing limitations here, these figures have been reproduced entirely in grey-scale, which can make it difficult to distinguish the overlays from the background. The color scales included with each figure refer to the overlay pixels only. We have attempted to reproduce the figures so that these pixels stand-out, but this has not been completely successful. The originals in color are available on line at `http://www.stat.cmu.edu/meetings/Bayes97/genfigs.html`.

References

Baker, J.R., Weisskoff, R.M., Stern, C.E., Kennedy, D.N., Jiang, A., Kwong, K.K., Kolodny, L.B., Davis, T.L. Boxerman, J.L., Buchbinder, B.R., Weeden, V.J., Belliveau, J.W. and Rosen, B.R. (1994). Statistical assessment of functional MRI signal change. In *Proceedings of the Society for Magnetic Resonance, Second Annual Meeting*, volume 2, 626. SMR.

Bandettini, P.A., Jesmanowicz, A., Wong, E.C. and Hyde, J. (1993). Processing strategies for time-course data sets in functional MRI ofthe human brain. *Magnetic Resonance in Medicine*, 30: 161–173.

Becker, W. (1989). Metrics. In R.H. Wurtz and M.E. Goldberg, editors, *The Neurobiology of Saccidic Eye Movements*, pages 13–39. Elsevier, New York.

Belliveau, J.W., Kennedy, D.N., McKinstry, R.C., Buchbinder, B.R., Weisskoff, R.M., Cohen, M.S., Vevea, J.M., Brady, T.J. and Rosen, B.R. (1992). Functional mapping of the human visual cortex by magnetic resonance imaging. *Science*, 254: 716–719.

Berman, R.A., Luna, B., McCurtain, B.J., Strojwas, M.H., Voyvodic, J.T., Thulborn, K.R. and Sweeney, J.A. (1996). fmri studies of human frontal eye fields [abstract]. In *Society for Neuroscience*, volume 22, p 1687.

Bronstein, A.M. and Kennard, C. (1985). Predicitve ocular motor control in Parkinson's disease. *Brain*, 108: 925–940.

Brown, M.A. and Semelka, R.C. (1995). *MRI: Basic Principles and Applications*. John Wiley and Sons, New York.

Bruce, C.J. and Goldberg, M.E. (1985). Primate frontal eye fields. I. single neurons discharging before saccades. *Journal of Neurophysiology*, 53: 603–635.

Carl, J.R. and Gellman, R.S. (1987). Human smooth pursuit: Stimulus-dependent responses. *Journal of Neurophysiology*, 57: 1446–1463.

Clementz, B.A. and Sweeney, J.A. (1990). Is eye movement dysfunction a biological marker for schizophrenia? A methodological review. *Psychology Bulletin*, 108: 77–92.

Cohen, J.D., Forman, S.D., Braver, T.S., Casey, B.J., Servan-Schreiber, D. and Noll, D.C. (1994). Activation of prefrontal cortex in a non-spatial working memory task with functional MRI. *Human Brain Mapping*, 1: 293–304.

Cohen, J.D., Noll, D.C. and Schneider, W. (1993). Functional magnetic resonance imaging: Overview nad methods for psychological research. *Behavior Research Methods, Instruments, Computers*, 25(2): 101–113.

Colby, C.L., Duhamel, J.R. and Goldberg, M.E. (1996). Visual, presaccadic, and cognitive activation of single neurons in monkey lateral intraparietal area. *Journal of Neurophysiology*, 75(5): 2841–2852.

de Boor, C. (1978). *A Practical Guide to Splines*. Springer-Verlag.

DiCiccio, T.J., Kass, R.E., Raftery, A. and Wasserman, L. (1997). Computing bayes Factors by combining simulation and asymptotic approximations. *J. Amer. Statist. Assoc.*, 903–915.

Eddy, W.F. (1997). Comment on Lange and Zeger. *Journal of the Royal Statisitical Society C*, 46: 19–20.

Eddy, W.F., Fitzgerald, M. and Noll, D.C. (1996). Improved image registration using Fourier interpolation. *Magn. Reson. Med.*, 36: 923–931.

Forman, S., Cohen, J.C., Fitzgerald, M., Eddy, W.F., Mintun, M.A. and Noll, D.C. (1995). Improved assessment of significant change in fucntional magnetic resonance fMRI: Use of a cluster size threshold. *Magn. Reson. Med.*, 33: 636–647.

Friedman, L., Jesberger, J.A. and Meltzer, H.Y. (1991). A model of smooth pursuit performance illustrates the relationship between gain, catch-up saccade rate, and catch-up saccade amplitude in normal controls and patients with schizophrenia. *Biological Psychiatry*, 30: 537–556.

Friston, K.J., Frith, C.D. and Frackowiak, R.S.J. (1994). *Human Brain Mapping*, 1: 69–79.

Friston, K.J., Jezzard, P. and Turner, R. (1994). Analysis of functional MRI time-series. *Human Brian Mapping*, 1: 153-171.

Friston, K.J., Holmes, A.P., Poline, J.B., Grasby, P.J., Williams, S.C.R., Frackowiak, R.S.J. and Turner, R. (1995). Analysis of fMRI time series revisited. *NeuroImage*, 2: 45–53.

Funahasi, S., Bruce, C.J. and Goldman-Rakic, .S. (1989). Mnemonic coding of visual space in the monkey's dorsolateral prefrontal cortex. *Journal fo Neurophysiology*, 61: 331–349.

Gaymard, B. and Pierrot-Deseilligny, C. (1990). Impairment of sequences of memory-guided sccades after supplementary motor area lesions. *Annals of Neurology*, 28: 622–626.

Genovese, C.R. (1997). Statistical inference in functional Magnetic Resonance Imaging. Technical Report 674, Department of Statistics, Carnegie Mellon University.

Gottlieb, J.P., MacAvoy, G. and Bruce, C.J. (1994). Neural responses related to smooth-pursuit eye movements and their correspondence with electrically elicited smooth eye movements in the primate frontal eye field. *Journal fo Neurophysiology*, 72(4): 1634–1653.

Green, P.J. (1995). Reversible jump MCMC computation and Bayesian moel determination. *Biometrika*, 82: 711–732.

Hastie, T.J. and Tibshirani, R.J. (1990). *Generalized Additive Models*. Chapman and Hall.

Haxby, J.V. Grady, C.L., Horwitz, B., Ungerleider, L.G., Mishkin, M., Carson, R.E., Herscovitch, P., Schapiro, M.B. and Rapoport, S.I. (1991). Dissociation of object and spatial visual processing pathways in human extrastriate cortex. *Neurobiology*, 88: 1621–1625.

Holzman, P.S., Proctor, L.R., Levy, D.L., Yasilo, N.J., Meltzer, H.Y. and Hirt, S.W. (1974). Eye-tracking dysfunctions in schizophrenic patients and their relatives. *Archives of General Psychiatry*, 31 143-151.

Just, M.A. and Carpenter, P.A. (1980). A theory of reading: From eye fixations to comprehension. *Psychological Review*, 87: 329–354.

Kennard, C. and Rose, F.C. (1988). *Physiological Aspects of Clinical Neuro-ophthalmology*. Chicago Year Book Medical Publishers, Chicago, IL.

Kwon, K.K., Belliveau, J.W., Chesler, D.A., I.E. Goldberg, Weisskoff, R.M., Poncelet, B.P., Kennedy, D.N., Hoppel, B.E., Cohen, M.S., Turner, R., Cheng, H., Brady, T.J. and Rosen, B.R. (1992). Dynamic magnetic resonance imaging of human brain activity during primary sensory stimulation. *Proc. Natl. Acad. Sci. U.S.A.*, 89: 5675.

Lange, N. and Zeger, S. (1997). Non-linear Fourier time series analysis for human brain mapping by functional magnetic resonance imaging. *Journal of the Royal Statistical Society C Applied Statistics*, 46: 1-29.

Lehmann, E.L. (1975). *Nonparametrics: Statistical Methods Based on Ranks*. Holden Day, Oakland, CA.

Leigh, R.J. and Zee, D.S. (1991). *The Neurology of Eye Movements (2nd ed.)*. F.A. Davis, Philadelphia, PA.

Lueck, C.J., Tanyeri, S., Crawford, T.J., Henderson, L. and Kennard, C. (1990). *Journal of Neurology, Neurosurgery and Psychiatry*, 53: 284–288.

Luna, B., Thulborn, K.R., Strojwas, M.H., McCurtain, B.J., Berman, R.A., Genovese, C.R. and Sweeney, J.A. (1998). Dorsal cortical regions subserving visually-guided saccades in humans: An fmri study. *Cerebral Cortez*, 8: 40–47.

Marr, D. (1982). *Vision*. W.H. Freeman and Co., New York.

Minshew, N.J., Sweeney, J.A. and Bauman, M.L. (1998). Neurologic aspects of autism. In D.J. Cohen and F.R. Volkmar, editors, *Handbook of Autism and Pervasive Developmental Disorders*, p. 344–369. John Wiley and Sons, second edition.

Minshew, N.J., Sweeney, J.A. and Furman, J.M. (1995). Evidence for a primary neocortical system abnormality in autism [abstract]. In *Society for Neuroscience Abstracts*, volume 21, page 293.13.

Mintun, M.A., Richle, M.E., Martin, W.R.W and Herscovitch, P. (1984). Brain oxygen utilization measured with 0-15 radiotrcers and positron emission tomography. *Journal of Nuclear Medicine*, 25: 177–187.

Munoz, D.P. and Wurtz, R.H. (1992). Role of the rostral superior colliculus in active visual fixation and execution of express saccades. *Journal of Neurophysiology*, 67: 1000–1002.

Ogawa, S., Tank, D.W., Menon, D.W., Ellermann, J.M., Kim, S., Merkle, H. and Ugurbil, K. (1992). Intrinsic signal changes acompanying sensory stimulation: Functional brain mapping using MRI. *Proc. Natl. Acad. Sci. U.S.A.*, 89: 5951–5955.

Pauler, D. (1996). *The Schwarz Criteion for Mixed Effects Models*, Ph.D. thesis, Carnegie Mellon Unviersity.

Petit, L., Clark, V.P., Ingeholm, J. and Haxby, J.V. (1997). Dissociation of sacccade-related and pursuit-related activation in human frontal eye fields as revealed by fmri. *Journal of Neurophysiology*, 77(6): 3386–3390.

Poline, J.B. and Mazoyer, B. (1994). Cluster analysis in individaul functional brain images: Some new techniques to enhance the sensitivity of activation detection methods. *Human Brain Mapping*, 2: 103–111.

Posner, M.I. and Raichle, M.E. (1994). *IMages of Mind*. Scientific American Library.

Rakic, P. (1995). The development of the frontal lobe: A view from the rear of the brain. In H.H. Jasper, Riggio, S., and P.S. Goldman-Rakic, editors, *Epilepsy and the Functional Anatomy of the Frontal Lobe*, pages 1-8. Raven Press, New York.

Rosenberg, D.R., Sweeney, J.A., Squires-Wheeler, E., Keshavan, M.S., Cornblatt, B.A. and Erlenmeyer-Kimling, L. (1997). Eye-tracking dysfunction in offspring from the New York high-risk project: Diagnostic specificity and the role of attention. *Psychiatry Research*, 66: 121–130.

Schervish, M.J. (1995). *Theory of Statistics*. Springer-Verlag.

Schlag, J. and Schlag-Rey, M. (1987). Evidence for a supplementary eye field. *Journal of Neurophysicology*, 57: 179–200.

Schwarz, G. (1978). Estimating the dimension of a model. *Ann. Stat.*, 6(2): 461–464.

Sommer, M.A. and Tehovnik, E.J. Reversible inactivation of macaque frontal eye field. *Experimental Brain Research*, in press.

Sweeney, J.A., Clementz, B.A., Haas, G.L. Escobar, M.D., Drake, K. and Frances, A.J. (1994). Eye tracking dysfunction in schizophrenia: Characterization of component eye movement abnormalities, diagnostic specificy, and the role of attention. *Journal of Abnormal Psychology*, 103: 222-230.

Sweeny, J.A., Luna, B.,Berman, R.A., McCurtain, B.J., Strojwas, M.H., Voyvodic, J.T., Genovese, C.R. and Thulborn, K.R. (1996a). fmri studies of spatial working memory [abstract]. In *Society for Neuroscience*, 22: 1688.

Sweeney, J.A., Mintun, Kwee, S., Wiseman, M.B., Brown, D.L., Rosenberg, D.R. and Carl, J.R. (1996b). A positron emission tomography study of voluntary saccadic eye movements and spatial working memory. *Journal of Neurophysiology*, 75(1): 545–468.

Talairach, J. and Tounoux, P. (1988). *Coplanar Stereotaxic Atlas of the Human Brain. Three-dimensional Proportional System: An Approach to Cerebral Imaging*. Thieme.

Thulborn, K.R. Personal communication.

Thulborn, K.R., Waterton, J.C., Matthews, P.M. and Radda, G.K. (1982). Oxygenation dependence of the transverse relaxation time of water protons in whole blood at high field. *Biochem. Biophys. Acta*, 714: 265–2770.

Tian, J.R. and Lynch, J.C. (1996). Functionally defined smooth and saccadic eye movement subregions in the frontal eye field of cebus monkeys. *Journal of Neurophysiology*, 7694): 2740–5753.

Tierney, L. (1994). Marov chains for exploring posterior distributions. *The Annals of Statistics*, 22(4): 1701–1727.

Tootell, R.B.H., Reppas, J.B., Kwong, K.K., Malach, R., Born, R.T., Brady, T.J., Rosen, B.R. and Belliveau, J.W. (1995). Functional analysis of human mt and related visual cortical areas using magnetic resonance imaging. *Journal of Neuroscience*, 15: 3215–3230.

Vardi, Y., Shepp, L.A. and Kaufman, L. (1985). A statistical model fo rpositron emission tomography. *J. Amer. Statist. Assoc.*, 80: 8–20.

Weaver, J.B., Saykin, A.Y., Burr, R.B., Riordan, H. and Maerlender, A. (1994). Principal component analysis of functional MRI of memory. In *Proceeding sof the Society for Magnetic Resonance, Second Annual Meeting*, p. 808. SMR.

Weisskoff, R.M. (1997). Personal communication.

Weisskoff, R.M., Baker, J.R., Belliveau, J., Davis, T.L., Kwong, K.K., Cohen, M.S. and Rosen, B.R. (1993). Power spectrum analysis of functionally-weighted MR data: What's in the noise? In *Proceedings of the Society for Magnetic Resonance in Medicine, Twelfth Annual Meeting*, p. 7. MRM.

Worsley, K.J. and Friston, K.J. (1995). Analysis of fMRI time series revisited - again. *NeuroImage*, 2:173–181.

Wurtz, R.H. and Goldberg, M.E. (1972). Activity of superior colliculus in behaving monkey. III. cells discharging before eye movements. *Journal of Neurophysiology*, 35: 575–586.

Wurtz, R.H. and Goldberg, M.E. (1989). *The Neurobiology of Saccadic Eye Movements*, Elsevier, New York.

Discussion

Jonathan Raz
Changying Liu
University of Michigan

Introduction

We applaud Chris on his excellent paper and presentation. We wish to enumerate some of the many admirable qualities in his work. First, Chris has performed an excellent job of demonstrating the use of Bayesian inference to answer questions that are difficult solve by frequentist methods. Second, many statisticians, including ourselves, would do well to emulate Chris' close collaboration with the investigators collecting the data to which his methods apply. Third, Chris has constructed a realistic statistical model with interpretable parameters, thus achieving his goal of addressing the scientific questions asked by the investigators. Fourth, Chris has demonstrated sophisticated computing skills in applying his statistical approach. Last, his written and verbal presentations were clear and carefully organized, and he included a powerful statement of the need for Bayesian methods in analysis of functional brain images.

Functional MRI data present special challenges to the statistician. The data may be viewed as consisting of two or three dimensional images at each of many time points, or as a multivariate time series with one series at each of many voxels. Thus, the data present what is inherently a time/space problem. As Chris clearly explains, a wide variety of sources contribute to the variability in the data. Moreover, the scientific questions that can and are asked of the data are varied and often difficult to address by classical methods.

Limitations of Existing Methodology

In our view, the current standard statistical approach to analysis of fMRI data is the Statistical Parametric Map (SPM), as developed by Karl Friston, Keith Worsley, and their colleagues (Friston et al., 1991, 1997; Worsley and Friston, 1997; Worsley et al., 1992). One of the important reasons for the wide-spread use of SPM is the free software available from the World Wide Web.

As a rough summary, here are the steps necessary to perform an SPM analysis of an fMRI data set.

1. Pre-process the data to register and filter images. A competitor to SPM in this necessarily complicated step is the FIASCO method (Eddy et al., 1996).
2. Convolve the stimulus series with a fixed hemodynamic response, which is treated as known by the investigators. (Several options are available in the software, including a delayed boxcar function and any user specified function). Let $x(t)$ denote the result at time t of this convolution.
3. At each voxel v, regress the fMRI time series $Y_v(t)$ on $x(t)$ and possibly other covariates.
4. Make an image ("statistical map") from t or F statistics.
5. Compute a threshold using the theory of extrema in random fields (Friston et al., 1991; Worsley et al., 1992).
6. Define areas of significant activation by the voxels in the statistical map that exceed the threshold.

Chris noted (Genovese, 1997) that the assumed spatial noise model may be unrealistic. We question the usefulness of this criticism, because *any* statistical approach requires a model of the spatial covariance structure, and any such model is open to criticism based on the presumed complexity of the true noise process. Furthermore, Keith Worsley (personal communication, October 15, 1997) argues that the SPM inference is typically robust to violations of the assumptions concerning the spatial covariance structure.

We certainly agree with Chris (Genovese, 1997) that the assumed hemodynamic response in existing methods is too simple. The SPM approach also is limited in that it does not allow the shape of the hemodynamic response to be estimated from the data. A substantial advantage of Chris' approach, as well as that of Lange and Zeger (1997), is that both the amplitude and shape of the hemodynamic response are represented by unknown parameters.

We point out two additional limitations of SPM, both of which are addressed by Chris' current or planned work. First, many investigators using fMRI seem to be concerned that registering and averaging across subjects leads to blurring of activation areas and thus to conservative tests. Investigators have complained to us that in some data sets an activation is significant in most of the subjects, but is not significant when evaluated using the average of registered images.

A second important issue was pointed out by Scott Zeger in a discussion of a presentation by Keith Worsley. Thresholded statistical maps are often interpreted as giving an estimate of an activation region, but this results in an estimated extent of activation that depends on the noise level.

The Proposed Model of fMRI Time Series

To better understand Chris' time series model, and to emphasize some of its features, we present the model in a form somewhat different from the form in Genovese (1997). Let $Y_v(t)$ denote the observed fMRI response at time t in voxel v. Since the experimental condition c can (and typically does) vary over time, we write the condition as $c(t)$. Let μ_v denote the baseline mean at voxel v, and $\gamma_{v,c(t)} \geq 0$ denote responsiveness (amplitude) of the hemodynamic response. An important feature of Chris' model is that the responsiveness can be identically zero at any voxel.

Let $\boldsymbol{\theta}_v$ denote a vector of spline parameters for voxel v, and let $b_1(t; \boldsymbol{\theta}_v)$, $b_2(t; \boldsymbol{\theta}_v)$, and $b_3(t; \boldsymbol{\theta}_v)$ denote the three fixed knot regression splines used to define the shape of the hemodynamic response. Then

$$\mu_v \, \gamma_{v,c(t)} \, [b_1(t; \boldsymbol{\theta}_v) \, b_2(t; \boldsymbol{\theta}_v) - b_1(t; \boldsymbol{\theta}_v) \, b_3(t; \boldsymbol{\theta}_v)]$$

represents the hemodynamic response. The three spline functions are expressed as functions of $\boldsymbol{\theta}_v$ in such a way that the spline coefficients can be interpreted as the "lag-on," "lag-off," "dip" and other quantities of interest to the investigator.

The drift at voxel v is defined by a variable or fixed knot regression spline $d_v(t; K_v, \boldsymbol{\kappa}_v, \boldsymbol{\delta}_v)$, where K_v is the scalar number of knots, $\boldsymbol{\kappa}_v$ is a vector of knots, and $\boldsymbol{\delta}_v$ is a vector of regression coefficients. A noise term $\varepsilon_v(t)$ represents additional variability not accounted for by the baseline, hemodynamic response, and the drift. The noise $\varepsilon_v(t)$ is assumed to be white.

In terms of these quantities, the time series model can be written as

$$\begin{aligned} Y_v(t) &= \mu_v\{1 + \gamma_{v,c(t)} \, [b_1(t; \boldsymbol{\theta}_v) \, b_2(t; \boldsymbol{\theta}_v) - b_1(t; \boldsymbol{\theta}_v) \, b_3(t; \boldsymbol{\theta}_v)]\} + \\ &\quad\; d_v(t; K_v, \boldsymbol{\kappa}_v, \boldsymbol{\delta}_v) + \varepsilon_v(t). \end{aligned}$$

An important feature of the model is that it contains interpretable parameters, notably the responsiveness γ and the parameters of the hemodynamic response $\boldsymbol{\theta}$. However, we believe the model of the shape of the hemodynamic response should be more general. In particular, some investigators have noticed an early dip in the hemodynamic response, and have ascribed considerable importance to it (Hu et al., 1997). While the proposed model could be extended to include additional parameters representing the early dip, we wonder if any low-dimensional parametric model will continue to be useful as additional features of the hemodynamic response are discovered and new experimental designs are developed. Possibly a more flexible and less parametric approach would be valuable in the long term. For example, we have had success using a simple linear filter with a finite number of unknown parameters (in engineering terms, an FIR filter) to represent the relationship between $Y_{t,v}$ and $c(t)$. Such an approach also was discussed in Nielsen et al. (1997).

The model of the drift is quite complicated in that it requires either a variable knot regression spline or a fixed knot regression spline with many knots and a roughness penalty. An alternative and possibly simpler model would absorb the drift into $\varepsilon_v(t)$ and allow high power at low frequencies in the noise.

We note that the responsiveness γ is scaled by the baseline μ. While we prefer to view the drift as low frequency random "noise", we note that if the drift is considered part of the "signal" (as the spline model seems to imply), then it would be natural to scale the responsiveness by $\mu + d$ rather than by μ alone.

Chris' model is designed specifically for fMRI experiments in which stimuli are presented in blocks of trials. Increasingly, experiments are being performed using "single trial" or "trial-based" designs (Zarahn et al., 1997; Dale and Buckner, 1997), in which the individual stimulus presentations are separated by substantial inter-stimulus intervals (ISIs). In many single-trial designs, the ISI is quite long [30 seconds in Zarahn et al. (1997)], and images are acquired every 1 or 2 seconds, so that the hemodynamic response is allowed to decay before the next stimulus is presented. We wonder if Chris' time series model is appropriate for representing the response in such designs.

In other single-trial designs (Dale and Buckner, 1997), the ISI is shorter than the duration of a response to a single stimulus, so that the acquired fMRI time series results from superimposed responses to several stimuli. Assuming additivity, we have shown that responses to the various stimulus types (i.e., various experimental conditions) are not identifiable without very strong prior assumptions, but that the differences between responses to two types of stimuli are identifiable. We have successfully estimated these differences using an FIR model.

In another type of experimental design, the stimulus would vary continuously as a function of time (Tootell et al., 1995). For example, the intensity of a stimulus could be a sinusoidal function of time. Chris' model is appropriate when the data are acquired in blocks with different types of continuously varying stimuli within each block, as in Tootell et al. (1995). Chris' model probably would not be appropriate for analysis of data acquired with continuously varying stimuli over the acquisition session, but we expect that an FIR model will perform well in that situation.

Conditional on the baseline, drift, and hemodynamic response, Chris' model assumes independent errors over both time and space. We note that future work on specifying the covariance structure will be very important to the usefulness of the model.

Bayesian Inference

The joint posterior distribution of the parameters is computed using several methods, including exact computation (ignoring the normalizing constant), Laplace approximation, and Monte Carlo approximation. By including sub-models with some γ's set identically to zero, and averaging over the posterior distributions in these submodels, Chris is able to estimate the posterior probability of zero response, and thus classify voxels as having a high or low probability of response.

A great advantage of the Bayesian approach is that it easily solves many problems that pose major difficulties in frequentist inference, such as: measuring the strength of evidence for a monotonic dose-response curve; estimating changes in the spatial extent of activation, with estimates of uncertainty; and estimating responsiveness, when responsiveness can be positive or identically zero.

As Chris points out, based on strict Bayesian reasoning, there is no selection bias from *a posteriori* choice of voxels of interest and no multiple comparisons problem. However, we believe that Bayesian procedures should be evaluated partly based on their frequentist properties. While computational barriers might rule out the possibility of performing a full-scale simulation study, we urge Chris to investigate the frequentist properties of his method in a stimulation study that uses a small image and a relatively simple version of his model.

We suspect that Chris' approach can result in good frequentist inference, but we are wary that the inference may depend heavily on the assumed spatial covariance structure and the prior distributions of the responsiveness parameters γ. Model-checking and sensitivity analysis will be important areas for future work.

Acknowledgments

We gratefully acknowledge helpful communications with Chris Genovese, Yingnian Wu, Bruce Turetsky, and Keith Worsley.

References

Dale, A.M. and Buckner, R.L. (1997). "Selective averaging of rapidly presented individual trials using fMRI," *Human Brain Mapping*, 5, 329-340.

Eddy, W.F., Fitzgerald, M., Genovese, C.R., Mockus, A., and Noll, D.C. (1996). "Functional imaging analysis software–computational Olio, in *Proceedings in Computational Statistics*, Prat, A., ed., Physica-Verlag, Heidelberg, 39–49.

Friston, K.J., Frith, C.D., Liddle, P.F., and Frackowiak, R.S.J. (1991). "Comparing functional (PET) images: The assessment of significant change," *Journal of Cerebral Blood Flow and Metabolism*, 11, 690–699.

Friston, K.J., Holmes, A.P., Poline, J.-B., Grasby, P.-J., Williams, S.C.R., Frackowiak, R.S.J., and Turner, R. (1995). "Analysis of fMRI time-series revisited," *Neuroimage*, 2, 45–53.

Genovese, C.R. (1997). "Statistical inference in functional magnetic resonance imaging," Technical Report 674, Carnegie Mellon Department of Statistics.

Hu, X., Le, T.H., and Ugurbil, K. (1997). "Evaluation of the early response in fMRI in individual subjects using short stimulus duration," *Magnetic Resonance in Medicine*, 37, 877-884.

Lange, N. and Zeger, S.L. (1997). "Non-linear Fourier time series analysis for human brain mapping by functional magnetic resonance imaging," *Applied Statistics*, 46, 1–29.

Nielsen, F.Å., Hansen, L.K., Toft, P., Goutte, C., Lange, N., Strother, S.C., Mørch, N., Svarer, C., Savoy, R., Rosen, B., Rostrup, E., and Born, P. (1997). "Comparison of two convolution methods for fMRI time series."*NeuroImage Supplement*, 5, S473.

Tootell, R.B.H., Reppas, J.B., Date, A.M., Look, R.B., Sereno, M.I., Malach, R., Brady, T.J., and Rosen, B.R. (1995). "Visual motion aftereffect in human cortical area MT revealed by functional magnetic resonance imaging," *Nature*, 375, 139–141.

Worsley, K.J., Evans, A.C., Marrett, S., and Neelin, P. (1992). "A three-dimensional statistical analysis for CBF activation studies in human brain," *Journal of Cerebral Blood Flow and Metabolism*, 12, 900–918.

Worsley, K.J. and Friston, K.J. (1997). "Analysis of fMRI time-series revisited – again," *Neuroimage*, 2, 173–181.

Zarahn, E., Aguirre, G. and D'Esposito, M. (1997). "A trial-based experimental design for fMRI," *Neuroimage*, 6, 122–138.

Discussion

Ying Nian Wu
University of Michigan

I congratulate the authors for this fine article.

The authors studied an important scientific problem. I was once involved in studying abnormalities in eye movements among schizophrenic patients and their relatives (Rubin and Wu, 1997), and I believe that understanding how brains work during eye movements will much enhance our understanding of schizophrenia. The first author, as a statistician, has clearly made a crucial contribution to this collaborative research by developing the statistical methodologies.

The article convincingly demonstrates the power of Bayesian modeling and inference to analyze complex signals such as fMRI images, and to address a wide range of scientific questions. The temporal model and the inference techniques may also be applied to other applications as well, such as fault detection in mechanical engineering, where the signatures of faults are to be captured among several sources of variations in the signals.

The following are some specific comments and questions.

The model

The temporal model for each voxel has one based line component describing the overall scale, and three other components, drift, activation, and noise, to characterize three sources of variations. Because the three sources of variations may confound with each other — a change in signal may be attributed to or be interpreted as any one of drift, noise, and activation, it is crucial to design the experiments and model the three variations carefully to differentiate them. The authors did a good job in this aspect. The low frequency drift is modeled by a spline with the smoothness enforced by the prior distribution. The activation has many repetitions, each with sufficient duration by experimental design, and the shape of activation is carefully parameterized, so that the periodic activation can be separated from low frequency drift and high frequency noise. The model assumes white noise. Extensions to colored noise or long tailed noise may be considered. About extensions of the temporal model, one has to be very cautious against the confounding effects.

The major inadequacy of the model is the assumption that each voxel follows an independent time series model, which may not adequately reflect either the physical reality or the goal of the research, which is to study functional connectivity, i.e., the spatial relationship, in the cortical circuits. The connectivity or the spatial relationship should be an unknown parameter to be incorporated into the model, and to be inferred from the posterior distribution, perhaps with the help of MCMC techniques.

Prior specification

The authors successfully incorporated the substantial prior knowledge about the four components of the model. It is important to study the sensitivity of the posterior conclusions to the prior specifications more carefully.

The prior distribution on the smooth drift and the specification of the noise model are important for extracting the activation. I am wondering whether we can learn some features of them using training data, such as signals obtained under the rest condition, so that we do not need to worry about the activation. We may extract the smooth drift curves using smoothing techniques under careful supervision, and learn about the parameters in the prior distribution on drift from these curves. Furthermore, we may want to see whether the prior distribution on drift can generate similar curves by sampling from it. We may also learn about the noise by properly detrending the drift.

Model checking

There are several important assumptions in the model, which should be rigorously checked, so that one can be confident about the scientific conclusions drawn using the model. Being able to capture some patterns that are missed by previous methods does not give much support to the model, because it is likely that the model captures some spurious features.

For the current situation where there are many possible extensions but no serious competing models, Rubin's posterior predictive check method (Rubin, 1984) may prove to be useful for checking the model assumptions and for suggesting worthwhile extensions. See Gelman, Meng, and Stern (1996) for recent extensions and applications of this method.

Experiments

The second experiment looks better than the first one because it better separates pursuit and saccade, albeit at the cost of magnitude and duration of the BOLD effect. The following are some questions.

Unlike smooth pursuit, saccade occurs at a random time with a very short duration, is the shape of activation described in the paper still adequate?

Since we are not sure about the intensity of saccade, is it possible to record the eye movement of the subject as angular displacement versus time, so that we can better interpret the results?

The authors mentioned that the target speed needs to be carefully chosen so that it is fast enough to engage the saccade system but is slow enough to get sufficient duration. Can we let the target move at a slow speed but with occasional "saccades" itself, to solve the conflict between saccade engagement and duration?

For studying working memory contribution in the frontal eye fields, besides manipulating the delay period, can we also manipulate the task difficulty? For instance, we may design several types of flashes with different colors and shapes, and design experiments where the subject needs to memorize more locations.

In summary, I think this work is a welcomed contribution to Bayesian statistics, neural sciences, and neural imaging, although more convincing scientific examples are needed for it to be accepted as the standard approach. I much appreciate the authors' good and hard work, and it is a great pleasure reading and commenting on this quality paper.

Additional references

Gelman, A., Meng, X.-L., and Stern, H. (1996) Posterior predictive assessment of model fitness via realized discrepancies (with discussion). *Statistica Sinica* **6**, 733–807.

Rubin, D. B. (1984) Bayesianly justifiable and relevant frequency calculations for the applied statistician. *Ann. of Statist.* **12**, 1151-1172.

Rubin, D. B. and Wu, Y. (1997) Modeling schizophrenic behavior using general mixture components, *Biometrics* **53**, 243-261.

Rejoinder

We would like to begin by thanking the discussants for their thoughtful comments and positive feedback on our approach. We would also like to thank the editors of this volume for the opportunity to respond. Our comments are organized by discussant.

Ying Nian Wu

Wu raises concerns about the spatial independence assumption in the model, as it is currently used. We share this concern, and we are currently working to incorporate spatial structure into the model. There are two issues here. The first is spatial dependence in the noise process. As we mentioned in the paper, there is evidence that non-local spatial correlations may be non-trivial for some data sets. These correlations appear to be caused by vascular and other physiological relationships among different regions of the brain, and they will be difficult to model. Nonetheless, the magnitude and importance of such correlations remains an unresolved empirical question. The second issue is spatial structure in the underlying response, similarities in the response across voxels. If we can capture these similarities, the model can borrow strength across voxels and improve the efficiency of inferences. The problem here is to identify regions that share such common structure; in current work, we are building these spatial relationships into the model as parameters at a deeper level in the hierarchy. MCMC is essential to this effort.

Wu notes also that the goal in our eye-movement studies is to assess the functional connectivity of the system. We believe that this is approachable by building on the model described here. Specifically, suppose we have identified functional regions with coherent responses to our tasks under study, nodes of the system so to speak. Then, we can model the temporal dynamics of activity for this network by assuming that these nodes interact by passing "activation" to other nodes in the network, perhaps at some lag. For example, we are considering a neural-network like model

$$A_r(t) = \gamma_r F(\mathcal{I}_r(t) + \sum_{1 \leq m \leq R} \beta_{rm} A_m(t - \tau_{rm})),$$

where $A_r(t)$ is the activation profile in node r at time t, $\mathcal{I}_r(t)$ is the input to node r, the β_{rm} are connectivity parameters between nodes that describe the strength of inhibition (if negative) or facilitation (if positive) along a single connection, the τ_{rm} are time-lags for that connection, and γ_r and the constrained but unknown function F describe how inputs to the node are combined. Here, we take the A_r's as the estimated activation profile from our temporal model and treat them as input data to the "neural network" model with appropriate uncertainties. These are used in a subsequent fit to estimate the connectivity parameters β_{rm} and τ_{rm}. The result is a description of the strength and timing of interaction among the nodes of the system.

Wu also makes an important point about extending the noise model within voxel: allowing arbitrary noise processes raises the possibility of confounds. For example, problems may arise in allowing the drift term when the noise model includes arbitrary temporal autocorrelations. As we evaluate different extensions to the noise model, some of the structure of the model may need to be revised or further constrained. Our first steps are to parameterize specific components of the noise spectrum, particularly those from physiological variations where we have strong constraints on the autocorrelation structure.

Wu suggests using training data to refine the priors and assess some of the empirical questions that remain unresolved. This is an excellent suggestion, and we are continuing to build a body of data sets with which to carry it out. Our software that implements these methods is being used by several groups, and we anticipate even wider usage with the newest version. This offers us several types of "training": a new library of data sets with which to evaluate empirical questions, check the model, and refine the model specification; and experience using the model both by itself and in comparison to other current methods. Regarding model checking, we appreciate the pointer to the methods of Rubin and Gelman et al.

Wu asks several good questions and points out intruiging alternatives regarding the experimental design and procedure for the eye-movement studies. First, although each saccade itself is rather short, many occur during each epoch, and as a result, the system "ratchets up" to a level where there is a detectable response even to the short saccades. Second, it would indeed be very informative to track the eye movements during the scan and use this data during the fit. There is an eye-tracker for just this purpose that will soon be available for use in our scanner, but various technical problems have delayed its availability for quite some time. In the interim, we have run many detailed eye-movement studies in the lab (where the eye tracker can be used without worries about magnetic fields) to understand how the eye movements manifest themselves. It turns out that the saccades have very nearly constant velocity and duration during these tasks. Third, the suggestion to allow the target to make jumps during the smooth movement in Experiment 3 seems like a very promising idea. The constraint on the subject's field of view remains, but this may allow us to better balance the contribution from the two processes. Lastly, we agree with Wu that it would be valuable to manipulate both task difficulty and the delay period in our studies of working memory involvement in the oculomotor system. We are pursuing both these avenues.

Jonathan Raz and Changying Liu

Raz and Liu question the relative benefits of parameteric versus non-parameteric modeling of the hemodynamic response function. We see value in both approaches and recognize, in particular, that non-parametric fitting of the response shape allows one to capture flexibly features that may not have been recognized *a priori*. However, we believe that the parametric approach has several advantages in this context. First, we want to apply strong constraints on the response, both to be consistent with empirical observation and to separate the response from the diverse other sources of variation in the data. A parametric model makes it easy to build such constraints, however complicated, into the structure of the response. This is not as easy to do with nonparametric methods because imposing general constraints on the function to be estimated can greatly complicate the procedures. As the constraints are weakened for computational convenience, additional uncertainty from confounding may creep into the estimates. Second, our focus is on inference about the response, and a key goal is to accurately assess our uncertainty about specific features of the response. Within a parametric model, computing these uncertainties is straightforward. In nonparametric function estimation, it can be challenging to construct suitable uncertainties (e.g., confidence envelopes) for the estimated function, especially since most smoothing methods are not linear. While we can more easily assess the uncertainty of nonlinear functionals of the estimated function, practical challenges arise even here. Third, many of the questions we are interested in answering relate to very specific features of the response, such as those regarding relative timing of activation in different voxels or regions or those gauging the relationship in the responses within and among voxels. What the parametric model lacks in flexibility it offers in interpretability because these questions can be expressed directly in terms of the model parameters. To address such questions nonparametrically is certainly feasible but requires that we use complicated nonlinear functionals of the estimated function.

Note that there is no need to maintain a low-dimensional parameterization if there are suitable constraints that can be applied, as in the FIR model that Raz and Liu discuss; the primary advantages of the low-dimensional model we use are interpretability and direct connection to features of interest. Also, with regard to the "early" dip, we should note that we can indeed incorporate it into the model. We have chosen not to do so at this point because it remains a controversial feature that has not been adequately explained and it has only been observed reliably at high magnetic field strength.

Raz and Liu raise two issues about the drift: that the spline model is more complicated than it needs to be and that the responsiveness scaling should perhaps include the drift magnitude. Regarding the former, while it is true that the spline model for the drift is more structured than other alternatives like the one Raz and Liu mention, the spline model is quite easy to work with and offers additional flexibility for dealing with special situations, e.g., a large movement artifact at a known time. Of course, other alternatives for the drift are reasonable as well. Regarding scaling of the response, we think of the drift as part of the noise, although the model handles it as a structured component of variation, and as such do not include it in the scaling of the response. Nonetheless, it is possible that this is not the best choice; indeed, the proper approach to scaling the response is an as yet unresolved empirical question.

Raz and Liu ask whether our approach is appropriate for designs with single trials rather than blocks of trials and designs with continuously varying stimuli. The model applies to single trial designs, even though the signal is often weaker than in the block designs. Our approach has been used successfully for single trial designs in several recent experiments by other researchers in the Pittsburgh area, not to mention our Experiment 3. The parameterized family we use for the response profile is flexible enough to capture the response shape even for very short stimuli. There is, however, a qualitative difference in the shape of responses to short and long stimuli. We are working to extend the model by allowing controlled variations in shape across conditions to handle the case where different conditions in the same experiment use very different time scales. We have not considered continuously varying stimuli, but it is an interesting idea. Our model would have to be modified somewhat to handle this.

We agree that model checking is an important part of the work that lays ahead. As we mentioned in response to Ying Nian Wu's comments, we are getting increasing amount of experience using this approach for diverse data sets as we build up a base of examples and as our software becomes more widely used. Nonetheles, the need for validation and model checking certainly has our attention.

Modeling Risk of Breast Cancer and Decisions about Genetic Testing

Giovanni Parmigiani
Donald A. Berry
Edwin Iversen, Jr.
Peter Müller
Joellen Schildkraut
Eric P. Winer

ABSTRACT Recent advances in the understanding of genetic susceptibility to breast cancer, notably identification of the BRCA1 and BRCA2 genes, and the advent of genetic testing, raise important questions for clinicians, patients and policy makers. Answers to many of these questions hinge on accurate assessment of the risk of breast cancer. In particular, it is important to predict genetic susceptibility based on easy-to-collect data about family history of breast and related cancers, to predict risk of developing cancer based on both family history and additional well recognized risk factors, and to use such predictions to provide women facing testing and preventive treatment decisions with relevant, individualized information.
In this paper we give an overview of several specific research projects that together address these goals. These studies are being carried out within the context of two interdisciplinary research programs at Duke University: the Specialized Program of Research Excellence (SPORE) in breast cancer and the Cancer Prevention Research Unit (CPRU) on improving risk communication. In both programs we found Bayesian modeling useful in addressing specific scientific questions, and in assisting individual decision making. We illustrate this in detail and convey how the Bayesian paradigm provides a framework for modeling the concerns and substantive knowledge of the diverse fields involved in this research. These fields include clinical oncology, human genetics, epidemiology, medical decision making, and patient counseling.

1 Background

1.1 Overview

Identifying and counseling individuals at high risk of developing cancer are essential components of cancer prevention efforts. In breast cancer, about 7% of the cases are believed to be due to inherited factors, and a family history of the disease has been the most powerful tool for identifying high-risk individuals. Recent years have marked exceptional scientific advancements in our understanding of inherited susceptibility to breast cancer. The most notable events are the identifications of the BRCA1 and BRCA2 genes. Inherited mutations of one of these genes confer to their carriers a much increased risk of developing breast cancer and also of ovarian, prostate, and other cancers. It has become possible, and increasingly common, to test for the presence of mutations of these genes.

While these new discoveries and the availability of BRCA1 and BRCA2 testing represent major opportunities both scientifically and clinically, they also raise important challenges. One of these is providing individualized counseling. It is increasingly recognized that there is a need for primary care clinicians to be able to assess familial risk factors for breast cancer and provide individualized risk information (Hoskins et al. 1995). The National Action Plan on Breast Cancer Working Group on Hereditary Susceptibility and the American Society of Clinical Oncology stress that "the decision whether to undergo predisposition genetic testing hinges on the adequacy of the information provided to the individual in regards to the risks and benefits of testing, possible testing outcomes, sensitivity and specificity of the test being performed, cancer risk management strategies, and the right of choice. Individuals should be provided with the necessary information to make an informed decision in a manner that they can understand (National Action Plan on Breast Cancer and American Society of Clinical Oncology 1997)."

In this case study, we give an overview of several related projects whose common goal is to develop a comprehensive approach to addressing individualized counseling of women who consider themselves at high risk of breast cancer. We discuss three aspects: assessing the probability of carrying an inherited genetic mutation based on family history of breast and related cancers; predicting the risk of developing cancer based on both family history and additional well recognized risk factors; using the results to provide women facing testing and preventive treatment decisions with relevant, individualized information.

In the remainder of this section we provide background information about known risk factors for breast cancer, and current options for genetic testing. In Section 2 we review existing models of breast cancer risk, and argue that there is a need for a new comprehensive risk model incorporating recent scientific advancements. We discuss our plans for developing such a model, and relay progress to date. We organize the exposition around two prediction problems:

1. (Section 3) Computing the probability that an individual is a carrier of an inherited mutation at BRCA1 or BRCA2 based on a detailed account of his or her family history;

2. (Section 4) Computing the probability that a woman will develop breast cancer based on a detailed account of her family history as well as other risk factors.

In Section 5 we discuss a decision analysis where this information is used to help women and their physicians in deciding about genetic testing. In Section 6 we illustrate the results using a specific example.

1.2 Risk factors for Breast Cancer

There are approximately 180,000 new cases of breast cancer each year in the United States. Approximately 45,000 women die of the disease annually (American Cancer Society 1996). Several factors are recognized to contribute to the risk of developing breast cancer, and are reviewed here.

Family history. The most highly predictive risk factors for breast cancer is a family history of breast and certain other cancers. The syndrome of hereditary breast cancer was initially characterized by the early age of onset, excess incidence of bilateral breast cancer, associations with other cancers, and vertical transmission (Lynch et al. 1984). Several epidemiologic studies have identified subgroups of women with a family history of breast cancer who are at an especially elevated risk. In general, a twofold to threefold increase in the risk of breast cancer has been associated with a family history of breast cancer in a mother or sister. Risk is higher for women with more than one first-degree relative, women with first-degree relatives who were diagnosed with breast cancer at an early age, and women with a first-degree relative who had bilateral breast cancer (Go et al. 1983; Goldstein et al. 1987; Claus et al. 1994; Claus et al. 1990a; Claus et al. 1990b). Using data from the Cancer and Steroid Hormone (CASH) study, Claus and colleagues demonstrated that the risk of breast cancer for first-degree relatives rose as the age of onset decreased. They reported the hazard ratio for a sister of a case with an affected mother and no additional affected sisters to be 5.9 (95% CI 3.9-8.9) when it was diagnosed at age 50, compared with 9.4 (95% CI 6.2-14.4) and 15.1 (95% CI 9.4-24.3) when the case was diagnosed at 40 and 30 years of age, respectively. The hazard ratios were even higher for those whose sister was affected. In addition, a family history of ovarian cancer has also been found to be associated with a 1.5 fold increase risk in breast cancer (Schildkraut and Thompson 1988).

These findings have led to formal genetic linkage analyses, which provided strong evidence for the existence of rare autosomal dominant inherited genes associated with increased susceptibility to breast cancer (Newman et al. 1988; Hall et al. 1990; Claus et al. 1990b). Based on data from the CASH study, the genetic model that best fit the breast cancer recurrence rates in first-degree relatives of the cases and controls specified a rare (gene frequency = 0.0033) allele that is dominantly inherited along with a high frequency of phenocopies or non familial cases (Claus et al. 1991). This result was in agreement with other studies (Newman et al. 1988; Bishop et al. 1988).

More recently, these discoveries have led to the identification of two breast cancer susceptibility genes. Approximately 2% of breast cancers and 10% of ovarian cancers are believed to occur in women who carry an inherited mutation at a gene called BRCA1, isolated on chromosome 17q21 (Miki et al. 1994; Futreal et al. 1994) A smaller fraction of breast cancer is attributable to inherited mutations at a gene called BRCA2, on chromosome 13q12-13 (Wooster et al. 1995). Mutations are rare, occurring in less than .2% of women (Ford and Easton 1995). But women with mutations are very likely to develop one or both of these cancers, and to develop them at relatively young ages (Easton et al. 1995; Schubert et al. 1997; Levy-Lahad et al. 1997; Easton et al. 1997; Struewing et al. 1997). This means that despite the mutations' rarity, they are likely to be present in families that have multiple occurrences of breast or ovarian cancer. The high incidence of breast and ovarian cancer among carriers suggests that family history of these diseases is a strong indicator of whether a mutation is present in the family, and therefore also whether a particular family member is a carrier.

Factors other than family history are also important. These are discussed next. The first two are related to each other and to overall estrogen exposure.

Reproductive factors. Generally speaking, the longer the reproductive lifespan, the greater the risk of developing breast cancer (Kelsey et al. 1993). Early age of menarche (La Vecchia et al. 1985; Brinton et al. 1988; Hsieh et al. 1990; Wang et al. 1992) late age at menopause (Hsieh et al. 1990; Pike et al. 1981; Ewitz and Duffy 1988; Kvale and Heuch 1988) and late age at first birth (Paffenberger et al. 1980; Bruzzi et al. 1988; Yuan et al. 1988) are all known to increase the risk of breast cancer. Although consistent, the relative risks associated with reproductive factors and breast cancer risk are weak, generally ranging below 2.0.

Oral contraceptive and estrogen use. Overall, oral contraceptives (OCs) have not been found to increase the risk of breast cancer. However, data have been inconsistent for long-term use of OCs. Some studies showed a modest increase with long-term use (Miller et al. 1989; Vessey et al. 1989; United Kingdom National Case-Control Study Group 1989; The WHO Collaborative Study of Neoplasia and Steroid Contraceptives 1990). In particular, published odds ratios range from 1.7 to 10.1 for long duration of OC use among women under age 35 at diagnosis (Miller et al. 1989; Vessey et al. 1989; Kay and Hannaford 1988). Other subgroups of users noted to be at increased risk include women under age 45 at diagnosis (Romieu et al. 1989), women with no children (The WHO Collaborative Study of Neoplasia and Steroid Contraceptives 1990; Prentice and Thomas 1987), and long-term use before age 25 (Lund et al. 1989; Olsson et al. 1989), or before a first birth (United Kingdom National Case-Control Study Group 1989; The WHO Collaborative Study of Neoplasia and Steroid Contraceptives 1990; Olsson et al. 1989). Estrogen replacement therapy has been found to be weakly associated with breast cancer risk among postmenopausal women (Colditz et al. 1995).

A recent large meta-analysis of oral contraceptives revealed that while there is a small increased risk of breast cancer within 10 years after stopping the use of oral contraceptive (RR (current use) = 1.24 (95% CI = 1.15-133) and RR (5-9 years after stopping) = 1.7 (95% CI = 1.02-1.12) there is not significant excess risk of having breast cancer diagnosed 10 or more years after stopping use (RR = 1.10 (95% CI = 0.96-1.05) (Collaborative group on hormonal factors in breast cancer 1996).

History of benign breast conditions. A positive history of benign breast disease has been associated with a two- to three-fold increase in risk of breast cancer (Kelsey and Gammon 1991; Dupont and Page 1985; Dupont and Page 1987). Although most studies suggest that the elevation in risk occurs primarily in women with epithelial proliferation or atypia in the biopsy, and that the risk directly increases with the degree of atypia, the exact relationship with breast cancer risk remains unclear (Krieger and Hiatt 1992; Dupont and Page 1985; Ebbs and Bates 1988; Palli et al. 1991). The wide range of normal variation in breast tissue has led many researchers and clinicians to question whether some conditions termed benign breast disease should be considered disease at all. Efforts to estimate the association between benign breast disease and breast cancer have been hampered by the lack of precision in defining benign breast disease and the absence of reliable incidence data. With the exception of family history, breast cancer and benign breast disease do not share any of the same risk factors (Dupont and Page 1987; Ernster et al. 1987). It remains unclear whether benign breast lesions represent precursors to malignant ones, whether they are markers of increased susceptibility to breast abnormalities, or whether they are indicators of other relationships.

Mammographic patterns and density. Previous studies support that there is a relationship between parenchymal pattern and the risk of breast cancer(Wolfe 1976; Saftlas et al. 1989; Boyd et al. 1984). A two to six-fold increase of breast cancer risk has been reported to be associated with various categorizations of patterns of mammographic features. In a recent study by Byrne and colleagues high-density parenchymal pattern effects were found to be independent of family history, age at first birth, alcohol consumption, and benign breast disease. Women with a breast density of 75% or greater had almost a fivefold increased risk of breast cancer compared with women with not visible density. This result suggests mammographic density may be one of the most important predictors of breast cancer to date.

Obesity. Obesity had been associated with decreased risk of breast cancer in premenopausal women, although results across studies are not consistent. This may be attributed to an increase in anovulatory cycles and a resulting decrease in absolute levels of estrogen and progesterone. Among post-menopausal women who are obese, the risk of breast cancer is increased (Kelsey and Gammon 1991; Newman et al. 1986; Swanson et al. 1988; Morabia and Wynder 1990). A likely mechanism is that obesity potentiates the conversion of plasma estrone levels. Excess levels of estrone are thought to promote breast cancer (Thomas 1984; Brinton et al. 1983).

Cigarette smoking and alcohol use. Cigarette smoking does not seem to be related to incidence of breast cancer, although one study found a relationship with breast cancer mortality (Calle et al. 1994). Some studies support an association between alcohol consumption and breast cancer (Schatzkin et al. 1987a; Yuan et al. 1988; Longnecker et al. 1988; Howe et al. 1991). While some studies show an increased risk for alcohol use, not all studies have been consistent (Chu et al. 1989; Rosenberg et al. 1982). Some studies have pointed to particular subgroups of women who are at increased risk following alcohol consumption: the association appears to be stronger among younger women (Williamson et al. 1987; Grunchow et al. 1985); and among premenopausal compared to postmenopausal women (Howe et al. 1991; Young 1989).

Race. The average annual age-adjusted incidence rate for breast cancer is lower among African-American women (92.8 per 100,00) as compared with white women in the U.S. (112.2 per 100,00) (Gloeckler-Reis et al. 1990). Age-specific incidence rates, however, vary by age and race. Among younger women, blacks have a higher rate than whites and among older women, whites have a higher rate than blacks, with the cross-over of rates occurring between the ages of 40 and 50 years (Gray et al. 1980; Janerich and Hoff 1982; Polednak 1986). Reasons for the age-specific differences are not known. Risk factors for breast cancer among African-American women are not well understood. However, the risk for breast cancer among blacks has been found to be associated with nulliparity or low parity (Mayberry and Stoddard-Wright 1992; Austin et al. 1979; Schatzkin et al. 1987b). Early age at menarche was not found to be associated with the risk of breast cancer in African-Americans (Mayberry and Stoddard-Wright 1992; Austin et al. 1979; Schatzkin et al. 1987b).

Lactation. The association between lactation and the risk of breast cancer is uncertain. A small reduction in risk has been reported in several studies (Byers et al. 1985; McTirnan and Thomas 1986; Katsouyanni et al. 1986; Yoo et al. 1992; United Kingdom National Case-Control Study Group 1993). Recently, Newcomb et al. (Newcomb et al. 1994) reported a slight reduction in risk of breast cancer among premenopausal women (RR = 0.78, 95% CI = 0.66-0.91), but not among postmenopausal women with breast cancer.

Exercise. The literature supports a modest inverse relationship between physical activity and breast cancer risk(Friedenreich and Rohan 1995; Mittendorf et al. 1995; McTiernanan et al. 1996), although not all studies have been consistent with this finding (Chen-Ling et al. 1997). A recent study by Thune, and colleagues which included 25,624 women, was the first prospective study to report an association between physical activity and a reduced breast cancer risk with an overall adjusted OR of 0.63 (95% CI = .42-.95).

1.3 Genetic Testing

The identification of the BRCA1 and BRCA2 genes allows for direct testing to ascertain whether there is a germline (inherited) mutation (Weber 1996). Genetic testing was initially offered in specialized clinics within academic medical centers, but more recently it has become available commercially (Nelson 1996; Oncormed, Inc. 1996; Myriad Genetics, Inc. 1996). The analysis is done on a small sample of blood. A comprehensive analysis, determining the full sequence of the protein-coding regions of the BRCA1 and BRCA2 genes is offered at a cost of $2,400 (Myriad Genetics, Inc. 1996). If a mutation is known to be present in a family, or if an individual belongs to a specific ethnic subgroup characterized by a small number of identified mutation, a mutation-specific test can be performed at considerably lower cost than a comprehensive analysis. Preliminary data indicate that the demand for testing may be high. In a sample of 121 first-degree relatives of ovarian cancer patients, 75% of the members indicated that they would "definitely" want to be tested (Lerman et al. 1994). Other studies have obtained similar results (Struewing et al. 1995). Thus it is appropriate to ask "What are the benefits and limitations associated with BRCA1 and BRCA2 testing?" "Who, if anyone, should be tested?"

The result of a test modifies the probability that a woman carries a mutation, and thus her knowledge of the risk of developing breast and/or ovarian cancer, and of the risk of transmitting a high susceptibility to her offspring. Such information may help some women decide to pursue a course of action (such as prophylactic mastectomy) to reduce their risk of developing cancer. If this measure reduces the risk of cancer, it might result in a net improvement in survival and also in overall quality of life, even after taking into account the negative quality of life aspects of the prophylactic intervention. In addition, women who are contemplating mastectomy and/or oophorectomy but who test negative, might benefit by avoiding surgery and its associated negative sequelae.

On the other hand, testing for BRCA1 and BRCA2 has technical limitations, and will not necessarily provide a definitive answer about whether a particular woman has a mutation in the BRCA1 or BRCA2 genes. Even comprehensive gene sequencing "may not be able to detect some kinds of mutations, such as large deletions or mutations that lie outside the protein-coding region of the gene" (Myriad Genetics, Inc. 1996). As a result, tests for BRCA1 and BRCA2 will have the potential for false-negatives. Albeit to a lesser extent, tests can also produce false-positives results. Also, among families with a genetic predisposition to breast cancer, a mutation in a yet to be identified gene may be the culprit and a test for BRCA1 and BRCA2 alone would not identify these cases. Thus a woman might well have a genetic predisposition, even if no mutations in BRCA1 or BRCA2 are found. Finally, although women with a mutation in BRCA1 or BRCA2 have higher rates of breast and ovarian cancer, it is not possible to say with certainty which women will develop cancer, and at what age.

In addition to these intrinsic limitations, testing is not without negative implications for a woman. A positive test outcome may limit a woman's eligibility for health insurance (Meissen et al. 1991; Schwartz et al. 1995) and may result in employment discrimination (Billings et al. 1992). Also, the decision to seek testing may affect, or be opposed by other family members.

The decision to undergo genetic testing is difficult, and appropriate counseling is necessary (National Advisory Council for Human Genome Research 1994; Hoskins et al. 1995; National Action Plan on Breast Cancer and American Society of Clinical Oncology 1997) An important component of counseling is an individualized evaluation of the probability that the woman carries a mutation and of the probability that she will ultimately develop breast cancer, based on information that can be easily ascertained during a counseling session.

2 Risk Models

Before discussing our approach to developing a comprehensive approach to addressing individualized counseling of women who consider themselves at high risk of breast cancer and/or are considering genetic testing, we review existing models and discuss some of their strengths and limitations.

2.1 *Models of breast cancer risk*

The *Gail et al.* Model. Currently, the "gold standard" for breast cancer risk prediction is a model developed by Gail and colleagues (Gail et al. 1989), which provides individualized estimates of absolute risk—the probability of develop breast cancer over a defined time period—for white females. This model has contributed substantially to place the due emphasis on absolute risk as the appropriate quantitative assessment for individualized counseling. It is currently the basis for selection into chemoprevention trials, such as the Breast Cancer Prevention Trial (Smigel 1992), and is used in clinical settings for patient counseling. Software and simple graphs are available to compute point estimates as well as confidence intervals of absolute risk (Benichou 1993; Benichou et al. 1996).

The empirical basis is given by the Breast Cancer Detection Demonstration Project (BCDDP). All subjects in the BCDDP consisted of women who were willing to have annual screening examinations. The overall approach is to model risk as a function of observable family history and other non-familial risk factors, without assuming a genetic model (Benichou 1995; Benichou and Gail 1995). Absolute risks is calculated by combining estimates of age–specific relative risk with estimates of baseline age–specific hazard rates derived from the BCDDP panel. Risk factors and associated estimates of relative risks were obtained through a case–control study of a subset of BCDDP participants. Breast cancer cases consisted of 2,852 white women newly diagnosed with in situ or invasive carcinoma between 1973 and 1980 at 28 participating centers. The control group consisted of 3,146 white women, also participating in the BCDDP, matched to cases by age, race, study center, date of entry into, and tenure in, the BCDDP screening program. Analyses focused on familial, reproductive, menstrual and medical history variables.

The four risk factors chosen for inclusion in the risk model were (1) a family history of breast cancer in first–degree relatives, (2) late age at first childbirth, (3) early menarche, and (4) multiple previous benign breast biopsies. Risk estimates are obtained from an unconditional logistic regression model with the four risk factors and a factor for age included as main effects and interactions between age and number of biopsies and age at first live birth and number of first–degree relatives diagnosed with breast cancer. This parameterization of risk factors partitions individuals into 216 risk categories. Relative risks are taken to be the estimated odds ratio in favor of the category in question relative to a baseline category. These age–specific relative risks are converted to absolute risks by combining them with age–specific estimates of baseline hazard obtained from the full BCDDP data set.

Limitations of this model include the representativeness of the BCDDP population, composed of women who volunteered to participate, and hence the generalizability of the results. Gail *et al.* (1989) note that composite age–specific hazards derived from BCDDP differ from population–based estimates such as those derived from the SEER Program, with relative risks (BCDDP to SEER) as high as

2.1 for the youngest age groups (20 to 39) and falling off monotonically to below 1.0 for the oldest age groups (70 to 79). Hence the Gail model is derived from a population at higher than average risk and is, as advertised, only accurate for predicting risk among other, similarly obtained, high risk populations. Three recent validation studies confirmed that the Gail model provides accurate predictions for women in regular screening but tends to exaggerate the risk for unscreened or sporadically screened women (Spiegelman et al. 1994; Gail and Benichou 1994; Bondy et al. 1994). Most U.S. women do not obtain yearly mammography.

Bondy *et al.* (1994) evaluated the Gail model using case-control data constructed from a study population consisting of women, aged 30-75, with at least one first-degree relative with breast cancer, who had participated in the American Cancer Society 1987 Texas Breast Cancer Screening Project, and who were observed over a five year period. Although Bondy et al. found that overall the model predicted the risk among women who adhered to American Cancer Society guidelines (O/E = 1.12; 95% CI =0.75-1.61), the model over-predicted risk among women who did not adhere to the guidelines as well as among younger women. Although there was no evidence of lack of fit in a case-control analysis of the variables from the Gail model, none of the variables achieved statistical significance in the Texas Breast Screening Project population. Bondy et al. concluded the model was appropriate only for those who receive annual mammography.

Spiegelman *el al.* (1994) also evaluated the performance of the Gail model in predicting breast cancer risk using prospectively gathered data from the Nurses' Health Study. This validation included women 61 years of age or younger. The validation study revealed that the model's performance was unsatisfactory for women aged 25-61 who do not participate in annual screening. Overall, 33% (95% CI = 28%-39%) excess prediction was found. A twofold excess prediction was observed among women who were premenopausal, among women with two or more first-degree relatives, and among women whose age at first birth was younger than 20 years. It was thought that the most likely explanation was the 30%-60% higher baseline incidence rate in the BCDDP population although this could explain some but not all of the observed over prediction. In addition, misspecification of the form of the Gail *el al.* model (additive rather than multiplicative) was considered a possible explanation. Other possibilities, including omission of important risk factors and secular trends in the incidence of breast cancer, were considered less likely to explain over-prediction.

Other non-genetic models of risk. Other models have addressed empirically the breast cancer risk of individuals with a family history of breast cancer, without assuming a genetic model. Ottman and colleagues (Ottman et al. 1983; Ottman et al. 1986) conducted a life-table analysis to estimate the cumulative risk of breast cancer at various ages for mothers and sisters of breast cancer patients in a population-based series. These cumulative risk estimates were then used to derive a probability of breast cancer within each decade between ages 20 and 70 for mothers and sisters according to age at diagnosis in the patients and whether the disease was unilateral or bilateral. Anderson and Badzioch (1985) estimated the probabilities of developing breast cancer for sisters of patients with three types of family histories: whose 1. mother, 2. sisters or 3. second-degree relatives had previous breast cancer.

The *Claus et al.* model. Claus et al. (1994) approached absolute risk prediction using the autosomal dominant genetic model developed in their previous work (Claus et al. 1991). The model was developed prior to the identification of the BRCA1 and BRCA2 genes and is based on a segregation analysis of the CASH study, a population-based, case-control study described in Section 4.1. The model applies to a high risk subset of the general population which included women with at least one female relative with breast cancer and is based on the number and types of relatives affected with breast cancer as well as the ages at which those relatives became affected. Analyses included only the mother and sisters of white breast cancer cases and controls. Genetic models were fit to the age-specific recurrence data of breast cancer cases and controls using a maximum likelihood approach. The best fitting model for the observed age-specific rates, an autosomal dominant major gene model, was used to generate the predictions. Under the autosomal dominant model, two age-at-onset distributions were examined: one for carriers of the abnormal allele and one for non-carriers, each of which had an estimate genotype-specific mean and variance. The parameters in the model included the gene frequency, and genotype-specific penetrance as well as the age-at-onset distribution means and variances. These estimates were used to calculate age-specific risks for a woman with one or two affected relatives. Unlike other models, including that of Gail *et al.* the Claus et al. model incorporates age-at-onset of the affected relatives. Risk factors, such as age at first birth and age at menopause, were not found to be important in the subset of women analyzed by Claus *et al.* The model by Claus et al. is consistent with subsequent linkage analyses of BRCA1 and BRCA2 families.

Recently, Houwing-Duistermaat and Van Houwelingen (1997) developed a generalization of the Claus model to handle an arbitrary number of relatives of any degree. They utilize the parameters estimated by Claus *et al.* as well as a first order approximation of the risk. This model does not account for the exact pedigree structure, as the first order approximation is based on ignoring relations between family members, and only preserving the degree of relationship with the proband.

2.2 *Models of carrier probability.*

The identification of BRCA1 and BRCA2 has generated interest in determining carrier probabilities based on family history. Shattuck-Eidens *et al.* (1995) provide a summary table of estimated probabilities of carrying a BRCA1 mutation for 11 categories of high risk women. Couch *et al.* developed a logistic regression model for predicting the probability of carrying a BRCA1 mutation based on summary information about family history. Predictors in their model include having family with a history of both breast and ovarian cancer; an average age of less than 55 years at the diagnosis of breast cancer in affected relatives; the presence of ovarian cancer; the presence of breast and ovarian cancer in the same relative; and Ashkenazi Jewish ancestry.

While useful, these models capture family history information using rather broad categories, such as number of affected relatives, sometimes giving insufficient detail for decision making. For example, a woman with given numbers of first- and second-degree relatives with breast and ovarian cancer can have a wide range in the probability of carrying a mutation, depending on the pedigree structure, the exact relationship of affected members, and family size.

2.3 *A Comprehensive Approach*

In this section we outline our plans for developing a model addressing both carrier probabilities and breast cancer risk. The overall approach is based on combining two components: a genetic model to predict genotype; a risk model explicitly including genotype as one of the risk factors. When predicting carrier status, the genotype is treated as the response variable; when computing absolute risk, the genotype is treated as a latent variable and integrated out. This approach permits direct incorporation of the rapidly accumulating body of knowledge about genetic inheritance of breast cancer, and grants a high level of resolution in incorporating features of family pedigrees. In addition to accurate incorporation of family history, our modelling efforts are motivated by several additional concerns.

Existing models are based on single datasets, collected for specific needs and not immediately extendible to general populations. The best way to develop a predictive model that is applicable in general populations, in the absence of a new and massive data collection, is to combine several high quality data sources. We focus on one cohort study and two case-control studies. Combining prospective and retrospective studies is a challenge: we approach it via a novel approach based on modeling directly the retrospective distribution of covariates in the case-control studies.

Existing models require input of a strictly specified set of covariates; in clinical settings, this may not always be feasible. We can add flexibility by enabling risk prediction based on incomplete covariate specifications, so that it is possible to use the model when covariates are not fully known. Retrospective modeling provides the basis for appropriate imputation of the missing covariates. Complete ascertainment of information about family history and numerous risk factors can be a substantial burden to both counselors and patients. A model that handles missing information can be used sequentially. An interview can start with the most important questions and be suspended at any time if desired.

Existing models selected risk factors based on a single data source. To exploit the large body of literature available on risk factors, some of which was summarized in Section 1, we convened an expert panel to assist with the selection of appropriate risk factors. This process was also helpful in identifying weaknesses of our data sets and in identifying published results that can be used to specify informative prior distributions on the parameters for which we expect insufficient or no information from out data sets.

In Sections 3 and 4 we discuss our modeling approach, breaking the exposition down into two components:

1. Prediction of BRCA1 and BRCA2 genotype status based on a woman's family history of breast and ovarian cancer in first- and second-degree relatives. Technically, this probability can be obtained from Bayes' rule, using family history as the evidence and gene prevalence as the prior. Bayesian methods are also used to formally incorporate uncertainty about some of the key inputs of the model, such as the age-specific incidence of disease for carriers, and overall prevalence. This model component is discussed in Section 3.

2. Prediction of onset of breast cancer, based on family history and other available risk factor information. This will use multiple studies. We begin by describing a strategy for producing absolute risk prediction combining case-control and prospective studies. We then show how to incorporate the genotype model component into the analysis. This model component is discussed in Section 4.

3 The Probability of Carrying a Mutation

3.1 Learning from Family History

In this section we discuss the calculation of the probability that an individual (the counseland) carries a germline mutation at BRCA1 or BRCA2, based on his/her family's history of breast and ovarian cancer. Our discussion is based on Berry *et al.* (1997) and Parmigiani *et al.* (1998). While the approach applies to both women and men, the counseland is usually female. In our current implementation, family history includes the counseland and her first- and second-degree relatives. For each member, we ascertain whether he or she has been diagnosed with breast cancer, and if so, age at diagnosis or, if cancer free, current age or age at death; similar data is obtained for ovarian cancer if the member is female.

We assume that individuals inherit two BRCA1/BRCA2 alleles, one from each parent, and that these alleles are either normal or mutated. We also assume that BRCA1 and BRCA2 mutations are inherited independently. More than 100 different mutations have been identified so far for both BRCA1 and BRCA2. We assume that, for each locus, the same increased risk of cancer is associated with being a carrier of every mutation. We only distinguish between individuals of Ashkenazi Jewish ancestry and others. We assume an autosomal dominant inheritance of mutations, empirically supported by the analysis of Claus *et al.* (1991). The implications of this assumption on the probability of carrying a mutation are as follows.

At each locus, an individual can have no mutations, one mutation or two mutations. Our model addresses the joint probability of a specific configuration of BRCA1 and BRCA2 genes, represented by a two-dimensional vector. We use the notation $P[\text{BRCA1} = i_1, \text{BRCA2} = i_2]$ to denote the probability of the counseland having i_1 mutated copies of BRCA1, $i_1 = 0, 1, 2$, and i_2 mutated copies of BRCA2, $i_2 = 0, 1, 2$.

Let the frequencies of mutations in the allele population be f_1 and f_2 for BRCA1 and BRCA2 respectively. In the absence of information about disease and family history, the probabilities that an individual inherits a given number of mutated copies of BRCA1 are:

$$\begin{aligned} P[\text{BRCA1=2}] &= f_1^2 \\ P[\text{BRCA1=1}] &= f_1(1-f_1) \\ P[\text{BRCA1=0}] &= (1-f_1)^2. \end{aligned} \qquad (1)$$

The probability that an individual carries at least one BRCA1 mutation is

$$\pi_1 = f_1^2 + 2f_1(1-f_1).$$

Of these, the fraction carrying two mutations is therefore

$$g_1 = \frac{f_1}{f_1 + 2(1-f_1)}.$$

Similar expressions hold for BRCA2. Joint probabilities can be obtained based on independence.

Our goal is to compute the joint probability distribution of the BRCA1 and BRCA2 genetic status variables of a woman given her family history. Indicate the family history by h. From Bayes' rule:

$$P[\text{BRCA1}, \text{BRCA2}|h] = \frac{P[\text{BRCA1}]P[\text{BRCA2}]P[h|\text{BRCA1}, \text{BRCA2}]}{P[h]}. \qquad (2)$$

This probability is posterior to family history ascertainment and prior to genetic testing. In genetics it is typically refereed to as "prior probability". The relevance of Bayesian calculations for genetic counseling has been recognized for a long time (Murphy and Mutalik 1969). The use of Bayesian techniques in software for determining the positive predictive power of family history has a precedent in Szolovits and Pauker (1992).

While the BRCA1 and BRCA2 variables can be assumed to be a priori independent, they will typically not be independent conditional on family history. Also, based on the joint probability distribution arising from (2) we can compute various summaries of interest in decision making as regards genetic testing. These include the probability that a woman is a carrier of either a BRCA1 or a BRCA2 mutation given her family history:

$$\pi^* = 1 - P[\text{BRCA1} = 0, \text{BRCA2} = 0|h],$$

the marginal probabilities that a woman is a carrier of BRCA1 given family history:

$$\begin{aligned} \pi_1^* &\equiv P[\text{BRCA1} = 1 \text{ or } 2|h] \\ &= \textstyle\sum_{i_1=1}^{2} \sum_{i_2=0}^{2} P[\text{BRCA1} = i_1, \text{BRCA2} = i_2|h] \end{aligned}$$

		Mother		
		0	1	2
Father	0	$1,0,0$	$\frac{1}{2},\frac{1}{2},0$	$0,1,0$
	1	$\frac{1}{2},\frac{1}{2},0$	$\frac{1}{4},\frac{1}{2},\frac{1}{4}$	$0,\frac{1}{2},\frac{1}{2}$
	2	$0,1,0$	$0,\frac{1}{2},\frac{1}{2}$	$0,0,1$

TABLE 1. *The probability of 0, 1 and 2 mutations in an offspring given any combination of the numbers of mutations in the parents.*

and the marginal probabilities that a woman is a carrier of BRCA2 given family history:

$$\begin{aligned}\pi_2^* &\equiv P[\text{BRCA2} = 1 \text{ or } 2|h] \\ &= \textstyle\sum_{i_1=0}^{2}\sum_{i_2=1}^{2} P[\text{BRCA1} = i_1, \text{BRCA2} = i_2|h].\end{aligned}$$

The likelihood function $P[h|\text{BRCA1, BRCA2}]$ in (2) is computing intensive because the conditioning event is the genotype of the counseland only; the genotypes of all other family members are unknown and need to be integrated out. This can be done by a chain conditioning. For example, in a nuclear family, we know the conditional probabilities of the parents' genotypes given the offspring and the conditional probabilities of the parents' phenotypes (their cancer status and age) given their own genotypes. The exact analytic expression for the integration of relatives' genotypes in pedigrees with first- and second-degree relatives is given in Parmigiani *et al.* (1998). We developed a C code to perform these calculations. Its core module can be used for any two genes, each leading to different penetrance functions for two different diseases.

Without giving a full treatment, we now review the building blocks that underlie this calculation: the probability of the genetic status of offspring given those of their parents; the probability of the genetic status of parents given those of their offspring; and the probability of disease outcome given genetic status. The first two are discussed next, the third is discussed in Section 2.4.

The probability of a genetic status configuration among offspring, given the genetic status of their parents can be computed based on random selection of the offspring's alleles from each parent's two alleles. We use the notation o for the offspring's genetic status, m for the mother's genetic status, and f for the father's genetic status. Each of these is a two-dimensional vector. The two coordinates represent the number of mutated alleles at BRCA1 and BRCA2, respectively. The set of possible values is $\{0, 1, 2\} \times \{0, 1, 2\}$. Because inheritance of BRCA1 and BRCA2 mutations is assumed to be a priori independent, we can focus on marginal distributions. Table 1 summarizes the probability of zero, one and two mutations of a gene in an offspring given any combination of the numbers of mutations in the parents. For example, the probability that an offspring has 0, 1 and 2 mutations, given that the mother has exactly one mutation and the father has exactly one mutation are $1/4$, $1/2$ and $1/4$.

The probability of any joint configuration of the parents' genetic status given that of an offspring requires a simple application of Bayes' rule. We begin by assuming independence between the genetic statuses of the two parents, which is tenable assuming random mating in the population. This implies

$$P[\mathsf{m},\mathsf{f}] = P[\mathsf{m}]P[\mathsf{f}],$$

Then, using Bayes' rule,

$$P[\mathsf{m},\mathsf{f}|\mathsf{o}] = \frac{P[\mathsf{m}]P[\mathsf{f}]P[\mathsf{o}|\mathsf{m},\mathsf{f}]}{\sum_{\mathsf{m}}\sum_{\mathsf{f}} P[\mathsf{m}]P[\mathsf{f}]P[\mathsf{o}|\mathsf{m},\mathsf{f}]},$$

where the conditional probabilities are those of Table 1.

3.2 *Uncertainty*

From expression (2) we see that inference about the genetic status of the counseland requires knowledge of mutation frequencies and of cancer rates for both carriers and noncarriers of mutations. These need to be evaluated for all family members whose history of breast and ovarian cancer (including no history) is available. These evaluations are based on empirical studies and so are uncertain. Uncertainty can be accommodated using a Bayesian approach as follows. We indicate by ξ the set of parameters indexing the cancer rate models, and by $p(\xi, f_1, f_2|\text{Published Data})$ the probability distribution expressing the uncertainty about (ξ, f_1, f_2) based on the published estimates.

We can then use $p(\xi, f_1, f_2|\text{Published Data})$ to compute the distribution of the random variables

$$P[\text{BRCA1}, \text{BRCA2}|h, \xi, f_1, f_2] = \frac{P[\text{BRCA1}, \text{BRCA2}|f_1, f_2]P[h|\text{BRCA1}, \text{BRCA2}, \xi, f_1, f_2]}{P[h|\xi, f_1, f_2]}, \quad (3)$$

with BRCA1=0, 1, 2 and BRCA2=0, 1, 2. This has the feature of examining separately the uncertainty deriving from the unknown genetic status of family members, which is integrated out, from uncertainty about population rates and prevalences. Alternatively, we can write the probability distribution of interest in the presence of estimation error by integrating out the unknown parameters:

$$P[\text{BRCA1}, \text{BRCA2}|h] = \int P[\text{BRCA1}, \text{BRCA2}|h, \xi, f_1, f_2] \, p(\xi, f_1, f_2|\text{Published Data}) df_1 df_2 d\xi. \quad (4)$$

We evaluate the integrand ratio exactly and we perform the integration by a simple Monte Carlo simulation.

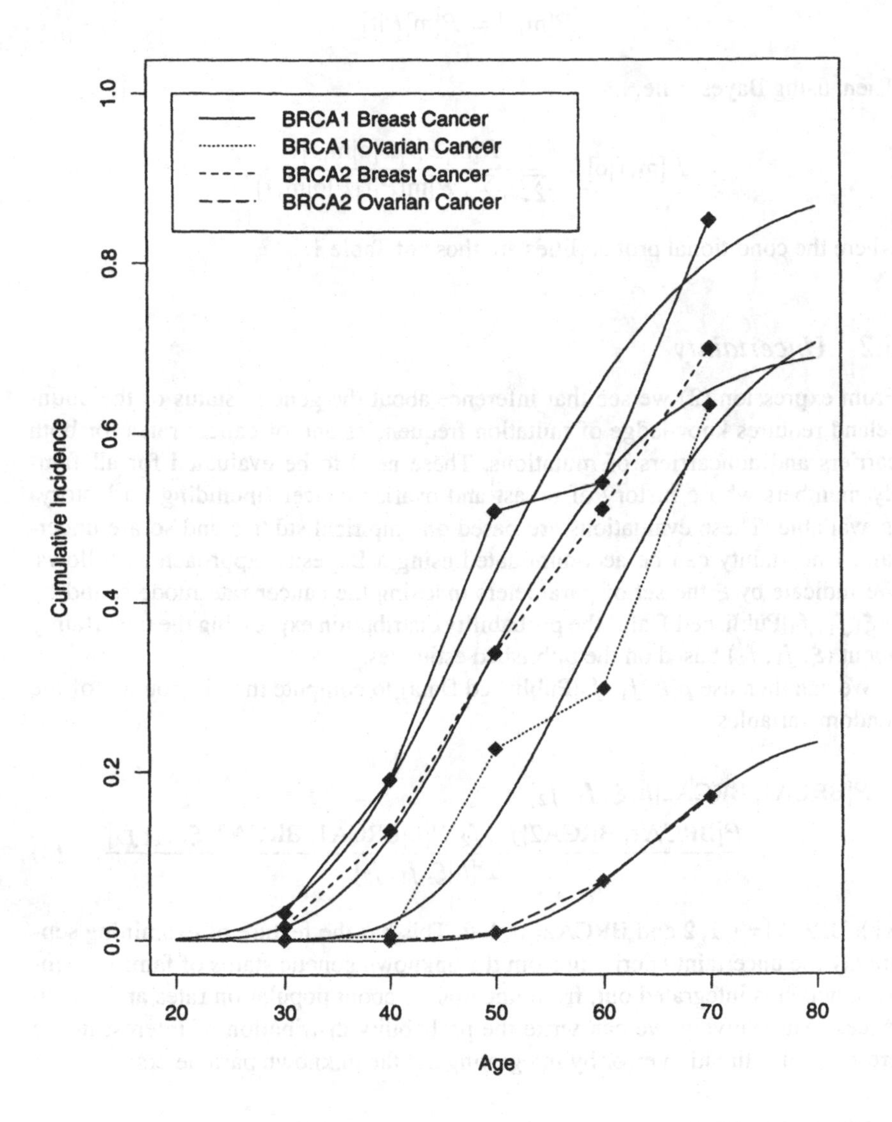

FIGURE 1. *Cumulative rates of breast (B) and ovarian (O) cancer for mutation carriers. The top panel refers to BRCA1 and is based on Easton, Ford and Bishop 1995; the bottom panel refers to BRCA2 and is based on Easton (1996) and Easton et al. (1997). Smooth lines indicate our interpolation, based on a three-parameter gamma c.d.f, with one of the parameters being the asymptote, which may be smaller than 1.*

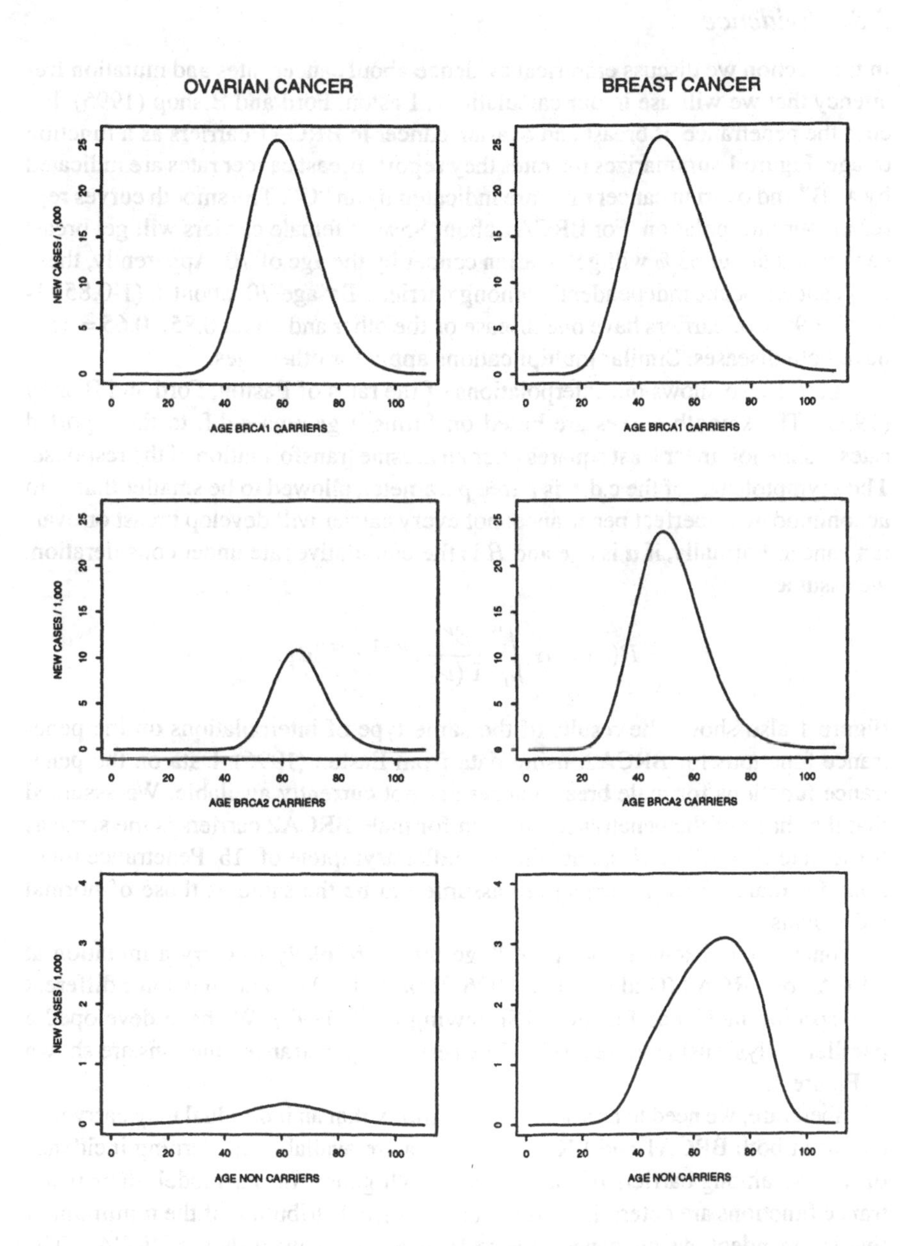

FIGURE 2. *Number of breast and ovarian cancer cases per year out of 1,000 born, for both BRCA1 and BRCA2 carriers and noncarriers. The scales for carriers and noncarriers are different, for better resolution. All else being equal, ovarian cancer is a stronger indication of a mutation than breast cancer and earlier breast cancer is a stronger indication than later breast cancer.*

3.3 Evidence

In this section we discuss empirical evidence about cancer rates and mutation frequency that we will use in our calculations. Easton, Ford and Bishop (1995) discuss the penetrance of breast and ovarian cancer in BRCA1 carriers as a function of age. Figure 1 summarizes the rates they report. Breast cancer rates are indicated by a "B" and ovarian cancer rates are indicated by an "O". The smooth curves represent our interpolation. For BRCA1 about 85% of female carriers will get breast cancer and about 65% will get ovarian cancer by the age of 70. Apparently, these two cancers occur independently among carriers. By age 70, about 1-(1-0.85)(1-0.65) = 95% of carriers have one disease or the other and about 0.85×0.65 = 55% have both diseases. Similar multiplications apply for other ages.

Figure 1 also shows our interpolations of the rates of Easton, Ford and Bishop (1995). The smooth curves are based on fitting a gamma c.d.f. to the reported rates, using nonlinear least squares after an arcsine transformation of the response. The asymptote (α) of the c.d.f. is a free parameter, allowed to be smaller than 1 to accommodate imperfect penetrance: not every carrier will develop breast or ovarian cancer. Formally, if a is age and R is the cumulative rate under consideration, we assume:

$$R(a) = \alpha \int_0^a \frac{\beta^\nu}{\Gamma(\nu)} x^{\nu-1} e^{-x\beta} dx.$$

Figure 1 also shows the results of the same type of interpolations on the penetrance functions for BRCA2, using data from Easton (1996). Data on the penetrance functions for male breast cancer are not currently available. We assumed that the shape of the penetrance function for male BRCA2 carriers is the same as for female BRCA2 carries, but with a smaller asymptote of .15. Penetrance functions for male BRCA1 carriers are assumed to be the same as those of normal individuals.

Women of Ashkenazi Jewish heritage are more likely to carry a mutation at BRCA1 or BRCA2 (Oddoux et al. 1996; Roa et al. 1996) and may have different age–specific incidence functions (Struewing et al. 1997). We have developed a parallel analysis using these results. The resulting penetrance functions are shown in Figure 3.

Albeit rare, we need to consider the possibility that an individual may carry mutations at both BRCA1 and BRCA2. No data are available concerning incidence of disease among carriers of mutations at both genes. In our model, their penetrance functions are determined as the cumulative distribution of the minimum of two independent events: cancer due to BRCA1 and cancer due to BRCA2. The results are not very sensitive to this assumption.

The disease history for a family member can include bilateral breast cancer. Estimation of rates of bilateral recurrence is difficult, because of the lack of good data. In the current version of our model, we approximate the rates for bilateral recurrences by assuming that the observed breast cancer rates $R(a)$ represent the distribution of the minimum of two independent events: cancer in the left and

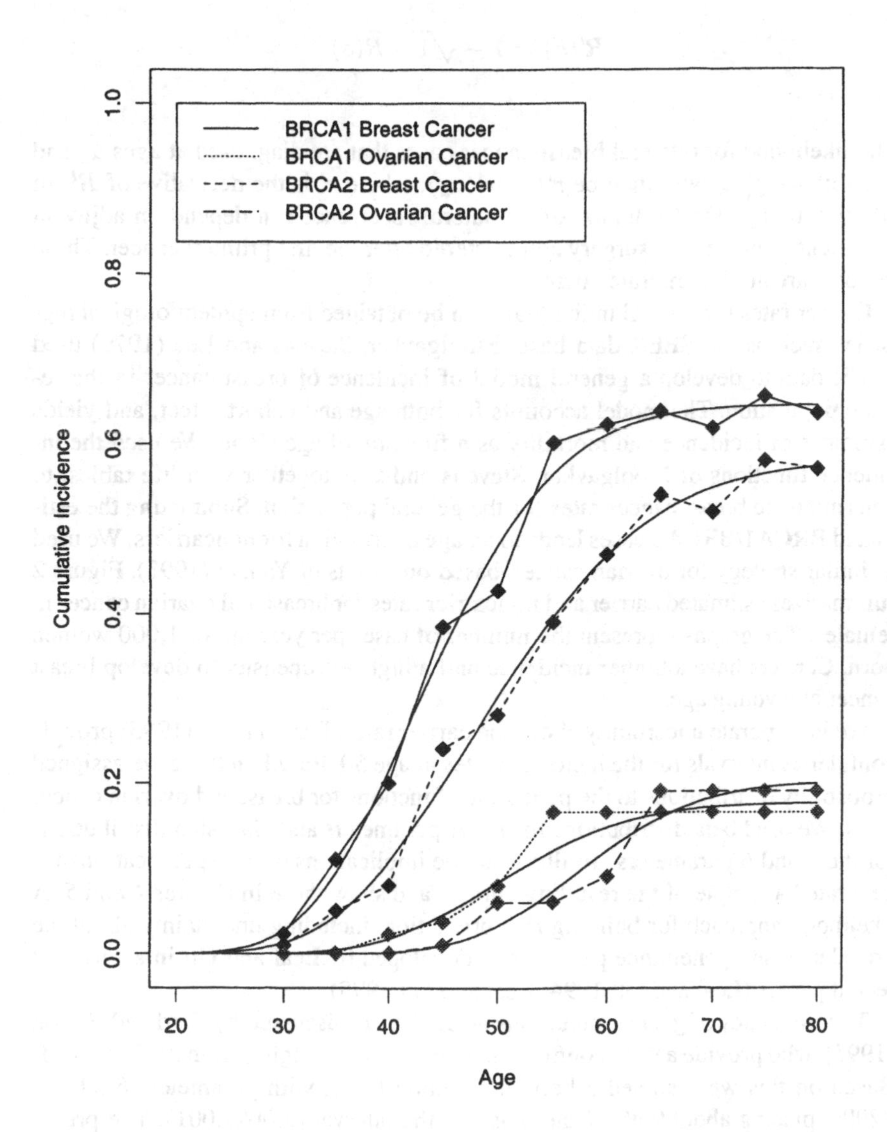

FIGURE 3. *Cumulative rates of breast and ovarian cancer for carriers of BRCA1 and BRCA2 mutation specific to the Ashkenazi Jewish population, as reported Struewing and colleagues (1997). Smooth lines indicate our interpolation, based on a three-parameter gamma c.d.f, with one of the parameters being the asymptote, which may be smaller than 1.*

cancer in the right breast. The cumulative rates of the time to cancer in a particular breast will then be:

$$R'(a) = 1 - \sqrt{1 - R(a)}$$

The likelihood for bilateral breast cancer cases that are diagnosed at ages a_1 and a_2, with $a_1 \leq a_2$ will then be $r'(a_1)r'(a_2)$, where r' is the derivative of R'. In addition to age, the likelihood of a bilateral recurrence can depend on adjuvant treatment or preventive surgery administered after the first primary cancer. These features are not incorporated here.

Cancer rates for normal individuals can be obtained from epidemiological registries such as the SEER data base. Moolgavkar, Stevens and Lee (1979) used SEER data to develop a general model of incidence of breast cancer in the female population. The model accounts for both age and cohort effect, and yields estimates of incidence and mortality as a function of age alone. We used the incidence functions of Moolgavkar, Stevens and Lee, together with life tables, to determine the breast cancer rates for the general population. Subtracting the estimated BRCA1/BRCA2 cases leads to an age distribution for noncarriers. We used a similar strategy for ovarian cancer, based on results of Yancik (1993). Figure 2 summarizes estimated carrier and noncarrier rates for breast and ovarian cancer in females. The graphs represent the number of cases per year out of 1,000 women born. Carriers have a higher incidence and a higher propensity to develop breast cancer at a young age.

We incorporate uncertainty about the carrier rates. Easton *et al.* (1995) provide confidence intervals for the incidence rates at age 50. Based on these we assigned probability distributions to the penetrance functions for breast and ovarian cancer rates. We used beta distributions for the α parameters and Gaussian distributions for the ν and β parameters. To illustrate the implications of our specifications we generated a sample of the resulting curves, and show these in Figures 4 and 5. A likelihood approach for building risk prediction, including uncertainty about the prevalence and penetrance parameters, developed by Leal and Ott in a series of recent papers (Leal and Ott 1994; Leal and Ott 1995).

The frequency of genetic mutations at BRCA1 is discussed by Ford and Easton (1995), who provide a 95% confidence interval for f_1 ranging from .0002 to .001. Based on this we assigned a beta distribution to f_1, with parameters 6.29 and 12000, placing about 95% of the mass on the interval $(.0002, .001)$. The prevalence of genetic mutations at BRCA2 is reviewed by Andersen (1996). Assessment of uncertainty are not available, so we assumed an amount of uncertainty comparable to that about the BRCA1 allele frequency. For individual of Ashkenazi Jewish origin, we use a beta distribution with parameters $(38, 3116 - 38)$ for BRCA1 and $(59, 3085 + 1255 - 59)$ for BRCA2(Oddoux et al. 1996; Roa et al. 1996). Cancer rates for the general population are based on large datasets and the uncertainty about their values can be ignored.

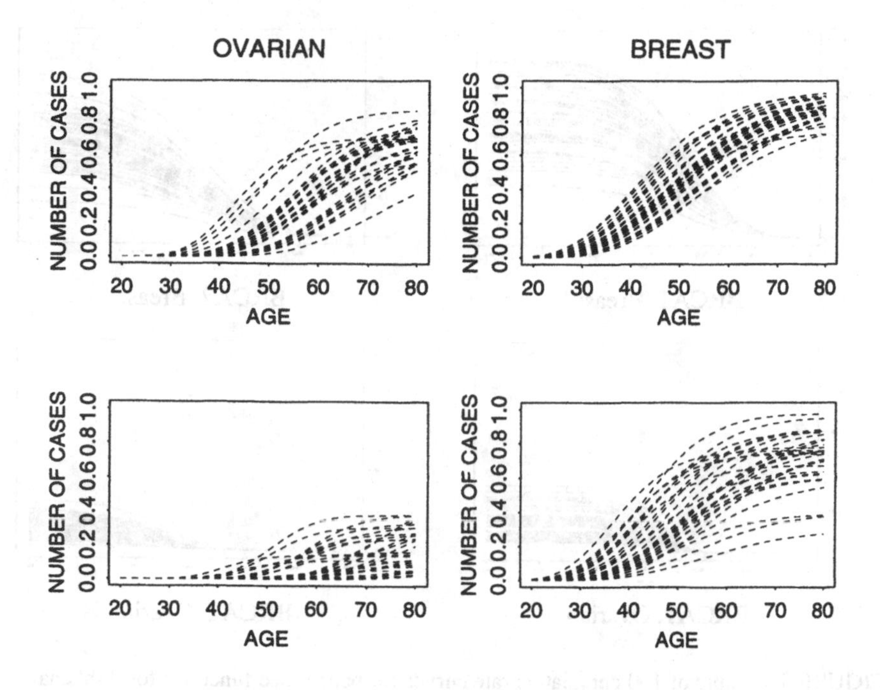

FIGURE 4. *Probability distribution of penetrance functions. Each graph is a sample of 30 penetrance curves. The top graphs refer to BRCA1 carriers; the bottom graphs to BRCA2 carriers. Curves are interpreted as in Figure 1. The uncertainty about the penetrance of the mutation is captured by variability in the total number of cases at age 80. The uncertainty about rates at young ages is high, in relative terms.*

3.4 Uses and limitations

Our model is currently being used in a randomized trial comparing tailored to standard printed materials in genetic counseling. Women in the tailored information arm receive estimates of their probability of carrying a mutation of BRCA1 or BRCA1, computed based on the model described here. The information is presented in graphical (pie chart), numerical, or verbal format depending on patient preferences. Counselees are provided with a range of probabilities reflecting uncertainty about the genetic parameters. The software performing the calculations is available to interested investigators for non-commercial purposes. Thus far it has been licensed to sites that are using it in counseling and in genetic epidemiology research. The pentrance and prevalences used by the model are stored in an external file, which is input to the program. In this way, it is convenient to create customized computations for subpopulations with different genetic parameters.

We considered BRCA1 and BRCA2 and regard all other breast cancer as being sporadic. This is a limitation of our model, as additional breast or ovarian cancer genes may exist (Håkansson *et al.* 1997, ,Vehmanen *et al.* 1997). Including other

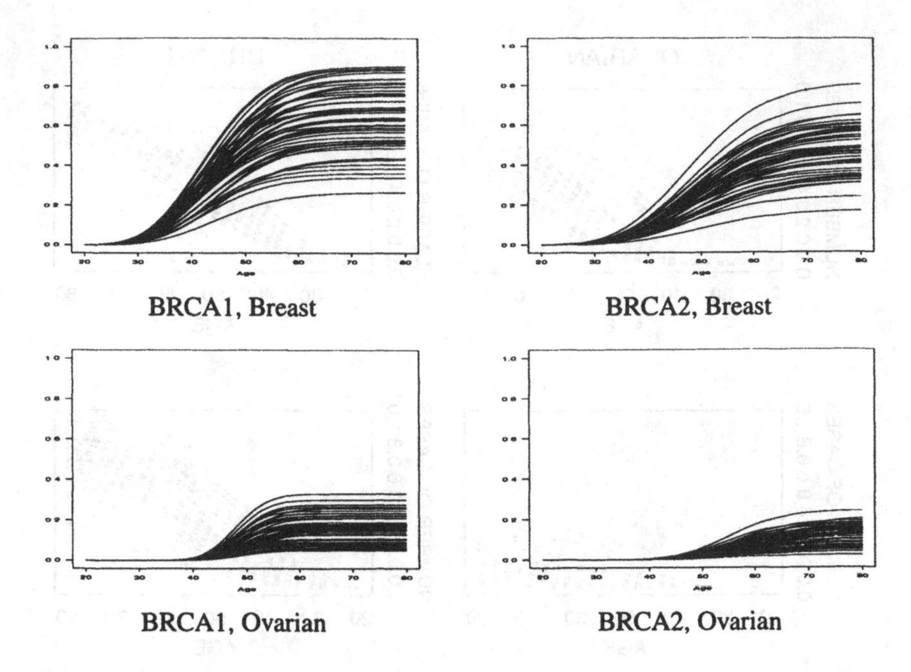

FIGURE 5. Sample of 100 cumulative rate curves for penetrance functions for Ashkenazi Jewish carriers.

genes and considering all other breast and ovarian cancer as being sporadic would give a different probability—usually smaller—that a woman carries BRCA1 and BRCA2. It would also give a different probability—larger—that a counseland carries at least one mutation at one or more breast cancer genes. Our procedure can be modified to include additional genes. However, such a modification would require information that is not yet available about the age-specific incidence of breast and ovarian cancer of such genes. Similar considerations apply to spontaneous mutations of BRCA1 and BRCA2 and interactions with environmental or other factors.

4 Absolute Risk of Breast Cancer

4.1 Data

Three datasets are available for the model development: the Cancer and Steroid Hormone (CASH) study, a study form the University of Wisconsin Comprehensive Cancer Center (UWCCC) often termed the Four State study, and the Iowa Women's Health Study (IWHS). Both the CASH study and the Four State study are multicenter, population-based, case-control studies. The IWHS is a cohort study. All three studies contain important risk factor information for breast cancer

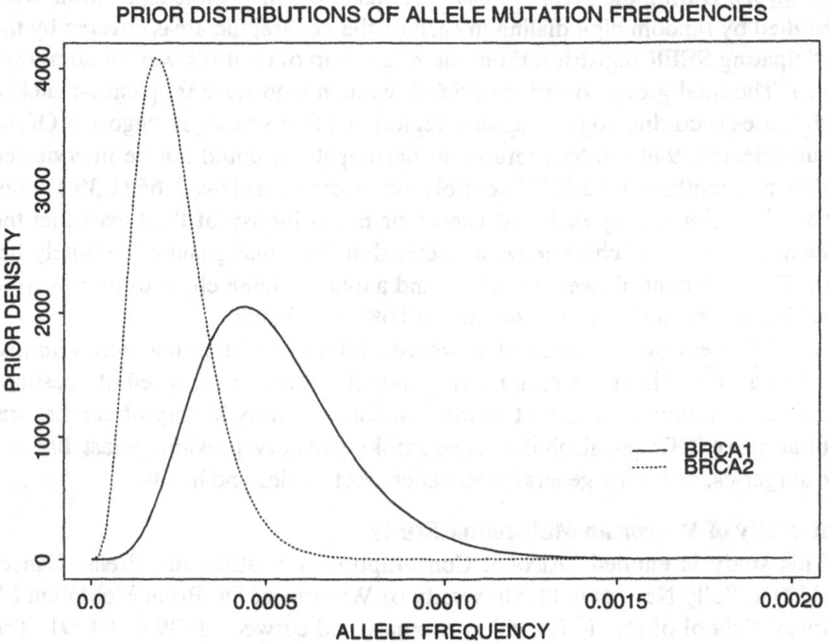

PRIOR DISTRIBUTIONS OF ALLELE MUTATION FREQUENCIES

FIGURE 6. *Probability distributions of the frequency of mutated alleles. The solid line is the distribution of f_1 while the dashed line is the distribution of f_2.*

and are described in detail below.

Cancer and Steroid Hormone (CASH) Study.

This Multicenter population-based case-control study was designed to examine the relationship between specific epidemiologic risk factors such as oral contraceptive use and breast cancer. From December 1, 1980 through December 31, 1982, eight centers of the SEER program of the National Cancer Institute identified cases of women with newly diagnosed breast cancer. The data from the CASH study is well documented and computerized in a SAS database. The data was provided by Dr. Marchbanks (Center for Disease Control and Prevention).

Cases. The eight geographic locations were the metropolitan areas of Atlanta, Detroit, San Francisco and Seattle, the states of Connecticut, Iowa, and New Mexico, and four urban counties of Utah. Breast cancer cases were women aged 20-54 years at diagnosis of a first primary breast carcinoma. Ninety-three percent of the eligible breast cancer cases were ascertained using rapid case ascertainment methods. Approximately 8-10 weeks elapsed between date of diagnosis and case ascertainment and 4,734 eligible cased were interviewed (80%). Most cases are white, although there are 490 (10%) black women in the breast cancer case group. Although breast cancer cases were histologically confirmed at the hospital of diagnosis, no independent pathology review was conducted for the CASH study.

Controls. During the period of case accrual, women eligible as controls were identified by random digit dialing in each of the geographic areas covered by the participating SEER registries. From these, a group of controls was randomly selected. The final group consisted of 5,698 women who were frequency-matched to the cases according to geographical region and five-year age categories. Of the group selected, 944 (16.5%) refused to participate or could not be interviewed within six months. Of the 4,754 controls who were interviewed, 66 (1.3%) either reported a prior history of breast cancer or had a biopsy of the breast that the respondent could not characterize as either definitely malignant or definitely benign. These 66 controls were excluded, and a total of 4,688 eligible controls were available for the analysis. Approximately 10% were black.

Subject Interview. A pretested structured 50-minute interview was administered by a trained interviewer in the respondent's home. It consisted of questions related to reproductive and past medical histories, family history of cancer, oral contraceptive (OC) and alcohol use, and smoking history, previous breast diseases and surgeries, and other general personal characteristics and habits.

University of Wisconsin Multicenter Study

This study is entitled "Alcohol Consumption, Lactation and Breast Cancer Risk' (Dr. Polly Newcomb PI, University of Wisconsin; Dr. Brian McMahon PI, Harvard School of Public Health) was conducted between 1989 and 1991. The specific aims of this study were to evaluate whether 1) cessation of alcohol intake before age 30 is associated with a lower risk of breast cancer, 2) alcohol intake before age 30 is associated more strongly with increased risk than consumption at later ages, and 3) increased duration of lactation is associated with a decreased risk of breast cancer among premenopausal women.

Cases. Women with breast cancer were identified from statewide tumor registries in Wisconsin, Maine, Massachusetts and New Hampshire. Cases were first screened for eligibility, and physicians were then contracted for permission to approach their patients (90% positive response). All female residents of Wisconsin, western Massachusetts, Maine, and New Hampshire who were diagnosed with breast cancer and were less than 75 years of age were eligible for the study. Cases were identified by each state's registry from April 1989 through December 1991. Eligibility was limited to case subjects with listed telephone numbers, driver's licenses verified by self-report (if less than 65 years of age), and known date of diagnosis. Of the 8,532 eligible cases subjects, 6,888 (80.7%) were available for the analysis.

Controls. Control subjects were selected by randomly sampling driver's license and Medicare beneficiary lists in each state. Of these age eligible subjects, 22% do not have a valid phone number or were deceased. Controls were required to have no history of breast cancer and to have a listed telephone number. Of the 11,319 potential controls 9,529 participated in the study for an overall response rate of 84.2%. The majority of the study subjects were from Wisconsin (60%). The majority of the study subjects reside in semi-rural areas; 95% of them are white, 2% African Americans and 3% of other races.

The data collection for this study was accomplished using a computer-assisted telephone interview (CATI) system developed at UWCCC. Reliability if the data collection had been demonstrated in reproducibility sub-studies which showed excellent concordance.

Iowa Women's Health Study Cohort

The Iowa Women's Health Study (IWHS) is a prospective study of mortality and cancer occurrence in older women. The cohort was selected from a January 1985 current drivers list obtained from the Iowa Department of Transportation. From this list, a total of 195,294 women between the ages of 55 and 69 were eligible. This sampling pool included approximately 94 percent of women in Iowa in this age range.

Study population. In December 1985 a 50% random sample of the eligible women was selected, yielding 99,929 women. Of these, 103 were immediately excluded because their mailing addresses were not in the state of Iowa. On January 16, 1986, the 99,826 women to be included in the study were mailed a 16-page questionnaire. Completed mailed questionnaires were received from 41,837 women, which, after excluding non-Iowa residents deceased individuals, and males, reflected a response rate of 42.7%. As of 1993, a total of 1,085 incident breast cancer cases had been diagnosed among the 37,105 women at risk for postmenopausal breast cancer. The 1994 breast cancer cases will be available in February 1996.

Vital status and current residence of non-respondents to the follow-up surveys were determined through the National Death Index and the National Change of Address Service, respectively. Out migration from Iowa is low; estimated to be about 1% annually in this cohort. The cohort is 99 percent white, and predominantly rural; 65 percent live in towns of less than 10,000 inhabitants. Approximately 81 percent have attained at least a high school education, 35 percent are current or former smokers, and 44 percent report regular consumption of alcoholic beverages.

Survey data. Epidemiologic data were obtained from a self-administered mailed questionnaire. These data include demographic characteristics, reproductive history, hormone use, lactation, occupation, body size, diet and exercise. Information on a variety of incident, self-reported diseases and current residence was collected through three follow-up questionnaires.

Follow-up. Follow-up for cancer occurrence in the cohort is accomplished using the State Health Registry of Iowa, part of the SEER program. Incident cases of cancer are identified through a computer program that matches Registry cases and study participants on names, zip code, birth date and social security number.

4.2 Selecting Predictors by Expert Consensus

The study questionnaires of each of the three datasets contain hundreds of questions and deal with a wide variety of risk factors. Our approach to selecting the candidate risk factors to be included in the model and the appropriate way of measuring them is based on the expert judgment of an Advisory Committee of investigators in breast cancer epidemiology and cancer control. Albeit in an informal way, our selection process incorporates the extensive literature on the subject and considers issues such as clinical plausibility of risk factors and publication bias (for example many experts ware investigators on landmark studies and could address the issue of whether certain effects were never found in their studies, or simply never analyzed).

The elicitation of expert opinion proceeded along several steps. We selected the experts, contacted them, and briefed them on the goals of our model. Interested experts became consultants of our project. Members of the Expert Advisory Committee include: Mitchell Gail (NIH); W. Douglas Thompson (Southern University of Maine); Elizabeth Claus (Yale University); Polly Marchbanks (Centers for Disease Control and Prevention); Polly Newcomb (University of Washington); Matthew Longnecker (National Institutes of Environmental Health Sciences); and Barbara Rimer (Duke University).

The next step was to seek the Committee's expertise to identify potentially interesting risk factors and group them by broad categories: family history; reproductive history; lifestyle; hormones and oral contraceptives; obesity and physical characteristics; benign breast disease and breast density and sociodemographic factors. For each category of risk factors we then convened a conference call with a subgroup of experts within our pool. The discussion was focussed on identifying: i) risk factors worth including in the model; ii) the appropriate way of measuring the risk factors and ascertaining the necessary information in a clinical setting; and iii) whether there are published studies whose results, if formally incorporated, would add significantly to the analysis.

Together with the experts, we agreed upon informal guidelines for inclusion of a risk factor. We focussed on factors with one or more of these features:

- a high relative risk;
- a high population attributable risk;
- the availability of supporting high-quality followup studies;
- ease of ascertainment in a clinical setting.

We begun by developing an encompassing list of candidate risk factors and organized 5 conference calls around related sets of factors. In many cases, we reached a consensus about whether a risk factor should be included. When agreement could not be reached or the literature was unanimously considered ambiguous, we deferred the decision about a risk factor to after the data analysis. Table 2 summarizes the risk factors that emerged from the panel's work.

4.3 Retrospective Modeling of Case Control Studies

The combination of case-control and cohort studies into a single prediction model poses interesting statistical challenges. Traditional analyses of case control stud-

Risk Factor	*Parameterization*
Family Pedigree	Ist and IInd degree relatives, BC and OC
Weight gain	Usual weight as adult (lbs) – weight at 18
Height	In feet
Body Mass	Quetelet's index (gm/cm^2) at adulthood
Surgical biopsies	Number of surgical biopsies performed
Needle biopsy	Never/ever had a needle biopsy
Biopsy	Never/ever had a biopsy
Benign breast disease	Never/ever had benign breast disease
Benign breast disease	Age of first diagnosis
Hyperplasia	No / minimal / Moderate or marked
Alcohol	Grams of alcohol consumed per day
Duration of OC use	In months
Recency of OC use	In months since last use
Latency of OC use	In months since first use
Duration of estrogen use	In months
Recency of estrogen use	In months since finished
Latency of estrogen use	In months since began
Age at menarche	In years
Age at menopause	In years
Natural menopause	Premenopausal / natural postmenopausal / surgical / other
Age at first pregnancy	In years
Parity	Number of pregnancies
Race	White / Black / Hispanic / Asian / Other

TABLE 2. *The risk factors used in our model and their parameterization. Pregnancies are counted if lasting longer than 6 months.*

ies focus on identifying important risk factors and on estimating relative risk, using maximum likelihood approaches (Breslow and Day 1980; Breslow 1996) and their extensions. Typically a particular parametric form for the prospective model is assumed. Consider the simple and common case of a binary response d indicating the occurrence of the disease under investigation. It is common to specify $p(d|\text{covariates})$ as a logistic regression with logistic slopes β. The appeal of this is that one can perform testing and inference directly on risk factors. The retrospective model $p(\text{covariates}|d)$ can be expressed in terms of the parameters β of the prospective model, using Bayes' rule. But under commonly adopted assumptions, the maximum likelihood estimates of the risk factors' effects can be derived by analyzing the data as though the prospective model were the actual sampling distribution (see Prentice and Pyke, 1979, and references therein).

Applications such as counseling and decision modeling require an assessment of absolute risk and therefore a more complete predictive analysis. In counseling asymptomatic individuals, for example, one needs to provide information about the risk of developing cancer in the next 5, 10, or 15 years, conditional on knowing that the counseland has not yet developed cancer. A full posterior and predictive analysis requires explicit consideration the retrospective model. One approach is to again specify a prospective model, derive the retrospective likelihood, and then proceed with posterior inference in the usual way. In our analysis we directly model the distribution of the covariates for cases and controls, following Müller et al. (1996). This achieves the immediate goals of providing a natural solutions for the problem of combining prospective and retrospective studies. In addition, it provides a framework for the imputation of missing variables and it facilitates computation and the Bayesian hierarchical inference in the case-control context. Implementation of hierarchical models based on a fixed prospective likelihood can be computationally prohibitive.

Bayesian approaches to case-control studies have precedents in Zelen and Parker (1986) who deal with fully Bayesian inference for categorical risk factors, Marshall (1988) and Ashby and Hutton (1996) who review issues and give examples and further references. Raftery and Richardson (1996) discuss Bayesian variable selection for case-control studies, using a prospective likelihood. Müller and Roeder (1997) discuss Bayesian inference in case-control studies with errors in variables and, while they constrain the model to take a specified prospective likelihood, they base their analysis on the full retrospective likelihood, using a mixture of normals for the unknown population distribution of the covariates.

In order to build a retrospective model one needs to specify the distributions $p(\text{Covariates}|d=0,\boldsymbol{\theta}_0), p(\text{Covariates}|d=1,\boldsymbol{\theta}_1)$, and $p(d=1|\boldsymbol{\theta}^*)$. Bayesian inference can then be based on the likelihood function:

$$\prod_{\text{controls}} p(\text{Covariates}_i|d=0,\boldsymbol{\theta}_0) \prod_{\text{cases}} p(\text{Covariates}_i|d=1,\boldsymbol{\theta}_1) \tag{5}$$
$$\times \prod_{\text{cohort}} [p(\text{Covariates}_i|d=1,\boldsymbol{\theta}_1)p(d=1|\boldsymbol{\theta}^*)]^{d_i}$$
$$[p(\text{Covariates}_i|d=0,\boldsymbol{\theta}_0)p(d=0|\boldsymbol{\theta}^*)]^{1-d_i}$$

The fundamental assumption underlying this model is that the control subjects in the case-control study are a sample for the same population as the prospectively collected cohort. Details of our specific proposal for implementing this approach in a single, age-matched study are discussed in Section 4.4. Müller *et al.* (1996) review results about relationships between prospective and retrospective models, and develop hierarchical extensions, that account for variability from study to study and from center to center within a study. This technique is further discussed in Section 4.4.

We need to model both categorical and quantitative variables. We specify a mixture of normals for the quantitative variables, and a multivariate probit model for the categorical variables given the quantitative variables. This results in a flexible non-linear regression approach for the prospective log-odds, replacing the linear log odds ratio traditionally used. In predicting risk of disease based on individual characteristics one often collects continuous variables, such as age or blood pressure. While broad discretization of these variables can be adequate for the purpose of identifying risk factors, counseling and decision making stress the need for smooth modeling that makes full use of continuous measurements. For example, it may not be easy for a patient to grasp the meaning of a model that predicts that, by turning 50, her risk of a disease increases by 50%. A smooth pattern may not only more be realistic, but also easier to communicate. In problems where continuous variables are present, linearity of log odds can represent a restrictive assumption and more general approaches need to be considered.

4.4 *Incorporating Pedigrees in the Analysis of Age-matched Case-control Studies*

Consider a single age-matched case-control study such as the CASH or UWCCC study. For each of n individuals, indicated by i, we have information about h_i, the relevant cancer history of her relatives. This notation differs from Section 3 where we use h to denote information about both the relatives and the counseland. We also have information about additional covariates, both categorical and continuous. It is convenient to separate quantitative from categorical variables. Let (z_i, x_i) denote the partition of the covariates into a $(p \times 1)$ subvector z_i of quantitative covariates and a $(q \times 1)$ subvector of categorical covariates x_i. Without loss of generality we will assume that all categorical covariates are binary. Here h_i refers to the family history of the individual sampled.

In addition, we have age and age-of-onset information. These studies use age-matched case-control designs and include all population cases in a specified age window. The age of onset of breast cancer a_i is known for all cases. Also, the age at interview y_i is known for all cases, and is set by design to be (at least approximately) the same as age of onset. Because controls are sampled to match the age distribution of cases, the age at interview of controls is fixed by design. Age of onset is unknown. Some individuals will never develop breast cancer and will have $a = \infty$ (in our implementation $a110$). Other individuals in the control group will develop cancer later. In the CASH study, the oldest controls are 54 years of age. The fraction of controls that will develop cancer later in life is on the order of 10% and cannot be ignored in the analysis. Information about cancer at later ages is borrowed from the other studies. As in the carrier probability model, age, in years, is treated as a discrete variable.

Based on this design, the contribution to the likelihood of the i-th subject is

$$p(x_i, z_i, a_i, y_i, h_i | A_i = Y_i, \boldsymbol{\theta}, \boldsymbol{\xi})$$

if she is a case while the contribution of the j-th subject is

$$p(x_j, z_j, h_j | Y_j = y_j, A_j > y_j, \boldsymbol{\theta}, \boldsymbol{\xi})$$

if she is a control, where $\boldsymbol{\theta}$ is the vector of all parameters and individuals are independent conditional on $\boldsymbol{\theta}$.

Traditional approaches for incorporating family history are based on including a subset of interesting summaries of the family history (such as the number of the first-degree relatives affected with breast cancer, or the average age of onset of affected family members) and using those as covariates. Our approach to incorporating covariates is based on assuming that family history influences the risk of cancer predominantly via the inheritance of susceptibility genes. We therefore introduce the vector g_i describing the unknown BRCA1 and BRCA2 genetic status of individual i. This will be treated as a latent variable. Given g, all other random variables are taken to be independent of h. The family history model of Section 3 provides us with the distributions $p(g_i|h_i, \xi)$ and $p(a_i|g_i)$.

To complete our probability model we specify the following conditional independence assumptions:

$$\begin{aligned} p(x_i, z_i|h_i, g_i, a_i, y_i, \theta, \xi) &= p(x_i, z_i|g_i, a_i, y_i, \theta) \\ p(h_i, g_i, a_i, y_i|\theta, \xi) &= p(h_i, g_i, a_i, y_i|\xi) \\ p(a_i, y_1|g_i, h_i, \xi) &= p(a_i, y_1|g_i, \xi). \end{aligned}$$

We then rewrite the contribution of the i-th case to the likelihood in terms of (5) as follows:

$$\frac{P(h_i|\xi)}{P(A=Y)} \sum_g p(x_i, z_i|g, a_i, y_i, \theta) p(g|h_i, \xi) p(a_i, y_i|g, \xi);$$

for the i-th control:

$$\frac{P(h_i|\xi)}{P(Y=y_i, A>y_i)} \sum_g \sum_{a>y_i} p(x_i, z_i|g, a, y_i, \theta) p(g|h_i, \xi) p(a, y_i|g, \xi).$$

Both denominators are independent of θ and can be ignored in deriving a posterior distribution. The genetic status vectors g_i and, for the controls, the age of onset a_i, are unobserved and need to be integrated out. This integration is performed within the MCMC algorithm for sampling from the posterior distribution of the parameters, by treating g_i and, when unknown, a_i as auxiliary variables (Higdon 1996).

For case-control studies with a simple binary response we model the covariate distributions for cases and controls separately, as in 5. In this context it is more natural to adopt the same strategy depending on whether the unobserved age of onset is finite or not. We define a model for each case: model 1 for individuals that eventually develop breast cancer ($a_i \leq 110$), and model 0, for individuals that don't. We will use $d = 0, 1$ as the model indicator, as in the simpler case control setting.

It now remains to specify the distribution

$$p(x_i, z_i|g_i, a_i, y_i, \theta) \tag{6}$$

In choosing distributional assumptions we face two competing desiderata. On one hand, technically simple structure and conjugate distributions are important to keep Markov chain Monte Carlo simulation feasible, especially with a view towards the additional hierarchical level which may be introduced to account for geographic or study-to-study variation. On the other hand, the model needs to allow for dependence among the covariates, including quantitative and categorical ones, multimodality, skewness etc.

With these considerations in mind we choose a mixture of normals for the quantitative covariates z_i and, conditional on z_i, a multivariate probit model for the categorical covariates x_i. We indicate by $Z_i = (z_i, 1)$ the covariate vector z_i augmented by a constant for the intercept. Our mixture of normals model for z_i is parameterized following the convenient specification used in Roeder and Wasserman (1997) and Mengersen and Robert (1996). We assume a common covariance matrix Σ_d, a common overall mean vector μ_d and offsets δ_{dm} for the terms $m = 1, \ldots, M$ of the mixture. The multivariate probit model for y_i given z_i is parameterized by the mean vector $m_d(z_i)$ and the precision matrix $V_d^{-1} = \lambda_d' \lambda_d$ for a multivariate normal latent variable x_i^*. The mean vector $m_d(z_i)$ is a regression on z_i, and λ_d is a lower triangular $(q \times q)$ matrix, constrained to have a unit diagonal to avoid likelihood non-identifiability. Without the unit diagonal constraint, one could inflate the k-th row and k-th column of V_d, and $m_d(z_i)$ by the same factor, without changing the likelihood.

The $(q \times p + 1)$ matrices κ_d describe the covariances between the continuous covariates and the latent covariates of the probit model. The parameterization with the additional factor λ_d^{-1} in the mean m_d of the latent normal parameters is chosen to allow resampling from complete conditionals (see Sections 4.2. and 4.3.). An alternative Bayesian model for the identified parameters of a multivariate probit is discussed by McCulloch, Polson and Rossi (1995).

In summary, the model specification is as follows. Subjects:

$$\begin{aligned} z_i | d_i = d, \boldsymbol{\theta}_d &\overset{ind}{\sim} \pi_{d0} N(\mu_d, S_d) + \textstyle\sum_{m=1}^{M} \pi_{dm} N(\mu_d + \delta_{dm}, S_d), \\ x_{ik} &= I(x_{ik}^* > 0), \\ x_i^* | z_i, d_i = d, \boldsymbol{\theta}_d &\overset{ind}{\sim} N(\underbrace{\lambda_d^{-1} \kappa_d Z_i}_{m_d(z_i)}, \underbrace{(\lambda_d \lambda_d')^{-1}}_{V_d}), \end{aligned} \tag{7}$$

Response specific parameters:

$$\begin{aligned} \mu_d &\sim N(\mu_H, V_H), \\ \delta_{dm} &\sim N(0, V_H), \\ S_d^{-1} &\sim W(s_H, [s_H S_H]^{-1}), \\ \pi_d &\sim \text{Dir}(\pi_H), \\ \kappa_{dkh} &\sim N(\kappa_{Hkh}, K_{Hkh}), \\ \lambda_{dkh} &\sim N(\lambda_{Hkh}, L_{Hkh}), \\ \lambda_{dkk} &= 1, \quad k = 1, \ldots, q, \\ \gamma &\sim \text{Be}(a_H, b_H), \end{aligned} \tag{8}$$

where $i = 1, \ldots, n$, $k = 1, \ldots, q$, $h = 1, \ldots, p + 1$, $d = 0, 1$ and $m = 1, \ldots, M$. A subscript H indicates fixed hyperparameters. Also, $W(s, S)$ is a Wishart distribution with scalar parameter s and matrix parameter S; $\text{Dir}(\pi)$ is a Dirichlet distribution; and $\text{Be}(a, b)$ is a Beta distribution.

Algorithms for sampling from the posterior and predictive distributions can be constructed using Markov chain Monte Carlo techniques. In this setting all full conditional distributions except those of $\boldsymbol{\xi}$ are available in closed form for block Gibbs sampling (Gelfand and Smith 1990).

4.5 *Prediction*

The main objective of our analysis is providing a predictive distribution of developing cancer for a cancer free individual. Formally, we would like to determine

$$p(a|y, h, x, z, A > y).$$

Using a sample $\boldsymbol{\theta}_m, \boldsymbol{\xi}_m, m = 1, \ldots, M$ from the posterior distribution of $(\boldsymbol{\theta}, \boldsymbol{\xi})$ we can use the expression:

$$p(a|y, h, x, z, A > y) = \sum_g \sum_m p(a|g, x, z, A > y, y, \boldsymbol{\theta}_m, \boldsymbol{\xi}_m) p(g|h, A > y, y, \boldsymbol{\xi}_m) \quad (9)$$

obtained by applying the conditional independence assumptions discussed earlier. The conditional probability $p(a|g, x, z, A > y, y, \boldsymbol{\theta}_m, \boldsymbol{\xi}_m)$ is relevant for the decision about testing, as it describes the states of knowledge that can be achieved if the test were performed. Its evaluation based on our modelling strategy requires an additional application of Bayes rule:

$$p(a|g, x, z, A > y, y, \boldsymbol{\theta}_m, \boldsymbol{\xi}_m) = \frac{p(x, z|g, a, y, \boldsymbol{\theta}_m) p(a|g, A > y, y, \boldsymbol{\xi}_m)}{\sum_{a>y} p(x, z|g, a, y, \boldsymbol{\theta}_m) p(a|g, A > y, y, \boldsymbol{\xi}_m)}. \quad (10)$$

In summary, we have presented a modeling strategy for determining absolute risk of cancer based on both case-control and prospectively collected data and incorporating detailed pedigree information. Estimation of this model can be implemented using MCMC, as discussed in Müller et al (1997).

5 Decision Making

The previous sections develop the prediction models necessary to address decisions related to breast cancer prevention in an patient-specific way. Risk prediction and carrier probability calculations can be presented to high-risk women as part of a individualized counseling session. In some cases, because of the complexity of choices related to cancer prevention, it can be important to provide women with additional quantitative support. For example, the decision about genetic testing depends on many factors, including the woman's risk of breast cancer as well as the effectiveness and cost of the testing procedure, the available prophylactic interventions, effectiveness and negative effects of these procedures, the impact of testing on other family members, the impact on the woman's ability to obtain insurance coverage and so on. In this section we consider a decision analysis of whether or not an individual should be tested for BRCA1 and BRCA2. Our discussion extends work of Tengs *et al.* (1997) Tengs *et al.* (1997) for BRCA1 alone.

5.1 Decision Model

Our approach to decisions about genetic testing is based on Bayesian sequential decision making. Two decisions are evaluated: The first is whether to be tested. The second is which prophylactic measure, if any, to choose. The first decision, whether or not to seek testing, depends on the outcomes that are expected from the second decision. Further, the outcomes expected with each prophylactic measure will depend upon the probability of being a carrier, perhaps modified by testing. We consider as outcome measure the expected quality-adjusted life years (QALYs). An "optimal" pair of decisions is one that maximizes the outcome measure of interest. If a woman does not elect testing, she still has some probability of being a carrier based on her family history and is left to choose a prophylactic measure (or not) without additional information. If she elects testing and tests positive for one of the two genes, then the probability that she carries a mutation is adjusted upward using Bayes rule. If she tests negative, the probability is adjusted downward.

After obtaining testing information, or not, she either elects to pursue no prophylactic measure, or chooses from among five groups prophylactic measures. The set of options that we considered are: 1. Bilateral Mastectomy (BM); 2. Oophorectomy at age 50 (OOP); 3. Bilateral Mastectomy followed by Oophorectomy at age 50 (BM+OOP); 4. Delayed Bilateral Mastectomy followed by Oophorectomy at age 50 (DBM+OOP); 5. No prophylactic measure. In option 4 we constrain the bilateral mastectomy to take place before the oophorectomy. We allow for the possibility that a woman who chooses to delay an intervention may develop breast or ovarian cancer during the interim.

Because BRCA1 and BRCA2 are associated with early cancer, women who do not develop breast or ovarian cancer in the interim will have a lower probability that they carry a mutation. Conversely, women who do develop cancer will have a higher probability of having a mutation. So for delayed interventions, we revise the post-test probability of having a mutation in light of new information about cancer status (breast cancer, ovarian cancer, or both) at the time of the first surgery and calculate post-surgical QALYs using this revised probability. We further assume that natural menopause occurs at age 50 and that after age 50 all women cease or decline hormone replacement therapy.

5.2 *Data sources*

Data on test accuracy, the likelihood of developing breast or ovarian cancer, and survival prospects were obtained from a variety of sources. We also included quality of life adjustments and the derivation of these estimates are also described below.

Test Accuracy. In the absence of good data-driven estimates of the reliability of testing, we consider two hypothetical tests designed to capture a range of possibilities. First, we consider an imperfect test with sensitivities of 0.85 and specificities of 0.99, for both BRCA1 and BRCA2, and independent errors for the two genes. This is consistent with expert judgment about commercially available testing procedures. Second, we consider a hypothetical perfect test with a sensitivities and specificities of 1. It is interesting to consider a range of test sensitivities because of our limited knowledge about it, and because a company can, to some extent, adjust the sensitivity of their test by looking at different sections of the genetic sequence. Further, any testing method (ASO, protein truncation, sequencing) is unlikely to identify a mutation which does not in fact exist so a high specificity is reasonable.

Intervention effects Prophylactic mastectomy and oophorectomy may reduce the probability of developing cancer, or delay its development. Unfortunately, information on the effectiveness of prophylactic mastectomy and oophorectomy is not yet available. To gather needed data on the likely effectiveness of these interventions, Tengs et al (1997) elicited estimates from breast cancer experts. The sample of experts consisted of all the MDs who are principal investigators (PI's) on projects of NCI-sponsored Specialized Programs of Research and Excellence (SPOREs) in breast cancer at six U.S. cancer centers. We mailed each expert a one-page questionnaire. In an accompanying letter we asked them to fax back their completed questionnaire. Of the 55 experts contacted, 18 responded for a response rate of 33%. It appears that some self-selection occurred, with those PIs who believed themselves more knowledgeable being more likely to respond. Further, following a suggestion in our letter, some of the PI's consulted with their colleagues in completing the questionnaire.

For each prophylactic intervention, the experts were asked to estimate the percent decrease in the life-time probability of developing breast or ovarian cancer that would be experienced by a 30-year-old woman who carries a BRCA1 mutation. We used the average of the responses as our estimate of the effectiveness of each intervention in our model. The expert estimates were elicited for women with a mutation. Although women without a mutation start with a lower risk, we assumed that prophylactic interventions are likely to have the same proportional benefit in preventing sporadic cancer, so the percent decrease in risk for non-carriers was assumed to be the same as for carriers. Further, although the percent decrease in risk was elicited for a life-time, we assumed that the cumulative probability of developing breast cancer at every age was reduced by the same percent. Recently Schrag *et al.* (1997) have published alternative expert-based estimates of effectiveness, with similar values.

Disease Onset For every combination of patient covariates, genetic status and prophylactic measure transition probabilities to the breast and ovarian states were estimated as in Section 3. The time to the first cancer event is computed assuming independence between breast and ovarian cancer incidence conditional on genetic status. This is consistent with the results in Easton *et al.* (1995).

Survival Information on the survival prospects for women who developed breast cancer was also taken from the SEER database. For example, age-specific five-year survival rates for women who were diagnosed with breast cancer ranged from 78% for women diagnosed before age 45 to 86% for women diagnosed after age 75. The expected length of life for ovarian cancer case is set to 4 years. Data on survival prospects following ovarian cancer were obtained from Rubin et al (1996) . We assume that women who develop both breast and ovarian cancer have the survival prospects of those with ovarian cancer alone.

5.3 *Quality of Life*

In deciding whether to seek testing, and whether to pursue some risk-reduction strategy upon learning the results, most women will consider changes in quality of life as well as quantity of life. However, the decision that maximizes survival may not maximize the quality of that survival. Tradeoffs are inevitable and are thus built into the model. For example, a woman might assign a year with breast cancer a quality of life weight of 0.75. Following the reasoning of the time-tradeoff method of utility elicitation (Torrance 1986; Sox et al. 1988), such a woman is assumed to be indifferent between living a full year with breast cancer and living 9 months cancer-free. The model allows for individualized health-related quality of life assessment. Specific values will be discussed in the context of the example. We combine the quality of life estimates associated with cancer status, Q1, and prophylactic measure, Q2, into a compound measure Q1*Q2. So, for example, a woman who develops ovarian cancer after prophylactic mastectomy would experience a 0.585 quality of life in subsequent years (i.e. 0.65 * 0.9 = 0.585), but a woman who remains cancer-free after mastectomy would experience a 0.9 quality of life (calculated as 1.0 * 0.9 = 0.9).

6 An Example

6.1 *Individual Information*

In this section we illustrate our approach as it applies to one specific, although fictional, case. Consider a 30-year-old woman who is currently free of cancer is contemplating testing for BRCA1. She wants to assure the best possible expected health outcome for herself. She has no children and does not plan to have children. She cares about both quantity and quality of life and is willing to make tradeoffs between them. She wants to make the decision that maximizes her expected quality-adjusted life-years (QALYs).

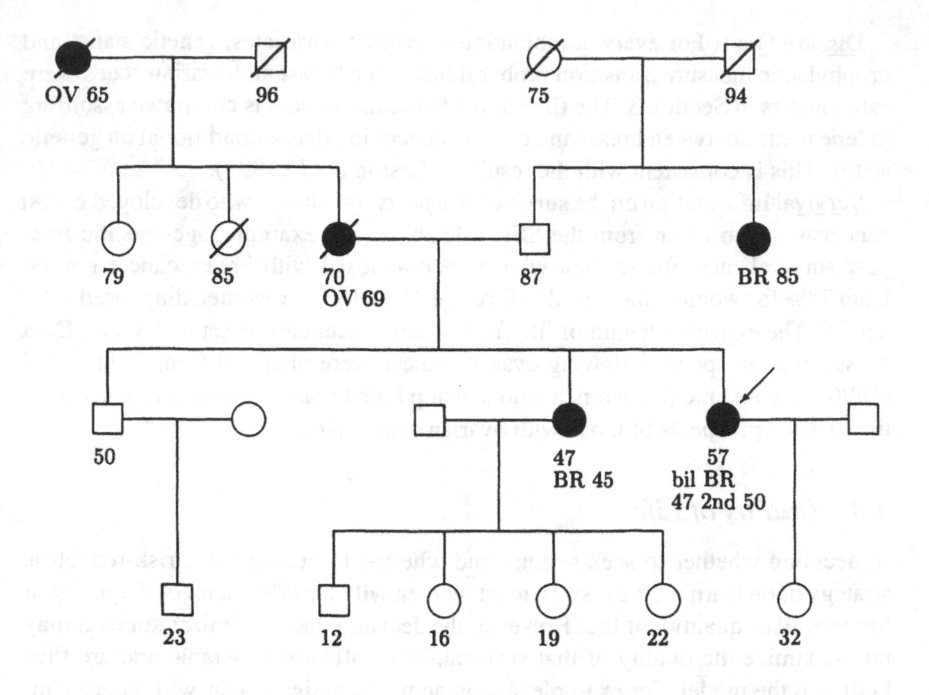

FIGURE 7. *The family history of Section 5. Age is shown for each family member, when known. This is either current age or age at death. If there is no indication of BC or OC, the member is free of both breast cancer and ovarian cancer.*

Her pedigree (based on a clinical case, but modified to maintain confidentiality) is shown in Figure 7. As customary in pedigree graphs, circles indicate females and squares indicate males. Offspring are below their ancestors. Horizontal lines connect mates and siblings, vertical lines connect parents and their children. The pedigree also shows the available cancer history and age information of family members. Breast cancer cases are indicated by BC and ovarian cancer cases by OC. Diagonal slashes indicate deaths. For cancer cases, the graph indicates age of diagnosis, for other individuals the graph indicates current age or age at death. The two cases can be handled in the same way in this context because the only relevant information for those family members is the length of time spent without either cancer. The arrows point to the individual to be counseled. For illustration, we assume that none of her great-grandparents are Ashkenazi Jews.

The software based on our model of Section 3 can only handle second degree relatives. To incorporate information from the whole pedigree, we can use compute the probability that the counseland's mother is a carrier. We have no information about the counseland's father, so we assume that he is randomly drawn from the general population, with a very small probability of being a carrier. Therefore the probability of the daughter being a carrier of the gene will be slightly over one half of the probability of her mother being a carrier. Focusing on the mother, based on the best estimate of the penetrance and prevalence parameters, her probability of carrying a mutation is 0.589 for BRCA1 and 0.377 for BRCA2. The probability of carrying a mutation at either gene is 0.966. In absence of information from the family of the counseland's father, the probability that the counseland carries a mutation is about 0.2945 for BRCA1 and 0.1885 for BRCA2

In addition to her strong family history, there are additional aspects of the counseland's life that make her at increased risk of cancer: she is using oral contraceptive and she had a previous breast biopsy. On the other hand her late age at menarche indicates a reduced risk. This results in a relative risk modifier of 1.52, with 95% probability interval $(1.15, 2.05)$.

As for quality of life, we assume she assigns a year with breast cancer the weight of 0.85, ovarian cancer 0.65, and both breast and ovarian cancer 0.60. Regarding following prophylactic measures, she assign to bilateral mastectomy a weight of .9 and to oophorectomy after menopause a weight of .98. The quality of life after mastectomy is a crucial element of her decision and we will show what happens by varying this value. A number of authors have reported that women assign the quality of life associated with unilateral mastectomy following breast cancer an average weight around 0.8 (Richardson 1991; Hall et al. 1992; Gerard et al. 1993). But this scenario differs from the present one in that it involves unilateral mastectomy as a treatment for breast cancer. So the value of .9 may be a reasonable one for many women.

6.2 Decision about Genetic Testing

Based on the individual information provided by the counseland we can compute the number of additional quality-adjusted life years that she can expect to live under each of the five strategies that we consider. In option 4, we explored various alternative delays for the bilateral mastectomy. Here we present results based on a three years delay. Delays up to 5 years may be optimal for our 30-year old woman.

The decision tree of Figure 8 summarizes the results. Because the optimal intervention differs depending on the pre-test probability of a mutation, she might benefit from testing. Because of the high probability of carrying a mutation, the option that maximizes QALYs in absence of testing is delayed bilateral mastectomy plus oophorectomy. Testing would adjust her probability up or down depending on the outcome. If she tested negative to both mutations, her probability would be revised down, but not to a sufficient extent to avoid any prophylactic intervention. In fact her preferred strategy would then be oophorectomy alone. On average, she can expect to gain 1.77 QALY's from the test.

By contrast, a perfect test would revise the probabilities to either 0 or 1. The resulting tree is in Figure 9. With a perfect test, the number of additional QALYs that a 30 year-old could expect, if she allowed the test to inform her decision, exceed the QALYs she could expect in the absence of testing, regardless of the pre-test probability of a mutation. For the woman considered here the expected gain is 2.75 QALYs. If she tests negative to both tests, her preferred choice is now no intervention. When using the imperfect test, if the outcome is a positive

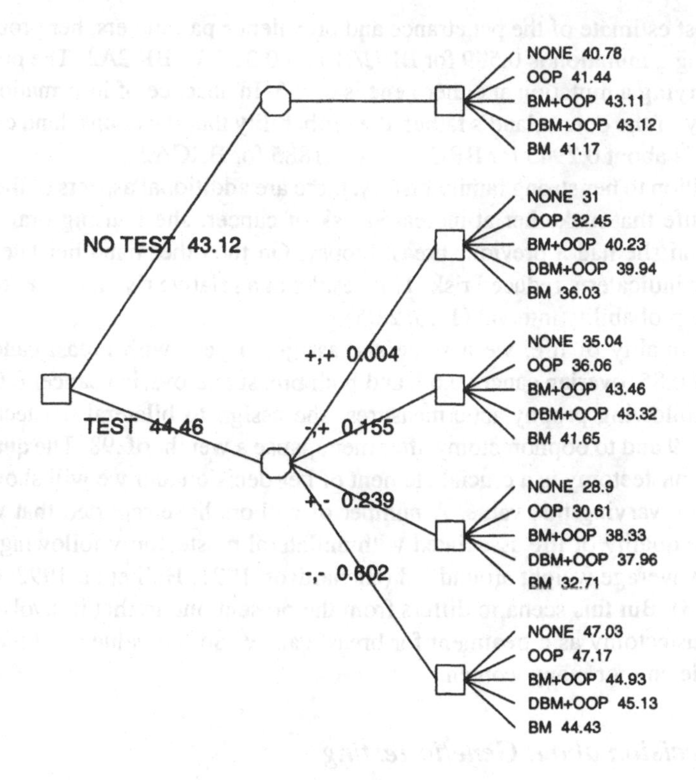

FIGURE 8. *Decision tree for testing for BRCA1 and BRCA2, for the woman of section 6.*

result for both BRCA1 and BRCA2, the woman is still not likely to carry both mutations. In fact, with the figures considered here, her posterior probability of carrying both genes is still lower than 2%. This explains why she appears to be better off if she tests positive for both genes (in which case she is still about twice as likely to be BRCA1 as BRCA2, but unlikely to be both) than if she tests positive for BRCA1 only (in which case she is most probably BRCA2). When using the perfect test, this situation does not occur, and testing positive to both genes is the worst scenario.

In general, QALYs with testing will exceed QALYs without testing for only those pre-test probabilities that can lead a woman to a different choice as a result of observing the outcome. Figure 10 summarizes the gain in expected quality adjusted life years as a function of pre-test probabilities of BRCA1 and BRCA2, using the imperfect test. To represent the problem in two dimensions, we assume, for the purpose of this plot only, that the probability of carrying both mutations is zero.

The decision trees of Figures 11 and 12 illustrate the effects of assuming a different quality of life adjustment from bilateral mastectomy. In Figure 11 the

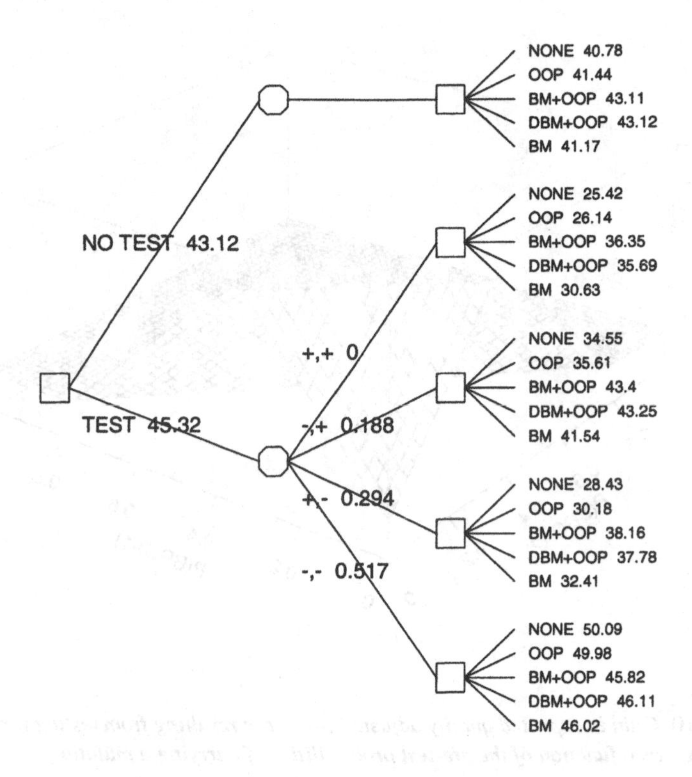

FIGURE 9. *Decision tree of testing for BRCA1 and BRCA2, for the woman of section 6, assuming a perfect test.*

quality adjustment of .98 is just sufficiently high that bilateral mastectomy followed by oophorectomy would be chosen even if the tests were negative. The test has no value for a women with such preferences. At the opposite end of the spectrum, in Figure 12 the quality adjustment of .6 is just sufficiently low that bilateral mastectomy is never considered. The preferred choice is always oophorectomy, and the test again is of no value. The same applies for all quality adjustments smaller than .6.

Finally, Figure 13 shows a the results of a probabilistic sensitivity analysis. This is performed by drawing jointly a sample from all input parameters, including the genetic parameters, the relative risk adjustment for patient covariates, and the hypothesized effects of prophylactic interventions. The probability distributions for the genetic parameters are based on the model of Section 3. The probability distribution used for the relative risk is the posterior distribution from the model of Section 4. As for the intervention effects, we used beta distributions modeled around the intervals given by the experts. For each set of parameters we solved the decision tree and calculated the expected gain from testing. For ease of inter-

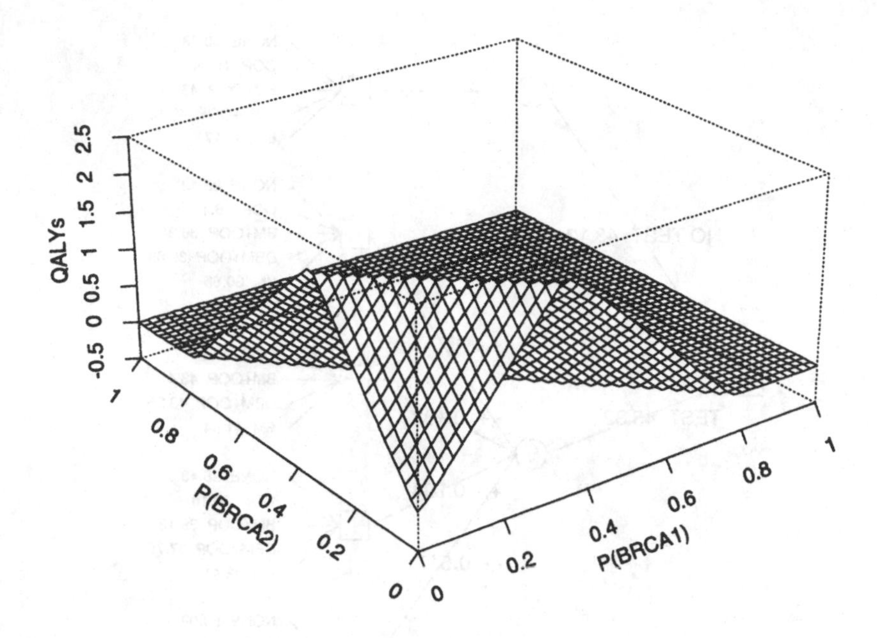

FIGURE 10. *Gain in expected quality adjusted life years resulting from testing for BRCA1 and BRCA2, as a function of the pre-test probabilities of carrying a mutation.*

pretation of the graph, we only graphed the value of testing for BRCA1. Similar results obtain when the probabilistic sensitivity analysis is extended to the whole surface on Figure 10.

7 Discussion

7.1 Issues and clinical implications of the decision analysis

A woman of average risk has only a 12 in 10,000 chance of being a BRCA1 mutation carrier and an even smaller chance of carrying a BRCA2 mutation. Our results show, not surprisingly, that women of average risk would derive virtually no benefit from testing for a mutation in BRCA1 under any of the circumstances explored here.

Women with a family history of early breast and/or ovarian cancer and therefore an elevated pre-test probability of carrying a mutation, might benefit from testing depending on a number of factors. These factors include the accuracy of the test; the exact risk of having a mutation; the prophylactic measure, if any, that will be pursued upon learning test results; the prophylactic measure, if any, that would be pursued in the absence of testing; the effectiveness of those measures; and

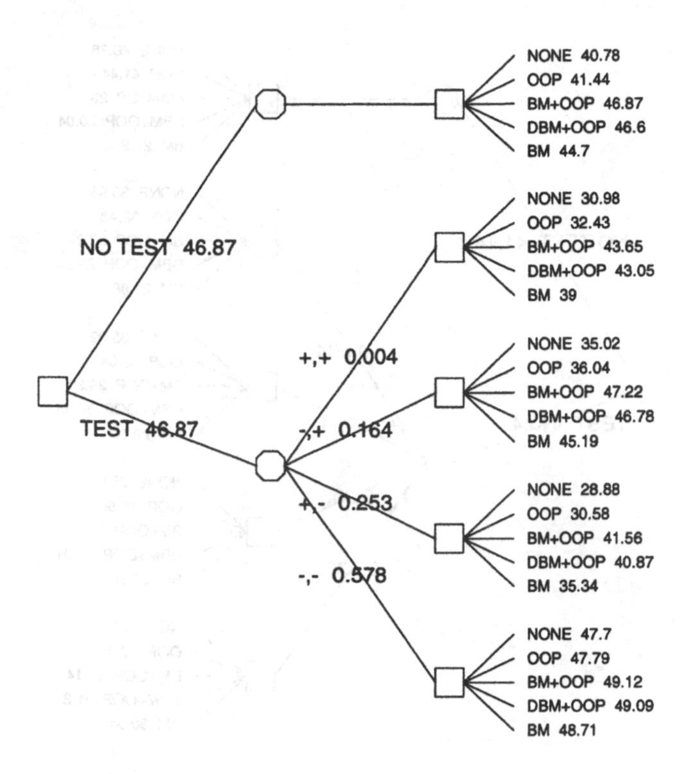

FIGURE 11. *Decision tree for testing for BRCA1 and BRCA2, for the woman of section 6, assuming a 2% loss of quality of life from bilateral mastectomy.*

individual preferences regarding quality of life.

When choosing between progressively more invasive measures ranging from no intervention to oophorectomy combined with mastectomy, a perfect test would have some value to all women. The gain that could be expected, however, varies considerably with the individual preferences regarding quality of life. In addition the value of testing must be weighed against other factors not considered in this model. For example if a woman's BRCA1 carrier status is recorded in her medical record, then she runs the risk that insurance companies might get this information and raise her premiums or even cancel her health, life, or disability insurance (Nelson 1995). Only a few states have some protection against genetic discrimination and no federal laws are yet in place.

Another key factor not considered here is cost. Testing is currently paid for out of pocket and it is expensive. From the societal perspective, it is important to determine whether testing is cost-effective, taking into account not only the cost of testing, but the cost of prophylactic measures as well as the avoided costs of cancer.

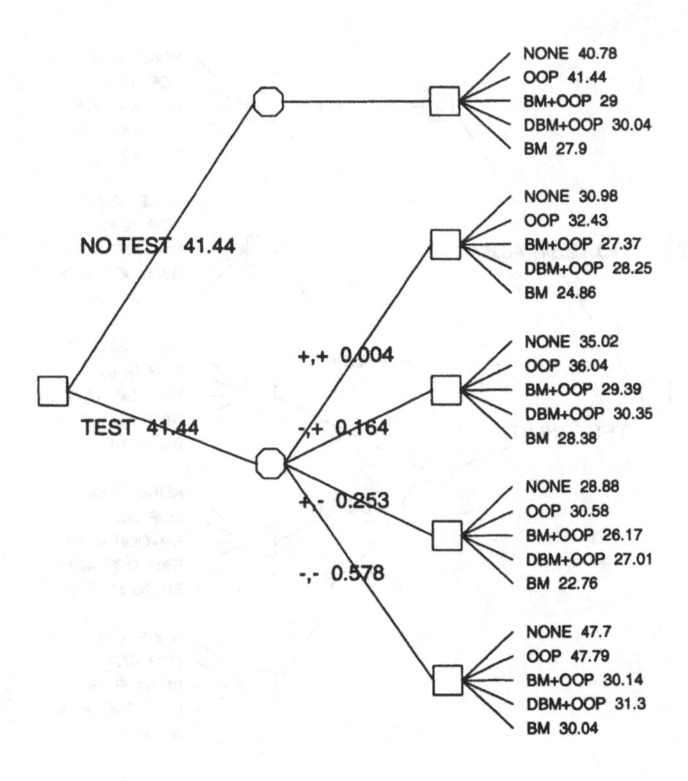

FIGURE 12. *Decision tree of testing for BRCA1 and BRCA2, for the woman of section 6, assuming a 40% quality of life loss from bilateral mastectomy. At a 30% loss, Bilateral mastectomy would still optimal for our woman if she tested positive.*

One important limitation in this analysis is that we assume a woman derives no satisfaction or discomfort from "just knowing" her BRCA1 and BRCA2 genotypes. Further, we assume that she derives no pleasure or pain from uncertainty about her genetic status. Some people may value information for its own sake even if it does not change their behavior. Others prefer not to know even if they would benefit medically from that knowledge. In general, testing is not without negative implications for a woman, as discussed in Section 1. Before BRCA1&2 testing becomes common place in clinical practice, it is critical that we consider how best to provide meaningful information that can help women decide if they indeed want to undergo testing.

This analysis reveals that testing for BRCA1 and BRCA2 may improve expected quality-adjusted survival for some women. The ideal candidate for testing is someone who is of moderate to high risk who is likely to decline all prophylactic measures in the absence of testing or if she tests negative, but whose values are such that she would consider pursuing a surgical option if she tested positive.

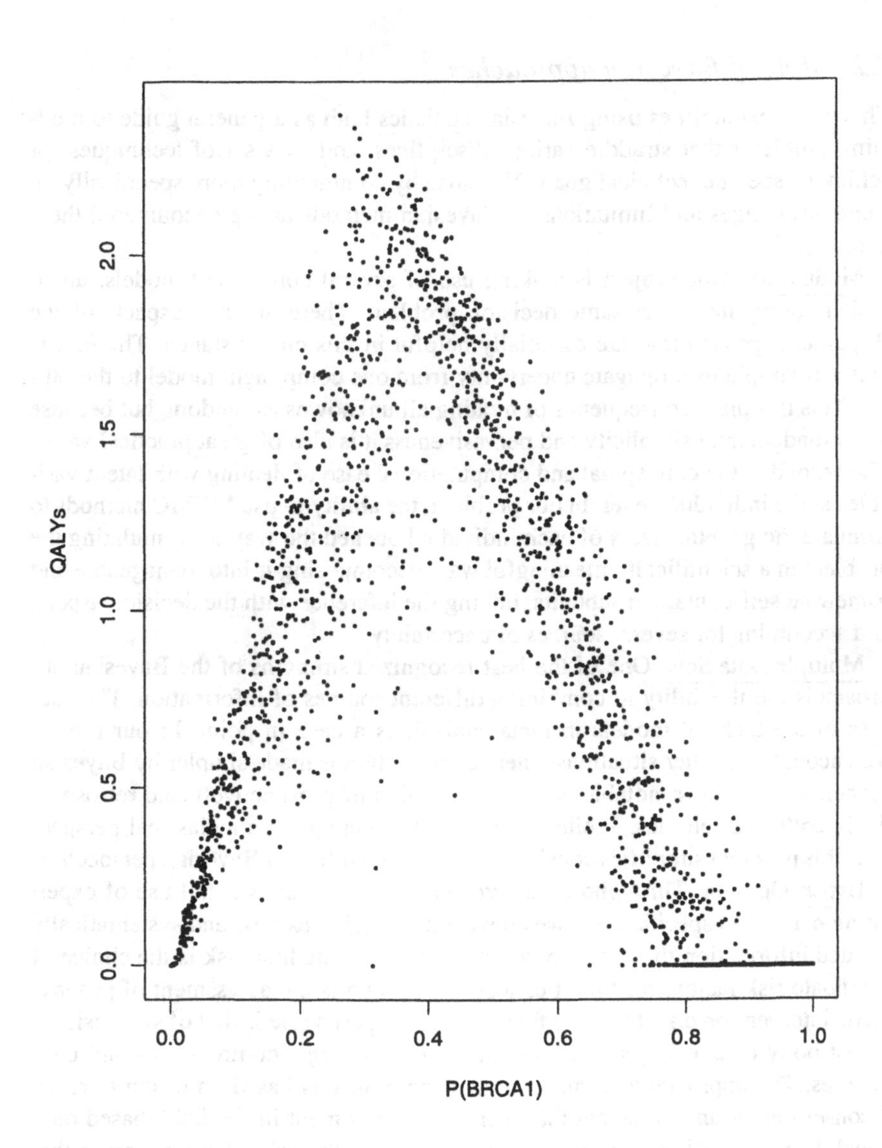

FIGURE 13. *Results of a probabilistic sensitivity analysis of the additional expected quality adjusted life resulting from testing. For ease of interpretation of the graph, we limited our sensitivity analysis to testing for BRCA1 only. Similar results obtain when the probabilistic sensitivity analysis is applied to the whole surface on Figure 10*

Testing is not recommend for women of average risk, as it is very unlikely to lead to risk-reduction measures that will substantially improve her survival and quality of life.

7.2 *Role of Bayesian approaches*

This paper exemplifies using Bayesian statistics both as a general guide to modeling problems that straddle various disciplines, and as a set of techniques for achieving specific technical goals. We close by commenting more specifically on some advantages and limitations of Bayesian methods as we encountered them here.

Modularity. Our project is making use of several component models, all of which contribute to the same decision problem. There are two aspects of the Bayesian approach that are especially helpful in this circumstance. The first is that it is simple to propagate uncertainty from one component model to the others. This is a plain consequence of treating all unknowns as random, but because of its fundamental simplicity and pervasiveness it is also of great practical value. The second is the conceptual and computational ease of dealing with latent variables at the individual level. In our problem, the ability to use MCMC methods to simulate the genetic status of each individual opened the way to formulating the problem in a scientifically meaningful way, decomposing it into manageable and somewhat self-contained subunits, linking the inference with the decision aspect, and accounting for several sources of uncertainty.

Multiple Data Sets. One of the best recognized strengths of the Bayesian approach is the flexibility in combining different sources of information. The success of hierarchical models in meta-analysis is a case in point. In our project we encountered other situations where combination is made simpler by Bayesian modelling. The most notable is the combination of prospectively and retrospectively collected samples. While extremely challenging from a classical perspective, this problem simplifies considerably when seen from a Bayesian perspective.

Expert Opinion. Throughout our project we made a substantial use of expert opinion. For two specific tasks we convened a panel of experts and systematically elicited information that was key to our modelling: the first task is the choice of candidate risk factors for breast cancer; the second is the assessment of prophylactic intervention effects. In the first case, the expert's role is that of synthesizing a vast body of existing studies, biomedical knowledge, controversies and conjectures. The approach to combining their opinion was based on trying to reach a consensus, or an agreement that there is disagreement in the field, based on a panel discussion of the issues. In the second case the role of the expert is that of shedding light into an almost uncharted territory. The approach to combining their opinions was based on eliciting separately their beliefs about certain key quantities and then combining them using a simple aggregation rule.

Limitations. In inference problems the main rate limiting step for Bayesian analyses is use of resources. For example, a hierarchical analysis of our case control studies, accounting for variability in sampling subregions, would be quite appropriate. As another example, pooling expert opinion is itself a problem that is amenable to a Bayesian treatment. Sometimes, A full Bayesian analysis is too labor intensive and has to be replaced with a simpler alternative, perhaps a simpler Bayesian analysis.

Other important issues arise in the context of the decision analysis. Genetic testing is a hard problem for women: it requires making trade-offs, and reaching a decision in spite of all the uncertainties. Expected utility is the most popular and best understood approach for doing this, and it is also very practical. Nonetheless, there are difficulties with it in this context. One set of problems stems from the fact that most women have multiple objectives, some of which are very difficult to translate into a single quantitative scale. Our approach here has been to single out an important and quantifiable measure, and leave out the intangibles. It can be very useful to have a measure of the gain in QALYs from testing, even if this is later to be qualitatively weighted against other aspects, such as the desire to "just know" whether a mutation is present. But it is clear that the full set of consequences to a woman cannot be reflected in QALYs. Another set of difficulties is related to the level of "numeracy" that is required to benefit from a decision analysis like the one presented here. Would a lay person be better off making her decision if presented with a well thought out list of advantages and disadvantages? We see this as a very important issue and are working with genetic counselors and patient advocates to make risk predictions and decision analysis as accessible and helpful as possible to lay people.

Acknowledgments

This work was funded in part by the NCI through the Specialized Program of Research Excellence (SPORE) grant in Breast Cancer at Duke University, P50 CA68438 and through the Cancer Prevention Research Unit (CPRU) at Duke university 1PO1–CA–72099–01, and by the NSF under grants DMS-9404151, DMS 9403818 and DMS-9305699. Omar Aguilar, Susan Paddock, Tammy Tengs and Luca Tardella contributed to the projects related to this article.

References

American Cancer Society (1996). *Cancer Facts and Figures*. American Cancer Society.

Anderson, D. E. and Badzioch, M. D. (1985). Risk of familial breast cancer. *Cancer*, 56:383–387.

Ashby, D. and Hutton, J. (1996). Bayesian epidemiology. In (Berry and Stangl 1996), pages 109–138.

Austin, H., Cole, P., and Wynder, E. (1979). Breast cancer in black american women. *Int J Cancer*, 24:541.

Benichou, J. (1993). A computer program for estimating individualized probabilities of breast cancer. *Computers and Biomedical Research*, 26:373–382.

Benichou, J. (1995). A complete analysis of variability for estimates of absolute risk from a population-based case-control study. *Biometrical Journal*, 37:3–24.

Benichou, J. and Gail, M. H. (1995). Methods of inference for estimates of absolute risk derived from population-based case-control studies. *Biometrics*, 51:182–194.

Benichou, J., Gail, M. H., and J, M. J. (1996). Graphs to estimate an individualized risk of breast cancer. *Journal of Clinical Oncology*, 14:103–110.

Berry, D. A., Parmigiani, G., Sanchez, J., Schildkraut, J., and Winer, E. (1997). Probability of carrying a mutation of breast-ovarian cancer gene BRCA1 based on family history. *J Natl Cancer Inst*, 89:9–20.

Berry, D. A. and Stangl, D. K., editors (1996). *Bayesian Biostatistics*, volume 151 of *Statistics: Textbooks and Monographs*. Marcel Dekker, New York, NY, USA.

Billings, P. R., Kohn, M. A., de Cuevas, M., Beckwith, J., Alper, J. S., and Natowicz, M. R. (1992). Discrimination as a consequence of genetic testing. *Am J Hum Gene*, 50:476–82.

Bishop, D. T., Cannon-Albright, L., McLellan, T., Gardner, E. J., and Skolnick, M. H. (1988). Segregation and linkage analysis of nine Utah brest cancer pedigrees. *Genetic Epidemiol*, 5:151–69.

Bondy, M. L., Lustbader, E. D., Halabi, S., Ross, E., and Vogel, V. G. (1994). Validation of a breast cancer risk assessment model in women with a positive family history. *J Natl Cancer Inst*, 86:620–625.

Boyd, N. F., O'Sullivan, B., Fishall, E., Simor, I., and Cooke, G. (1984). Mammographic patterns and breast cancer risk: methodologic standards and contradictory results. *J Natl Cancer Inst*, 72:1253–9.

Breslow, N. E. (1996). Statistics in epidemiology: The case-control study. *Journal of the American Statistical Association*, 91:14–28.

Breslow, N. E. and Day, N. E. (1980). *Statistical Methods in Cancer Research*. International Agency for Research on Cancer, Lyon.

Brinton, L. A., Hoover, R., and F, F. J. (1983). Reproductive factors in the aetiology of breast cancer. *Br J Cancer*, 47:757–762.

Brinton, L. A., Schairer, C., Hoover, R. N., *et al.* (1988). Menstrual factors and risk of breast cancer. *Cancer Inves*, 6:245–254.

Bruzzi, P., Negri, E., La Vecchia, C., *et al.* (1988). Short term increase in risk of breast cancer after full term pregnancy. *BMJ*, 297:1096–1098.

Byers, T., Graham, S., Rzepka, T., and J, M. (1985). Lactation and breast cancer: evidence for a negative association in premenopausal women. *Am J epidemiol*, 121:664–674.

Byrne, C., Schairer, C., Wolfe, J., Parekh, N., Salane, M., Brinton, L., Hoover, R., and Haile, R. (1995). Mammographic features and breast cancer risk: effects with time age and menopause. *J Natl Cancer Inst*, 87:1622–9.

Calle, E. E., Miracle-McMahill, H. L., Thun, M. J., and Heath, Jr, C. W. (1994). Cigarette smoking and risk of fatal breast cancer. *Am J Epidemiol*, 139(10):1001–1007.

Chen-Ling, C., White, E., Malone, K. E., and Daling, J. R. (1997). Leisure-time physical activity in relation to breast cancer among young women (washington, united states). *Cancer Causes and control*, 8:77–84.

Chu, S. Y., Lee, N. C., Wingo, P. A., *et al.* (1989). Alcohol consumption and the risk of breast cancer. *Am J Epidemiol*, 130:867–877.

Claus, E. B., Risch, N., and Thompson, W. D. (1990a). Age of onset as an indicator of familial risk of breast cancer. *American Journal of Epidemiology*, 131:961–972.

Claus, E. B., Risch, N., and Thompson, W. D. (1990b). Using age of onset to distinguish between subforms of breast cancer. *Ann Hum Genet*, 54:169–177.

Claus, E. B., Risch, N., and Thompson, W. D. (1991). Genetic analysis of breast cancer in the cancer and steroid hormone study. *American Journal of Human Genetics*, 48:232–242.

Claus, E. B., Risch, N., and Thompson, W. D. (1994). Autosomal dominant inheritance of early-onset breast cancer: Implications for risk prediction. *Cancer*, 73:643–651.

Colditz, G. A., Hankinson, S. E., Hunter, D. J., *et al.* (1995). The use of estrogens and progestins and the risk of breast cancer in postmenopausal women. *NEJ*, 332:1589–1593.

Collaborative group on hormonal factors in breast cancer (1996). Breast cancer and hormonal contraceptives: collaborative reanalysis of individual data on 53,297 women with breast cancer and 100,239 women without breast cancer from 54 epidemiological studies. *Lancet*, 347:1713–27.

Couch, F. J., DeShano, M. L., Blackwood, M. A., Calzone, K., Stopfer, J., Campeau, L., Ganguly, A., Rebbeck, T., Weber, B. L., Jablon, L., Cobleigh, M. A., Hoskins, K., and Garber, J. E. (1997). BRCA1 mutations in women attending clinics that evaluate the risk of breast cancer. *N Engl J Med*, 336:1409–15.

Dupont, W. D. and Page, D. L. (1985). Risk factors for breast cancer in women with proliferative disease. *New England J Med*, 312:146–51.

Dupont, W. D. and Page, D. L. (1987). Breast cancer risk associated with proliferative disease, age at first birth, and family history of breast cancer. *Am J Epidemiol*, 125:769–779.

Easton, D. F., Ford, D., and Bishop, D. T. (1995). Breast and ovarian cancer incidence in BRCA1-mutation carriers. *Am J Hum Genet*, 56:265–271.

Easton, D. F., Steele, L., Fields, P., Ormiston, W., Averill, D., Daly, P. A., McManus, R., Neuhausen, S. L., Ford, D., Wooster, R., Cannon-Albright, L. A., Stratton, M. R., and Goldgar, D. E. (1997). Cancer risks in two large breast cancer families linked to BRCA2 on chromosome 13q12-13. *Am J Human Genetics*, 61:120–128.

Ebbs, S. R. and Bates, T. (1988). Breast cyst type does not predict the natural history of cyst disease or breast cancer risk. *Br J Surg*, 75:702–704.

Ernster, V. L., Wrencsh, M. R., Petrakis, N. L., *et al.* (1987). Benign and malignant breast disease: initial study results of serum and breast fluid analyses of endogenous estrogens. *JNCI*, 79:949–960.

Ewitz, M. and Duffy, S. W. (1988). Risk of breast cancer in relation to reproductive factors in denmark. *Br J Cancer*, 58:99–104.

Ford, D. and Easton, D. F. (1995). The genetics of breast and ovarian cancer. *Br J Cancer*, 72:805–812.

Friedenreich, C. M. and Rohan, T. E. (1995). Physical activity and risk of breast cancer. *Eur J Cancer Prev*, 4:145–151.

Futreal, P. A., Liu, Q., Shattuck-Eidens, D., Cochran, C., Harshman, K., Tavtigian, S., Bennett, L. M., *et al.* (1994). BRCA1 mutations in primary breast and ovarian carcinomas. *Science*, 226:120–122.

Gail, M. H. and Benichou, J. (1994). Epidemiology and biostatistics program of the national cancer institute. *J Natl Cancer Ins*, 86(8):573–575.

Gail, M. H., Brinton, L. A., Byar, D. P., Corle, D. K., Green, S. B., Schairer, C., and Mulvihill, J. J. (1989). Projecting individualized probabilities of developing breast cancer for white females who are being examined annually. *J Natl Cancer Inst*, 81:1879–1886.

Gelfand, A. E. and Smith, A. F. M. (1990). Sampling-based approaches to calculating marginal densities. *Journal of the American Statistical Association*, 85(410):398–409.

Gerard, K., Dobson, M., and Hall, J. (1993). Framing and labeling effects in health descriptions: Quality adjusted life years for treatment of breast cancer. *Journal of Clinical Epidemiology*, 46:77–84.

Gloeckler-Reis, L. A., Hankey, B. F., and Ewards, B. K., editors (1990). *Cancer Statistics Review 1973-87*, volume 90-2789. US Department of Health and Human Services, NIH, Bethesda, MD.

Go, R. C. P., King, M. C., Bailey-Wilson, J., *et al.* (1983). Genetic epidemiology of breast cancer and associated cancers in high risk families I segregation analysis. *JNCI*, 71:455–461.

Goldstein, A. M., Haile, R. W. C., Marazita, M. L., *et al.* (1987). A genetic epidemiologic investigation of breast cancer in families with bilateral breast cancer I. segregation analysis. *J Natl Cancer Inst*, 78:911–918.

Gray, G. E., Henderson, B. E., and Pike, M. C. (1980). Changing ratio of breast cancer incidence rates with age of black females compared with white females in the united states. *Journal National Cancer Institute*, 64:461–463.

Grunchow, H. W., Sobocinski, D. A., Barboniak, J. J., *et al.* (1985). Alcohol consumption, nutrient intake, and relative body weight among us adults. *Am J Clin Nutr*, 42:289–295.

Hall, J., Gerard, K., Salkeld, G., and Richardson, J. (1992). A cost utility analysis of mammography screening in australia. *Soc Sci Med*, 34:993–1004.

Hall, J. M., Lee, M. K., Newman, B., Morrow, J. E., Anderson, L. A., Huey, B., and King, M. C. (1990). Linkage of early-onset familial breast cancer to chromosome 17q12. *Science*, 250:1684–1689.

Higdon, D. (1996). Auxiliary variable methods for markov chain monte carlo with applications. Discussion Paper DP96-17, Institute of Statistics and Decision Sciences, Duke University.

Hoskins, K. F., Stopfer, J. E., Calzone, K. A., Merajver, S. D., Rebbeck, T. R., Garber, J. E., and Weber, B. L. (1995). Assessment and counseling for women with a family history of breast cancer a guide for clinicians. *JAMA*, 273:577–585.

Houwing-Duistermaat, J. J. and Van Houwelingen, J. C. (1997). A family history score for breast cancer. *Genetic Epidemiology*, 14:530.

Howe, G. R., Roharli, A., *et al.* (1991). The association between alcohol and breast cancer risk: evidence form the combined analysis of six dietary case-control studies. *Int J Cancer*, 47:707–710.

Hsieh, C.-C., Trichopoulos, D., Katsouyanni, K., *et al.* (1990). Age at menarche, age at menopause, height and obesity as risk factors for breast cancer: associations and interatctions in an international case-control study. *Int J Cancer*, 62:1625–1631.

Iversen, Jr, E.S., Parmigiani, G. and Berry, D. (1998). Validating Bayesian prediction models: a case study in genetic susceptibility to breast cancer. In

Case Studies in Bayesian Statistics, Volume IV, pp. 321-338, Springer-Verlag.

Iversen, Jr. E.S., Parmigiani, G., Berry, D., and Schildkraut, J. (1997a). A Markov Chain Monte Carlo approach to survival models with detailed family history. *Genetic Epidemiology*, 14:531.

Iversen, Jr. E.S., Parmigiani, G., Berry, D., and Schildkraut, J. (1997b). Genetic susceptibility and survival: Application to breast cancer. Discussion Paper DP997-38, Institute of Statistics and Decision Sciences, Duke University.

Janerich, D. T. and Hoff, M. B. (1982). Evidence for a crossover in breast cancer risk factors. *American Journal of Epidemiology*, 116:737.

Katsouyanni, K., Trichopoulos, D., Boyle, P., *et al.* (1986). Diet and breast cancer: a case-control study in greece. *Int J Cancer*, 38:815–820.

Kay, C. R. and Hannaford, P. C. (1988). Breast cancer and the pill - a further report from the royal college of general practitioners oral contraceptive study. *Br J Cancer*, 58:675–80.

Kelsey, J. L. and Gammon, M. D. (1991). Epidemiology of breast cancer. *Epidemiologic Reviews*, 12:228–240.

Kelsey, J. L., Gammon, M. D., and John, E. M. (1993). Reproductive factors and breast cancer. In Kelsey, J. L., editor, *Epidemiologic Reviews*, volume 15 of *Epidemiologic Reviews*, pages 36–47. The Johns Hopkins University School of Hygiene and Public Health, Baltimore, MD.

Krieger, N. and Hiatt, R. A. (1992). Risk of breast cancer after benign breast disease: variation by histologic type, degree of atypia, age at biopsy, and length of follow-up. *American Journal of Epidemiology*, 135:619–31.

Kvale, G. and Heuch, I. (1988). Menstrual factors and breast cancer risk. *Cancer*, 62:1625–1631.

La Vecchia, C., Decarli, A., and Di Pietro, F. (1985). Menstrual cycle patterns and the risk of breast disease. *Eur Journal Cancer*, 21:417–422.

Leal, S. M. and Ott, J. (1994). A likelihood approach to calculating risk support interval. *American Journal of Human Genetics*, 54(5):913–917.

Leal, S. M. and Ott, J. (1995). Variability of genotype-specific penetrance probabilities in the calculation of risk support intervals. *Genetic Epidemiology,*, 12:859–862.

Lerman, C., Daly, M., Masny, A., and Balshem, A. (1994). Attitudes about genetic testing for breast-ovarian cancer susceptibility. *J Clin Onco*, 12:843–50.

Levy-Lahad, E., Catane, R., Eisenberg, S., Kaufman, B., Hornreich, G., Lishinsky, E., Shohat, M., Weber, B. L., Beller, U., Lahad, A., and Halle, D. (1997). Founder BRCA1 and BRCA2 mutations in ashkenazi jews in israel: Frequency and differential penetrance in ovarian cancer and in breast-ovarian cancer families. *Am J Hum Genet*, 60:1059–1067.

Longnecker, M. P., Berlin, J. A., Orza, M. J., *et al.* (1988). A meta-analysis of alcohol consumption in relation to risk of breast cancer. *J Am Med Assoc*, 260:652–656.

Lund, E., Meirik, O., Adami, H.-O., *et al.* (1989). Oral contraceptive use and premenopausal breast cancer in sweden and norway: possible effects of different pattern of use. *Int J Epidemiol*, 18:527–532.

Lynch, H. T., Albano, W. A., Heieck, J. J., Mulcahy, G. M., Lynch, J. F., Layton, M. A., and Danes, B. S. (1984). Genetics, biomarkers and control of breast cancer: a review. *Ca Genet Cytogenet*, 13:43–92.

Markus, J.N., Watson, P., Page, D.L., Narod, S.A., Lenoir, G.M., Tonin, P., Linder-Stephenson, L., Salerno, G., Conway, T.A., and Lynch, H.T. (1996) Hereditary breast cancer: pathobiology, prognosis, and BRCA1 and BRCA2 gene linkage. *Cancer*, 77:697–709.

Marshall, R. J. (1988). Bayesian analysis of case-control studies. *Statistics in Medicine*, 7:1223–1230.

Mayberry, R. M. and Stoddard-Wright, C. (1992). Breast cancer risk factors among black women and white women: Similarities and differences. *Am J Epidemiol*, 136:1145–1156.

McCulloch, R. E., Polson, N. G., and Rossi, P. (1995). Bayesian analysis of the multinomial probit model with fully identified parameters. Technical report, University of Chicago, GSB, `http://gsbrem.uchicago.edu/Papers/WPS.html`.

McTiernanan, A., Stanford, J. L., Weiss, N. S., Daling, J. R., and Voigt, L. F. (1996). Occurrence of breast cancer in relation to recreational exercise in women age 50-64 years. *Epidemiology*, 7:598–604.

McTirnan, A. and Thomas, D. B. (1986). Evidence for a protective effect of lactation on risk of breast cancer in young women: results from a case-control study. *Am J Epidemiol*, 124:353–358.

Meissen, G. J., Mastromauro, C. A., Kiely, D. K., McNamara, D. S., and Myers, R. H. (1991). Understanding the decision to take the predictive test for huntington disease. *Am J Med Genet*, 39(4):404–410.

Mengersen, K. and Robert, C. (1996). Testing for mixtures: a Bayesian entropic approach. In Bernardo, J. M., Berger, J. O., Dawid, A. P., and Smith, A. F. M., editors, *Bayesian Statistics 5*. Oxford University Press, Oxford.

Miki, Y., Swenson, J., Shattuck-Eidens, D., Futreal, P. A., Harshman, K., Tavtigian, S., *et al.* (1994). A strong candidate for the breast and ovarian cancer susceptibility: gene BRCA1. *Science*, 266:66–71.

Miller, D. R., Rosenberg, L., Kaufman, D. W., *et al.* (1989). Breast cancer before age 45 and oral contraceptive use: new findings. *Am J Epidemiol*, 129:269–280.

Mittendorf, R., Longnecker, M. P., Newcomb, P. A., Dietz, R. W., Greenberg, E. R., Bogdan, G. F., Clapp, R. W., and Willet, W. C. (1995). Strenous physical activity in young adulthood and risk of breast cancer (united states). *Cancer Causes and Control*, 6:347–353.

Moolgavkar, S. H., Stevens, R. G., and Lee, J. A. H. (1979). Effect of age on incidence of breast cancer in females. *Journal of the National Cancer Institute*, 62:493–501.

Morabia, A. and Wynder, E. L. (1990). Epidemiology and natural history of breast cancer. *Surg Clin N Am*, 70:739–752.

Müller, P., Parmigiani, G., Schildkraut, J. M., and Tardella, L. (1996). A Bayesian hierarchical approach for combining case-control and cohort studies. Discussion Paper DP96-29, Institute of Statistics and Decision Sciences, Duke University, http://www.isds.duke.edu/.

Müller, P. and Roeder, K. (1997). A Bayesian semiparametric model for case-control studies with errors in variables. *Biometrika*, 84.

Murphy, E. A. and Mutalik, G. S. (1969). The application of Bayesian methods in genetic counseling. *Human Heredity*, 19:126–151.

Myriad Genetics, Inc. (1996). Myriad genetic homepage. http://www.myriad.com.

National Action Plan on Breast Cancer and American Society of Clinical Oncology (1997). Hereditary susceptibility to breast and ovarian cancer: An outline of the basic fundamental knowledge needed by all health care professionals. http://www.napbc.org/napbc/hsedcurr.htm.

National Advisory Council for Human Genome Research (1994). Statement on use of DNA testing for presymptomatic identification of cancer risk. *JAMA*, 271:785.

National Cancer Institute: Surveillance, Epidemiology, and End Results (SEER) Program (1997). Seer homepage. http://www-seer.ims.nci.nih.gov.

National Surgical Adjuvant Breast and Bowel Project (1998). Breast cancer prevention trial shows major benefit, some risk. Press Release, http://www.nsabp.pitt.edu/PR-040698.htm.

Nelson, N. (1995). Caution guides genetic testing for hereditary cancer genes. *J Natl Cancer Inst*, 88:70–2.

Nelson, N. J. (1996). Caution guides genetic testing for hereditary cancer genes. *JNCI*, 88(2).

Newcomb, P. A., Storer, B. E., Longnecher, M. P., *et al.* (1994). Lactation and a reduced risk of premenopausal breast cancer. *N Engl J Med*, 330:81–87.

Newman, B., Austin, M. A., Lee, M., *et al.* (1988). Inheritance of human breast cancer: evidence for autosomal dominant transmission of high-risk families. *Proc Natl Acad Sci USA*, 85:3044–3048.

Newman, S. C., Miller, A. B., and Howe, G. R. (1986). A study of the effect of weight and dietary fat on breast cancer survival time. *Am J Epidemiol*, 123:767–774.

Oddoux, C., Struewing, J. P., Clayton, C. M., Neuhausen, S., Brody, L. C., Kaback, M., Haas, B., Norton, L., Borgen, P., Jhanwar, S., Goldgar, D., Ostrer, H., and Offit, K. (1996). The carrier frequency of the BRCA2 6174delT mutation among Ashkenazi Jewish individuals is approximately 1%. *Nature Genetics*, 14:188–190.

Olsson, H., Moller, T. R., and Ranstam, J. (1989). Early oral contraceptive use and breast cancer among premenopausal women. *JNCI*, 81:1000–1004.

Oncormed, Inc. (1996). Oncormed homepage. `http://www.pslgroup.com/dg/oncormed.htm`.

Ottman, R., Pike, M. C., King, M. C., *et al.* (1986). Familial breast cancer in a population-based series. *American Journal of Epidemiology*, 123:15–21.

Ottman, R., Pike, M. C., King, M. C., and Henderson, B. E. (1983). Practical guide for estimating risk for familial breast cancer. *Lancet*, pages 553–558.

Paffenberger, Jr, R. S., Kampert, J. B., and Chang, H.-G. (1980). Characteristics that predict risk of breast cancer before and after menopuase. *American Journal of Epidemiology*, 112:258–68.

Palli, D., Del Turco, M. R., Simoncini, R., *et al.* (1991). Benign breast disease and breast cancer: a case-control study in a cohort in italy. *Int J Cancer*, 47:703–706.

Parmigiani, G., Berry, D. A., and Aguilar, O. (1998). Determining carrier probabilities for breast cancer susceptibility genes BRCA1 and BRCA2. *American Journal of Human Genetics*.

Pike, M., Henderson, B. E., and Casagrande, J. T. (1981). The epidemiology of breast cancer as it relates to menarche, pregnancy, and menopause. In Pike, M. C., Siiteri, P. K., and Welsch, C. W., editors, *Hormones and breast cancer*, pages 3–18, Cold Spring Harbor, NY. Cold Spring Laboratory.

Polednak, A. (1986). Breast cancer in black and white women in new york state. *Cancer*, 58:807–815.

Porter, D.D., Cohen, B.B., Wallace, M.R., Smyth, E., Chetty, U., Dixon, J.M., Steel, C.M., and Carter, D.C. (1994). Breast cancer incidence, penetrance and survival in probable carriers of BRCA1 gene mutation in families linked to BRCA1 on chromosome 17q12-21. *British journal of Surgery*, 81:1512–1515.

Porter, D.E., Dixon, M., Smyth, E., and Steel, C.M. (1993) Breast cancer survival in BRCA1 carriers. *Lancet*, 341:184–185.

Prentice, R. L. and Pyke, R. (1979). Logistic disease incidence models and case-control studies. *Biometrika*, 66:403–412.

Prentice, R. L. and Thomas, D. B. (1987). On the epidemiology of oral contraceptives and disease. *Adv Cancer Res*, 49:285–401.

Raftery, A. E. and Richardson, S. (1996). Model selection for generalized linear models via glib: Application to nutrition and breast cancer. In (Berry and Stangl 1996), pages 109–138.

Richardson, J. (1991). Economic assessment of health care: Theory and practice. *Australian Economic Review*, 0:4–21.

Rimer, B.K. and Glassman, B. (1997). Tailoring communications for primary care settings. *Methods of Information in Medicine*, 2.

Roa, B. B., Boyd, A. A., Volcik, K., and Richards, C. S. (1996). Ashkenazi Jewish population frequencies for common mutations in BRCA1 and BRCA2. *Nature Genetics*, 14:185–187.

Roeder, K. and Wasserman, L. (1997). Practical bayesian density estimation using mixtures of normals. *Journal of the American Statistical Association*, 92.

Romieu, I., Willett, W. C., Colditz, G. A., *et al.* (1989). Prospective study of oral contraceptive use and risk of breast cancer in women. *JNCI*, 81:1313–1321.

Rosenberg, L., Shapiro, S., , and Stone, D. (1982). Epithelial ovarian cancer and combination oral contraceptives. *Journal of the American Medical Association*, 247:3210–3212.

Rubin, S. C., Benjamin, I., Behbakht, K., Takahashi, H., Morgan, M. A., LiVolsi, V. A., *et al.* (1996). Clinical and pathological features of ovarian cancer in women with germ-line mutations in brca1. *N Engl J Med*, 335:1413–1416.

Saftlas, A. F., Wolfe, J., Hoover, R. N., Brinton, L. A., Schairer, C., Salane, M., *et al.* (1989). Mammographic paremchymal patterns as indicators of breast cancer risk. *Am J Epidemiol*, 189:518–26.

Schatzkin, A., Jones, Y. D., Hoover, R. N., *et al.* (1987a). Alcohol consumption and breast cancer in the epidemiologic follow-up study of the first National Health and Nutrition Examination Survey. *N Engl J Med*, 316:1169–1173.

Schatzkin, A., Palmer, J., Rosenberg, L., *et al.* (1987b). Risk factors for breast cancer in black women. *JNCI*, 78:213–217.

Schildkraut, J. M. and Thompson, W. D. (1988). Familial ovarian cancer: a population-based case-control study. *Am J Epidemiol*, 128:456–466.

Schrag, D., Kuntz, K., Garber, J., and Weeks, J. (1997). Decision analysis - effects of prophylactic mastectomy and oophorectomy on life expectancy among women with BRCA1 or BRCA2 mutations. *N Engl J Med*, 336:1465–1471.

Schubert, E. L., Lee, M. K., Mefford, H. C., Argonza, R. H., Morrow, J. E., Hull, J., Dann, J. L., and King, M.-C. (1997). BRCA2 in american families with four or more cases of breast or ovarian cancer: Recurrent and novel mutations, variable expression, penetrance, and the possibility of families whose cancer is not attributable to BRCA1 or BRCA2. *Am J Hum Genet*, 60:1031–1040.

Schwartz, M., Lerman, C., Daly, M., Audrain, J., Masny, A., and Griffith, K. (1995). Utilization of ovarian cancer screening by women at increased risk. *Cancer Epidemiol Biomarkers Prev*, 4(3):269–273.

Scully, R., Chen, J., Plug, A., Xiao, Y., Weaver, D., Feunteun, J., Ashley, T., and Livingston, D.M. (1997). Association of BRCA1 with Rad51 in mitotic and meiotic cells. *Cell*, 88(2):265–275.

Shattuck-Eidens, D., McClure, M., Simard, J., Labrie, F., Narod, S., Couch, F., and Hoskins, K. (1995). A collaborative survey of 80 mutations in the BRCA1 breast and ovarian cancer susceptibility gene implications for presymptomatic testing and screening. *JAMA*, 273:535–541.

Smigel, K. (1992). Breast cancer prevention trial takes off. *Journal of the National Cancer Institute*, 84:669–670.

Sox, H. C., Blatt, M. A., Higgins, M. C., and Marton, K. I. (1988). *Medical Decision Making*. Butterworth and Heinemann, Boston.

Spiegelman, D., Colditz, G. A., Hunter, D., and E, H. (1994). Validation of the Gail et al model for predicting individual breast cancer risk. *J Natl Cancer Inst*, 86:600–607.

Struewing, J. P., Hartge, P., Wacholder, S., Baker, S. M., Berlin, M., McAdams, M., Timmerman, M. M., Brody, L. C., and Tucker, M. A. (1997). The risk of cancer associated with specific mutations of BRCA1 and BRCA2 among Ashkenazi jews. *New England Journal of Medicine*, 336:1401.

Struewing, J. P., Lerman, C., Kase, R. G., Giambarresi, T. R., and Tucker, M. A. (1995). Anticipated uptake and impact of genetic testing in hereditary breast and ovarian cancer families. *Cancer Epidemiology, Biomarkers and Prevention*, 4:169–173.

Swanson, C. A., Jones, Y. D., Schatzkin, A., *et al.* (1988). Breast cancer risk assessed by anthropometry in the nhanes epidemiologic follow-up. *Cancer Res*, 48:5363–5637.

Szolovits, P. and Pauker, S. (1992). Pedigree analysis for genetic counseling. In Lun, K. C. *et al.*, editors, *MEDINFO-92 Proceedings of the Seventh Conference on Medical Informatics*, pages 679–683, New York. Elsevier.

Tengs, T. O., Winer, E. P., Paddock, S., Aguilar, O., and Berry, D. A. (1997). Testing for the BRCA1 breast-ovarian cancer susceptibility gene: A decision analysis. Discussion paper, Institute of Statistics and Decision Sciences.

Tengs, T. O., Winer, E. P., Paddock, S., Aguilar, O., and Berry, D. A. (1998). Testing for the BRCA1 and BRCA2 Breast-Ovarian Cancer Susceptibility Gene: A Decision Analysis. *Medical Decision Making*.

The WHO Collaborative Study of Neoplasia and Steroid Contraceptives (1990). Breast cancer and combined oral contraceptives: Results from a multinational study. *Br J Cancer*, 61:110–119.

Thomas, D. B. (1984). Do hormones cause breast cancer? *Cancer*, 53:595–604.

Thune, I., Brenn, T., Lund, E., and Gaard, M. (1997). Physical activity and the risk of breast cancer. *N Engl J Med*, 336:1269–1275.

Torrance, G. W. (1986). Measurement of health state utilities for economic appraisal: A review. *J Hlth Econ*, 5:1–30.

United Kingdom National Case-Control Study Group (1989). Oral contraceptive use and breast cancer risk in young women. *Lancet*, 1(8645):973–982.

United Kingdom National Case-Control Study Group (1993). Breast feeding and risk of breast cancer in young women. *BMJ*, 307:17–20.

Vessey, M. P., McPherson, K., Villard-Mackintosh, L., *et al.* (1989). Oral contraceptives and breast cancer: Latest findings in a large cohort study. *Br J Cancer*, 59:613–617.

Wang, Q.-S., Ross, R. K., Yu, M. C., Ning, J. P., Henderson, B. E., and T, K. H. (1992). A case-control study of breast cancer in tianjin, china. *Cancer Epidemiol Biomarkers Prev*, 1:435–439.

Weber, B. (1996). Genetic testing for breast cancer. *Science & Medicine*, 3:12–21.

Williamson, D. F., Forman, M. R., Binkin, N. J., *et al.* (1987). Alcohol and body weight in united states adults. *Am J Public Health*, 77:1324–1330.

Wolfe, J. N. (1976). Breast parenchymal patterns and their changes with age. *Radiology*, 121:545–52.

Wooster, R., Bignell, G., Lancaster, J., Swift, S., Seal, S., and Mangion, J. (1995). Identification of the breast cancer susceptibility gene BRCA2. *Nature*, 378:789–92.

Yancik, R. (1993). Ovarian cancer. age contrasts in incidence, histology, disease stage at diagnosis, and mortality. *Cancer*, 71:517–523.

Yoo, K. Y., Tajima, K., Kuroishi, T., *et al.* (1992). Independent protective effect of lactation against breast cancer: a case-control study in japan. *Am J Epidemiol*, 135:726–733.

Young, T. B. (1989). A case-control study of breast cancer and alcohol consumption habits. *Cancer*, 64:552–558.

Yuan, J.-M., Yu, M. C., Ross, R. K., *et al.* (1988). Risk factors for breast cancer in chinese women in Shanghai. *Cancer Research*, 48:1949–1953.

Zelen, M. and Parker, R. A. (1986). Case-control studies and Bayesian inference. *Statistics in Medicine*, 5:261–269.

S. Greenhouse
George Washington University

1. What is the nature of the Quality of Life questions asked. Certain QOL standardized questionaires could reasonably be used to compare different groups in a study even if the groups differ from the population used for standardization. But what meaning can be ascribed to the responses of an individual, say from a different culture or environment, so that the response is helpful in making a decision with regard to testing.

2. I question the procedure of forming a Likelihood which combines a probability distribution of a covariate, x, in a cohort study with a frequency (i.e. a non-probability) distribution of x in a case-control study.In the former case, P(x) is a sampling distribution in the population, whereas in the case-control study, "P(x)" is the frequency of x among the selected cases and among the selected controls. In the event that the controls are selected at random from the population then P(x) may approximate the distribution of x in the general population. A further problem with the formulation is the specification of the parameters in the cohort study and in the case-control study. I assume that 'theta' represents either a relative risk (which is appropriate for the cohort study) or an odds ratio (appropriate for the case-control study). But how does the distribution of the covariate x involve either of these parameters. P(x—D=1) may for example be N(mu, sigma) with no reference to 'theta'.

3. You point out that mutations in the BRCA gene seem to be more likely to occur in younger women. It is well known that the tumor in younger women is more aggressive than in older women and that the tumor in younger women is probably of shorter duration than the age of tumor detected in older women. Could these observations be related to and perhaps effected by the finding that mutations occur more frequently in younger women.

4. I don't recall ever reading that BRCA genes are sex linked. Assuming they are not then why so few breast cancers in males (1 per million in 1986?) Either there are fewer mutations in males or else these genes express themselves in other tumors in males. Interesting research questions.

Discussion

Larry Kessler
FDA/CDRH

I take great pleasure in congratulating the authors for presenting a comprehensive analysis and a useful model in a very important area. The comments below are divided into two sections; the first includes general comments about the paper, and the second section, with more specific comments, is further subdivided into those related to clinical or epidemiologic aspects of the paper followed by comments specifically targeted to the decision model part of the paper

The authors have tackled an important problem which becomes even more important as the entire field of genetic testing grows. It would be of interest to know whether the modeling strategy developed by Parmigiani et al. appears readily generalizable to other genetic tests in other disease situations. Their particular effort is especially important given the prominence that breast cancer plays in the current health care arena. The National Cancer Institute estimates between 1 in 8 and 1 in 10 women alive today will at some point in their lifetime be diagnosed with breast cancer and, as noted in the paper, women who carry the BRCA1 or BRCA2 gene may have in excess of a 50% risk to contract this disease.

The characterization of the problem into two major pieces, the probability that a woman is a BRCA1/2 carrier given her family history or pedigree and the probability that the woman will eventually be diagnosed with breast cancer given the pedigree and available risk factor information is both natural and leads to an appropriate Bayesian framework. What does this Bayesian perspective offer? It appears that it offers an convenient structure for combining data from disparate sources; a suitable structure for the use of expert opinion in several places in the model; and a full posterior and predictive analysis for this particular problem. Do we lose anything with a Bayesian analysis? Perhaps, we lose useful and conventional variance measurements and associated confidence intervals, which have become relatively standard in clinical applications. Perhaps, because of this, the product of the Parmigiani model could present a communication problem between modelors and their intended audience.

Turning to more specific comments, this particular effort combines different data sets which brings along with it certain assumptions not explicitly stated. For example, are all the data sets drawn from some theoretical superpopulation? Are the quality of the data roughly equivalent? The authors could choose to weight the contributions of the various data sets, either by a conventional approach using the inverse of the variance information in the data or by a combination of variability measures and expert judgment.

Breast cancer screening could affect a number of aspects of the analysis and it is not mentioned very much at all in the paper. For example, some of the underlying studies used by Parmigiani were developed during a rapid rise in the prevalence of breast cancer screening with mammography in the United States. This rise would have an effect on a variety of information from these studies, including incidence and prevalence of the disease as well as survival for women detected with breast cancer. A second effect is the possible influence of screening on the age of onset of the disease. And, finally, a woman presenting herself for possible genetic testing will have a different risk of developing the disease depending on her current mammographic history. A woman age 40 never having had a mammogram will appear to have a different risk level than a woman who has had a series of mammograms with negative findings.

A number of analyses of breast cancer have come to the conclusion that this is a heterogeneous disease. In fact, as the authors point out, women who carry the BRCA1 or BRCA2 gene tend to be at dramatically higher risk for earlier onset of breast cancer. It is possible that breast incidence rates in this country in fact reflect a combination of two or more different diseases, both resulting in similar appearing clinical manifestations of breast cancer. Would such an occurrence affect the ways in which the authors have modeled the data?

Not mentioned by the authors but perhaps included in their analysis, "no history of breast cancer in the family" may really mean no "known" history in the family. Individuals may become deceased from competing causes prior to having had a full chance of becoming a breast cancer case and, as such, there may be some biases in the model because this is not taken into account. In some data described by the authors and then fitted by several models, the incidence of breast cancer appears to have a flattened asymptote as women age. Almost all epithelial cancers increase without such an asymptote in virtually every cancer registry in the world. Why do these data show an asymptote for their fitted models? It is quite possible that a flattening of the breast cancer rates reflects either a period or a cohort effect but not the true incidence of breast cancer at older ages.

Turning to comments concerning the decision model, it is clear that some aspects of this model need careful reconsideration. For example, quality of life estimates are most certainly age dependent. An additional year of life at age 70 will be of far different value to women currently at age 30, 40, 60 or 70. In fact, years of life in the future are almost always discounted in such analyses to correct for time preferences of such quantities. Although this is fairly common practice, it does make communicating information from such decision models more difficult for the clinician laywomen to whom this model is directed. One possible option from the branches of the decision tree appears possibly overlooked.

Early and frequent breast cancer screening, including mammography, clinical breast examination, and breast self examination, could be an additional option for a woman who may test positive for one of the two genetic mutations. However, as with several of the fundamental aspects of the model, a great deal of uncertainty of information exists related to how an aggressive screening regimen would affect the probability of diagnosis and mortality from genetically transmitted breast cancer. Two critical pieces of information are lacking and they will certainly have dramatic influences on the model: genetic test performance, and the effect of the surgical prophylactic interventions suggested on breast cancer mortality.

This particular clinical situation presents the quintessential paradigm for psychological effects of the applications of such a model, for example framing effects will certainly influence how a woman appreciates risks and possible interventions with various decision strategies.

There may be interesting regulatory ramifications of the building of such a model. To the degree that such a model will be used as an adjunct to counseling it can probably receive widespread use with no barriers. However, should this or similar models be used in the direct diagnosis and or therapeutic decision making for women, this kind of model and essentially the software around it may be considered a medical device. Medical devices are required to go through either a premarket approval process or a market notification procedure to alert the FDA to the availability of products on the market. The policies of the Center for Devices and Radiological Health concerning software such as this are still developing at this time.

I would like to conclude these comments by calling attention to the need for validation of this particular model. The authors have recognized this important feature of the model. My desire to validate the outcome of such a model is meant as a compliment to the fine work of the authors, for the validation of such a model is perhaps the penultimate step to widespread utility, the goal of the best of applied modelers.

Discussion

Nozer D. Singpurwalla
The George Washington University

I have not read the paper I am discussing, nor have I heard its presentation at the CMU Workshop. However, my exposure to *retrospective* versus *prospective* modeling was triggered by an inspiring presentation by Peter Mueller at the Bernoulli Society Meetings in Calcutta, India, in December 1997. This was followed up by a discussion with Peter in February 1998 at Duke. I thank Peter for his contribution to my appreciation of the problem, and give below what I think is a completely subjectivistic approach for its solution. Peter and Giovanni Parmigiani, in an e-mail message to Constantine Gatsonis, have labeled the approach "radical;" hopefully, it is not "radically wrong!"

In what follows, the following convention is used:

- $P(A; B)$: is the (subjective) probability of event A, in the light of history (to include data) B;
- $P(A \mid B)$: is the (subjective) conditional probability of A, were one to know B.

Statement of Problem: One needs to assess $P(Y_{n+1} = y_{n+1}; x_{n+1}, (\underline{x}, \underline{y}))$, where Y_{n+1} is the response to a pre-selected stimulus x_{n+1}, and $(\underline{x}, \underline{y}) = ((x_1, y_1), \ldots, (x_n, y_n))$, with x_i being the observed stimulus to a pre-selected response $y_i, i = 1, \ldots, n$.

Proposed Approach: By the law of total probability

$$P(Y_{n+1} = y_{n+1}; x_{n+1}, (\underline{x}, \underline{y})) = \int P(Y_{n+1} = y_{n+1} \mid \underline{\theta}; x_{n+1}, (\underline{x}, \underline{y})) \cdot P(\underline{\theta}; x_{n+1}, (\underline{x}\underline{y})) d\underline{\theta},$$

for some vector of parameters, say θ.

Assumption 1: $Y_{n+1} \perp (\underset{\sim}{x}, \underset{\sim}{y})$ given x_{n+1} and (were you to know) $\underset{\sim}{\theta}$; "$\perp$" denotes independence.

Thus $P(Y_{n+1} = y_{n+1} \mid \underset{\sim}{\theta}; x_{n+1}, (\underset{\sim}{x}, \underset{\sim}{y})) = P(Y_{n+1} = y_{n+1} \mid \underset{\sim}{\theta}; x_{n+1})$.

Assumption 2: (A probability model for Y_{n+1} given x_{n+1} and $\underset{\sim}{\theta}$):

Any subjectively specified choice can be made, and for purposes of discussion, assume that

$$(Y_{n+1} \mid \underset{\sim}{\theta}; x_{n+1}) \sim \aleph(\alpha + \beta x_{n+1}, \sigma^2),$$

where $\underset{\sim}{\theta} = (\alpha, \beta, \sigma^2)$, and "$\aleph$" denotes the Gaussian distribution.

Assumption 3: $P(\underset{\sim}{\theta}; x_{n+1}, (\underset{\sim}{x}, \underset{\sim}{y})) = P(\underset{\sim}{\theta}; (\underset{\sim}{x}, \underset{\sim}{y}))$; i.e. x_{n+1} being pre-selected, it does not provide any information for specifying out uncertainty about θ.

The task therefore is to evaluate $P(\underset{\sim}{\theta}; (\underset{\sim}{x}, \underset{\sim}{y}))$, where now, in retrospective studies, it is the y_i's that are pre-selected and the corresponding x_i's, the observed entities.

To proceed further, consider the following probability statement:

$$P(\underset{\sim}{\theta} \mid X_i; y_i, i = 1, \ldots, n) \propto P(X_1, \ldots, X_n \mid \underset{\sim}{\theta}; y_i, i = 1, \ldots, n) P(\underset{\sim}{\theta}),$$

by Bayes' law, where $P(\underset{\sim}{\theta})$ is the prior on $\underset{\sim}{\theta}$.

Assumption 4:

$$\begin{array}{lll} X_1 \perp X_2 & , \ldots, X_n, & \text{given } \underset{\sim}{\theta} \text{ and } \underset{\sim}{y}, \\ X_2 \perp X_1, X_3 & , \ldots, X_n, & \text{given } \underset{\sim}{\theta} \text{ and } \underset{\sim}{y}, \ldots, \\ X_n \perp X_1 & , \ldots, X_{n-1}, & \text{given } \underset{\sim}{\theta} \text{ and } \underset{\sim}{y}. \end{array}$$

Thus

$$P(\underset{\sim}{\theta} \mid X_i; y_i, i = 1, \ldots, n) \propto \Sigma_{i=1}^{n} P(X_i \mid \underset{\sim}{\theta}; \underset{\sim}{y}) P(\theta).$$

But $X_i = x_i, i = 1, \ldots, n$ have been observed, and so each term $P(x_i \mid \underset{\sim}{\theta}, \underset{\sim}{y})$ is the *likelihood* of $\underset{\sim}{\theta}$ holding X_i fixed at x_i, and $\underset{\sim}{y}$ (which is pre-selected), for $i = 1, \ldots, n$.

From a subjectivistic viewpoint, one is free to choose the likelihood $P(x_i \mid \underset{\sim}{\theta}, \underset{\sim}{y})$ via any suitable strategy that one feels appropriate, as long as it reflects one's level of support provided by fixed quantities x_i and $\underset{\sim}{y}$, for the various values of $\underset{\sim}{\theta}$. The only requirement here is that likelihood be specified for the same $\underset{\sim}{\theta}$ that appears in the probability model for Y_{n+1}; i.e., *likelihood identifiability*. An implication of the foregoing comment is that one does not need to have assessed a probability model for X_i (given $\underset{\sim}{\theta}$ and a pre-selected y_i), in order to specify a likelihood. Thus, to address the problem of retrospective modeling the specification of a joint probability model for $(X_i, Y_i), i = 1, \ldots, n, n+1$ is *not essential*. All that one needs is a probability model for Y_{n+1} given θ and the pre-selected x_{n+1}, and any subjectively specified (or expert opinion based) likelihood function for $\underset{\sim}{\theta}$ given the observed x_i and the corresponding pre-selected y_i. This facility, namely that of divorcing the specification of a likelihood from the specification of a probability model is not available to those whose view of probability is relative frequency.

Suppose then, that the likelihood for $\underline{\theta}$, having observed an x_i and given its corresponding (pre-selected) $y_i, i = 1, \ldots, n$ is chosen to be of the form

$$\prod_{i=1}^{n} \frac{1}{\sqrt{2\pi\sigma}} \exp\left[-\frac{1}{2}\left(\frac{x_i - (\alpha + \beta y_i)}{\sigma}\right)^2\right],$$

then

$$P(\theta; (\underline{x}, \underline{y})) \propto \prod_{i=1}^{n} \frac{1}{\sqrt{2\prod\sigma}} \exp\left[-\frac{1}{2}\left(\frac{x_i - (\alpha + \beta y_i)}{\sigma}\right)^2\right] P(\underline{\theta}).$$

Thus, for prediction under the retrospective model, all of the above can be packaged as

$$P(Y_{n+1} = y_{n+1}; x_{n+1}, (\underline{x}, \underline{y})) \propto \int \frac{1}{\sqrt{2\prod\sigma}} \exp\left[-\frac{1}{2}\left(\frac{y_{n+1} - (\alpha + \beta x_{n+1})}{\sigma}\right)^2\right] \cdot$$
$$\prod_{i=1}^{n} \frac{1}{\sqrt{2\prod\sigma}} \exp\left[-\frac{1}{2}\left(\frac{x_i - (\alpha + \beta y_i)}{\sigma}\right)^2\right] P(\alpha, \beta, \sigma) d\underline{\theta},$$

and that is all there is to it!

Discussion

Steven J. Skates
Massachusetts General Hospital

First, let me congratulate Parmigiani and colleagues on an excellent and scholarly paper exemplifying the application of Bayesian approaches to provide a timely answer to an important medical problem. The timeliness is emphasized by the fact that the software implementing this solution is already in use at various medical sites, assisting women worried about risk of mortality from breast cancer in making decisions about genetic testing and prophylatic medical interventions. This discussion focuses on the statistical modeling involved and why I believe Parmigiani et. al.'s approach gives more accurate answers than standard approaches such as straightforward logistic regression. Additional comments address decision analyses based on expected quality adjusted life years.

Statistical modeling

This paper derives methods for computing two probabilities:

1. Pr(carrying defective gene | family history)

2. Pr(Breast cancer | family history & other risk factors)

In a simpler situation, such as comparing the success rate of two treatments in a randomized trial, probabilities are calculated by simply counting proportion of success within each arm of the trial. However, the probabilities required to assess risk of breast cancer are conditioned by high dimensional "events", instead of a simple one dimensional comparison with two levels (treatment A vs. B). That is, family history is complicated when fully recorded, and there are multiple known risk factors for breast cancer. High dimensions naturally imply data are sparse, so "noise" or natural variation is a crucial component to the problem and must be represented in the model for an accurate description. The "curse of dimensionality" thus arises, which is nothing more than asserting that the real world is complicated, or in this situation, that patients are individuals requiring individualized answers. I feel that statisticians have a responsibility for noisy complicated problems and need to develop methods to extract information as efficiently as possible. Parmigiani et. al. have made significant advances in accomplishing such an aim for assessing risk of breast cancer.

Typical answers for evaluating $Pr(Y|X_1, X_2, X_3, \ldots)$ is to estimate θ in an additive model of the form:

$$fPr(Y|X_1, X_2, X_3, \ldots) = \theta_0 + \theta_1 X_1 + \theta_2 X_2 + \cdots \quad (2)$$

or even generalized additive models of the form

$$fPr(Y|X_1, X_2, X_3, \ldots) = \theta_0 + \theta_1 f_1(X_1) + \theta_2 f_2(X_2) + \cdots \quad (3)$$

where $f, f_1, f_2, \ldots$ are well chosen functions. The resulting estimates of θ are used to calculate probabilities for future combinations of $X's$. However, there is a HUGE assumption in these models; namely, additivity. The size of this assumption dramatically increases with the dimensionality of X; it assumes that the first addition function $+_1$ holds across the full support of $X_2, X_3, \ldots$, that $+_2$ is similarly appropriate across the support of $X_1, X_3, \ldots$, and that any combination of $+_i$ functions are appropriate across the complement of $+$ functions. Such a collection of assumptions is surely too much to expect to hold. Furthermore, just the sheer number of assumptions make for a daunting task in attempting to determine if they hold, and for assessing the implications if they fail to hold. Reducing the dimensionality where additivity is assumed allows the possibility of exploring a smaller dimensional space for adequacy of the assumptions and understanding the impact if the assumptions are not met. In classical applied mathematics, much effort has been expended in re-expressing functions of 3 or more variables as a function of functions of two or fewer variables, due to this very problem of dimensionality. Statisticians should also expend a great deal of effort into reducing dimensionality where possible, such as by incorporating scientific knowledge or using more flexible statistical models, before resorting to traditional additive models, which somehow "magically" replace commas with plus signs and $\theta_i's$.

This paper presents two methods to counter the problem of dimensionality and the assumption of additivity :

1. Use of conditional independence, and
2. Flexible multivariate distributions

The interdisciplinary team enables both methods to be incorporated, allowing effective inclusion of biological knowledge to reduce dimensionality, and recent advances in statistics to surmount the traditional assumption of additivity.

The first method incorporates the biological understanding that given the genetic status of the individual, the risk of breast cancer is independent of family history. That is, familial breast cancer risk is mostly genetically conferred, and knowing the genetic status of the individual provides all the information contained in the family history about the risk. This method assumes that environmental factors do not interact with genetic risk factors for breast cancer. Given the weak effect of most environmental factors so far discovered, this seems to be a reasonable assumption. Using this conditional independence substantially reduces the dimensionality of family history.

Other risk factors are either quantitative, denoted by Z_i (weight gain, no. of surgical biopsies, duration of OC use, etc.), or categorical, denoted by X_i (needle biopsy, hyperplasia, race, etc.). The distribution of environmental risk factors (Z, X) initially depends on genetic status, age at onset, and age at interview. However, given that environment and genetic factors are assumed independent, the dependence on genetic status drops. Also, the dependence on age at onset is simplified to dependence on whether the woman eventually contracts breast cancer or not. Finally, the age at interview is assumed not to affect the distribution of risk factors, although exposure to hormonal replacement therapy for example may not yet be recorded for a woman interviewed at a pre-menopausal age. This simplification is surely minor. Therefore, the distribution of (Z, X) is modeled as two distributions between women who eventually contract breast cancer and those who do not. The two distributions are flexible mixtures of multivariate Gaussian distributions, with the distribution of dichotomous variables X modeled as a latent component of the overall multivariate Gaussian distribution. The mixture of multivariate Gaussian distributions overcomes the strong assumption of additivity that would be present in both a single Gaussian distribution (probit regression model) or a logistic regression model. From these models and the combined data sources, posterior distributions are derived for parameters describing the genetic penetrance and the contribution of environmental risk factors to development of breast cancer.

Decision Making

The resulting statistical models are then combined to develop a predictive distribution of age of onset of breast cancer (age greater than 110 implying the patient never gets the disease) given family history, environmental risk factors, and age at "interview" or current age. These predictive distributions can then be used - creatively - to inform patients in genetic counseling sessions.

Decision making is explored by examining years of life saved and quality adjusted years of life for five interventions, including genetic testing with or without prophylatic mastectomy and/or oophorectomy, and no intervention. This exploration relies on the predictive distributions derived in the modeling section, and on estimates of operating characteristics of the genetic test, improvement in survival given prophylatic interventions, and estimates of quality of life with the disease or following interventions. An example of a 30 year woman with a family history is explored, bringing to the fore the power of Bayesian modeling and decision making, when real life choices and consequences, quantified by a believable approach, are presented.

Critique and Comments

I have a few minor suggestions and some comments, more for future work than anything else.

I believe the power of this approach derives from the solid foundations the authors have presented. Therefore, I would like to see more emphasis on how each conditional independence assumption is justified, for it is on these solid bricks that the structure is built, instead of simply asserting a list of conditional independence assumptions.

Given the power of the modeling, it would be insightful if a standard analysis, such as a logistic regression model, were used to derive risk estimates of breast cancer given environmental risk factors and a simplified genetic risk factor. This would represent some quantification of family history, as is seen in standard epidemiological approaches. Having a standard model would then enable differences in predictive risk levels between the two approaches to be highlighted. Finally, examination of the source data sets to determine which model gave better predictions, and how ubiquitous this was, would potentially complete the case empirically for deriving powerful models for the problem of breast cancer risk. It may also shed some light on where, why, and to what extent, traditional additive models fail.

Traditional models have the advantage of a simpler summary for presentation and transmission to other users. The question arises as to how best to publish results of such analyses in a journal. It may be that the Web is a more appropriate vehicle by which to furnish results of the modeling and analysis. The other advantage of a traditional model is interpreting results, and in effect, learning about risk factors for breast cancer. It is not immediately clear what we have learned from this analysis which builds on a mixture of multivariate Gaussian distributions?

The statistical modeling could be improved further, in that some of the environmental variables may lend themselves to dimension reduction. For example, variables connected to oral contraceptives (OC) are presumably collected as a proxy for OC exposure. Therefore deriving a single OC exposure variable from the multiple OC variables may lead to improved models and a reduction in dimension.

The decision making section gives a paradoxical example of where the quality adjusted life years is more if the woman tests positive for both BRCA1 and BRCA2 than if she tests positive only for BRCA1. The authors indicate this paradox derives from the imperfect operating characteristics of the test, where in the absence of empirical data they suggest a specificity of 99%. While this may seem like a high specificity, I would suggest that for a genetic test, it would be quite low, leading to a non-intuitive situation as given above. A more realistic specificity may be 99.9%, that is, a genetic abnormality would be mistakenly identified once per 1,000 tests on women with normal BRCA1 and BRCA2 genes. At this level of specificity I would hazard a guess that the paradox disappears. Thus obtaining empirical data on the operating characteristics of the test could be an important next step in refining the decision making program.

Finally, a question arises as to whether expected (quality adjusted or not) life years saved is the appropriate metric for decision making; in particular, is the expectation a sufficient summary on which to base a decision. Given that people can be risk adverse, risk neutral, or risk seeking, saving 30 years of life with a probability of 0.1 may be a very different scenario from saving the last three years of life, say from 78 to 80 with a probability of 1.0. I would like to suggest presenting the predictions as a distribution of years of life saved, or distributions on years of life lived under the five different decisions. Given the statistical modeling and computing power employed so far in this project, this should be very feasible. More input from genetic counseling experts as to how to present these results would also be very helpful.

Conclusion

Parmigiani and colleagues have presented a very powerful methodology to provide answers to patients facing important medical decisions. The power stems from interdisciplinary collaboration, and incorporating realistic assumptions and biological knowledge into their statistical modeling, which I believe, overcome the major assumption of many traditional analyses. Their Bayesian approach is helpful in correctly propagating uncertainty and quantifying expert opinion. However, I would not be surprised if a frequentist justification could be derived which would give models and results close to those obtained herein. For me, the advantage of the Bayesian approach is that it avoids distractions arising in multivariate frequentist inference, such as ancillarity (or M-ancillarity), marginal likelihoods, and uniformly most powerful tests, and instead allows one to focus on accurate statistical modeling where most of the "payoff" lies. I believe that if statisticians do not focus more of their energies on modeling and providing realistic incisive answers, then our discipline could easily become overshadowed by competition from approaches with names such as neural networks, data mining, knowledge engineering, and artificial intelligence. This paper presents a very competitive statistical modeling solution based on solid foundations for a medical decision making problem, sets a high standard for other statisticians to follow, and I feel, serves statistics, medicine, and ultimately patients better than approaches with Madison Avenue like names.

References

Almeida, M. and Gidas, B. (1993). A Variational Method for Estimating the Parameters of MRF from Complete or Incomplete Data, *Annals Appl. Prob.* **3**, 103-136.

Heyman, M.A., Payne B.D., Hoffman, J.I.E. and Rudolph, A.M. (1977). Blood flow measurements with radionuclide-labled particles. *Progress in Cardiovascular Diseases* **10**, 55-79.

Rejoinder

Individualized prediction of the risk of developing a disease is likely to become increasingly important, especially in relation to genetic susceptibility. As we continue in our efforts to develop better risk prediction models, the discussants' encouragement and insightful comments will be very helpful.

An aspect of our analysis that was touched upon by three of the discussants is the role of quality of life adjustments in making decisions about genetic testing. We first comment on this issue, and then address each discussant's specific issues, starting with the invited discussions.

There seems to be no question that decisions about genetic testing and prophylactic interventions involve, among other aspects, a trade-off between length and quality of life. Two issues emerged. The first concerns the most helpful way of presenting quantitative information to help women address this trade-off. In our view it is intructive to explore the implications of a formal decision analysis. The most important contribution of this analysis is to assist with the multi-stage aspect of the testing decision, which is otherwise difficult to conceptualize.

In practice, a formal decision analysis could be carried out by eliciting individual values, converting them into quality of life measures, and then providing women with the results. Alternatively, individuals could be provided with results of an analysis based on quality of life assessments elicited from other women who experienced the various health states involved. This may imply cross-cultural or cross-generational comparisons, which, as Greenhouse indicates, are potentially problematic. However, women sometimes inquiry explicitly about the quality of life experienced by others. At the opposite extreme, patients can be provided with the probabilistic ingredients of the decision analysis. Then they can be counseled to address the trade-off at a more intuitive level than the formal decision analyses. This approach can be valuable, especially in decision problems that do not involve multiple decisions over time. For example Schrag and colleagues (1997) suggest it in the context of decision–making about prophylactic mastectomy.

The second issue related to quality of life are the limitations of providing expectations of life length, or of quality adjusted life length (Kessler, Skates). Problems here arise because the conditions on which maximization of expected QALYs is predicated, and in particular risk neutrality with respect to length of life, are not easily met. We agree that an interesting alternative is to look at the whole distribution of QALYs. However, in practical counseling situations, the meaning of this distribution may be difficult to convey. Also, a woman would be faced with the problem of understanding and choosing from several distributions, which is difficult. The alternative of measuring risk aversion is also a challenging task.

A key aspect in meaningful use of the tools of risk prediction and decision analysis is proper education and counseling. As Kessler notes, counselors can be critical in framing the risk information using the appropriate level and approach. Ideally, not only the risk predictions, but also the presentation could be individualized to a patient's individual values, attitudes and level of instruction (Rimer and Glassman 1997).

Kessler

We are definitely optimistic about generalizing of our approaches to other inheritable diseases. In fact, applications to colon and ovarian cancer are under way.

Uncertainty assessment is not emphasized in the paper as much as it perhaps should have been. While we abandon conventional asymptotic methods for deriving confidence intervals, we do obtain the full predictive distribution of the absolute risk over time. This can be used to derive probability intervals that are more likely to give realistic uncertainty assessments than conventional asymptotic approaches. Predictive distributions are interesting and important, and we were remiss in not illustrating them. However, our sensitivity analysis incorporates this uncertainty, as well as others, and converts it into a scale that is directly relevant for decision making.

We agree that combining different data sets always requires strong assumptions. While there are study-specific effects in our model, an analysis more systematically acknowledging study-to-study differences would have been stronger. With only three studies, and a highly parameterized model, implementing standard hierarchical models could be arduous, and, as Dr. Kessler suggests, a combination of formal analysis and expert judgment could prove more reliable. One way in which expert judgment can help is in specifying the likely degree of heterogeneity among studies.

Indeed, screening is very important. The Gail model (Gail et al. 1989) is based on a regularly screened cohort. One of the motivations for our own work has been to develop a tool that can be applied more generally. However, we have no information about who was being screened or with what frequency in the three studies. Regarding your related point, a critical aspect of previous screening history is currently captured by our model via the benign breast disease variables.

Our predictions are different for "no family history" and "no known family history". Relatives who are free of disease at age A are factored in the calculations and contribute the information that "there was no disease at or before age A" —a censored observation. Therefore young unaffected relatives and relatives who died at young age of competing causes provide little evidence for or against susceptibility, while long-lived unaffected relatives provide evidence against susceptibility. Our age-at-onset distributions for noncarriers are based on the SEER empirical distributions, and reflect an increasing incidence at all ages. However, as you note, the cumulative distribution of the age of onset for the Ashkenazi Jewish population does flatten around age 60, suggesting that susceptible individuals either develop breast cancer by age 60 or never. This finding has not been independently confirmed in other studies as yet. A number of hypotheses have been advanced to explain it, including relationships with menopausal status and hormonal exposure.

Increased mammography has been suggested as a potentially useful strategy for high risk women and could have been included in our decision analysis. However, it is unclear whether increased mammography is of benefit for BRCA1&2 carriers, and there is even some suggestion that it may be harmful, because of a possible interaction with mutations in the RAD51 gene (Scully et al. 1997). We could have incorporated an expert assessment of the benefit, but it would have been small, and sufficiently similar to the "do nothing" arm that conclusions about the value of genetic testing would have been unaltered. The Breast Cancer Prevention Trial (BCPT) recently released results showing "a 45 percent reduction in breast cancer incidence among the high-risk participants who took tamoxifen, a drug used for the past two decades to treat breast cancer" (National Surgical Adjuvant Breast and Bowel Project 1998). While it is not clear how benefits are distributed among carriers and noncarriers, another interesting alternative would be to include tamoxifen as one of the options.

We agree very strongly that risk prediction models, whether simple or complex, call for external validation. We are working on validating various components of our model. Elsewhere in this volume (Iversen et al. 1998) we report on some of the methodological issues that arose during our effort. The issue of validation is closely related to the "regulatory ramification" you mentioned. External validation could play a systematic role in guaranteeing the quality of risk prediction models. A number of interesting methodological issues arise.

Skates

We too are concerned about the dimensionality of the model. In the genetic model component, the complexity of the model is supported by the simple building blocks of Mendelian inheritance. With regard to other covariates, not only the magnitude of the effect, but also the functional form is to be inferred from the data, and that is substantially more challenging. With the powerful modeling techniques available today, it is simple to fit the data well; the real issue is generalizability. We emphasized three strategies: 1) breaking down complex problems into simpler, scientifically meaningful components; 2) incorporating expert opinion; and 3) validating on external data sets. The first two are exemplified in the paper, while the third is in progress.

The conditional independence of genetic and other risk factors is a problematic assumption. Our approach has been to require independent evidence in order to include effects in the model. This does not mean that we believe that there are no interactions between genes and environment. Why is cancer penetrance incomplete? What are the factors that determine whether a carrier will develop breast cancer? Are these inherited or environmental? It is conceivable that environmental components play a big role, but we have no information wether they do or not. Including interactions into a counseling tool would be excessive specualtion.

The architecture of our model is complex and unquestionably hard to communicate. Our view, however, is that the resulting predictions should be just as easy to communicate to users as absolute risk predictions obtained using simpler and less accurate approaches. Regarding logistic regression parameters and relative risks, those are important and understandable to statisticians, but are they relevant to a lay person in decision making? And should they be? Imagine being told that by performing some simple and painless preventive task for 20 minutes every day you can reduce risk of developing cancer at site X by 3 times, compared to an individual with no risk factors, or with no known risk factors. Is such relative risk information sufficient for you to make a decision? Or would you also want to know how likely you are to get cancer at site X?

Dr. Skates comments on the apparent paradox that the prognosis for a woman testing positive to both genes is better that that of a woman testing positive to BRCA1 only. For a woman who already tested positive to BRCA1, testing positive to BRCA2 could be good news: if the test is not perfectly specific, the chance of carrying both genes, which assuming independence is a priori less than one in a million, is still quite small. However, there is now a chance that cancer in her family is due to a slightly less lethal gene (BRCA2), so that her life expectancy is a little higher. But if the test had perfect specificity, so that we could be sure that she has mutations in both genes, her life expectancy would of course be worse. As indicated by Dr. Skates, the apparent paradox would be resolved.

Greenhouse

We agree that risk prediction based on retrospective modeling could have poor performance if case-control studies are not representative. For our model, we sought both prospective and case-control studies. The IOWA study is the only prospective study we have been given access to. However, both of the case-control studies included in our model are population–based, which lends some credibility to our assumption of generalizability. Because this is one of the most important assumptions in the model, we look forward to assessing its legitimacy in external validations using prospectively collected data. We apologize for the confusion created by our unorthodox parameterization. θ is the ensemble of parameters characterizing the retrospective distributions. So the elements of θ are neither relative risks nor odds ratios, but simply the μ and σ in Dr. Greenhouse's example, plus more.

The mutations of BRCA1 and BRCA2 considered in this paper are inherited. This means that carriers have had mutated genes since birth. While it may be true that on an overall basis tumors developed at a younger age tend to be more aggressive and lead to a worse prognosis than tumors developed later in life, we don't know whether this applies to tumors that result from germline mutations of BRCA1 and BRCA2. Markus et al. (1996) compare breast cancer tumor pathobiology and survival prognosis between sporadic cases, gene-linked or suspected BRCA1 cases, and suspected non-BRCA1 heritable cases presenting to a high-risk clinic. Both of the latter two groups had better survival rates than the sporadic group, with survival among BRCA1 cases better than among non-BRCA1 cases. Porter et al. (1993,1994) study 35 probable carriers from eight BRCA1-linked families and find evidence of improved survival. These studies are small and not population-based as the subjects were recruited from high–risk breast cancer clinics. In Iversen et al. (1997) we analyzed the CASH data using family history as a proxy for genotype and found that BRCA2 carriers may have a substantially more favorable prognosis than non-carriers, controlling for age and several other factors.

Indeed, BRCA1 and BRCA2 are not sex linked. The reason why fewer male carriers than female carriers develop breast cancer is still unknown. Answering this question may uncover important aspects of the function of the gene, which is still being investigated. It is plausible that estrogen is involved, and estrogen is sex-dependent.

Singpurwalla

We agree that likelihood and prospective inference can be based on different probability models. It is important that the model used for retrospective likelihood and prospective inference share common parameters, otherwise there is no learning. We propose to derive both from one underlying model $p(x, d)$, which is a natural approach, but certainly not the only one. In a sense our approach accommodates what Dr Singpurwalla proposes. By defining a single encompassing model $p(x, d)$ we imply a model $p(x|d)$ for the likelihood and a model $p(d|x)$ for the predictive distribution.

The Bayesian approach to population pharmacokinetic/pharmacodynamic modeling

Jon Wakefield
Leon Aarons
Amy Racine-Poon

ABSTRACT It is one of the principal aims of drug development to discover, for a particular agent, the relationship between dose administered, drug concentrations in the body and efficacy/toxicity. Understanding this relationship leads to the determination of doses which are both effective and safe. Population pharmacokinetic/pharmacodynamic models provide an important aid to this understanding.
Pharmacokinetics considers the absorption, distribution and elimination over time of a drug and its metabolites. Pharmacokinetic data consist of drug concentrations along with (typically) known sampling times and known dosage regimens. A dosage regimen is defined by a route of administration and the sizes and timings of the doses. Pharmacodynamics considers the action of a drug on the body. Pharmacodynamic data consist of a response measure, for example blood pressure, a pain score or a clotting time, and a known dosage regimen. Population data arise when these quantities are measured on a group of individuals, along with subject-specific characteristics (covariates) such as age, sex or the level of a biological marker. When identical doses are administered to a group of individuals large between-individual variability in responses is frequently observed. The mechanisms which cause this variability are complex and include between-individual differences in both pharmacokinetic and pharmacodynamic parameters. The general aim of population studies then is to isolate and quantify the within- and between-individual sources of variability. The explanation of between-individual sources of variability in terms of known covariates is important as it has implications for the determination of dosage regimens for particular covariate-defined subpopulations.
In this chapter we describe the drug development process from a population pharmacokinetic/pharmacodynamic perspective. In particular we describe how the nature of the statistical analysis and the models that are used are modified as the type of data and the aims of the study change through the various phases of development. The Bayesian approach to population modeling is particularly appealing from a biological perspective as it allows informative prior distributions to be incorporated. These priors may arise from previous studies and/or from medical/biological considera-

tions. From an estimation standpoint a Bayesian approach is preferable because of the difficulties which a classical approach encounters due to the large numbers of parameters, the nonlinearity of the subject-specific models which are typically used and the large numbers of variance parameters.
We illustrate the population approach to drug development by describing a number of studies which were carried out by Ciba for a particular anti-clotting agent. We also present a detailed analysis for one of the studies.

1 Introduction

It is the aim of drug development to discover safe and efficacious doses for clinical use. Berry (1990) provides a description of the drug development process, and in particular the types of trial that are undertaken. Population pharmacokinetic/-pharmacodynamic (PK/PD) models provide a valuable aid to the drug development process by identifying sources of, and quantifying the remaining, variability in drug concentrations and response measures. 'Population' here makes explicit the fact that we wish to gain an understanding of the dose/concentration/response relationship across different groups as defined by covariates such as sex, age and weight. Statistically, population PK/PD models fit within the framework of nonlinear hierarchical models. These models contain many layers of assumptions and it is difficult to check the appropriateness of these assumptions, particularly those which concern unobservable quantities. Consequently it is essential that the modeling be informed by medical/pharmacological information, both in the form of the deterministic/stochastic parts of the model and also via priors for model parameters. In this paper we aim to describe how population PK/PD modeling is carried out in practice. We illustrate the approach using a study of the anti-coagulant drug REVASC^{TM}. This study was carried out in Phase II of the drug's development; we also describe five other studies which were carried out in Phase I and informed the analysis of the main study.

The structure of this chapter is as follows. In Section 2 we provide an overview of the population approach and drug development. In Section 3 we outline the role of PK/PD studies in drug development and in Section 4 describe the statistical models which are used for the PK/PD data of a single individual. In Section 5 we explicitly consider population models from a drug development perspective and in Section 6 describe a three stage hierarchical model which may be used for the analysis of population data. In Section 7 we describe a number of population PK/PD studies which were carried out for REVASC^{TM} and present a detailed analysis of one of the studies in particular. We finish by describing those areas of population PK/PD for which research is still required.

2 Overview of the population approach and drug development

The aim of a clinical development program for a new drug is to provide relevant information on the safety and efficacy of the compound so enabling the prescribing physician to optimally treat individual patients. Frequently, however, the dosing recommendations that emerge from such studies are found inappropriate and when individual dose adjustment is needed, the recommendations provided may be insufficiently informative to allow the adjustment to be undertaken in an optimal manner.

Pharmacokinetics has been defined as what the body does to a drug, and pharmacodynamics as what a drug does to the body. A PK model of a drug attempts to relate drug dosage to drug concentration, usually measured in blood or plasma, or to drug excretion, usually in urine. A pharmacodynamic model attempts to relate drug concentration, ideally at the site of action of the drug but more usually in blood, to some pharmacological effect. The aim of PK/PD modeling therefore is to combine the PK model of a drug with a pharmacodynamic model in order to relate dose to effect in a quantitative manner. To this end PK/PD modeling allows the separation of the factors that influence the inter-individual variability in pharmacokinetics from those that influence pharmacodynamics. Specifically the identification of these influential individual-specific variables (covariates) provides the basis for individual dose selection and hence better therapy. Covariates which are likely to affect PK/PD parameters include demographic variables (such as age and sex), biological information (such as the values of physiological markers), genetic information (phenotype and genotype), comedications, environmental factors and disease states. Figure 1 shows a schematic representation of the relationship between dose, concentration and response.

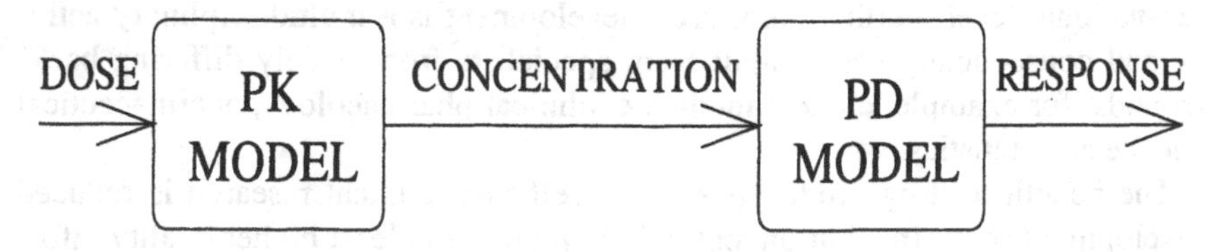

FIGURE 1. Schematic representation of the dose/concentration/response relationship.

An understanding of the dose-effect relationship is basic to the process of identifying those doses to be used in phase III clinical trials and in clinical practice. It is derived in a piece-wise fashion from the various phases of drug development. In establishing a dose-effect relationship, it is essential to also define the concentration-effect relationship. Appropriate mathematical models and the methodology to estimate the model parameters and to identify relevant covariates, that influence inter-individual variability in drug response, are required for both dose-concentration and concentration-effect modeling. In this sense dose-effect and PK/PD modeling are important tools in drug development. The size of study necessary to detect important covariate effects currently has not been established and will depend on the variability and magnitude of the parameter-covariate relationship, on the sampling design and attained compliance levels.

In early drug development, the use of clinical outcomes, especially in healthy volunteers, is often not feasible and intermediate or surrogate outcomes have to be chosen to replace the real clinical endpoint. A primary motivation for the use of a surrogate endpoint is the possible reduction in sample size or trial duration that can be expected when a rare or distal outcome is replaced by a more frequent or proximate outcome. Such reductions have important cost implications, and in some cases may be pivotal in regard to the trial feasibility. Other motivations include the possibility that true outcome measurement may be unduly invasive, uncomfortable or expensive. Furthermore, outcome events close in time to the treatment or intervention activities under study may be more readily interpreted than are more distal outcomes such as study subject death, which may be also confounded by secondary treatments or competing risks. This motivation, however, seems to relate more to the choice of principal endpoint than to the issue of defining a replacement, or surrogate, for a selected endpoint.

The history of drug development contains many examples of drugs that entered the market with suboptimal dosing regimens, usually erring on the side of initially recommended doses being considerably higher than necessary (Temple, 1989). So far as suboptimal dosing regimens are concerned an important issue is the needless over-exposure of patients to drug, with avoidable dose-dependent side-effects and hazard to patients. The potential exists, when initially-selected doses are too high, that dose-dependent adverse reactions might trigger the conclusion that a potentially valuable drug has too high a risk-benefit ratio to remain in the marketplace. Loss of a potentially valuable therapeutic agent can have adverse public health consequences. Drug development today is an international business and in particular, drug registration is being organized increasingly at a pan-national level. Furthermore, drug development is a multidisciplinary activity and consequently needs input from specialists from widely differing backgrounds: for example, clinical medicine, clinical pharmacology, pharmaceutical science and statistics.

The benefit to drug producers of more efficient clinical research is reduced development costs. In addition, potentially more, or at least higher quality information will be obtained on the drug. This will have direct effect on the consumer and the community as end-costs may well be reduced and the patient is more likely to obtain optimum therapy. Therefore there is a potential knock-on effect, in that quality of life is likely to be improved and time in hospital may be reduced. Consequently the efficient design and performance of clinical trials during the development of a new drug is in the interest of both the developer and consumer. Therefore society as a whole benefits, in that medicines are designed in an optimal manner and patients should achieve the maximum benefit from a drug with the minimum risk.

3 Pharmacokinetics/pharmacodynamics in drug development

In Section 2 it was stated that the aim of PK/PD modeling was to relate dose to effect in a quantitative manner. Broadly speaking drug development is divided into a preclinical phase and a clinical phase. The preclinical phase includes drug discovery and early development, mainly involving safety assessment studies in animals. Pharmacokinetic studies are carried out in at least two animal species, typically rat and dog. These studies will involve the development of assay methodology and the assessment of absorption, distribution and elimination of the drug in the animal. Preliminary formulation work is undertaken to define the formulations to be used in man. Extensive toxicology testing is initiated – studies which may be still ongoing during the clinical development phase – and the metabolic fate of the drug is defined from in vitro and in vivo studies. One of the major aims of the preclinical program is to ensure safety before first administration to man of the IND (Investigational New Drug) and indeed to define the doses to be administered in the first clinical studies.

Although the definitions vary from company to company and depend on the therapeutic area, the clinical development program is divided into three broad phases – I, II and III. Phase I studies involve the first administration of the IND to man and can be defined as tolerability studies. In general these studies will be carried out in normal, healthy (usually male) young volunteers. However in fields like oncology, because of the toxicity of the agents, initial studies will be in patients. Initial pharmacokinetic studies will involve single rising dose protocols followed by multiple dose studies. Bioavailability assessment and studies investigating the influence of food, comedication, gender and age may also be performed. These studies are usually performed in small groups (6 to 24) of individuals under careful experimental control. Typically, between 12 to 15 blood samples will be taken per subject per study, depending on the total amount of blood drawn per subject during the whole study.

Phase II studies are usually the first patient studies and one of their primary goals is to define the dose-response relationship. The response may be a surrogate marker, as clinical outcome, for example survival time, may be too difficult to implement in study designs at this stage. As with phase I studies, phase II studies tend to be in small groups of patients and are carried out under carefully controlled experimental conditions. Pharmacokinetic studies will again involve single and multiple dose administration, primarily to assess differences between the patient population and a normal, young volunteer population. In addition the effect of disease states, such as renal and hepatic insufficiency are studied at this stage. Other pharmacokinetic and pharmacodynamic studies performed in phase II include drug interaction studies, which are increasingly being directed from preclinical information, and comparative studies with competitive products. Often a Phase II study will collect sparse concentration/response data on all patients. Alternatively the majority of the centres participating in the study will collect sparse data on patients whilst the remainder will collect full profiles. This is because of logistical constraints on the centres.

Phase III studies are large scale clinical trials in the target population designed to demonstrate efficacy and tolerability of the drug. The dynamic marker is now a clinically relevant measure such as mortality or morbidity. Given that the primary purpose of these studies is the demonstration of efficacy, in the past relative little pharmacokinetic and pharmacodynamic information has been gathered during this phase. However now there is a growing realisation that it is important to collect pharmacokinetic and pharmacodynamic information during phase III. Logistically, however, it is difficult to obtain a large number of blood samples or dynamic measures in each patient. Consequently relatively sparse pharmacokinetic and pharmacodynamic data - sometimes only one sample per subject - is obtained during phase III studies. In addition phase III studies are frequently not carried out under the same degree of experimental control as phase I and II studies, and consequently the quality of data obtained during phase III may not be of the same standard as that obtained earlier in the development program, though the assay may have been improved. At the end of the phase III program, the company will apply to a regulatory authority for a NDA (New Drug Application) which will allow the drug to be marketed. Some data are collected post-marketing in what are called phase IV studies but in general any pharmacokinetic or pharmacodynamic studies are carried out by clinical investigation sites at this stage, rather than the company.

4 Mathematical models for individual PK/PD data

4.1 Compartmental models for PK data

In this section we describe the class of models that are used to model individual drug concentrations and responses. We let $y(t)$ denote the concentration of an individual at time t. The drug is introduced into the body via a particular route of administration. Common routes include intravenous bolus, intravenous infusion, oral or subcutaneous. An *intravenous bolus* is an instantaneous introduction of drug directly into the blood stream via an injection, an *intravenous infusion* is a constant introduction of drug directly into the blood stream over some specified period, and a *subcutaneous dose* is an injection beneath the skin and is, consequently, not directly into the bloodstream. After introduction the drug undergoes the processes of absorption, distribution and elimination. These processes are assumed to give rise to concentrations which vary with time in such a way that they may be modeled via a sum of exponentials form. A conceptual way of viewing the way in which these models arise is by considering the body as being modeled as a series of homogeneous compartments or pools. Gibaldi and Perrier (1982) describe the rationale behind this modeling and the types of model which may be appropriate for different drugs/routes of administration. For example a three-compartment model may nominally consist of blood, soft tissue and deep muscle compartments. Within each of these compartments the drug is assumed to have identical kinetic behaviour. The flow between compartments is typically described by a series of linear differential equations. For a p-compartment system denote the amount of drug in compartment i by $x_i(t)$, $i = 1, ..., p$. Then the rate of change of drug in compartment i may be modeled by

$$\frac{dx_i}{dt} = \sum_{j=0}^{p}(k_{ji}x_j - k_{ij}x_i) \tag{1}$$

where k_{ij} denotes the rate constant associated with flow from compartment i to compartment j. Compartment 0 denotes the outside environment from where the drug is administered and to where drug is eliminated. Under this model the rate of change of drug is assumed to be proportional to the amount of drug in the donor compartment. In general the rate of change of drug may be modeled via some other form. For example the rate of change may be taken to be independent of the amount of material in the donor compartment to give a zero-order equation. Alternatively the rate of flow from compartment i to compartment j may be described by a Michaelis-Menten type relationship:

$$\frac{-k_{ij1}x_i}{k_{ij2} + x_i}.$$

Note that for small x_i this equation is approximately linear whilst for larger values the rate of flow approaches a constant value. An advantage of a linear set of equations such as (1) is that it is straightforward to obtain expressions for the amounts of drug in the different compartments as a function of time. For other forms for which analytical solutions do not exist numerical integration is required.

For general PK modeling the compartmental system is a convenient visualisation but the compartments have no physiological interpretation. For physiological analyses, which typically use a large number of compartments to model the various organs and tissues of the body and only obtain data on a small number of individuals, greater interpretation is possible. However, parameter identification is limited due to the fact that samples can, in general, only be taken from one or two compartments. Gelman, Bois and Jiang (1996) consider the Bayesian approach to physiological modeling.

We now describe in some detail the simple two-compartment model illustrated in Figure 2. Compartments one and two may nominally be thought to represent blood and tissue, respectively. We suppose that a single intravenous bolus of drug of size D is given at time zero. We then have

$$\frac{dx_1}{dt} = k_{21}x_2 - k_{12}x_1 - k_{10}x_1 \tag{2}$$

and

$$\frac{dx_2}{dt} = k_{12}x_1 - k_{21}x_2. \tag{3}$$

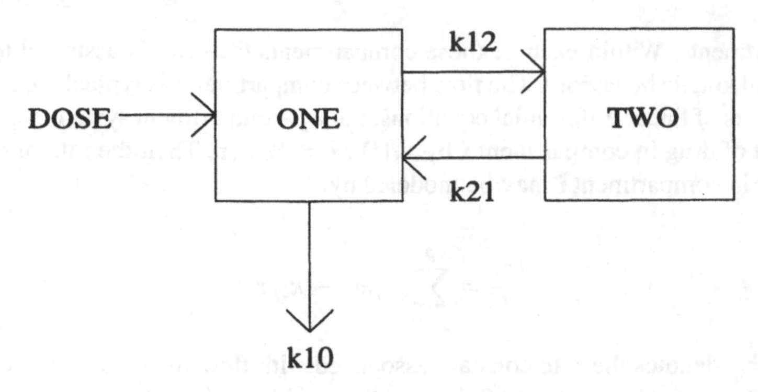

FIGURE 2. The two-compartment model with dose introduced into the plasma compartment.

Solving these two equations subject to $x(0) = D$ gives

$$x_1(t) = D\{A\exp(-\lambda_1 t) + (1-A)\exp(-\lambda_2 t)\}$$

and

$$x_2(t) = \frac{Dk_{12}}{(\lambda_2 - \lambda_1)}\{\exp(-\lambda_1 t) - \exp(-\lambda_2 t)\}$$

where

$$\lambda_1, \lambda_2 = (k_{10} + k_{12} + k_{21} \pm [(k_{10} + k_{12} + k_{21})^2 - 4k_{10}k_{21}]^{1/2})/2$$

and $A = (\lambda_1 - k_{21})/(\lambda_1 - \lambda_2)$. Figure 3 shows the concentrations of drug in each of the compartments versus time for particular choices of (k_{10}, k_{12}, k_{21}). We note that the amount of drug initially increases in compartment two during the distribution phase before elimination becomes the dominant process. Absorption into compartment one is immediate here because the drug was introduced as a bolus.

We typically observe drug *concentrations* of drug in compartment one and so we introduce a further parameter V_1 which is the *volume* of compartment one. Note that V_1 is a nominal volume in that its value may be much greater than the volume of blood in the body. This may be due to the binding of the drug to plasma proteins and/or compartment one containing well-perfused tissues, such as the heart and lungs. For this reason this parameter is often referred to as the *apparent volume of distribution*. We then have $y(t) = x_1(t)/V_1$ to give

$$y(t) = \frac{D}{V_1}\{A\exp(-\lambda_1 t) + (1-A)\exp(-\lambda_2 t)\}.$$

It is clear from the form of $y(t)$ that it is possible to work with many parameterizations. For example one may choose to work with the rate constants and volume as used above. There are certain parameters which are of fundamental interest

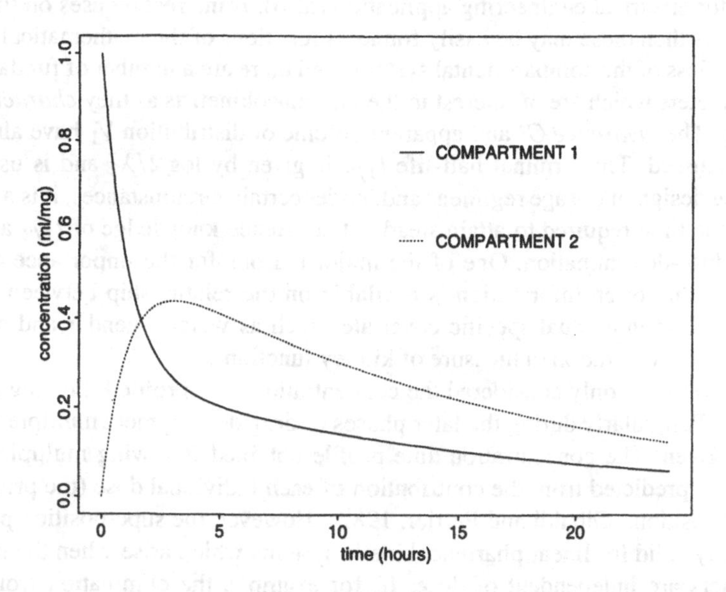

FIGURE 3. Concentration/time relationship for the system represented in Figure 2 with $D = 1, V = 1, k_{10} = 0.2, k_{12} = 0.5$ and $k_{21} = 0.3$.

and for which pharmacokinetists have prior knowledge. In particular *clearance* parameters define the volume of a compartment which is cleared of drug in unit time. For the model defined by (2) and (3) there are two clearance parameters, the clearance from the central compartment $Cl = V_1 k_{10}$ and the distribution clearance $Cl_d = k_{12}V_1 = k_{21}V_2$. The parameter Cl is defined in general (that is for any compartmental system) as D/AUC where AUC denotes the area under the concentration/time curve. Hence a *pharmacokinetic* parameter set would be given by (V_1, V_2, Cl, Cl_d) where each of the parameters is greater than zero.

The most obvious parameterization from a *statistical* perspective would consider the set $(A_1, A_2, \lambda_1, \lambda_2)$ where

$$y(t) = D[A_1 \exp(-\lambda_1 t) + A_2 \exp(-\lambda_2 t)]$$

with $A_1, A_2 > 0$ to ensure positive concentrations. Identifiability is imposed by assuming $\lambda_1 > \lambda_2 > 0$. The disadvantage of using this parameterization is that the parameters which the pharmacokineticist has prior information on, namely Cl and V_1, are nonlinear transforms of the mathematical set and hence it is not so straightforward to carry out covariate modeling (see Section 6.1).

The identifiability issues become worse as the number of compartments increases. It is in phase I studies that more complex PK modeling is carried out, and in these studies, as described in Section 3, covariate modeling is rarely carried out due to the homogeneity of the individuals. Therefore the mathematical parameterization is appealing here. Godfrey (1983) describes in detail a number of issues, including identifiability, relating to compartmental systems (not just for

PK but for electrical engineering applications also). If interest focuses on the PK parameters then these may be easily found as functions of the mathematical set.

Regardless of the compartmental system used there are a number of fundamental parameters which are of interest to the pharmacokinetists as they *characterise* the drug. The *clearance* Cl and apparent volume of distribution V_1 have already been discussed. The terminal half-life $t_{1/2}$ is given by $\log 2/\lambda_2$ and is used to guide the design of dosage regimens and, under certain circumstances, it is a measure of the time required to attain steady-state. Hence knowledge of $t_{1/2}$ allows simple dose-determination. One of the major reasons for the importance of Cl and V_1 is that often information is available on the relationship between these quantities and individual-specific covariates such as weight, gender and serum creatinine concentration (a measure of kidney function).

We have so far only considered the concentration-time profile following a single dose. Particularly during the later phases of drug development multiple doses will be given. The concentration/time profile obtained following multiple dosing can be predicted from the contribution of each individual dose (the principle of superposition, Gibaldi and Perrier, 1982). However, the superposition principle is only valid for linear pharmacokinetic systems which arise when the model parameters are independent of dose. If, for example, the elimination from the central compartment is described by a Michaelis-Menten relationship then this principle does not hold. If PK parameters change with either dose or time, the concentration-time profile at steady-state cannot simply be predicted by superposition and derived quantities such as AUC and maximum concentration increase either less than or greater than proportionally with dose. Phenytoin provides an example of a drug for which the AUC and maximum concentration relationship with dose increases greater than linearity. As described in Section 3 it is one of the aims of Phase I of drug development to explore the 'dose proportionality' issue. Therapeutically, drugs which display nonlinear pharmacokinetics can be difficult to prescribe.

We shall let θ denote the p_θ-dimensional vector of PK parameters, whether this be the pharmacokinetic or the mathematical parameterization. We assume that each of the elements of θ is defined on the whole real line. So for example for the two compartmental model described above with the mathematical parameterization we will have $\theta = (\log A_1, \log A_2, \log(\lambda_1 - \lambda_2), \log \lambda_2)$. The model for the concentration as a function of time t arising from a particular compartmental system and dosage regimen (times and sizes) will be denoted $f_1(D, \theta, t)$. The subscript 1 acknowledges the hierarchy we shall develop in Section 6: here we are modeling a first-stage variable, namely the concentration.

Both the compartmental system and differential equations described above are obviously huge simplification of the body and the processes which act on a drug molecule. However, the resultant concentration-time models have been found empirically to mimic the true concentrations seen.

As mentioned in Section 3, in Phase I of drug development a small number of (usually) healthy volunteers, typically less than 20, are given single doses and subsequently a large number (relative to later phases) of blood samples are taken in order to determine drug concentrations. Traditional, or non-compartmental PK analyses have used simple numerical methods such as the trapezium rule to estimate the AUC (and hence Cl). Other summary measures are similarly computed. During this phase the individuals in general form a very homogeneous group. Due to the abundance of data per individual the statistical modeling at this stage can often utilise two- or three-compartment models. The number of compartments is generally chosen as the maximum number which can be fitted. As drug development proceeds the individuals become more heterogeneous and the data per individual becomes sparser. Consequently, traditionally, simpler compartmental systems are assumed (though see the discussion of Sections 4.3 and 7.3).

4.2 PK error models

We wish to model the differences between the observed concentrations y and the modeled concentrations f_1. These error terms, which describe the difference, will represent not only assay precision but also model misspecification. Particularly at the later stages of drug development this latter quantity is likely to be substantial. The compartmental systems described above are large simplifications. Also the environment within which the trials are performed is important. Early trials will be carried out within the drug company and so recorded sampling times and administered doses are likely to be accurate, in a hospital environment this is far less likely. Since the model misspecification is likely to vary smoothly as a function of time one might expect the errors to be correlated but such models have rarely been fitted due to the sparsity of data. Davidian and Giltinan (1995) describe such models.

Within the PK literature a frequently-used error model is

$$y = f_1(D, \theta, t) + \epsilon^y$$

with the error terms ϵ^y taken as independent, zero mean normal random variables with variance $f_1(D, \theta, t)^\gamma \sigma_y^2$ with $\gamma \geq 0$. Wakefield (1996a) uses such a model and treats γ as an unknown parameter. Davidian and Giltinan (1993) discuss various error models. The case $\gamma = 2$ results in a constant coefficient of variation which often mimics assay precision. An alternative to this model which produces an approximate constant coefficient of variation, and is both statistically and biologically appealing, is given by

$$\log y = \log f_1(D, \theta, t) + \epsilon^y$$

with the ϵ^y now taken as independent zero mean normal random variables with variance σ_y^2.

It is standard practice in analytical chemistry to set a lower limit of quantification below which concentrations are discarded and treated as missing. One common strategy is to accept only concentrations which have a coefficient of variation (based on a calibration curve) below 20%. Note that although zeros are recorded for those concentrations below the lower limit of quantification this limit is typically known and so these observations may be treated as censored in the conventional way (Wakefield and Racine-Poon, 1995). Ignoring these zeros leads to estimation bias since the observations are clearly not missing at random.

4.3 PD models

Pharmacodynamic responses can either be quantal or continuous. An example of a quantal response is pain score recorded in an analgesic trial. For example Sheiner, Beal and Dunne (1997) analyse data recorded on a four-point scale using a mixed-effects approach with a cumulative logit model.

Here we shall concentrate on continuous responses since these are relevant to the application which we shall describe in Section 7. Let $z(t)$ denote the response (or some transform of the response such as the logarithm) measured at time t. A general model is then given by

$$z = g_1(\theta, \phi, t) + \epsilon^z \tag{4}$$

with the ϵ^z's taken as independent zero mean normal random variables with variance σ_z^2. So the modeled response depends not only on a p_ϕ-dimensional vector of PD parameters ϕ, but also on the PK parameters θ since the concentration of drug is assumed to influence the resultant response. Here and throughout the paper we use $p(\cdot)$ as a generic notation for a probability density function. The distribution of y and z is given by

$$p(y, z|\theta, \phi) = p(y|\theta)p(z|\theta, \phi).$$

We are therefore making the assumption that z is conditionally independent of y, given θ. In particular this means that the errors in concentrations and responses at the same time point are independent.

Equation (4) represents an ideal analysis in which inference for the PK parameters will also be influenced by the response data. Often when PK/PD data are jointly modeled this relationship is simplified. Suppose for example that the response at time t depends directly on the true concentration at time t, $f_1(D, \theta, t)$. The easiest way to model this relationship would be to simply plug in the observed concentrations. This approach does not acknowledge that the concentrations are measured with error and so we are faced with an errors-in-variables problem. If it were assumed that the relationship between response and concentration were linear then, from standard errors-in-variables results (Carroll, Ruppert and Stefanski, 1995), one might expect the coefficient describing the strength of this relationship to be *attenuated*. If the response/concentration relationship is nonlinear then it is not clear what the effect of the errors-in-variables will be. A more refined procedure would be to obtain estimates of the PK parameters $\hat{\theta}$ and then regress the response on $f_1(D, \hat{\theta}, t)$. The disadvantage of this is that the PD data will not inform the estimation of θ and the variability in the PD parameters will be underestimated since the variability in $\hat{\theta}$ has not been acknowledged. One possibility here would be to allow fluctuations about the observed concentration y via a simple errors-in-variables approach, without explicitly considering a PK model. In some instances one of these more simplistic approaches may be followed when we *a priori* know that our PK and PK/PD models have been greatly simplified due to sparsity of data. For example when we use a one-compartment PK model for a two-compartment system we know that the high peak concentrations will be systematically underestimated whereas concentrations near the end of the administration period (trough concentrations) will be systematically overestimated. In this case fitting the full PK/PD model simultaneously will lead to bias in the PD parameter estimation. Similarly when the PK/PD link or PD model is oversimplified bias will be expected when the joint PK/PD model is used.

Another consequence of taking one of the simpler approaches is that it is not so straightforward to determine the response consequences of specific doses since we have not carried out joint estimation.

Often the pharmacological response is not directly related to concentration but lags behind the concentration/time profile. This can occur either when the drug acts as a precursor which gives rise to the response or when the site of action of the drug is not blood but a tissue into which the drug must distribute before eliciting an effect. A discussion of models used to handle this situation is given in Holford and Sheiner (1981).

5 Population PK/PD

Population pharmacokinetics can be defined as the study of the variability in outcome between individuals when standard dosage regimens are administered, the outcome usually being the plasma concentration-time profile. It is of interest to both quantify the variability of this response within the population and to account for that variability in terms of patient specific variables, such as age, sex, weight, disease state, etc. The current interest in population pharmacokinetics stems from the concern that the pharmacokinetics of new drugs are not studied in relevant populations, that is, patients likely to receive the drug, at an early enough stage in the drug development program. In particular the United States Food and Drug Administration (FDA) (Temple, 1983; Temple, 1985), and others (Abernathy and Azarnoff, 1990), are concerned that the pharmacokinetics of a new drug should be studied in elderly populations 'so that physicians will have sufficient information to use drugs properly in their older patients' (Food and Drug Administration, 1989). The obvious time to collect PK information on the target population is during large-scale clinical trials carried out during Phase III of the drug development program. However, because of logistic and ethical reasons, it is improbable that intensive experimentation can be carried out on each and every patient. At best one could hope for one or two blood samples per patient. Therefore traditional PK analysis, which involves the determination of an individual's PK parameters, is untenable (see Section 4.1). Instead data analysis techniques that focus on the central tendency of the PK information and are capable of utilizing very sparse data have to be employed. Population pharmacokinetics has come to mean the design, execution and analysis of PK studies involving sparse data, although the data analysis techniques can be applied to data obtained from conventional PK studies. The label *population pharmacokinetics* is perhaps unfortunate but it does convey the sentiment that interest is focused on the population rather than the individual.

The implementation of a population approach within the drug development program is the subject of much debate (Colburn, 1989). It has been suggested that a 'population screen' be employed in which blood samples are taken from a wide range of individuals so that, essentially, the concentration-time profile is covered within the population (Sheiner and Benet, 1985). The advantages of such an approach are that data are collected in the target population, an assessment of the variability within the population is obtained and, hopefully, the factors that control that variability may be discovered. Although the goals are indisputable, much concern has been expressed about the logistics of implementation of a population approach during Phase III of the drug development program. A common statement that is made is 'garbage in, garbage out'. It should be pointed out that this criticism can be made of any poorly designed or executed study, not just population studies. However there are particular problems associated with Phase III studies due to the fact that they are in general multicentre and in many cases in an outpatient setting. Compliance and accurate timing of both dosing and sampling are clearly critical issues. At present there are virtually no guidelines on experimental design, both in terms of sample timing and subject numbers, particularly within subgroups. Similarly, we have no idea of the power of the approach to detect important inter-subject differences and overall there is no hard data on the cost-to-benefit ratio. At present we are still on the learning curve.

To date most population modeling in phase III has been concerned with pharmacokinetic data, although some mixed effects modeling has been carried out on continuous response variables obtained from small scale controlled clinical trials (Pitsiu, Parker, Aarons and Rowland, 1993). Sheiner, Beal and Dunne (1997) provide an elegant application of mixed effects modeling to pain score data obtained from an analgesic study. The current status of population PK/PD is simply a reflection of the lack of PK/PD modeling in drug development. However with the growing realization of the importance of PK/PD modeling to drug development, we can expect to see an increase in the activity of population PK/PD and population PD.

We finally note that information obtained from population analyses during drug development may also be used after the drug has been marketed for the *individualization* of dosage regimens. Such on-line therapeutic drug monitoring is often carried out in a hospital environment when patients with an acute condition require careful tuning of doses. Typically only sparse data are available and so prior information, in the form of estimates of the parameters of the population distribution, are required. Vozeh and Steimer (1985) describe this technique and Wakefield (1994, 1996*b*) Bayesian sampling-based approaches.

6 Statistical aspects of population PK/PD

6.1 A three stage hierarchical model

It is natural to model population PK/PD data hierarchically since this allows the variability in concentrations/responses to be separated into within-individual and between-individual components. The joint PK/PD aspect is important since one can see how much of the variability in responses is due to variability in concentrations. At the first stage of the hierarchy *within-individual* modeling of the PK/PD data are carried out. The data of each individual are assumed to follow the same underlying functional form (that is f_1 and g_1) but the parameters of these forms take different values for different individuals. At the second stage the *between-individual* differences are modeled by assuming that the individual PK and PD parameters arise, after accounting for individual-specific covariates, from a common probability distributions. An excellent review of the statistical aspects of population PK/PD modeling is given by Davidian and Giltinan (1995), Yuh *et al* (1994) provide a bibliography of population PK applications and methodological approaches, and Steimer *et al* (1994) give an introduction to population PK/PD ideas in drug development.

First stage model

Let y_{ij} denote the concentration (in blood, plasma or urine) of individual i at time t_{ij}, $i = 1, ..., N$, $j = 1, ..., n_i$. Then the model for concentration is given by

$$\log y_{ij} = \log f_1(D_i, \theta_{ij}, t_{ij}) + \epsilon_{ij}^y$$

where ϵ_{ij}^y are independent and identically distributed as $N(0, \sigma_y^2)$ and θ_{ij} represent the PK parameters of individual i at time t_{ij}. We index by both i and j because at the second stage of the hierarchy we will model the PK parameters as functions of individual covariates which may be time-varying.

Let z_{ij} denote the response of individual i at time t_{ij}, $i = 1, ..., N$, $j = 1, ..., n_i$. For notational simplicity, and because it is the case in the studies described in the next section it is assumed that the PK and PD data were collected from the same individuals and at the same time points. In general this is not a requirement, PK and PD data on the same individual need not be collected at the same time points. In fact some individuals may just contribute PK or PD data though to learn about the PK/PD relationship there must be some individuals with both types of data. We may also have PD data which is a cumulative response over some time period.

The response model is given by

$$z_{ij} = g_1(\theta, \phi_{ij}, t_{ij}) + \epsilon_{ij}^z$$

where ϵ_{ij}^z are independent and identically distributed as $N(0, \sigma_z^2)$.

Second stage model

Let μ_θ and Σ_θ represent location and scale parameters for the PK parameters and μ_ϕ and Σ_ϕ represent location and scale parameters for the PD parameters. In the most general case μ_θ and μ_ϕ are vectors of length p_θ and p_ϕ, respectively and Σ_θ and Σ_ϕ are matrices of dimension $p_\theta \times p_\theta$ and $p_\phi \times p_\phi$, respectively. In some cases, particularly with sparse data, some elements of θ and/or ϕ may be taken to be fixed effects, thereby reducing the dimensionality of Σ_θ and/or Σ_ϕ. In other instances the off-diagonal elements may be taken to be zero. A priori, however, enough empirical evidence has been gathered to follow the most general model – simplifications result when the data do not inform about particular elements of θ and ϕ. In such cases a Bayesian may stay with the general model and place informative priors on parameters for which there is little information in the data. In Section 7 such a model is used.

Let X_{ij} denote a vector of individual-specific covariates measured at time t_{ij}. The relationship between the PK parameters and covariates is then modeled as

$$\theta_{ij} = f_2(\mu_\theta, X_{ij}) + \delta_i^\theta.$$

The 'error' terms δ_i^θ model the difference between the PK parameters of individual i at time t_{ij} and that predicted by the covariate model f_2. Note that δ_i is independent of time, that is j. In general the function f_2 is taken to be linear, that is $f_2(\mu_\theta, X_{ij}) = X_{ij}\mu_\theta$, see for example Davidian and Gallant (1993) and Wakefield (1996*a*).

The choice of which components of X to include for each element of θ is a very difficult problem. It is essentially a multiple regression in which the dependent variable θ is multivariate and unobserved. Covariate selection in population PK analyses has often proceeded via a forward selection procedure, with graphical plots driving the inclusion of covariates (Maitre, Buhrer, Thomson and Stanski, 1991), and analytical approximations being used to evaluate the likelihood of the population parameters. Using these plots to drive a forward selection procedure is hazardous in several ways. It is difficult to decide on which of a discrete or a continuous covariate is to be included first. Adding more than one covariate at a time is also dangerous since the covariates themselves are correlated. Choosing whether a particular covariate should be included for different elements of the θ vector is also difficult. A covariate may be needed for volume only (say) but may appear to be necessary for clearance also because of correlation in the random effects distribution. The problems of forward selection are well-documented, Miller (1990). The approach described above, which is typical of those taken in population PK analyses, has a number of additional disadvantages. The procedure is computationally expensive as each of the analyses requires a numerical minimization to be carried out. The test statistic has only an asymptotic distribution under the null hypothesis and the statistic itself is being calculated via an analytic approximation. It is difficult to assess the impact of these approximations for a particular dataset. Simulating the test statistic under the null hypothesis in order to obtain a Monte Carlo significance level is computationally prohibitive. Most importantly, the significance of the covariates is only being judged in a *statistical*, and not in a *clinical*, sense. For example suppose the statistical significance levels of two regressors are of similar size. This does not imply that each has equal clinical significance. The regression coefficient for a particular covariate may be statistically significant but may have little effect on the clinical aim. Conversely a regressor which does not attain statistical significance may have a clinical impact. Wakefield and Bennett (1996) consider this problem in a specific application when the design of a dosage regimen was the objective.

We argue that the choice of covariates is driven by the aim of the analysis and prior knowledge, both from previous studies with the same or similar drugs and from pharmacokinetists. To this end it is relevant to recall the meaning of the components of θ. Although the mathematical parameterization is the most convenient for overcoming identifiability problems, it is not the parameterization which is natural for covariate modeling. In particular, pharmacokinetists frequently have prior beliefs concerning the effect of the covariates on clearance and volume. In general the logarithms of clearance and volume are not linear functions of the mathematical parameters and so a general strategy is difficult to determine.

The relationship between the PD parameters and covariates is similarly modeled as

$$\phi_{ij} = g_2(\mu_\phi, X_{ij}) + \delta_i^\phi.$$

The interpretation/difficulties with this modeling are similar to those for the PK modeling.

We now consider the joint distribution of $\delta^\theta, \delta^\phi$. The basic assumption we make is that

$$p(\delta^\theta, \delta^\phi | \mu_\theta, \Sigma_\theta, \mu_\phi, \Sigma_\phi) = p(\delta^\theta | \mu_\theta, \Sigma_\theta) p(\delta^\phi | \mu_\phi, \Sigma_\phi).$$

A number of approaches have been suggested for the modeling of $p(\delta^\theta | \mu_\theta, \Sigma_\theta)$. The choice of this distribution is far less obvious than the choice of the distribution at the first stage for which assay data are available. At the second stage the distribution of the PK parameters in their untransformed form, after conditioning on covariates, has often been found empirically to be skewed. The obvious choice is a p_θ-dimensional lognormal distribution. This was used implicitly by Beal and Sheiner (1982) and explicitly by Lindstrom and Bates (1990). Wakefield, Smith, Racine-Poon and Gelfand (1994) used a log student distribution in an attempt to robustify the modeling.

Figure 4 represents a directed graph of the joint PK/PD model of a single individual.

There is always the possibility that a covariate which is an important predictor has not been measured, however. This will result in a mixture distribution. Consequently a number of authors have avoided a specific parametric family. Mallet (1986) proposed to allow the distribution to be completely general and obtained a nonparametric maximum likelihood (NPML) estimate (which is discrete). The NPML method suffers from a number of disadvantages. The first stage variance, σ_y^2, is assumed known and no standard errors are produced. No estimates of uncertainty are produced by the method and so it is not clear whether interesting

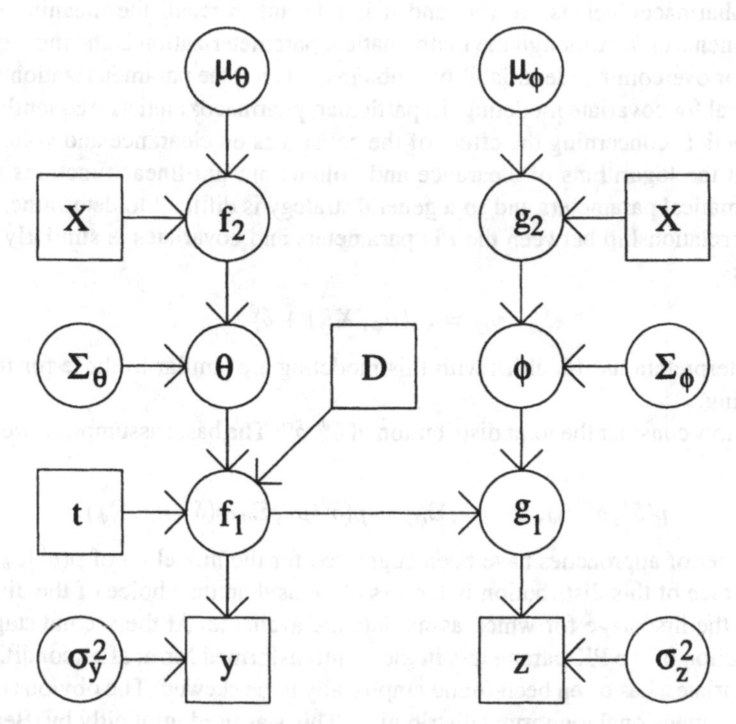

FIGURE 4. Directed graph for the population PK/PD model. Observable quantities are placed in square boxes, *X* represents individual-specific covariates, *D* dose, *t* time, *y* concentration and *z* response.

features such as bimodalities are real or merely artifacts of the data. Mentré and Mallet (1994) extended the original method to allow a completely general specification at this stage and obtain a nonparametric maximum likelihood estimate of the relationship between θ and X by assuming that the joint distribution of (θ, X) is being estimated. The difficulty with this is that a probability distribution of high dimension is being estimated with very sparse data.

Davidian and Gallant (1993) propose a semiparametric method in which the assumption of smoothness is made at the second stage and the random effects are modeled as arising from a distribution that is represented as a normal times a polynomial.

Wakefield and Walker (1997), Muller and Rosner (1997), and Walker and Wakefield (1998) take a Bayesian nonparametric approach using Dirichlet process priors.

Third stage model

At the third stage of the hierarchy priors are specified for the population parameters μ_θ, μ_ϕ, Σ_θ, Σ_ϕ, σ_y^2 and σ_z^2. We first note that those elements of μ_θ, μ_ϕ that correspond to nonlinear parameters at the first stage, proper priors are required in order to guarantee propriety of the posterior distribution. For the same reason we also need proper priors for Σ_θ and Σ_ϕ. For computational convenience (see next Section) normal prior distributions $N(c_\theta, C_\theta)$, $N(c_\phi, C_\phi)$ are assumed for μ_θ, μ_ϕ, respectively; Wishart distributions $W(r_\theta, (r_\theta R_\theta)^{-1})$, $W(r_\phi, (r_\phi R_\phi)^{-1})$ are assumed for Σ_θ^{-1}, Σ_ϕ^{-1}, respectively; and gamma distributions $Ga(a_\theta^2/A_\theta, a_\theta/A_\theta)$, $Ga(a_\phi^2/A_\phi, a_\phi/A_\phi)$, for σ_y^{-2} and σ_z^{-2}, respectively. The values c_θ, c_ϕ, R_θ, R_ϕ, a_θ and a_ϕ represent prior guesses for the relevant parameters and C_θ, C_ϕ, r_θ, r_ϕ, A_θ and A_ϕ represent the precision of these estimates.

6.2 *Implementation details*

We let $y_i = (y_{i1}, ..., y_{in_i})$, $z_i = (z_{i1}, ..., z_{in_i})$, $y = (y_1, ..., y_N)$, $z = (z_1, ..., z_N)$, $\delta^\theta = (\delta_1^\theta, ..., \delta_N^\theta)$ and $\delta^\phi = (\delta_1^\phi, ..., \delta_N^\phi)$. In general interest will focus upon the population parameters μ_θ, Σ_θ, μ_ϕ, Σ_ϕ. The posterior for these quantities is given by

$$p(\mu_\theta, \Sigma_\theta, \mu_\phi, \Sigma_\phi \mid y, z) = \int p(\mu_\theta, \Sigma_\theta, \mu_\phi, \Sigma_\phi, \sigma_y^2, \sigma_z^2 \mid y, z) d\sigma_y^2 d\sigma_z^2$$

where

$$\begin{aligned} p(\mu_\theta, \Sigma_\theta, \mu_\phi, \Sigma_\phi, \sigma_y^2, \sigma_z^2 \mid y, z) \quad &\propto \quad p(y, z \mid \mu_\theta, \Sigma_\theta, \mu_\phi, \Sigma_\phi, \sigma_y^2, \sigma_z^2) \\ &\times \quad p(\mu_\theta, \Sigma_\theta, \mu_\phi, \Sigma_\phi, \sigma_y^2, \sigma_z^2) \qquad (1) \end{aligned}$$

and, utilizing the conditional independencies discussed previously and represented in Figure 4,

$$p(y, z \mid \mu_\theta, \Sigma_\theta, \mu_\phi, \Sigma_\phi, \sigma_y^2, \sigma_z^2) = \prod_{i=1}^{N} \int p(y_i \mid \delta_i^\theta, \mu_\theta, \sigma_y) \times$$

$$p(z_i \mid \delta_i^\theta, \mu_\theta, \delta_i^\phi, \mu_\phi, \sigma_z) \times p(\delta_i^\theta \mid \Sigma_\theta) p(\delta_i^\phi \mid \Sigma_\phi) d\delta_i^\theta d\delta_i^\phi.$$

Due to the nonlinear nature of the first stage PK and PD models these integrals are analytically intractable. Racine-Poon and Wakefield (1998) provide a review of estimation techniques for population PK modeling. Within the pharmaceutical industry the package NONMEM (Beal and Sheiner, 1993) is often used. Within this package various analytical approximations are available for calculating the required integrals and the resultant likelihood is then numerically maximised. In particular the First Order (FO), First Order Conditional Estimation (FOCE) and Laplacian methods may be used. The FO method takes a linearisation about the population mean whilst the FOCE is essentially the method of Lindstrom and Bates (1990). Pinheiro and Bates (1995) compare various analytical and numerical approximations. Inference is made through asymptotic arguments. Unfortunately it is, in general, difficult to assess the adequacy of the approximation and the appropriateness of the asymptotics.

Mallet's NPML method is computationally fast but the numerical maximisation is sensitive to the starting ranges specified for the parameters. The parameters of the normal times polynomial form of Davidian and Gallant (1993) are estimated by maximum likelihood. Again the numerical maximization is difficult and it is recommended that a 'wave of starting values be used'. The likelihood is evaluated using numerical or Monte Carlo integration. These and other techniques are described in Davidian and Giltinan (1995) and Vonesh and Chinchilli (1997). Wakefield and Walker (1997) compare the methods of Mallet and Davidian and Gallant with a nonparametric Bayesian method. For the latter convergence and prior sensitivity are important issues.

The first Bayesian approach to population PK modeling is due to Racine-Poon (1985) using an approximation to the first stage model that allows an EM-type algorithm to be used. Markov chain Monte Carlo (Smith and Roberts, 1993) was first used in the context of population PK models by Wakefield, Smith, Racine-Poon and Gelfand (1994) who used a Gibbs sampler. For the priors described in the previous section and the first stage models outlined in Section 4 all of the required conditional distributions assume known forms apart from the random effects. Wakefield, Smith, Racine-Poon and Gelfand (1994) used the ratio-of-uniforms method (Wakefield, Gelfand and Smith, 1991) for generating from these distributions whilst Gilks, Best and Tan (1995) used the Metropolis Adaptive Rejection Sampling (MARS) algorithm (see also Gilks, Neil, Best and Tan, 1997). Both of these methods are only practically feasible for univariate conditional distributions which can lead to slow mixing. Subsequently a Hastings-Metropolis (Metropolis *et al*, 1953, Hastings, 1970) step has been used to generate from the multivariate conditional (Bennett, Racine-Poon and Wakefield, 1996). This is the approach that was used for the analysis in Section 7.

7 Population PK/PD of Recombinant Hirudin

7.1 Pharmacology of hirudin

REVASC^{TM} (recombinant desulfatohirudin, produced in yeast by recombinant DNA technology), is a selective thrombin inhibitor nearly identical in protein structure to the natural leech anticoagulant, hirudin variant 1. Preclinical studies showed REVASC^{TM} to be a highly specific thrombin inhibitor and a potent anticoagulant. REVASC^{TM} demonstrated positive results in various preclinical models of venous, arterial and foreign surface thrombosis, as well as arterial thrombolysis. In vitro, the dose-response curve is shallower than that of unfractionated heparin (a competitor drug) enabling it to be used over a much greater concentration range (more than two orders of magnitude) and to higher levels of anticoagulation (up to three times control APTT (activated partial thromboplastin time). Based upon these results, REVASC^{TM} was recommended for clinical development for prevention of venous and arterial thrombosis and reduction of reocclusion during and after thrombolysis. In acute and chronic toxicological studies, REVASC^{TM} displayed no evidence of systemic toxicity. Dose is limited by the extension of the pharmacological activity, that is inhibition of blood coagulation.

In a biotransformation study in six healthy male volunteers following a single intravenous infusion of 3mg/kg/h for 6 hours, two metabolites were detected in trace amounts in the urine. They represented only about 7% of the administered dose. No other metabolites were detected. The observed data suggested that REVASC^{TM} is eliminated and metabolized in the kidney. Total unitary excretion of intact REVASC^{TM} amounted to 40–50% of the total dose. The apparent renal clearance was approximately 80 mL/min. The difference between the renal clearance and the creatinine clearance (120 mL/min) may be accounted for by resorption. The non-renal clearance, about 20% of the administered dose, was essentially due to this resorption process and an additional undefined non-renal clearance. After intravenous bolus administration, the steady-state volume of distribution of REVASC^{TM} was about 18L. It may be concluded from this small steady-state volume that REVASC^{TM} does not enter cells but stays in the circulation and extracellular space. The central volume of distribution is about 5L, approximately the plasma volume. This also suggested that a minimum of two compartments is required to model the pharmacokinetics of REVASC^{TM}. The pharmacological effect of REVASC^{TM} is based on its binding to thrombin. The thrombin-hirudin complex, which is much too large to be cleared by renal filtration, is present in trace amounts in plasma in healthy volunteers. The binding to thrombin is therefore not a major route of elimination for healthy volunteers.

Human studies demonstrated that after both intravenous and subcutaneous administration, areas under the plasma concentration curves are dose proportional.

Mean terminal elimination elimination half-lives were similar (2–2.5 hr), and dose-independent. Clearance is primarily renal and dose-independent (2.1–2.5 ml/min/kg). Similarly, the steady state volume of distribution is dose-independent (0.25 L/kg). Following intravenous administration, plasma concentrations are best described by a three exponential model. The absolute bioavailability of subcutaneous doses is approximately 100% and half-lives were similar for the subcutaneous and intravenous routes. Total REVASC^{TM} plasma clearance is reduced in the elderly (1.62 ml/min/kg) compared to young subjects. REVASC^{TM} is assayed by a sandwich ELISA method, based on the recognition of the analyte by a monoclonal capture antibody and a polyclonal signal antibody. The assay measures free drug, *i.e.* not drug bound to thrombin.

Phase I clinical trials in normal volunteers and patients demonstrated that the APTT response to REVASC^{TM} was dose-dependent and consistent after intravenous or subcutaneous dosing. The details of the studies which demonstrated these findings are given in the next section. APTT is strongly correlated with REVASC^{TM} plasma levels. There is no evidence for delay in the onset of anticoagulant effect or from an extended duration of activity beyond the actual presence of REVASC^{TM} in plasma.

We first describe each of the study objectives and designs, and then the models that were used for analysis.

7.2 *Study description*

In this section we describe five different phase I studies that were carried out for REVASC^{TM}. The first study is reported in Wakefield and Racine-Poon (1995) whilst studies 2–5 are described in Racine-Poon and Wakefield (1996). We then describe a Phase II study upon which we shall concentrate.

Study 1

This study consisted of four groups of four volunteers, each of whom received on day 1 of the study an intravenous bolus injection of REVASC^{TM}. Each group received one of the doses 0.01, 0.03, 0.05 and 0.1mg/kg. Approximately 28 hours later they received an intravenous bolus followed by a constant rate intravenous infusion of heparin over 24 hours during which they received a second intravenous bolus injection of REVASC^{TM} at the same doses as given on day 1. Hence the concentration of heparin was held constant over the second administration of REVASC^{TM}. Heparin also inhibits blood clotting forming and it was envisaged that the two drugs may be co-administered and so it was of interest to see whether there was any interaction between them. Blood samples were taken at 0 hours, immediately before the bolus of drug, and between 0.08 hours and 24 hours subsequently. Sixteen blood samples were taken in total from each individual on each of days 1 and 3. For this study, and each of the studies we now describe, the plasma concentration of REVASC^{TM} of each sample was determined along with the clotting measure APTT.

Study 2

The aim of this study was to investigate the absorption characteristics of the subcutaneous administration route. To this end it is of interest to determine whether the PK model and the PK/PD relationship were altered by the administration route. In terms of the pharmacokinetics it was of interest to see whether the distribution and elimination phases are the same for the two administration routes and also to estimate the *bioavailability*. This quantity is defined as the proportion of the administered dose which is absorbed into the bloodstream and is obviously of importance when decisions concerning the size of the administered dose are being considered. The bioavailability can only be estimated when concentration data are available from the intravenous route *and* from the alternative route of administration (subcutaneous here). This is because we do not measure absolute levels of drug and from concentrations alone the total amount of drug that has been absorbed cannot be determined. Sixteen young healthy male volunteers were administered each of the intravenous and subcutaneous doses in a two-period randomized crossover open (that is not 'blind') study. Note that to carry out a blind trial here two injections would have to be given at each administration time, one intravenous and one subcutaneous. Eight of the volunteers received, on separate days, doses of 0.3mg/kg of body weight, one intravenous and one subcutaneous, while the remaining eight volunteers each received 0.5mg/kg in both forms. Notice, therefore, that the doses depend on the weight of the individual. On each of the days of administration blood samples were taken during a 24-hour period following administration. Figure 5 shows the concentration/time data for four of the individuals and both routes of administration.

Study 3

The purpose of this study was to investigate the possible modification of the PK profile and the PK/PD relationship by the covariate age. The study was carried out with 12 elderly volunteers who had a normal range of renal function. Each of the six volunteers received a single subcutaneous application. Six of the volunteers received a dose of 0.3mg/kg ,and the remaining volunteers received 0.5mg/kg. Blood samples were taken for a 24-hour period after application.

Study 4

The purpose of this study was to investigate the PK and PK/PD relationship after repeated subcutaneous dosing. Eight healthy volunteers received subcutaneous dosing twice daily for six consecutive days. Four of the volunteers received 0.3mg/kg every 12 hours, whereas the remaining four received 0.5mg/kg every 12 hours. Two blood samples were taken in each administration period, one immediately before administration (trough) and one 3 hours after administration (peak). On the seventh day three samples were collected following the morning dose. This study was reported in Verstraete *et al* (1993).

Study 5

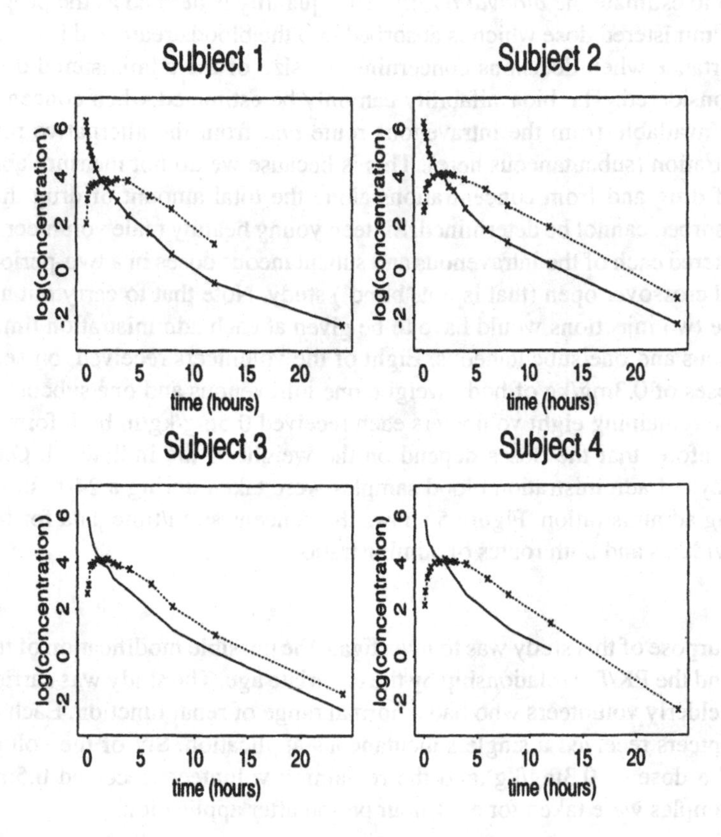

FIGURE 5. Concentration/time data for four individuals from Study 2 who received doses of 0.3mg/kg. The solid line denotes concentrations from the intravenous bolus route, the dashed from the subcutaneous.

The purpose of this study was to investigate the PK and PK/PD relationship at equilibrium. If an intravenous infusion is continued indefinitely, a constant concentration/time profile results as the amount of drug entering the body at any instant of time is equal to the amount being eliminated from the body. In practice, equilibrium is reached after some finite time that depends on the elimination rate constant. Eight young male volunteers received a constant infusion over 72 hours, four at a rate of 0.2mg/kg per hour and four at 0.3mg/kg per hour. Blood samples were taken during the infusion period and for 12 hours after the infusion was complete.

Study 6

In this phase II dose-finding trial, 301 orthopaedic patients undergoing total hip replacement therapy were administered subcutaneous doses twice daily, following the operation, with REVASC^{TM}. More details of this study may be found in Eriksson *et al* (1996). The principal aim of the trial was to find an appropriate dose for later trials. Consequently each of the patients received repeated administration of one of the doses 10, 15 or 20mg. Patients in another arm of the trial received unfractionated heparin as an active comparator treatment; we do not consider these patients in this analysis. Four days after the operation two blood samples were taken after the morning dose and the concentration of REVASC^{TM} was determined. By this time *steady-state* concentration levels had been attained. This means that the amount of drug absorbed daily is equal to the amount eliminated. In the trial protocol it was specified that the samples should be taken approximately 2 and 10 hours after the dose. Due to the hospital environment the actual times, which were recorded, varied quite considerably about these nominal, that is protocol pre-determined, values. The nominal 12 hour doses ranged between 10 and 14 hours. The 2 and 12 hour measurements correspond, approximately, to peak and trough concentrations. Figure 6 shows all of the (log) concentration data plotted versus time. We can see the spread in the sampling times around 2 and 12 hours. We see that there is a tendency for the concentrations corresponding to the 20mg/10mg doses to be high/low, but there is a large amount of variability.

At each of the sampling times the APTT was also determined, along with an additional baseline measure before the randomization. These data are plotted in Figure 7.

Patient specific covariates were also recorded. These were: dose (mg), weight (kg), height (cm), age (years), gender, smoking and serum creatinine concentration (mg/dl).

For this study the clinical outcomes of primary interest were incidence of deep vein thrombosis and safety, not the clotting measure APTT though the latter was measured. APTT is linked to bleeding, which is one potential well-known complication of anticoagulant drugs such as heparins and REVASC^{TM}, and to efficacy and so to illustrate PK/PD modeling we shall investigate the relationship between individual PK and PD parameters and covariates and the links to safety/efficacy.

Table 1 summarises the 6 studies which we consider here and illustrates how the study design (numbers of subjects and observations) changes through drug

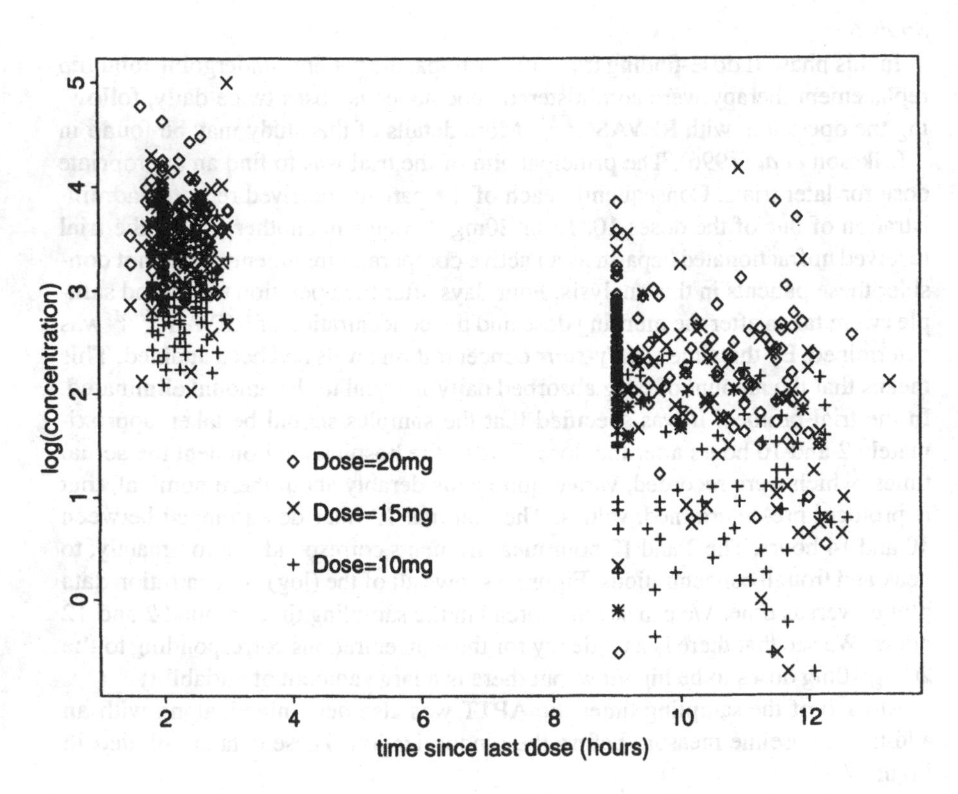

FIGURE 6. Concentration/time data for the 301 individuals of study 6.

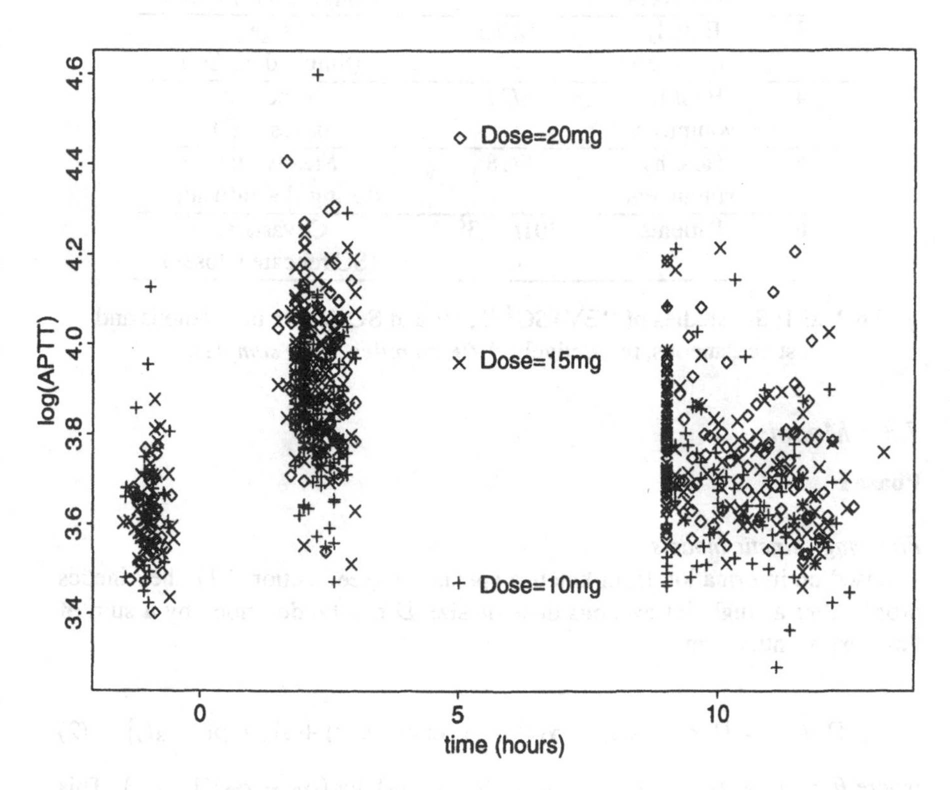

FIGURE 7. APTT/time data for the 301 individuals of Study 6. The times of the APTT measurements before time zero are nominal, that is planned, times, they are pre-dose measurements.

development.

Study	Study population	No. of subjects/ samples	Aim of study, effect of:
1	Healthy volunteers	16 and 16/16	Coadministration (Single dose IV bolus)
2	Healthy volunteers	16 and 16/16	Absorption route (Single dose IV/SC)
3	Elderly volunteers	12/16	Age (Single dose SC)
4	Healthy volunteers	8/27	Repeated doses (SC)
5	Healthy volunteers	8/18	Steady-state dosing (IV Infusion)
6	Patients	301/2[a],3[b]	Covariates (SC repeated doses)

TABLE 1: Six studies of REVASC^{TM}, IV and SC denote intravenous and subcutaneous, respectively. [a] *PK samples,* [b] *PD samples.*

7.3 Models

Phase I

First stage kinetic models

Based on information from kinetics specialists (see Section 7.1), the kinetics profile after a single intravenous dose of size D can be described by a sum of three exponential terms,

$$f_1(D, \theta, t) = D \times [A_1 \exp(-\alpha_1 t) + A_2 \exp(-\alpha_2 t) + A_3 \exp(-\alpha_3 t)] \quad (2)$$

where $\theta = (\log A_1, \log A_2, \log A_3, \log(\alpha_1 - \alpha_2), \log(\alpha_2 - \alpha_3), \log \alpha_3)$. This parameterization ensures identifiability via the constraint $\alpha_1 > \alpha_2 > \alpha_3 > 0$, and positive predicted concentrations via $A_1, A_2, A_3 > 0$. Equation (2) implies dose proportionality (see Section 3), an assumption that was validated in previous studies in which different doses were administered to the same volunteers in a randomized order.

For the infusion route of administration which was used for Study 5 the disposition model described by (2) was convolved with a constant rate of input to produce predicted concentrations (Gibaldi and Perrier, 1982). No additional parameters are required in this case.

A first-order absorption model was assumed for the subcutaneous data. The concentration after a single dose is therefore given by:

$$f_1(D,\theta,t) = F_a D k_a \quad \times \quad \left[A_1 \frac{\exp(-\alpha_1 t)}{(k_a - \alpha_1)} + A_2 \frac{\exp(-\alpha_2 t)}{(k_a - \alpha_2)} + A_3 \frac{\exp(-\alpha_3 t)}{(k_a - \alpha_3)} - B\exp(-k_a t)\right] \quad (3)$$

where $0 < F_a < 1$ is the fraction of the dose absorbed, $k_a > 0$ is the absorption rate constant and

$$B = \frac{A_1}{(k_a - \alpha_1)} + \frac{A_2}{(k_a - \alpha_2)} + \frac{A_3}{(k_a - \alpha_3)}$$

For this model we have $\theta = (\text{logit}F_a, \log k_a, \log A_1, \log A_2, \log A_3, \log(\alpha_1 - \alpha_2), \log(\alpha_2 - \alpha_3), \log \alpha_3)$.

Note that F_a can only be estimated when both intravenous and subcutaneous doses are given to the same individual, which is true for Study 2. Note also that the data from a subcutaneous dose alone would have been insufficient to estimate all three distribution phases due to the confounding with the absorption phase. From Figure 5 it is possible from the intravenous data to identify three straight lines corresponding to the three phases of distribution; for the subcutaneous data this is not possible, however. With the additional intravenous data all three phases may be estimated, however.

For the repeated dosing study (Study 4) the principle of superposition was used to sum the contributions of single doses of the form (3), so the predicted concentrations were of the form

$$\sum_{s=0}^{m} f_1(D,\theta,t+s\tau)$$

where $m+1$ doses are administered prior to time t and τ is the dosing interval.

Let y_{ijk} denote the jth concentration on the ith individual with $k = 1$ representing the intravenous experiments and $k = 2$ the subcutaneous experiment. The error models were taken to be of the form

$$\log y_{ijk} = \log f_1(D_i, \theta_i, t_{ijk}) + \epsilon^y_{ijk}$$

where ϵ^y_{ijk} are independent and identically distributed as $N(0, \sigma^2_{yk})$. We use different error variances because although the same assay technique is used (see Section 7.1), we would expect greater model misspecification for the subcutaneous experiment as the assumption of first order absorption is only approximately true.

First stage dynamic models

To aid in the identification of the effect/concentration relationship the observed APTT measurements were plotted against drug concentrations. The consecutive observations were joined to identify whether *hysteresis loops* were evident (Holford and Sheiner, 1981). Hysteresis essentially relates to the constancy over time of the effect/concentration relationship. The absence of hysteresis indicates that regardless of the time the effect corresponding to a given concentration is constant. Such loops may occur for a number of reasons, for example when the effect lags behind the concentration which could occur when the site of action is not the bloodstream. In this case an effect compartment model (Holford and Sheiner, 1981) may be assumed. No such loops were evident and an instantaneous relationship between drug concentration and dynamic effect was therefore assumed (see Racine-Poon and Wakefield, 1996, Figure 2). In Wakefield and Racine-Poon (1995) a linear relationship was assumed between APTT and concentration. This was later refined and the reciprocal of APTT was modeled via an inhibition sigmoid Emax model (Holford and Sheiner, 1981). This model is given by:

$$g_1(\theta, \phi, t) = APTT_0 \left[\frac{f_1(d, \theta, t)^{1/2} + IC_{50}^{1/2}}{IC_{50}^{1/2}} \right] \quad (4)$$

where $APPT_0 > 0$ is the baseline APTT (that is, the APTT when no drug is present) and $IC_{50} > 0$ is a parameter that corresponds to the concentration which would be required to produce 50% inhibition. We take $\phi = (\log APTT_0, \log IC_{50})$. Initially the Hill coefficient (the power within equation 4) was estimated for the 16 individuals of Study 2 using maximum likelihood. Its value was found to be close to 0.5 and subsequently it's value was fixed at this value. The fit of this model to the data of Studies 3–5 confirmed that this was reasonable.

Let z_{ij} denote the jth APTT measurement on the ith individual. The error model was taken to be of the form

$$\log z_{ij} = \log g_1(\theta_i, \phi_i, t_{ijk}) + \epsilon_{ij}^z$$

where ϵ_{ijk}^z are independent and identically distributed as $N(0, \sigma_z^2)$.

Second stage kinetic model

The initial analyses did not include covariates. Once the analysis of Study 3 had been completed, estimates of the random effects of the PK parameters were plotted with plotting symbols representing the young/old covariate (the only covariate for these phase I studies). No differences between the two groups were detected. A t distribution was taken as the population distribution.

Second stage dynamic model

Again a no-covariate analysis was carried out. For Study 3 when the random effects were plotted it was found that the $\log APTT_0$ and $\log IC_{50}$ parameters were both much lower for the elderly volunteers. Consequently we allowed different locations for the population distributions of the young and old. A normal distribution was taken as the population distribution.

Third stage priors

For both the PK and the PD parameters relatively non-informative priors were assumed for all of the population parameters.

Phase II

First stage kinetic model

The sparsity of the data here lead to the fitting of a simple one compartment model with first-order absorption and elimination (Section 4.1); we discuss the implications of this simplification later in this section. Under the principle of superposition the predicted concentration is the sum of the contributions from the previous two doses which are of size D. The first stage form is therefore given by

$$f_1(D,\theta,t) = \frac{Dk_a}{V(k_a - Cl/V)} \sum_{l=1}^{2} \left(\frac{\exp(-Cl/V(t-t_l))}{1-\exp(-Cl\Delta_l/V)} \right. \quad (5)$$

$$\left. - \frac{\exp(-k_a(t-t_l))}{1-\exp(-k_a\Delta_l)} \right) \quad (6)$$

where t_l is the time since dose l was given, $l = 1, 2$, Δ_l is the dosing interval for the two doses and $\theta = (\log Cl, \log V, \log k_a)$. The planned dosing intervals were $\Delta_1 = 10$ hours and $\Delta_2 = 14$ hours though the actual times varied for different patients. As with the Phase I data lognormal errors were assumed.

One of the primary objectives here is to obtain accurate estimates of clearance and volume for the patient population. One aim then is to investigate the relationship between these parameters and the covariates. Clearance and volume are also important because they may be used to determine maintenance and loading doses for a patient. When multiple doses are given the loading dose is given by

$$\text{Loading Dose} = \frac{\bar{C} \times V}{F} \quad (7)$$

where $\bar{C}$ represents the desired average concentration, V the volume and F the bioavailability. The dosing rate as a function of clearance is given by $\text{DR}(Cl) = \bar{C} \times Cl/F$ where Cl is the clearance. Then, for a known dosing interval Δ, we have

$$\text{Maintenance Dose} = \text{DR}(Cl) \times \Delta. \quad (8)$$

Given the importance of clearance and volume we wished to examine the effect on the estimation of these quantities of using a simplified model (recall for the phase I data a three compartment model was used). To this end ten simulations of the following kind were carried out. Using the three compartment model and population parameter estimates from Studies 1–5 we simulated data of the same design as Study 6 (301 individuals with two time points each). The 'true' clearance and volume are then functions of the seven parameters of the three compartment model. These were compared with the estimates obtained from the simulated data when analysed using a one compartment model. It was found that both the clearance and volume were well-estimated but the peak and trough concentrations were under- and over-estimated, respectively. This is reflected in Figure 8 which shows a plot of the residuals versus fitted concentrations for the Study 6 data. We see the effect of assuming a simple one compartment model – low (trough) concentrations are over-estimated whilst high (peak) concentrations are under-estimated. Figure 9 shows the (unstandardised) residuals from the PK analysis. These and all other residuals were calculated by simply substituting the posterior means of the parameters into $f_1(D, \theta, t)$, strictly we should evaluate the posterior distribution of the residual itself.

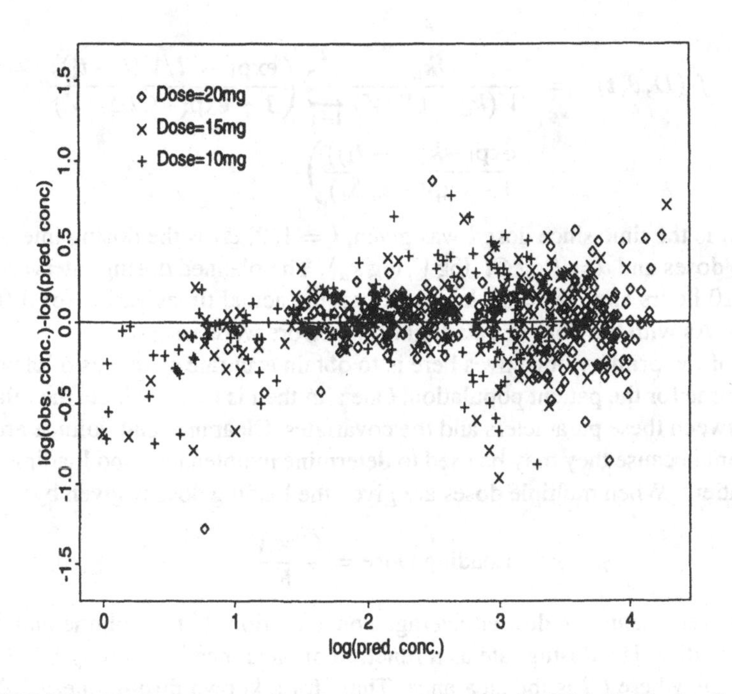

FIGURE 8. Residuals versus predicted concentrations for the 301 individuals of study 6.

First stage dynamic model

The model (4) was used to model the APTT though now we substitute *fitted* concentrations into the relationship, i.e. we do not account for the uncertainty in these fitted values. We do this because the concentration data are very sparse and we believe that the PD parameter estimation could be biased due to the feedback from the PK data. If we had complete faith in our PK/PD link model this would not be a problem but here the model is empirically-derived. Again lognormal errors were assumed.

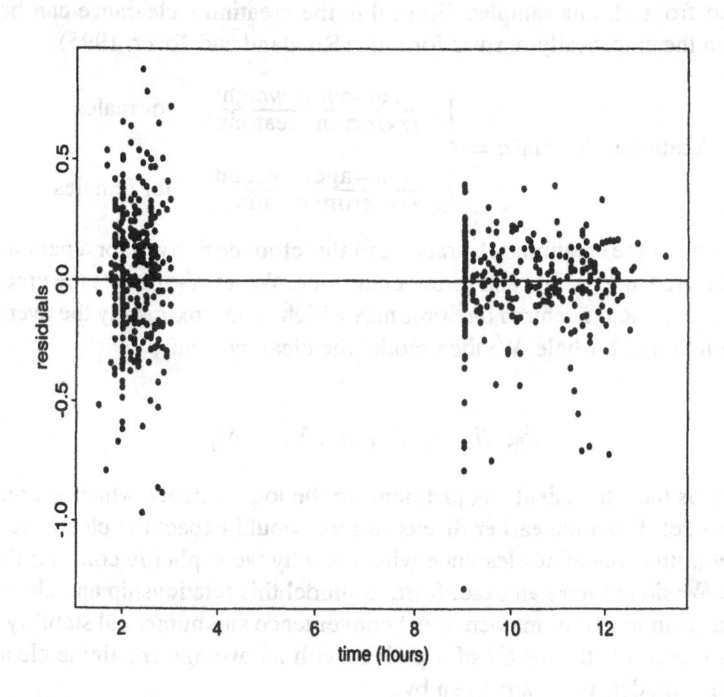

FIGURE 9. Residuals versus time for the population PK analysis of Study 6.

Second stage kinetic model

Recall from Section 6 that we have

$$\theta_i = f_2(\mu_\theta, X_i) + \delta_i^\theta.$$

where here $\theta_i = (\log k_{ai}, \log Cl, \log V_i)$ and we have suppressed the subscript j because none of the variables were time-varying. As noted in Section 6 it is very difficult to carry out covariate modeling, in particular because the dependent variable θ_i is multivariate and unobserved. For these data we have two random effects and six covariates and hence $(2^6)^2 = 4096$ potential models, even before we consider transforms of the covariates such as body surface area, which is a function of height and weight and is sometimes used as a predictor of volume.

To form the covariate model for θ_i we use biological information as far as possible. Here we describe an initial model, in the next Section further relationships will be investigated using graphical techniques. We know that REVASC^{TM} is eliminated primarily through the kidneys via filtration (Section 7.1). In fact we know from studies in healthy volunteers in which urine is collected that the drug is 98% renally eliminated. Hence if we could obtain the filtration rate of the kidneys, as determined by the creatinine concentration in urine this would give us a highly informative covariate for the clearance. Unfortunately this covariate is logistically difficult to measure so instead the serum creatinine concentration is measured from plasma samples. From this the creatinine clearance can be estimated via the empirically-derived formula (Rowland and Tozer, 1995):

$$\text{Creatinine Clearance} = \begin{cases} \frac{(140-\text{age})\times\text{weight}}{72\times\text{serum creatinine}} & \text{for males} \\ \\ \frac{(140-\text{age})\times\text{weight}}{85\times\text{serum creatinine}} & \text{for females.} \end{cases}$$

An estimate of the creatinine clearance can therefore be derived for a patient from their measured age, weight and serum creatinine. We let X_{1i} denote the creatinine clearance of patient i centred on 80ml/min which is approximately the average in the population as a whole. We then model the clearance via:

$$\log Cl_i = \mu_{\theta 0} + \mu_{\theta 1} X_{1i} + \delta^{\theta}_{1i} \tag{9}$$

where δ^{θ}_{1i} is the i-th patients 'adjustment' to the log clearance which is predicted by the model. From the earlier discussion we would expect the clearance to increase with the creatinine clearance which is why we explicitly consider this parameter. We do not have an exact form to model this relationship and choose the loglinear form in (9) for mathematical convenience and numerical stability. Note that $\mu_{\theta 0}$ represents the $\log Cl$ of a patient with an average creatinine clearance. Alternative models to (9) are given by:

$$Cl_i = \mu_{\theta 0} + \mu_{\theta 1} X_{1i} + \delta^{\theta}_{1i}$$

and

$$Cl_i = \mu_{\theta 0} \times X_{1i}^{\mu_{\theta 1}} + \delta^{\theta}_{1i},$$

both of which have been used in the PK literature. Within the observed limit of concentration/creatinine clearance one cannot assess which is the better model from a purely statistical perspective. This choice will be essential, however, if one wishes to extrapolate beyond the range of the observed data. Hence if such extrapolation is required the choice of model must be based on pharmacological/biological information. In the patient population it is not the case that 98% is renally eliminated. More thrombin is produced, hirudin binds to thrombin, and is then eliminated through the liver (hepatic elimination). Hence we have a significant non-renal route of elimination. We will return to this in the analysis stage when we address sensitivity.

We turn now to the volume parameter. On simple physiological grounds we expect the volume to increase with weight or body mass. We denote the weight of the i-th patient, centred by 80kg, by X_{2i} and again fit the simple loglinear model

$$\log V_i = \mu_{\theta 2} + \mu_{\theta 3} X_{2i} + \delta^{\theta}_{2i} \qquad (10)$$

where δ^{θ}_{2i} is the i-th patients 'adjustment' to the log volume which is predicted by the model.

For the third parameter k_a there is very little information in the absorption phase since there were no early sampling times. An initial model

$$\log k_{ai} = \mu_{\theta 4} + \delta^{\theta}_{3i}$$

was assumed with very tight priors being placed on $\mu_{\theta 4}$ and the variance of the δ^{θ}_{3i}. The variance on the prior for $\mu_{\theta 4}$ corresponded to a change in k_a of $\pm 5\%$. The posterior for $\mu_{\theta 4}$ from this analysis was located at an unreasonably high value, however, because even with a strong prior the data were too sparse too discount the intravenous model which corresponds to an infinite k_a. The same behaviour occurred when k_a was treated as a fixed effect, again with a tight prior. Two final analyses were carried out. In the first of these k_a was allowed to take the single value $\exp(-1.2)$. This latter was chosen from Studies 2–4 where the posterior mean of the population $\log k_a$ was -1.2. In the second analysis k_a was allowed to take one of five discrete values, centred on $\exp(-1.2)$ and with a spread of $\pm 5\%$.

We assume that the pair corresponding to volume and clearance $\delta^{\theta}_i = (\delta^{\theta}_{1i}, \delta^{\theta}_{2i})$ arise from the zero mean bivariate normal distribution with variance-covariance matrix Σ^{θ}.

Second stage dynamic model

We initially assume no covariate relationships, ie

$$\begin{aligned} \log APTT_{0i} &= \mu_{\phi 0} + \delta^{\phi}_{1i} \\ \log IC_{50i} &= \mu_{\phi 1} + \delta^{\phi}_{2i} \end{aligned} \qquad (11)$$

with $\delta^{\phi}_i = (\delta_{1i}, \delta_{2i}) \sim N(0, \Sigma^{\phi})$. Although age was used as a covariate in earlier studies, for those studies we did not have measures of creatinine clearance (which are functions of age). Consequently the need for age in the covariate model for the PD parameters could have been due to the fact that we did not account for creatinine clearance in the PK second stage model.

Third stage kinetic model

Here we specify prior distributions for $\mu_{\theta} = (\mu_{\theta 0}, \mu_{\theta 1}, \mu_{\theta 2}, \mu_{\theta 3})$, Σ^{θ} and σ^2_y.

Studies 1–5 were all in healthy volunteers and a three compartment model was used. Extrapolation from healthy volunteers to patients is a risky enterprise. REVASC^{TM} binds to thrombin and after an operation more thrombin is produced by the body. This leads to less free drug being present in the plasma and so increases both the apparent volume of distribution and the clearance.

If we have a compartment model with more than one compartment then there are two volume parameters, the volume of the central compartment and the volume at steady-state. At steady-state all of the compartments are in equilibrium and the volume relates total drug in the body to the concentration in the central compartment (which is the same as the concentration in all other compartments since we are at equilibrium). With a three compartment model there is a specific formula to evaluate the volume at steady state, Gibaldi and Perrier, (1982), p. 215. The clearance can also be readily calculated as the dose divided by the area under the concentration/time curve.

On the basis of the studies using the three compartment model we can obtain prior distributions for $\mu_{\theta 0}$ (clearance intercept) and $\mu_{\theta 2}$ (volume intercept) based on predictions from the Phase I studies. There were also three additional studies which we have not described in which weight and creatinine clearance were measured. Hence prior estimates of the regressors describing these relationships in (9) and (10) were also obtained. This was done by first simulating a large number of individuals from the posterior distribution of the population parameters. For each of these individuals we calculate the clearance and the volume at steady-state and the mean and the variance/covariance matrices of these quantities was then evaluated to give $c_\theta = (2.41, 0.0101, 3.56, 0.095)$. The variance of the prior C_θ was taken to be a diagonal matrix with diagonal elements 0.09. The prior estimate for $\mu_{\theta 0}$ corresponds to a clearance value of 11litres/hour for the patient population. The prior estimate of Σ_θ, R_θ was taken to be a diagonal matrix with diagonal elements 0.04. These values correspond to a coefficient of variation on clearance and volume of 20% which is typical for studies such as these. We take $r_\theta = 2$, which is the smallest value that gives a proper prior, and choose $A_\theta = a_\theta = 0$, with 602 observations there are sufficient information in the data to estimate σ_y^2.

Third stage dynamic model

As a prior mean for μ_δ we use studies 1–5 to obtain the value $c_\delta = (3.5, 5.0)$ with C_δ diagonal with variances 1.0. We take R_δ as diagonal with diagonal elements 0.04 which again corresponds to 20% coefficient of variation for $APTT_0$ and IC_{50}.

7.4 Summary of analyses of studies 1–5

Studies 1–5 provided various information which was subsequently used in the analysis of Study 6. These studies were therefore vital to the analysis of Study 6 because the data in that study were very sparse. In particular it was found that the bioavailability was equal to one and there was dose proportionality so the principle of superposition could be assumed. The instantaneous PK/PD relationship was also found to be appropriate for the three compartment model. As described above the third stage priors for Study 6 were also informed by the earlier studies.

7.5 *Analysis of Study 6*

The ratio of APTT post-administration to baseline APTT may be taken as a surrogate for the possibility of a bleed. In particular it was desirable to keep this ratio to less than 1.5 and in the observed data of this study this was always the case.

We first examine the adequacy of the model via various diagnostics. Figure 10 displays the first stage residuals, $\log z_{ij} - \log g_1(\theta, \phi, t)$, plotted versus time. We see that residuals at the first time are slightly more positive whilst at the final time point they are more negative. One possible reason for this is the bias in the predicted concentrations. To investigate this we carried out an analysis in which the observed concentrations were used to regress the response on but this made little difference to the residual plots or the posterior distributions of the population parameters. Another possibility to explain this lack of fit would be to allow the Hill coefficient to be an unknown parameter rather than the value of 0.5 which was that used for the healthy volunteer Studies 2–5.

Figure 11 plots the first stage PD residuals versus the individual covariates – patterns in this plot might indicate that we have missed a second stage covariate relationship. Here there appear to be no patterns.

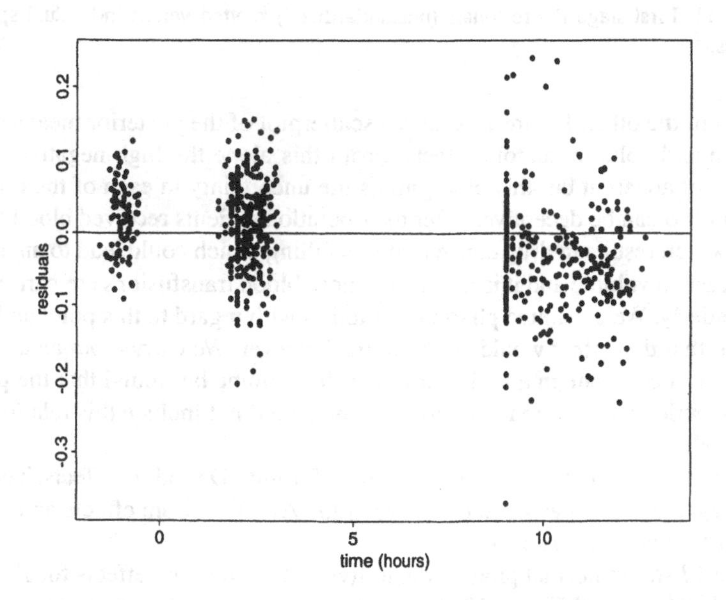

FIGURE 10. First stage PD residuals (unstandardized) plotted versus time.

Figures 12 and 13 display the posterior means of the PK random effects δ^{θ}_{i1}, $\delta^{\theta}_{i2}, i = 1, ..., 301$ versus the individual covariates. There appears to be an association between age and both the clearance and volume random effects. There is negative correlation (posterior median -0.71, 90% highest posterior interval -0.44,-0.85) in the population distribution of clearance and volume, however, so inclusion of age for one of these may lead to a disappearance of the apparent as-

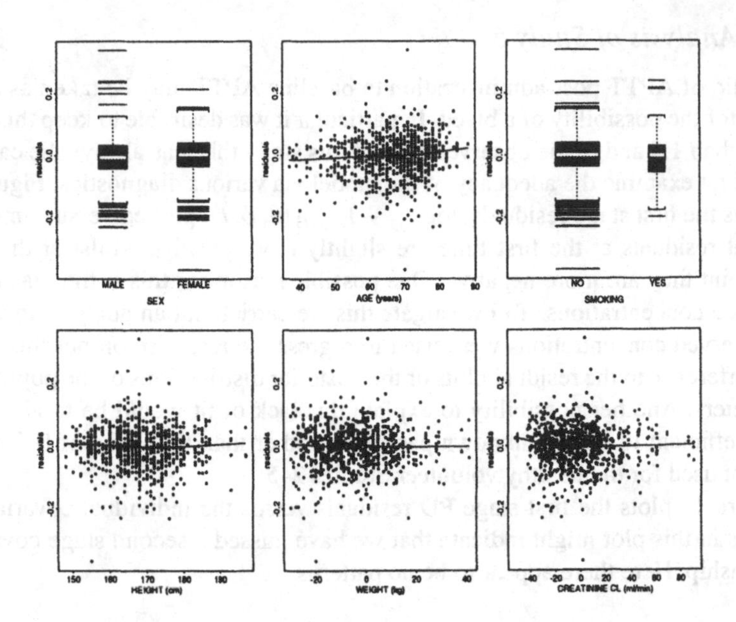

FIGURE 11. First stage PD residuals (unstandardized) plotted versus individual specific covariates.

sociation in the other. Figure 14 shows a scatterplot of the posterior means of the clearance and volume random effects. From this alone the high negative correlation is not apparent but this plot ignores the uncertainty in each of the random effects and so can be deceptive. After the operation patients received blood transfusions which results in drug concentrations falling which could lead to an apparent increase in volume and it is likely that more blood transfusions are carried out for the elderly. We consulted pharmacokinetists with regard to this point and they indicated that this effect would not be large, however. We carried out an analysis with age as a covariate in a loglinear model for volume but found that the posterior distribution was centred close to zero and so did not include this relationship in our model.

Figures 15 and 16 show the covariate plots for the PD random effects. There is some suggestion of an association between the APTT random effects and height but otherwise no obvious patterns.

Figure 17 shows normal plots of the individual PD random effects for $APTT_0$ (left) and IC_{50} (right) Figure 18 shows a bivariate plot of the posterior means of these same quantities. The normality assumption appears viable although there is some evidence of skewness in the second stage distribution of the $ATTT_0$ random effects. These plots should be interpreted with caution, however, as estimates and not observations are being plotted (Lange and Ryan, 1989).

Figure 19 shows the posterior distributions of the PK population mean parameters. Each of these distributions are well-behaved apart from $\mu_{\theta 3}$ which is highly skewed. We found that the Markov chain was slow mixing and a large number

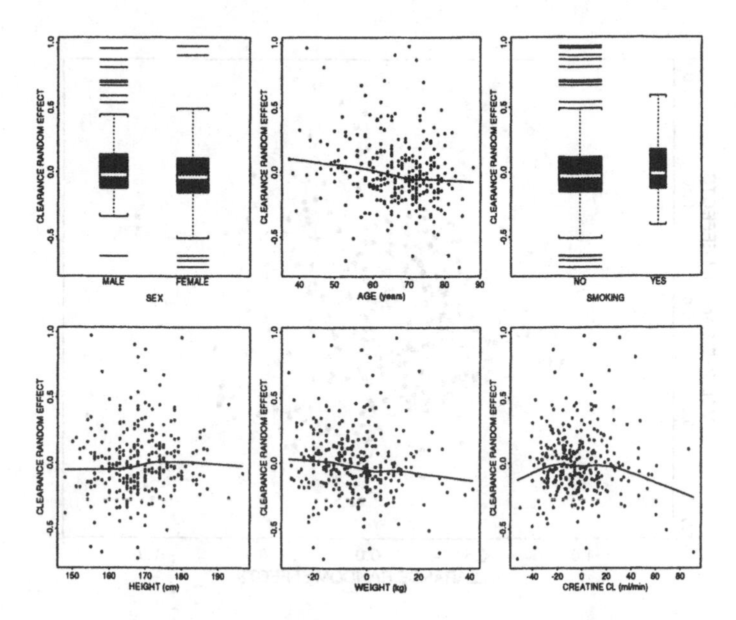

FIGURE 12. Second stage residuals (random effects) for clearance plotted versus individual specific covariates.

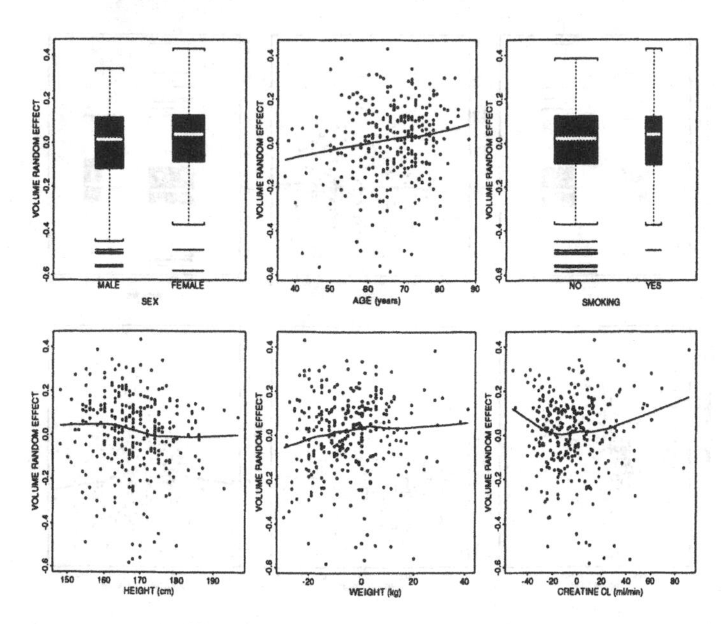

FIGURE 13. Second stage residuals (random effects) for volume plotted versus individual specific covariates.

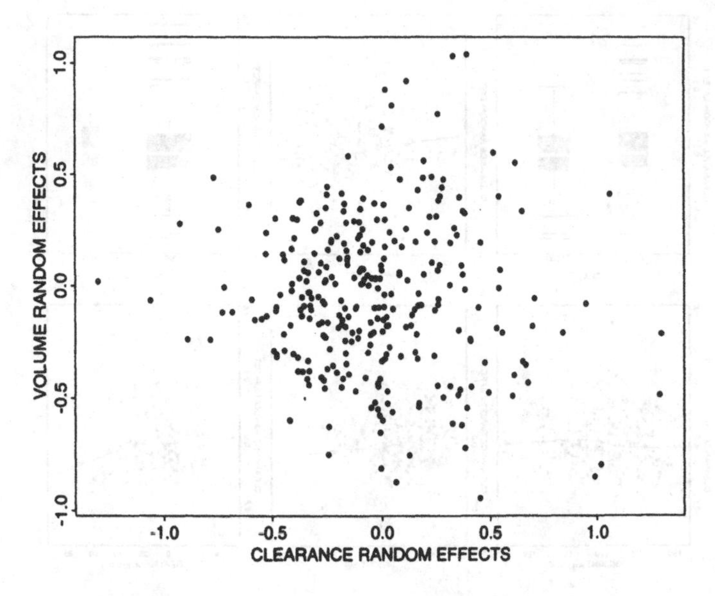

FIGURE 14. Posterior means of clearance random effects plotted versus posterior means of volume random effects.

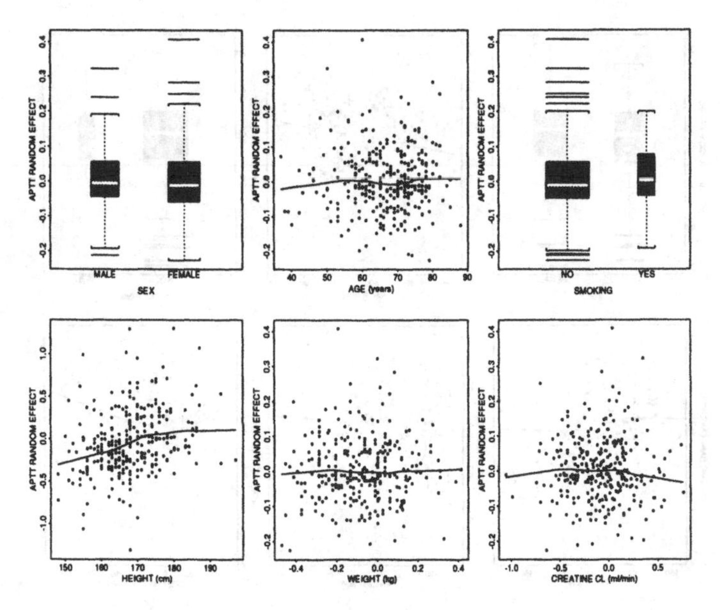

FIGURE 15. Second stage residuals (random effects) for $APTT_0$ plotted versus individual specific covariates.

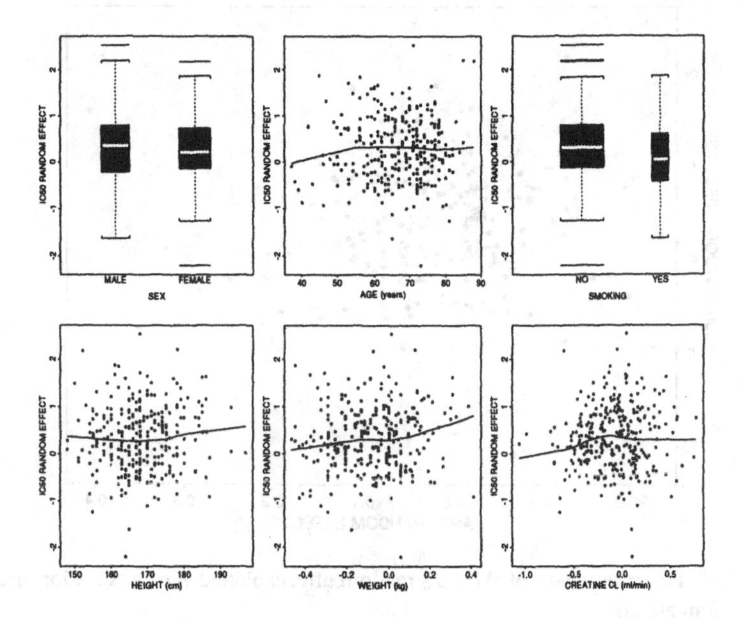

FIGURE 16. Second stage residuals (random effects) for IC_{50} plotted versus individual specific covariates.

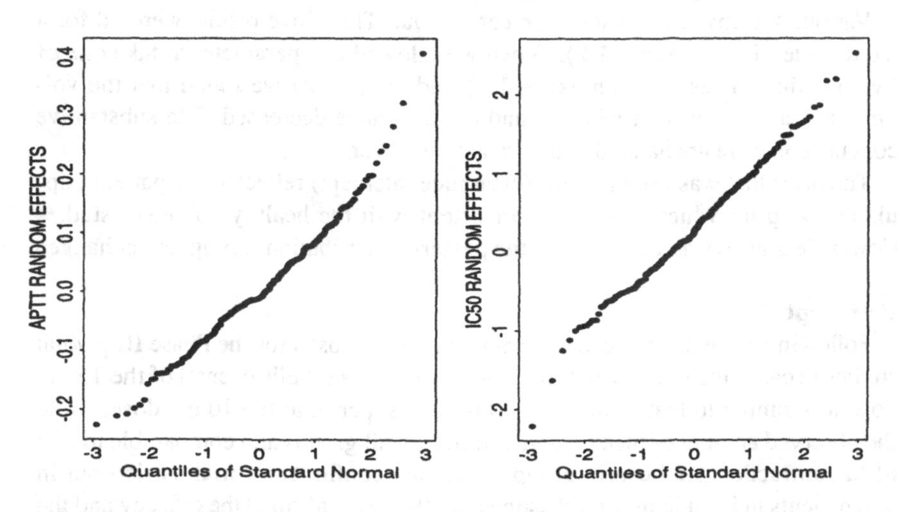

FIGURE 17. Normal scores plots for $APTT_0$ random effects (left) and IC_{50} random effects (right).

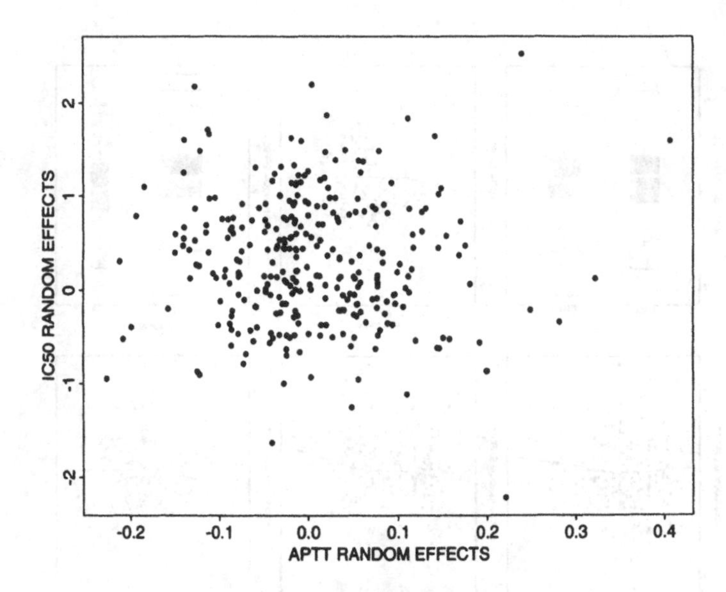

FIGURE 18. Posterior means of $APTT_0$ random effects plotted versus posterior means of IC_{50} random effects.

of iterations were required to obtain reliable inference. The posterior means of these four parameters can be used within a simple plug-in approach for formulas (7) and (8) in order to determine loading and maintenance doses for new patients with measured weight and creatinine clearance.

Various sensitivity analyses were carried out. The above results were all for a fixed value of $k_a = \exp(-1.2)$. When we allowed this parameter to take one of five possible values between $\exp(-1.1)$ and $\exp(-1.3)$ we found that the volume increased with increasing k_a and the clearance decreased. The substantive conclusions were unchanged in this range, however.

The prior that was used for $\mu_{\theta 0}$ (clearance intercept) reflected the patient population. A prior which was more consistent with the healthy volunteer studies (lower clearance) was also used but the posterior distribution was again unchanged.

Postscript

Following Study 6, the dose of 15mg b.i.d. was chosen for the Phase III pivotal studies because the efficacy (i.e. the rate of thromboembolic events) of the 15 mg dose was similar to that of the 20mg dose and superior to the 10 mg dose, while the observed amount of bleeding was similar in all groups and comparable to that of the unfractionated heparin group. A second confirmatory trial carried out in 400 patients in Scandinavia (Eriksson *et al*, 1997*a*) confirmed the efficacy and the safety of the 15mg dose regimen. Finally, another large-scale study performed in 2000 patients (Eriksson *et al*, 1997*b*) successfully demonstrated the efficacy and the safety of the REVASC^{TM} 15mg b.i.d. compared to a low-molecular-weight

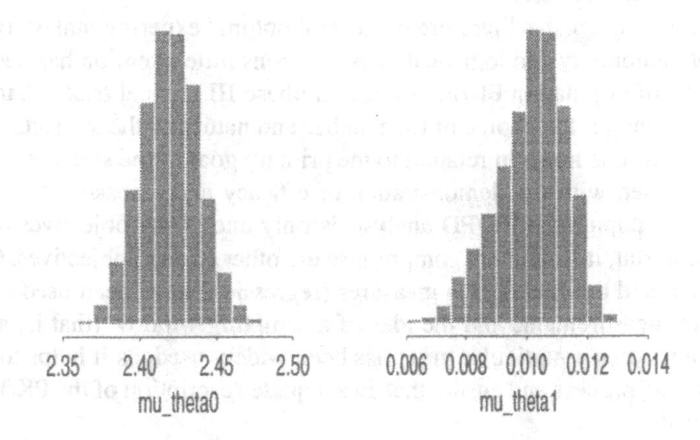

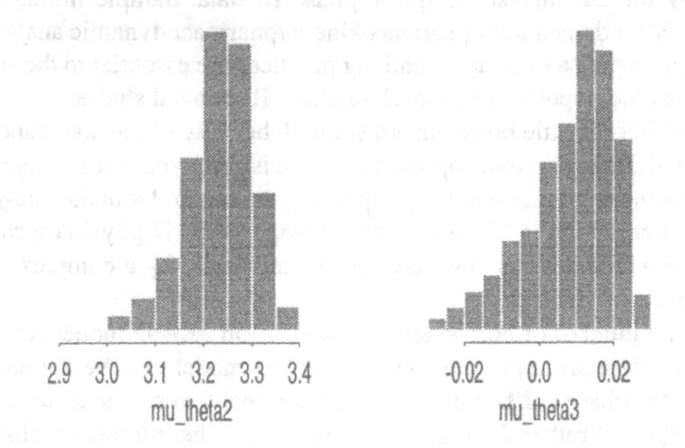

FIGURE 19. Posterior distribution for PK population mean parameters, $\mu_{\theta 0}$ is the clearance intercept, $\mu_{\theta 1}$ is the regressor for clearance/creatinine clearance relationship, $\mu_{\theta 2}$ is the volume intercept and $\mu_{\theta 3}$ is the regressor for volume/weight.

heparin. REVASC^{TM} is now approved is the European Community.

8 Future directions

There are a number of outstanding methodological problems associated with population PK/PD analyses.

There is an extensive literature on general optimal experimental design but because of complexity and logistical considerations little attention has been paid to the design of population PK/PD studies in phase III clinical trials (Aarons *et al*, 1996). In general, the choice of the number and nature of the subjects in a phase III clinical trial is made in relation to the primary goal of the study, which is usually concerned with the demonstration of efficacy and assessment of safety. In fact since a population PK/PD analysis is only one of the objectives of a phase III clinical trial, it should not compromise the other (major) objectives. Computer simulation and optimal design measures (regression) have been used to plan the timing of measurements and the idea of a sampling window (that is, a range of times rather than a particular time) has been widely used, as it helps to structure the sampling process and ensure that an adequate description of the PK/PD profile is obtained.

Some prior knowledge of the PK and PD models and covariate relationships is necessary for the analysis of sparse phase III data. Sample timing and dosing history is fundamental to pharmacokinetic/pharmacodynamic analysis. Consequently, good data and sample handling practices are essential to the successful application of the population approach to phase III clinical studies.

In phase I there is little borrowing of strength because of the abundance of data per individual. Hence it would appear that there is some room for designing trials with fewer sampling times and fully exploiting the hierarchy in the analysis.

Another area which is likely to grow in importance is physiological models in which the whole body is described more realistically by a complex system of compartments.

There are a number of components of population PK/PD models for which it may be more appropriate to use errors-in-variables modeling. The sampling times, particularly for Phase II/III studies are often nominal times – the actual scheduled times are reported rather than the actual times (see the number of observations reported at 9 hours in Figure 9) so a Berkson model (Carroll, Ruppert and Stefanski, 1995) may be appropriate. Wang and Davidian (1996) have examined this issue. Many second stage covariates are measured with error or are the result of formulas such as that for creatinine clearance in Section 7.3. Finally when a reliable joint PK/PD model cannot be obtained it may be preferable to use the observed concentrations and acknowledge the uncertainty in these values using an errors-in-variables approach.

The multi-stage hierarchical models that are used for population data contain many layers of assumptions and there has been little work on assessing their adequacy (see Hodges, 1998). Flexible models such as the nonparametric and semi-parametric techniques described in Section 6.2 are important. Experience using these techniques is required but their use, at least as exploratory tools, is likely to be valuable though any physiological information that is known should obviously be incorporated.

Covariate selection remains an important problem as the detection of associations between PK and PD parameters and individual characteristics has important implications for the determination of dosage regimens.

For population studies the important principle underlying each of design, prior specification, assessment of model adequacy and model selection is that each of these procedures must be informed by biological/pharmacological information and not by statistical methodology alone.

References

Aarons, L., Balant, L.P., Mentré, F., Morselli, P.L., Rowland, M., Steimer, J.-L. and Vozeh, S. (1996). Practical experience and issues in designing and performing population pharmacokinetic/pharmacodynamic studies, *Eur. J. Clin. Pharmacol.*, **49**, 251–254.

Abernathy, D.R. and Azarnoff, D.L. (1990). Pharmacokinetic investigations in elderly patients. Clinical and ethical considerations. *Clin. Pharmacokinet.* **19**, 89–93.

Beal, S.L. and Sheiner, L.B. (1982). Estimating population kinetics. *CRC Critical Reviews in Biomedical Engineering*, **8**, 195–222.

Beal, S.L. and Sheiner, L.B. (1993). *NONMEM User's Guide*, University of California, San Fransisco.

Bennett, J.E., Racine-Poon, A. and Wakefield, J.C. (1996). MCMC for nonlinear hierarchical models. In *Markov Chain Monte Carlo Methods in Practice* (eds. W.R. Gilks, S. Richardson and D.J. Spiegelhalter), 339–357. London: Chapman and Hall.

Berry, D.A. (1990). Basic principles in designing and analyzing clinical studies. In *Statistical Methodology in the Pharmaceutical Sciences* (ed. D.A. Berry), 1–55. Marcel-Dekker, Inc. New York and Basel.

Carroll, R.J., Ruppert, D. and Stefanski, L.A. (1995). *Measurement Error in Nonlinear Models.* Chapman and Hall, London.

Colburn, W.A. (1989). Controversy IV: population pharmacokinetics, NONMEM and the pharmacokinetic screen; academic, industrial and regulatory perspectives. *J.Clin.Pharmacol.*, **29**, 1–6.

Davidian, M. and Gallant, A.R. (1993). The non-linear mixed effects model with a smooth random effects density. *Biometrika*, **80**, 475–88.

Davidian, M. and Giltinan, D.M. (1993). Some simple estimation methods for investigating intra-individual variability in nonlinear mixed effects models. *Biometrics*, **49**, 59–73.

Davidian, M. and Giltinan, D.M. (1995). *Nonlinear Models for Repeated Measurement Data.* Chapman and Hall, London.

Eriksson, B.I., Kalebo, P., Zachrisson, B., Ekman, S., Kerry, R. and Close, P. (1996). Prevention of deep vein thrombosis with recombinant hirudin CGP 39393 in hip prosthesis surgery. Evaluation of three dose levels of recombinant hirudin in comparison with unfractionated heparin. *Lancet*, **347**, 635–39.

Eriksson, B.I., Ekman, S., Lindbratt, S., Baur, M., Torholm, C., Kalebo, P. and Close, P. (1997*a*). Prevention of deep vein thrombosis with recombinant hirudin – results of a double-blind multicenter trial comparing the efficacy of desirudin (Revasc) with that of unfractionated heparin in patients having a total hip replacement. *J. Bone J. Surg. (Am)*, **79A**, 326–33.

Eriksson, B.I., Wille-Jorgensen, P., Kalebo, P., Mouret, P., Rosencher, N., Bosch, P., Baur, M., Ekman, S., Bach, D., Lindbratt, S. and Close, P. (1997*b*). Recombinant hirudin, desirudin, is more effective than a low-molecular-weight heparin, enoxaparin, as prophylaxis of major thromboembolic complications after primary total hip replacement. To appear in *New England Medical Journal.*

Food and Drug Administration (1989). Guideline for the study of drugs likely to be used in the elderly. Washington, DC.

Gelman, A., Bois, F.Y. and Jiang, J. (1996). Physiological pharmacokinetic analysis using population modeling and informative prior distributions. *Journal of the American Statistical Association*, **91**, 1400–12.

Gibaldi, M. and Perrier, D. (1982). *Drugs and the Pharmaceutical Sciences, Volume 15 Pharmacokinetics, Second Edition*. Marcel Dekker.

Gilks, W.R., Best, N.G. and Tan, K.K.C. (1995). Adaptive rejection Metropolis sampling within Gibbs sampling. *Appl. Statist.* **44**, 455–472.

Gilks, W.R., Neal, R.M., Best, N.G. and Tan, K.K.C. (1997). Corrigendum to 'Adaptive rejection metropolis sampling within Gibbs sampling'. *Applied Statistics*, **46**, 541–2.

Godfrey, K.R. (1983). *Compartmental Models and their Applications*. London, Academic Press.

Hastings, W. (1970). Monte Carlo sampling-based methods using Markov chains and their applications. *Biometrika*, **57**, 97–109.

Hodges, J.S. (1998). Some algebra geometry for hierarchical models, applied to diagnostics. *Journal of the Royal Statistical Society, Series B*, **60**, 497–536.

Holford, N.H.G. and Sheiner, L.B. (1981). Understanding the dose-effect relationship: Clinical application of pharmacokinetic- pharmacodynamic models. *Clinical Pharmacokinetics*, **6**, 429–453.

Lange, N. and Ryan, L. (1989). Assessing normality in random effects model. *Annals of Statistics*, **17**, 624–642.

Lindstrom, M. and Bates, D. (1990). Nonlinear mixed effects model for repeated measures data. *Biometrics*, **46**, 673–87.

Maitre, P., Buhrer, M., Thomson, D., and Stanski, D. (1991). A three-step approach to combining Bayesian regression and NONMEM population analysis. *Journal of Pharmacokinetics and Biopharmaceutics*, **19**, 377–84.

Mallet, A. (1986). A maximum likelihood estimation method for coefficient regression models. *Biometrika*, **73**, 645–656.

Mentré, F. and Mallet, A. (1994). Handling covariates in population pharmacokinetics, *International Journal of Bio-Medical Computing*, **36**, 25–33.

Metropolis, N., Rosenbluth, A., Rosenbluth, M., Teller, A., and Teller, E. (1953). Equations of state calculations by fast computing machines. *J. Chemical Physics*, **21**, 1087–91.

Miller, A.J. (1990). *Subset Selection in Regression*, Chapman and Hall, London.

Muller, P. and Rosner, G.L. (1997). A Bayesian population model with hierarchical mixture priors applied to blood count data. *Journal of the American Statistical Association*, **92**, 1279–92.

Pinheiro, J. and Bates, D. (1995). Approximations to the loglikelihood function in the nonlinear mixed effects model. *Computational and Graphical Statistics*, **4**, 12–35.

Pitsiu, M., Parker, E.M., Aarons, L. and Rowland, M. (1993). Population pharmacokinetics and pharmacodynamics of warfarin in healthy young adults, *Eur.J.Pharm.Sci.*, **1**, 151–157.

Racine-Poon, A. (1985). A Bayesian approach to nonlinear random effects models. *Biometrics*, **41**, 1015–1024.

Racine-Poon, A. and Wakefield, J.C. (1996). Bayesian analysis of population pharmacokinetic and instantaneous pharmacodynamic relationships. In *Bayesian Biostatistics*, (ed. D. Berry and D. Stangl). Marcel-Dekker.

Racine-Poon, A. and Wakefield, J.C. (1998). Statistical methods for population pharmacokinetic modelling. *Statistical Methods in Medical Research*, **7**, 63–84.

Rowland, M. and Tozer, T.N. (1995). *Clinical Pharmacokinetics: Concepts and Applications, Third Edition.* Williams and Wilkins.

Sheiner, L.B., Beal, S.L. and Dunne, A. (1997). Analysis of non-randomly censored ordered categorical longitudinal data from analgesic trials (with discussion). *Journal of the American Statistical Association*, **92**, 1235–55.

Sheiner, L.B. and Benet, L.Z. (1985). Premarketing observational studies of population pharmacokinetics of new drugs. *Clin.Pharmacol.Ther.*, **38**, 481–487.

Smith, A. and Roberts, G. (1993). Bayesian computation via the Gibbs sampler and related Markov chain Monte Carlo methods. *J Roy Statist Soc, Series B*, **55**, 3–23.

Steimer, J., Vozeh, S., Racine-Poon, A., Holford, N., and O'Neill, R. (1994). The population approach: rationale, methods and applications in clinical pharmacology and drug development. In *Handbook of experimental pharmacology*, (eds. P. Welling and H. Balant). Springer Verlag.

Temple, R. (1983). Discussion paper on the testing of drugs in the elderly. Washington, DC: Memorandum of the Food and Drug Administration of Department of Health and Human Service.

Temple, R. (1985). Food and Drug Administration's guidelines for clinical testing of drugs in the elderly. *Drug Information Journal*, **19**, 483–486.

Temple, R. (1989). Dose-response and registration of new drugs. In *Dose-response relationships in Clinical Pharmacology*. Eds. Lasagne, L., Emill, S. and Naranjo, C.A. Amsterdam, Elsevier, p145–167.

Verstraete, M., Nurmohamed, M., Kienast, J. *et al* (1993). Biological effects of recombinant hirudin (GP 39393) in human volunteers. *J. Amer. Coll. Cardiol.*, **22**, 1080–1088.

Vonesh, E.F. and Chinchilla, V.M. (1997) *Linear and Nonlinear Models for the Analysis of Repeated Measurements*. New York, Dekker.

Vozeh, S. and Steimer, J.J. (1985). Feedback control methods for drug dosage optimization: concepts, classifications and clinical applications. *Clinical Pharmacokinetics*, **10**, 457–476.

Wakefield, J.C. (1994). An expected loss approach to the design of dosage regimens via sampling-based methods. *The Statistician*, **43**, 13–29.

Wakefield, J.C. (1996*a*). The Bayesian analysis of population pharmacokinetic models. *J. Amer. Statist. Assoc.*, **91**, 62–75.

Wakefield, J.C (1996*b*). Bayesian individualization via sampling based methods. *Journal of Pharmacokinetics and Biopharmaceutics*, **24**, 103–31.

Wakefield, J.C. and Bennett, J.E. (1996). The Bayesian modeling of covariates for population pharmacokinetic models. *Journal of the American Statistical Association*, **91**, 917–927.

Wakefield, J.C., Gelfand, A.E. and Smith, A.F.M. (1991). Efficient generation of random variates via the ratio-of-uniform method. *Statist. Comput.*, **1**, 129–133.

Wakefield, J.C. and Racine-Poon, A. (1995). An application of Bayesian population pharmacokinetic/pharmacodynamic models to dose recommendation. *Statistics in Medicine*, **14**, 971–86.

Wakefield, J.C., Smith, A.F.M., Racine-Poon, A. and Gelfand, A.E. (1994). Bayesian analysis of linear and non-linear population models using the Gibbs sampler. *Appl. Statist.*, **43**, 201–221.

Wakefield, J.C. and Walker, S.G. (1997). Bayesian nonparametric population model: formulation and comparison with likelihood approaches. *Journal of Pharmacokinetics and Biopharmaceutics*, **25**, 235–53.

Walker, S.G. and Wakefield, J.C. (1998). Population models with a nonparametric random coefficient distribution. To appear in *Sankhya, Series B*.

Wang, N. and Davidian, M. (1996). A note on covariate measurement error in nonlinear mixed effects models. *Biometrika*, **83**, 801–812.

Yuh, L., Beal, S., Davidian, M., Harrison, F., Hester, A., Kowalski, K., Vonesh, E., and Wolfinger, R. (1994). Population pharmacokinetic/ph- armacodynamic methodology and applications: a bibliography. *Biometrics*, **50**, 566–675.

Discussion

Frédéric Y. Bois
Lawrence Berkeley National Laboratory, Berkeley

I would first like to congratulate the authors for a very clear and complete presentation of the state of the art in Bayesian population pharmacokinetics. I would warmly recommend its reading to statisticians interested in the subject matter and its idiosyncrasies, and to all pharmacologists curious to understand how real-life clinical data can be analyzed with pharmacokinetic models. Reviews of this quality are in my opinion essential tools for building interdisciplinary teams in academic, regulatory, or industrial environments. It may be no accident that it has itself been written by a multitalented group ...

Incidentally, a major advantage of the Bayesian approach is that it is more natural to many clinicians. That may well be linked to the frequentation of patients: at the time of diagnosis we never wonder whether the subject is "statistically significantly" ill with pneumonia, but rather what are the chances of pneumonia versus competing possibilities. The development of numerical tools able to deal with the computational complexity of population pharmacokinetics is in this respect a major advance.

My only concerns are with the potential pitfalls of the modeling philosophy which consists in using simpler models as data per individual become sparser, while is it known that a complex model is in fact more scientifically reasonable. The authors have been careful in checking that no major bias (at least of clearance and volume of distribution) was introduced when going from a three-compartment to a one-compartment model. Yet, there may still remain more subtle problems with the proposed estimation. It seems also that hirudin is a nicely behaved drug for that matter. Things may not go that well with other compounds, and we should be very cautious. The first problem with that expedient is potential estimation bias. Figure 1 shows what can happen when concentration kinetic data generated by an underlying two-compartment process are fitted with a one-compartment model (with first-order absorption). With only two data points, the solution is unique. The estimated peak concentration is 1.3 times the true peak. Other data points along the true two-compartment kinetic curve could have led to underestimation of the peak, or worse errors. I did not look for an exaggerated figure, and actual drugs can certainly exhibit worse behavior. Peak levels of drugs in the body tend to be associated with acute toxic effects and their correct estimation is a basic requirement of formulation quality control.

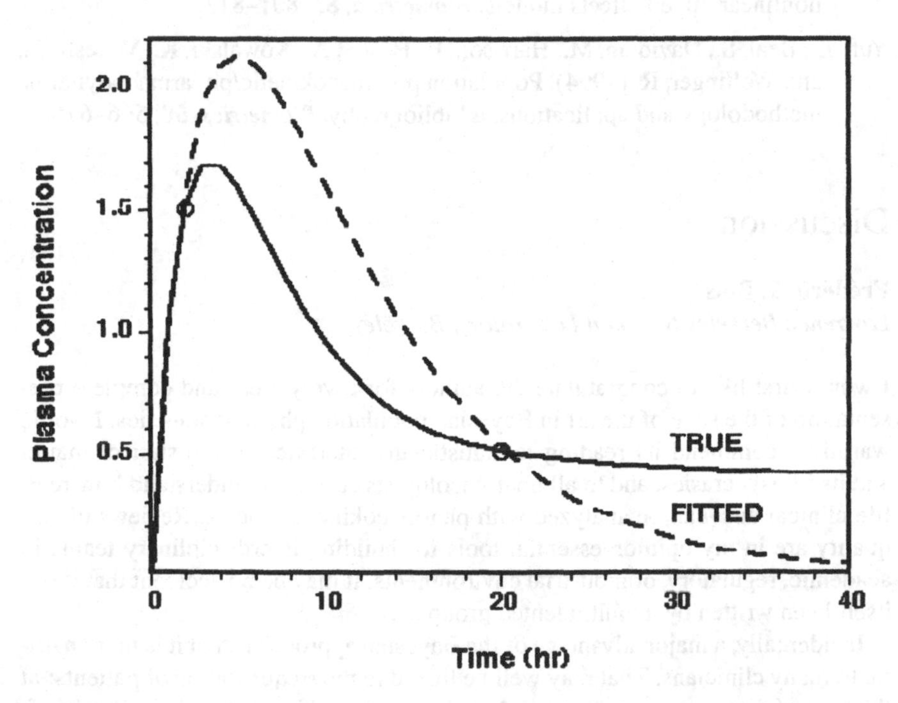

FIGURE 1: Potential estimation bias in the peak plasma concentration of a drug when a one-compartment model (dashed line) is fitted to data from an actual two-compartment kinetic process (solid line).

The second problem I see concerns scale rather than location. In some situations, population variability can be greatly overestimated if the wrong model is fitted. See Figure 2: for all four simulated subjects, exactly the same two-compartment kinetics actually apply for drug X; yet, because of different sampling schedules, very different peak concentrations are obtained when fitting a one-compartment model. The problem disappears if the same sampling schedule is adopted for all subjects, but a particularity of most population studies (which can be construed as an advantage) is that samples can be obtained at different times in different subjects . . .

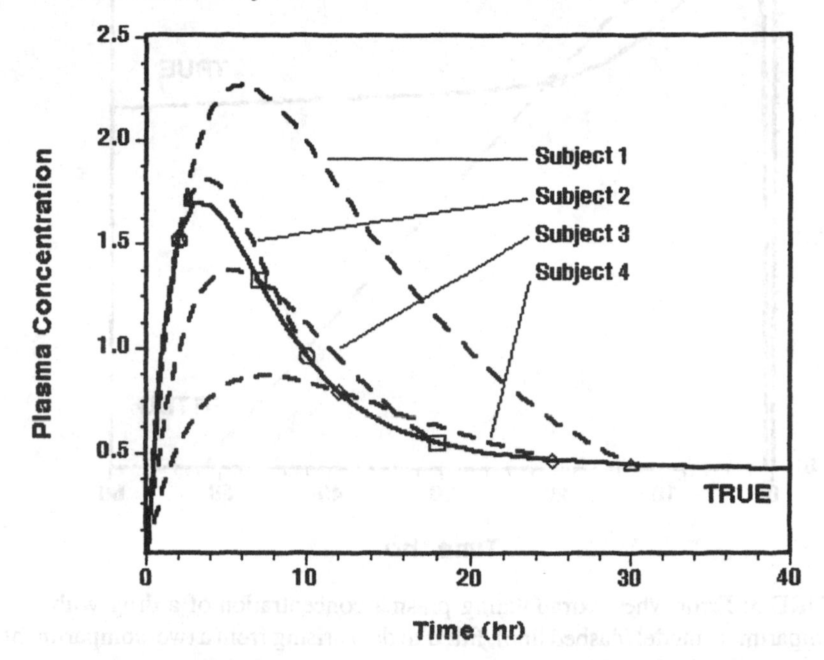

FIGURE 2: Estimation bias in the variance of peak plasma concentrations of a drug when a one-compartment model (dashed lines) is fitted to individual data arising from actually the same two-compartment kinetic process (solid line).

Further problems, not met by the authors, can arise when extrapolating the fitted model outside the data range. This is a standard pitfall, exemplified in Figure 3. The plasma concentration after 40 hours will be very badly predicted by the small, wrong, model. Indeed, one would never ever attempt such extrapolations, except maybe when looking at posterior predictions in a Bayesian context, e.g. for dosing adjustments.

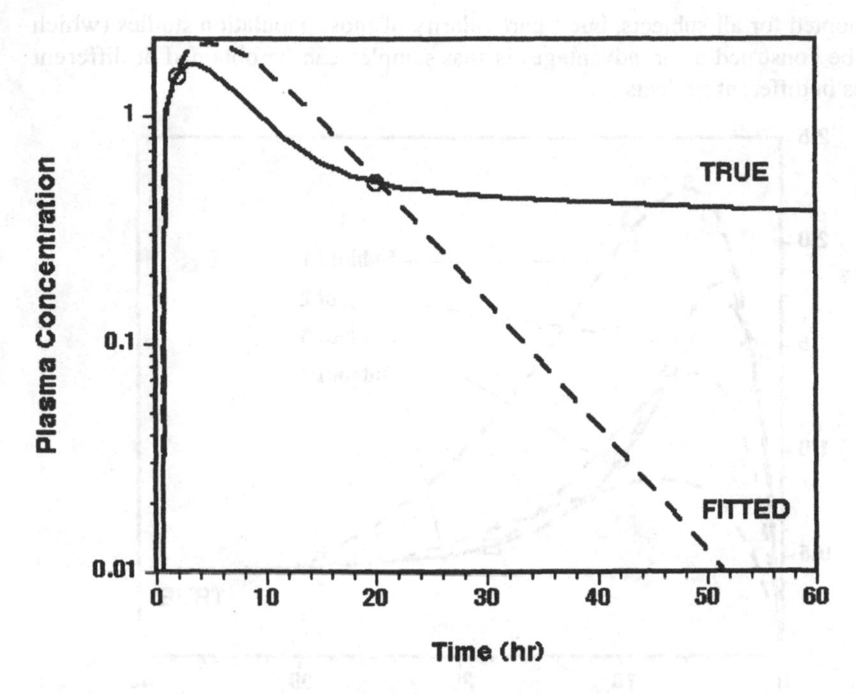

FIGURE 3: Error when extrapolating plasma concentration of a drug with a one-compartment model (dashed line), fitted to data arising from a two-compartment kinetic process (solid line).

More philosophically, when juggling models, I would be afraid of introducing some confusion in the booming field of hirudin pharmacokinetics. Are the kinetics mono-, bi-, or tri-compartmental? I am not sure myself. Maybe this is an arcane and hair-splitting issue of quality assurance in science. Yet, I can easily imagine a clinician stumbling, after a quick and dirty literature search, on a population study of hirudin and learning with delight that a one-compartment model has been "successfully" fitted to the data. S/he takes a pocket calculator and computes the plasma concentration remaining after 40 hours and ... see Figure 3. We may maintain that models are application-specific, that they have no validity outside the particular purpose they were designed for, but that is a grim and pessimistic perspective for pharmacokinetics as a research activity. I would rather favor building trust in the models we use.

On another philosophical plane, I would argue that forgetting the past data and experience, when starting an analysis of hirudin population kinetics, is not an epitome of the Bayesian approach. I wish we could do better. Naively, assuming in particular that the US FDA did not exist, we might ask what the phase I pharmacokinetic studies are good for, if they don't inform much of the analyses in phase II. The authors might advance two excuses for erring on the side of economy: they avoid identifiability problems, and they may not have in fact good prior information on their population of patients. These two issues are linked. If prior information about the parameters of a three-compartment model were strong, identifiability problems resulting from very sparse data would be much reduced (even if at the price of some reparameterizations). The problem rests mostly with the issue of extrapolating prior information. I would actually add that question as one of the challenges for future work in the area of population pharmacokinetics. The problem does not concern the extrapolation of the model structure; a three-compartment model, at least, is certainly adequate for hirudin kinetics in patients. The obstacle is the non-applicability of some of the parameter distributions derived from healthy individuals to patients. I would argue that part of the problem is due to the inability of the models used, even with three-compartments, to make full use of *a priori* physiological information. A more complete model could incorporate a better description of elimination processes, describe the fixation of hirudin to plasma proteins, and identify the various compartments. Physiological models can be quite refined in their description of drug kinetics. They are harder to compute but they allow the use of strong informative distributions and ease the extrapolations by modeling the source of the differences between subjects. Obviously, I have the easy part in just suggesting further work, but the authors themselves underline in their discussion the potentials of physiological modeling and do point to current research in the area. There are, from the practicing pharmacokineticist point of view, terrible difficulties in parameterizing and fitting a physiological model, "it just can't be done" as someone told me early in my Ph.D. work. I think that recent development in Bayesian statistics do give us a way to go forward, provided we learn to forget some of our classical habits.

Discussion

Marie Davidian
North Carolina State University

The authors are to be congratulated for a clear, well-written, and practically relevant exposition on a topic that has both generated much interest and engendered much confusion. Appreciation for the role of pharmacokinetics, pharmacodynamics, and their interplay in the drug development process among statisticians has been increasing. However, many of these statisticians familiar with PK/PD have acquired their knowledge through serendipitous experience or individual initiative. For the broader population of statisticians in industry and especially those working exclusively in Phase III clinical trials, my experience is that, although there is an appreciation that PK/PD is somehow important, knowledge of its role in the overall process is a bit vague. Compounding the problem is that there are few self-contained accounts that clarify the role of PK/PD in a way accessible to the statistician. This article fulfills this need well; thus, regardless of religious preference (Bayesian vs. frequentist) these statisticians will find it to be a useful resource. I look forward to being able to hand it to my students interested in a realistic account of PK/PD analysis and associated statistical methods in drug development.

Praising the broad utility of the article is nice, but what of the statistical aspects? Unfortunately, those hoping for some anti-Bayesian polemics from one whose work in this area has been decidedly frequentist will be disappointed, as I believe that taking a Bayesian perspective in the current context is entirely sensible. Thus, the focus of my comments is not on statistical underpinnings but rather on the critical issues in population PK/PD analysis that I believe transcend the mere choice of statistical philosophy. I concentrate on this because I believe that, ultimately, one of the main contributions of the work of Drs Wakefield, Aarons, and Racine-Poon, henceforth WAR-P, is their fair and balanced focus on these issues.

As made abundantly clear by the authors, both statistical and subject-matter considerations in PK/PD analysis are characterized by a myriad of assumptions. On the subject-matter side, the choice of structural models is dictated by assumptions on the underlying physiological processes that are, mostly of necessity, gross simplifications. Depending on the sparsity of the data at hand, this choice may be subject to further simplifying assumptions in order to make fitting feasible. A prominent feature of the situation is that the processes of PK and PD occur at the *individual* level, while, for drug development, interest focuses mostly on the *population*. Thus, the natural statistical framework within which to represent data on PK/PD from several subjects is that of a hierarchical model. Tailoring this model to fit the needs of the particular situation involves a number of additional assumptions. At the very least, one must specify the nature of intra- and inter-subject variation. For studies involving identification of the sources of this variation, like the authors' Study 6, the analyst is faced with the additional challenge of incorporating subject-specific covariate information into the model; this involves assumptions about the functional form of the relationship between meaningful parameters like drug clearance and covariates, for which there is usually little scientific guidance. These parameters are of course unobservable, exacerbating the challenge.

Once the basic model framework has been established, the statistical paradigm within which implementation will be carried out is chosen. The most widely-used methods are "frequentist." The marginal likelihood is almost always analytically intractable due to the nonlinearity of the structural model, forcing numerical evaluation of the required integrals. The most popular inferential strategies seek to avoid this complication by appealing to a linearization that allows the likelihood to be approximated; variants of this approach are implemented in the package NONMEM (Beal and Sheiner, 1993), whose widespread availability and focus on PK/PD makes the methods favored by pharmacokineticists. Fitting proceeds treating the approximation as exact (an assumption), and inference on model parameters is carried out by invoking asymptotic approximations (a further assumption). It is certainly possible to avoid approximations by carrying out the integration numerically (e.g. Davidian and Gallant, 1993); inference will still involve an appeal to large sample results. As demonstrated by WAR-P the Bayesian approach introduces a third "hyperprior" stage, which incorporates further distributional assumptions, possibly representing realistic prior information from previous studies. The necessary integrations may be carried out by appealing to MCMC techniques, thereby eliminating the need for linear approximation. However, this comes at the expense of possible sensitivity to the distributional assumptions that most often are fully specified at each model stage, including priors, thus, representing a source of concern to frequentists in the same way asymptotic approximations are to Bayesians. Luckily, my experience has been that, when the data contain sufficient information, both frequentist and Bayesian approaches yield similar conclusions, which is as it should be, as Dr Wakefield and I once encountered firsthand. While analyzing the same data set, we noted that our respective results exhibited an almost eerie correspondence given our different assumptions, techniques, and software, right down to comparison of my large-sample confidence intervals to the quantiles of his posteriors, that could not be attributed to a mutual hallucinatory experience induced by the previous night's session at the local pub.

Before I stray into the details of such sessions, back to the point. To summarize the population PK/PD analysis endeavor, this type of modeling and inference in general is not an enterprise to be undertaken lightly. Before one even begins to contemplate the details of a statistical analysis, one must recognize that the basic model framework relies on numerous assumptions, some that may be justified by subject-matter considerations or on the basis of empirical evidence, others that may be a matter of convenience or the subjectivity of the analyst. That done, the inferential strategy chosen carries with it further, generally unverifiable assumptions. This choice may be dictated by the political leanings of the analyst; however, in practice, it is usually dominated by the availability of accessible software.

My main message is thus. For PK/PD problems, complex hierarchical models represent the appropriate framework for analysis, and I believe that the Bayesian perspective, allowing information from previous studies to be incorporated in a natural way throughout the development process, holds considerable promise. This modeling is of course not new; population PK/PD analysis via frequentist techniques has been carried out for almost two decades with success by pharmacokineticists, some of whom, although not trained formally in statistics, have an intuitive feel for how things work. Among these individuals, such as the pioneer of population PK/PD analysis, Lewis Sheiner, and his colleagues, the complexity of the modeling and limitations of what may be gleaned from real data are well-appreciated. However, regardless of whether frequentist methods, approximate or exact, or Bayesian methods are used, I am concerned that, as popularity of the approach increases, the complexity, the broad array of assumptions, both subject-matter and statistical, and the numerous pitfalls of implementation, may be under-recognized by the broader population of consumers, both statisticians and pharmacokineticists. The FDA has already issued a draft guidance on population PK/PD analysis, encouraging more widespread undertaking of these analyses in the pharmaceutical industry. There is thus an urgent need to make these users aware of the limitations involved; this would be beneficial to all users, even those with practical experience. In their paper, WAR-P have done a commendable job of realistically warning the reader at each instance where an assumption has been made and of its possible consequences and highlighting the need for relevant sensitivity analyses. It is important that the considerable enthusiasm among statisticians and scientists be tempered by realism; this cautious stance stems not only from what I have encountered when fielding questions from individuals new to the area, but also from my own painful learning process.

To be a little more specific, I would like to point out briefly some pitfalls in greater detail. First, subject-matter issues aside, these models are *hard*. The nonlinearity of the structural model is in itself a major complication, raising questions of parameterization, as noted by the authors. Incorporation of subject-level covariates in the second stage model leads to additional parameters to be fitted, and it is natural within such a complex framework to wonder whether lack of identifiability of some parameters may be an unintended consequence of this modeling, which may be impossible to determine by inspection. Although the classical Bayesian view would hold that identifiability is not a problem for a Bayesian analysis, in practical situations this is simply not true if one is interested in more than a simply hollow academic exercise. If one's goal is to get sensible, useful answers to real questions, it is a key issue in this application, regardless of paradigm. Lack of identifiability may rear its head through unstable implementation, or may lead to physically implausible solutions. *Near* lack of identifiability may be more insidious, and may lead to modeling assumptions that, although nonsensical from a biological perspective, may be required to achieve stability. For example, when the variation in one individual-specific PK/PD parameter is small relative to the others, this may dictate treating that parameter as if it is fixed in the population, which is certainly not realistic. Recognizing such situations is tricky, and determining how best to handle them while holding true to the science is trickier still. These issues and others like them may not be fully appreciated; whether this poses major consequences for the qualitative inferences drawn may or may not be serious in any given problem, but it is prudent to be concerned. Moreover, they may manifest themselves in or affect different methods in different ways, making validity of results method-dependent, a feature that will likely go unnoticed unless different methods are tried.

Speaking of implementation, both frequentist and Bayesian approaches are difficult. For the former, with which I am more familiar, complex, high-dimensional optimization or equation-solving is required, even if approximations are introduced. In my experience, one can never be too skeptical about the results; objective functions may be riddled with local maxima or minima, so finding the "true" solution may require appeal to numerous starting values. The validity of common approximations, particularly for the computation of standard errors and confidence intervals, may be poor. Even for exact likelihood methods, the relevance of asymptotic theory in finite samples is not guaranteed. How these factors conspire with those mentioned above may be impossible to sort out or interpret. I have noted a tendency for fascination with some of the most complex procedures whose operating characteristics are the least understood, such as those involving less restrictive assumptions on the distribution of the random effects. I was once witness to a presentation of an analysis where a mixture of *three* normal distributions for the random effects (yes, the mixing proportions were estimated) was fitted to PK data on *seventeen* subjects; it is scary to contemplate how these various problems may have afflicted this endeavor...

A strength of the authors' account is their willingness to point out the corresponding issues in the Bayesian implementation. My comments here represent the impressions of an observer who has little experience with the nuts and bolts, so are of necessity somewhat superficial, but I believe the spirit is correct. The obvious comparison with computational issues for frequentist approaches is with implementation through the use of Markov chain Monte Carlo techniques. I say this somewhat tongue and cheek, but it is almost as though a religious movement has developed over the application of these methods. In this context, my fear is that, just as with frequentist approaches that allow flexible, nonparametric estimation of the random effects distribution, there is a widespread enthusiasm for MCMC among individuals who may not have studied them with same depth as the authors. This is no doubt in part a consequence of the facts that the premise seems (deceptively) simple and the implementation involves some pretty-fancy sounding stuff, a combination which tends both to fascinate and be unwittingly misunderstood. The possibility for confusion is indeed ripe; I am sure that Drs Wakefield and Racine-Poon the perhaps extreme case of one individual who was certain that the MCMC implementation was equivalent to the "EM algorithm."

It is clear that the performance of these methods in difficult problems such as population PK/PD are still not entirely understood (not that they are for frequentist methods, either). For example, as the authors remark with respect to their analysis of Study 6 in Section 7.5, "We found that the Markov chain was slow mixing and a large number of iterations were required to obtain reliable inference." This statement is a bit vague (how "large?" what constitutes "slow?" "reliable?"), but I do not fault the authors for that, as they were speaking to an audience with some familiarity with the methods. This statement does convey the message that, even among the experienced, the attitude is still cautious, and interpretation of the behavior observed not entirely clear. Another obvious frequentist target is sensitivity to prior specification; WAR-P are again to be commended for stressing the need for investigating investigation of this issue. A concern is again that, for less experienced users, these issues may not be given their full day in court, leading to the Bayesian version of analyses similar to the three-normal mixture affair above.

Although my tone comes off as pessimistic, in reality I have positive expectations for the future. There are other applications, such as analysis of HIV dynamic data from AIDS clinical trials, where hierarchical nonlinear methods are just beginning to be applied; thus, both in the population PK/PD arena and these other areas, there are new challenges, such as issues of measurement error in covariates, missing covariates, and informatively missing responses. The Bayesian approach has a certain advantage over frequentist methods for these problems, as introduction of these complications into the analysis is more straightforward. As we continue to learn about performance in practice and as appreciation for what can and cannot be accomplished realistically with real data becomes more widespread, these analyses will hopefully one day be a standard component of the statistician's repertoire. Thankfully, researchers like WAR-P here to offer balanced guidance along the way.

References

Beal, S.L. and Sheiner, L.B. (1993). *NONMEM User's Guide*, University of California, San Francisco.

Davidian, M. and Gallant, A.R. (1993). The nonlinear mixed effects model with a smooth random effects density, *Biometrika* **80**, 475-488.

Discussion

S. Greenhouse
George Washington University

Dr. Aarons gave us an excellent review of the long haul necessary to develop a new drug almost always taking a number of years. He refers to the large amount of information that becomes available at the time a phase three clinical trial is planned. Now the drug they were discussing, hirudin, is an anticoagulant and therefore must have some adverse effects of bleeding. A possible long term effect is stroke. Clearly it is difficult for the drug company to do long term studies but does anyone think about the issue of how to estimate potentially long term severe adverse effect? Somehow one is uneasy with the response that that is the reason we need to do phase three clinical trials.

Rejoinder

We would first like to thank all of the discussants for their comments. We consider their comments under a number of headings.

Simplified Model

We essentially agree with the comments of Professor Bois concerning the use of a simplified model and find the examples he presents enlightening. Ultimately the effects, and therefore the adequacy, of the simplified model will be scenario-specific and in particular will depend on the objectives of the study. In our case study, *clinically*, a one-compartment model is adequate for the safe and efficacious use of the drug. Clearance is not poorly estimated here and the clearance controls the dosing rate. From our simulations and Professor Bois' examples it is clear that the peak and trough levels are less-well predicted by the simple model. This was also found for the drug bismuth by Bennett, Wakefield and Lacey (1997) where, in addition, some systematic discrepancies were found for the clearance parameter.

We note that for extravascular administration, due to the finite time for absorption, the peak/trough swings are dampened down making the use of the simple model less of a clinical problem. Extrapolation is probably not as great a problem for chronic (i.e. multiple) dosing.

Phase I studies are tolerability studies and provide baseline information on the pharmacokinetics of a drug. They therefore guide the design of phase II studies by providing the basis of the PK model and parameter estimates. Phase II studies are generally carried out in a patient population which differs from the young, healthy population that is considered in phase I. Consequently the PK parameter values, but usually not the model, differ between phase I and phase II studies. Due to the sparsity of data in most phase II studies one must either assume a simplified model (as we did), or impose a strong prior distribution. The latter is difficult, however, because of the aforementioned differences between the phase I and II groups. Ideally, as suggested by Professor Bois, one would resort to the underlying physiology to predict how the parameters would change between the two groups. However the information necessary to make these predictions only becomes available during the phase II and III programs; the phase I population is too homogeneous to provide the necessary range in the covariates. Nevertheless it may be possible to utilize pre-clinical information and data on related drugs to facilitate this prediction in a Bayesian manner.

Finally it should be mentioned that there are many examples in clinical practise where a reduced model is used routinely and successfully to guide therapy. A good example is the antibiotic gentamicin for which a one-compartment model is used for dosing but strictly the model has at least two-compartments (Evans *et al*, 1980). We re-iterate than in practice it is always necessary to show that the simpler model is adequate for its purpose.

In general the combination of information from different studies/experiments is an outstanding problem. The perceived wisdom currently seems to be that unless the data are directly comparable, in which case one would combine the datasets anyway, then one should use the posterior distributions from previous studies as priors, but, in the sense of conservatism, with increased variances. Clearly further work is needed in this area.

In fact the reason that we did not use a joint PK/PD model was because we did not sufficiently believe in our first-stage PK model. Bennett and Wakefield (1998) consider in detail the data of Study 6 and compare the effect of: using the observed concentrations; using the fitted concentrations obtained from a one-compartment model; jointly estimating the PK/PD relationship with a one-compartment model; and using the observed concentrations with the errors in these being acknowledged via an errors-in-variables model.

A closely-related problem is one of population meta-analysis. Initial work (Wakefield and Rahman, 1998) indicates that this is feasible but the exchangeability of studies is a very strong assumption which needs very careful consideration, particularly beyond phase I.

Identifiability

We agree with Professor Davidian concerning the difficulties of parameterization in the face of model identifiability. Even in the case of the simple 'flip-flop' model (Gibaldi and Perrier, 1982) the usual approach is to ensure identifiability by assuming the parameterization $(\log V, \log(k_a - k_e), \log Cl)$, if it is believed that $k_a > k_e$, or parameterization $(\log V, \log(k_e - k_a), \log Cl)$, if it is believed that $k_e > k_a$. We note that already we face difficulties if we wish to then regress $\log k_a$ on individual-specific covariates at the second stage. Real difficulties arise if we obtain data from a population within which k_a and k_e are of similar size. In this case a mixture distribution at the second stage may provide a solution.

Implementation

Professor Davidian is right to point out that the implementation of the Bayesian approach is not without its difficulties. The BUGS software (Spiegelhalter, Thomas, Best and Gilks, 1994) is currently being extended to handle population models within a menu-driven environment, but convergence issues remain, particularly if the model that is used is inadequate in some respect, or when one is faced with near-identifiability problems.

Safety

There is no simple answer to Professor Greenhouse's contribution, the possibility of long-term and infrequent adverse events are difficult to deal with. These possibilities must be considered case-case by ethics committees and regulatory bodies that oversee the conduct of clinical trials, such as the FDA. The experience with related compound is an important consideration. We also note that safety monitoring continues after the drug has been granted a marketing licence. Safeguards have improved but are not infallible.

Additional References

Bennett, J.E., Wakefield, J.C. and Lacey, L.F. (1997). Modeling of trough plasma bismuth concentrations. *Journal of Pharmacokinetics and Biopharmaceutics*, **25**, 79-106.

Bennett, J.E. and Wakefield, J.C. (1998). Errors-in-variables in joint PK/PD modeling. *Manuscript under preparation.*

Evans, W.E., Taylor, R.H., Feldman, S. Crom, W.R., Rivera, G., Yee, G.C. (1980). A model for dosing gentamicin in children and adolescents that adjusts for tissue accumulation with continuous dosing, *Clinical Pharmacokinetics*, **5**, 295-306.

Spiegelhalter, D.J., Thomas, A., Best, N.G. and Gilks, W.R. (1994). *BUGS: Bayesian Inference Using Gibbs Sampling, Version 3.0.* Cambridge: Medical Research Council Biostatistics Unit.

Wakefield, J.C. and Rahman, N.J. The combination of population pharmacokinetic studies. *Submitted for publication.*

CONTRIBUTED PAPERS

Longitudinal Modeling of the Side Effects of Radiation Therapy

Sudeshna Adak
Abhinanda Sarkar

ABSTRACT One of the most important aspects of treating cancer patients with radiation therapy is to find an optimal dose regimen that cures without causing undue side effects. A recent clinical trial undertaken by the Eastern Cooperative Oncology Group made a study of toxic reactions in patients receiving hyperfractionated accelerated radiation therapy (HART). 28 lung cancer patients were observed over a four week time period during which the radiation therapy was being given for the first 16 days. At the end of each week, the patient was examined for side effects and the binary outcome (toxic reaction/no toxic reaction) was recorded. The primary focus of this analysis is to determine how pre-treatment clinical factors (e.g. age, stage of lung cancer and ambulatory status) affect the week-specific risk of experiencing a toxic reaction.

We develop a multivariate logistic regression model for longitudinal binary data which looks at the regression of the marginal probabilities on the covariates. The model allows for nonstationarity in the regression coefficients over time in order to reflect the reduced risk of toxicity over time. Correlations in the measurements over time are modeled by the inclusion of "interaction" terms which are present according to whether a corresponding complete subgraph is present in a random graph. The data is then allowed to determine the dependence structure by using the posterior distribution of the graph. The resulting structure is recursively incorporated into determining the posterior distribution of the regression coefficients. The Bayesian methodology for longitudinal binary data developed in this paper allows a data-driven choice of the correlation structure in the repeated measures and simultaneously identifies the factors that influence the risk of side effects over time.

1 Introduction

Radiation therapy has been found to be the mainstay of treatment and the only chance of cure for patients with locally advanced, inoperable "non-small cell lung cancer". It is estimated that there are approximately 80,000 new patients in this category every year in the U.S. alone (American Cancer Society, 1996). The standard radiation dose recommended for these patients is a total of 50-60 Gy, delivered over six weeks, once daily for five days a week. (1 Gy, pronounced "gray", is equivalent to 1 joule of radiation energy absorbed per kilogram.) However, approximately 80% of the patients suffer a relapse and additionally, the cancer spreads to other parts of the body in the majority of the patients (Withers *et al.*, 1988). Newer approaches to treating this disease include hyperfractionated radiation and accelerated regimens. Hyperfractionation schemes give multiple small doses daily instead of a conventional larger single daily dose. An accelerated regimen reduces the overall duration of treatment to two-three weeks in comparison to the six weeks of a more traditional course of radiation therapy.

Trial 4593, a phase II trial, was conducted by the Eastern Cooperative Oncology Group (ECOG) to test the feasibility, toxicity and efficacy of a *hyperfractionated accelerated radiation therapy (HART)* regimen. Combining hyperfractionation with acceleration allows a much higher dose of the radiation therapy to be delivered while reducing the overall duration of the treatment. This is expected to reduce the chances of a relapse. However, a major concern is that patients may experience acute side effects and ultimately die of serious complications as a direct result of the radiation.

One of the objectives of trial 4593 was to evaluate the risk of side effects resulting from an accelerated regimen of radiation therapy. It is important for both the physician and patient to know the risk of developing a toxic reaction to the radiation — both when a reaction is likely to occur and how long can it be expected to last. Patients in trial 4593 were assessed every seven days over a four week period to detect toxicities. This paper carries out a detailed analysis of the *longitudinal binary outcomes* (side effects/no side effects) observed during trial 4593 over a four week time period. We propose a Bayesian methodology to investigate if pre-treatment clinical data such as ambulatory status of the patient, stage of lung cancer and age of the patient affect the week-specific risk of developing toxic reactions to radiation therapy.

The next section contains a brief description of trial 4593 and the hyperfractionated accelerated radiation therapy or *HART*. We also describe the side effects and the procedure for quantifying and measuring them. Section 3 describes the model specification and the priors for the model parameters. Section 4 briefly outlines the computational procedures used to sample from the posterior distributions. Section 5 contains the results for our case study data and summarizes our conclusions about trial 4593.

2 Radiation induced toxicity

2.1 *Patients and Methods*

Radiation therapy was delivered to a total dose of 57.6 Gy using a hyperfractionated accelerated schedule. Treatment began on a Monday and finished on the third Tuesday, with weekend breaks. This amounted to a total of 12 planned treatment days over 16 elapsed days. Three fractions were delivered daily and the minimum interval between fractions was 4 hours. The first and third fraction of each day consisted of large AP-PA fields encompassing the primary tumor and draining lymphatics with 1-1.5 cm margins. The fraction size for these fields was 1.5 Gy. The second fraction utilized a lateral photon field and encompassed the primary tumor and involved nodes. This fraction size was 1.8 Gy. Attempts were made to design the radiation field so as to minimize the volume of esophagus treated without compromising the margins around the tumor or the spinal cord.

2.2 *Toxic Effects of Radiation Therapy*

Radiation therapy uses high-energy x-rays, electron beams or radioactive isotopes to shrink and destroy malignant growths or cancers. The life cycle of a cell, cancerous or healthy, has four phases, one of which is the dividing or *mitotic* stage. This is the point at which radiation is most effective. When radiation hits a dividing cancerous cell, the cell either dies or sustains an injury that prevents it from dividing, or if the cell divides, the two cells cannot divide further. Since cancerous cells are rapidly dividing all the time, they will be most affected by the radiation. However, this also implies that radiation will affect actively dividing healthy cells such as those present in the skin, mucous membranes lining the esophagus (gastrointestinal tract) and bone marrow. Radiation does not distinguish between tumor cells and normal cells. Even though great care is taken to shield healthy tissue during treatment, some side effects are unavoidable.

When the thoracic region is irradiated, the mucous membrane lining the esophagus is destroyed. The most common side effects seen from this are: sore throat or difficulty swallowing, poor appetite and weight loss. This combination of symptoms of radiation toxicity is often referred to as *esophagitis* (Cox (1986)). It is standard clinical practice to assess a combined grade of esophagitis toxicity (Soffer *et al.*, 1994) in patients receiving radiotherapy. It is necessary to systematically assess the patients during and after treatment in order to detect esophagitis. To this end, ECOG developed an *Acute Esophagitis Toxicity Criteria* for assessing an overall toxicity grade. The esophagitis grade was based on points given by the physician on examination of the patient. A detailed description of the toxicity assessment tool is given in Table 1 below.

Once esophagitis has developed, the patient requires care designed to prevent and decrease the risk of further complications and also care designed for the palliation of symptoms. While esophagitis is graded on a scale of 0(none) to 4(lethal), there were no cases of lethal esophagitis in trial 4593. So, the presence/absence of detectable toxicity is of clinical relevance and accordingly the esophagitis score has been dichotomized. (Combined score of 0-3 points: No detectable esophagitis). A more sophisticated analysis involving the ordinal esophagitis grade was not considered feasible with the limited number of patients.

2.3 *The Data*

- **Thirty patients were enrolled on the trial 4593. Two of the patients enrolled were not considered evaluable for response to treatment. (Both patients died before response to treatment could be assessed). These two patients have been excluded from all analysis of response and survival data from this trial (Mehta *et al.*, 1997). In keeping with this practice of ECOG, we have chosen to exclude these patients from our analysis of esophagitis toxicity as well.**

- **Longitudinal Binary Responses:** For 28 patients, binary data indicating the presence/absence of detectable esophagitis was collected on a weekly basis. Complete esophagitis data is available for a period of four weeks.

- **Covariates:** Eligibility criteria for trial 4593 included: age of patient ≥ 18 years; performance status (patient fully active or ambulatory and able to do work of a light or sedentary nature); biopsy proven unresectable stage IIIA or IIIB non-small cell lung cancer (see chapter 13, Murphy *at al.*, 1995). It is anticipated that these pre-treatment clinical data are the main factors to affect the possibility of radiation complications.

Table 1: ECOG Acute Esophagitis Toxicity Criteria

Assessment Criteria		Points
	None	0
Sore Throat	Mild	1
(Dysphagia)	Moderate	2
	Severe	3
	None	0
Analgesia	Topical Analgesia	1
Required	Non-narcotic	2
	Narcotic	3
	Normal	0
Nutrition/	Soft diet/liquids	1
hydration	NG tube feeding	2
	IV hydration	3
	None	0
Weight	$\leq$5% of baseline weight	1
Loss	5-10% of baseline weight	2
	$>$10% of baseline weight	3

ECOG Esophagitis Score is based on the sum of points.
0-3 points: No detectable esophagitis; $>$ 3 points: Toxic Reaction.

Figure 1 shows the pattern of toxicity incidence over the four weeks in relation to age. Tables 2 and 3 give the number of toxicity incidences over the four weeks classified by ambulatory status and stage of lung cancer respectively.

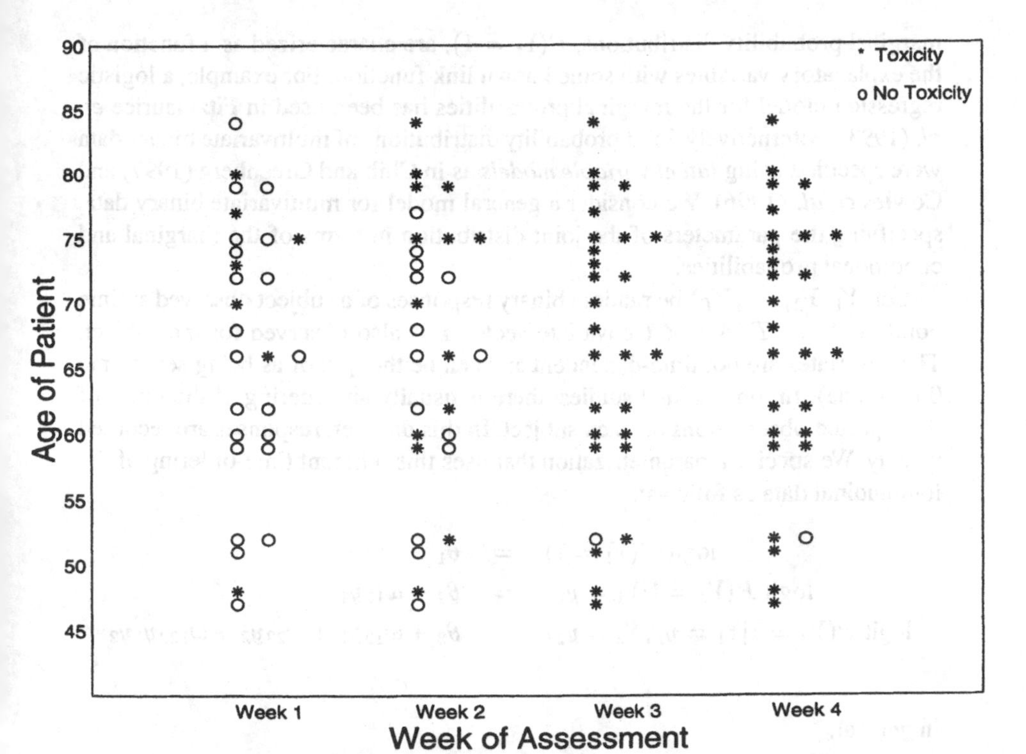

Figure 1: Plot of Toxicity Incidence

Table 2: Number of Toxicity Incidences

Ambulatory Status	Week 1	Week 2	Week 3	Week 4
Fully Active	1	3	7	8
Can do light work	6	11	20	19

Table 3: Number of Toxicity Incidences

Stage of Lung Cancer	Week 1	Week 2	Week 3	Week 4
Stage IIIA	2	4	12	12
Stage IIIB	5	10	15	15

3 The multivariate logistic model for longitudinal binary data

3.1 *Likelihood Specification*

Various models have been proposed to analyze the effect of covariates on longitudinal binary data. In *transitional models* (as in Bonney (1987)), the conditional probability distribution of the subjects' current status given their past history, $P(Y_t = 1|y_s, s < t)$, is linked to the covariates. In *marginal models*, the marginal probability distributions, $P(Y_t = 1)$, are characterized as a function of the explanatory variables with some known link function. For example, a logistic regression model for the marginal probabilities has been used in Fitzmaurice *et. al.* (1993). Alternatively, joint probability distributions of multivariate binary data were specified using *latent variable models* as in Chib and Greenberg (1997) and Cowles *et. al.* (1996). We consider a general model for multivariate binary data, specifying the parameters of the joint distribution in terms of the marginal and conditional probabilities.

Let $(Y_1, Y_2, \ldots, Y_T)$ be random binary responses of a subject observed at time points $1, 2, \ldots, T$. A $p \times 1$ covariate vector x is also observed for the subject. The covariates are not time-dependent and can be thought of as being set at time 0 (baseline). In longitudinal studies, there is usually an ordering of the times of the repeated observations on each subject. In this data set, responses are recorded weekly. We specify a parametrization that uses this inherent time-ordering of the longitudinal data as follows:

$$\begin{aligned} \text{logit } P(Y_1 = 1) &= \theta_1 \\ \text{logit } P(Y_2 = 1 | Y_1 = y_1) &= \theta_2 + \omega_{12} y_1 \\ \text{logit } P(Y_3 = 1 | Y_1 = y_1, Y_2 = y_2) &= \theta_3 + \omega_{13} y_1 + \omega_{23} y_2 + \omega_{123} y_1 y_2 \\ &\cdots \end{aligned}$$

In general,

$$\begin{aligned} \text{logit } P(Y_t = 1 \quad | \quad & Y_1 = y_1, Y_2 = y_2, \cdots, Y_{t-1} = y_{t-1}) \\ &= \theta_t + \sum \omega_{s_1, s_2, \ldots, s_k, t} y_{s_1} y_{s_2} \cdots y_{s_k} \end{aligned} \tag{1}$$

where the summation is over all $s_1, s_2, \ldots, s_k$ which are subsets of the set $\{1, 2, \ldots, t-1\}$. A similar conditional formulation has been used in Bonney (1987).

The marginal distribution of Y_t is Bernoulli with $\mu_t = P(Y_t = 1)$. The question of clinical relevance in this trial is how the pre-treatment covariates affect toxicity over time. The regression parameters are introduced via the marginal logistic regression

$$\text{logit } \mu_t = x' \beta_t. \tag{2}$$

We thus have time-dependent parameters which allows for the covariate effects to change over time. One of our eventual objectives is to look for this type of nonstationarity. For a given value of the parameters ω and $(\beta_1, \beta_2, \cdots, \beta_T)$, we can explicitly calculate $(\theta_1, \theta_2, \cdots, \theta_T)$ from solutions to polynomial equations. Details of this computation are provided in Section 4. The regression parameters (β) and the "interaction" parameters (ω) thus completely specify the likelihood, $P(Y = y | \beta, \omega)$.

This is regarded as a marginal specification as the covariates enter into the joint distribution of responses via the marginal expectations and regression parameters can be given their standard interpretation. This is in the spirit of Fitzmaurice *et al.*, 1993). On the other hand, the ω's are contrasts of conditional logits (log-odds) which can be explicitly computed in our case. For example,

$$\begin{aligned}
\omega_{12} &= \text{logit } P(Y_2 = 1|Y_1 = 1) - \text{logit} P(Y_2 = 1|Y_1 = 0) \\
\omega_{13} &= \text{logit } P(Y_3 = 1|Y_1 = 1, Y_2 = 0) - \text{logit} P(Y_3 = 1|Y_1 = 0, Y_2 = 0) \\
\omega_{23} &= \text{logit } P(Y_3 = 1|Y_1 = 0, Y_2 = 1) - \text{logit } P(Y_3 = 1|Y_1 = 0, Y_2 = 0) \\
\omega_{123} &= \text{logit } P(Y_3 = 1|Y_1 = 1, Y_2 = 1) - \text{logit } P(Y_3 = 1|Y_1 = 1, Y_2 = 0) \\
&+ \text{logit } P(Y_3 = 1|Y_1 = 0, Y_2 = 0) - \text{logit } P(Y_3 = 1|Y_1 = 0, Y_2 = 1) \\
&\ldots
\end{aligned}$$

These parameters determine the correlation structure in the longitudinal binary responses. They can also be interpreted in terms of log odds-ratios conditional on the past history. This circumvents some of the concerns raised in Diggle *et al.*, 1994), Chapter 8 of interpretation of such log odds-ratios conditional on all other outcomes, both past and future.

3.2 *Priors for regression coefficients*

We need to specify priors on the β's and the ω's. These will reflect our differing objectives for the two classes of parameters and the differing nature of the prior information.

The marginal distribution of the binary outcomes are linked to the covariates through a logistic regression as defined in equation (2). We specify a central Gaussian prior for β_1, the regression parameters for the first time point. The prior variance is $k_1(X'X)^{-1}$ where X is the $n \times p$ covariate matrix for the set of n subjects. Ibrahim and Laud (1991) showed that this prior behaves as a noninformative prior for large values of k_1. The prior for the regression parameter β_t is defined conditionally given the previous β's:

$$\beta_t | (\beta_1, \beta_2, \ldots, \beta_{t-1}) \sim N(\beta_{t-1}, k_2 I). \tag{3}$$

Note that this means that the regression parameters are not independent over time in the prior but that they evolve as a Markov chain. Harrison and Stevens (1976) discuss use of such Markovian priors.

3.3 *Hierarchical priors using graphs*

One of the objectives of this analysis is to identify good models for the correlation structure of the data in terms of parsimony and performance. So, we consider models in which some of the ω's, the "interaction" terms in equation (1) can possibly be zero. We choose priors for the ω's that are a mixture of a central Gaussian and a positive mass at zero, i.e. we have a positive prior probability that an interaction term can be degenerate at zero. This prior belief about the ω's reflects the flexibility in the choice of models. There are $(2^T - T - 1)$ interaction terms in the specification of the likelihood and hence $2^{(2^T - T - 1)}$ possible models. We restrict this very large class by considering only models induced by a graph.

Consider a graph G with T vertices labeled $1, 2, \ldots, T$. The interaction $\omega_{s_1, s_2, \ldots, s_k}$ is non-zero if and only if the subgraph $(s_1, s_2, \ldots, s_k)$ is complete in G. A subgraph is said to be *complete* if all pairs of vertices in the subgraph are joined by edges. Thus each selection of edges corresponds to a model in which only those interactions are present for which the subgraphs are complete. Christensen (1990) discusses such graphs in Chapter 4. It is to be noted that we do not interpret such graphs as conditional independence structures. This separates our use of graphs from the more elaborate treatment in (for example) Cox and Wermuth (1993).

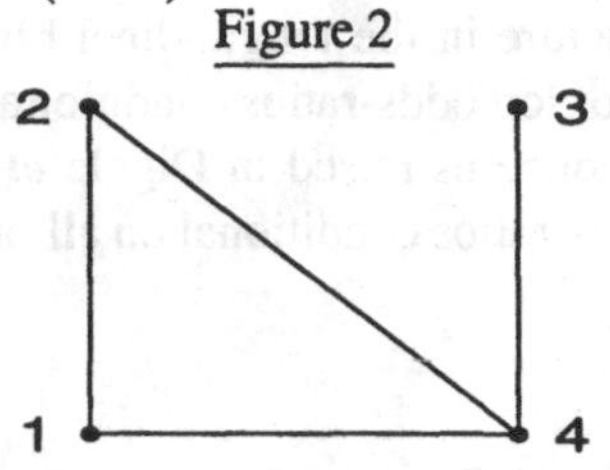

An example may help clarify this. Suppose $T = 4$ and that the graph in Figure 2 was selected. The interaction parameters which are allowed to be non-zero are ω_{12}, ω_{14}, ω_{24}, ω_{124}, and ω_{34}. The remaining six interaction terms are set to be zero.

Note that this means that a higher-order interaction is present if and only if all lower-order interactions involving the same time points are also present. There are $2^{T(T-1)/2}$ such models as compared to $2^{(2^T - T - 1)}$ models if we allow each interaction to be freely present or absent.

We now specify a prior probability distribution on graphs of T vertices as follows. Choose each edge independently according to a prior distribution on the edges. Two possible mechanisms for specifying the prior probabilities for each edge are:

Exchangeable Prior: Each edge is chosen with probability p, irrespective of the vertices joined. For example, the graph of Figure 2 has prior probability $p^4(1-p)^2$.

Serial Correlation Prior: Alternatively, the edge joining vertices s and t are selected with probability $p^{|s-t|}$. In this prior, the two-factor interactions between close time points are more likely to be in selected in the prior than those between more widely separated time points. So, the graph of Figure 2 has prior probability $p^7(1-p)(1-p^2)$.

We chose $p = 0.5$ for all our computations reported in Section 5. The larger the value of this prior parameter the more likely it will be that complex models with many interaction terms will be included in the model. If $p = 1$, the full model will always be chosen. At the other extreme, $p = 0$ allows for only the independent model where there is no intertemporal correlation. For any $p \in (0, 1)$, all $2^{T(T-1)/2}$ models are allowed by the prior. The prior beliefs about the models is transformed by the data into a posterior opinion of the models. It is in this sense that we allow the data to choose the appropriate dependence structure using formal Bayes factors.

Once we have selected which interactions are present, the prior is completely specified by iid $N(0, \sigma^2)$ distributions for the selected interactions. The complete prior on the interaction terms is thus hierarchical and is a mixture of discrete and continuous parts. We choose σ^2 large in order to keep the continuous part of the prior noninformative. The discrete part of the posterior can be easily interpreted, as explained in the previous paragraph, and will be one of our interests.

4 Sampling from the Posterior

Under the model introduced in Section 3, the posterior distributions $[\beta|y]$ and $[\omega|y]$ do not yield closed analytical expressions. In this section, we describe a Gibbs sampler (Gelfand and Smith, 1990) for the calculation of all the relevant distributions and related summary statistics.

- *Likelihood Computation:* In order to evaluate the likelihood for a given set of values for the parameters, $[\beta, \omega]$, we need to compute θ_{it}, $i = 1, \ldots, n$; $t = 1, \ldots, T$. The parameter θ_{it} is θ_t in equation (1) evaluated for the i-th subject. The θ_{it}'s can be recursively obtained by the following system of equations:

$$\theta_{i1} = x_i'\beta_1$$

For $t > 2$,

$$\text{logit}\, \mu_{it} = \log \frac{\mu_{it}}{1 - \mu_{it}} = x_i'\beta_t$$

$$\mu_{it} = \sum P(Y_{it} = 1 | Y_{i(<t)} = y_{i(<t)}).P(Y_{i(<t)} = y_{i(<t)})$$

where $Y_{i(<t)} = (Y_{i1}, Y_{i2}, ..., Y_{i(t-1)})$ and the summation is over all 2^{t-1} possible values of $y_{i(<t)}$. Thus, it follows from equation (1) that

$$\begin{aligned} \mu_{it} &= \sum \frac{P(Y_{i(<t)} = y_{i(<t)})}{1 + exp(-\theta_{it} + \sum \omega_{s_1, s_2, \ldots, s_k, t} y_{is_1} y_{is_2} \cdots y_{is_k})} \\ &= \sum_{s=1}^{2^{t-1}} \frac{a_{is}^{(t)}}{1 + d_s^{(t)} u} \qquad (4) \\ \text{where } \theta_{it} &= -\log(u) \end{aligned}$$

and where $a_{is}^{(t)}$'s come from $P(Y_{i(<t)})$ and hence depend only on $(\theta_{is'}, s' < t)$ and on ω's.

Equation 4 is a polynomial equation in u whose coefficients can be determined given the values of the parameters $[\beta, \omega]$. Thus, the likelihood can be evaluated by finding the solutions to these polynomial equations.

- Recall that the prior distribution of β is multivariate normal but that of the ω's is a mixture that puts a positive mass at zero. To incorporate the mixture prior, we introduce an auxiliary binary random vector Z, which is of the same dimension as ω. A particular interaction term ω is included in the model if the corresponding coordinate of Z is 1 and the term is excluded otherwise. We can define

$$\omega = Z * \alpha \tag{5}$$

where $*$ is component-wise multiplication. The (mixture) prior on ω is equivalent to the (discrete) graphical prior on Z and the (continuous) iid $N(0, \sigma^2)$ prior on all components of α.

- We implement a Gibbs sampler to sample from the full conditional of distribution $[\beta, \alpha, Z|y]$. Samples from the posterior distribution $[\omega|y]$ can then be obtained using equation (5). Each realization of Z corresponds to a graphical model. The introduction of α allows us to avoid specifying distinct parameter spaces for each model. This circumvents the need to specify "pseudopriors" as in Carlin and Chib (1995).

- At each iteration of the Gibbs sampler, we first draw a sample from the full conditional distribution $[Z|y, \beta, \alpha]$. This is followed by updating β's and α's by sampling from their full conditional distributions.

- The support of Z is restricted to the $2^{T(T-1)/2}$ models induced by the graphs described in Section 3.3. The posterior distribution can be exactly evaluated at each of these $2^{T(T-1)/2}$ points as follows:

$$\pi(z|\beta^*, \alpha^*, \mathrm{y}) = \frac{\pi(z) \times P(\mathrm{y}|\beta^*, \alpha^*, z)}{\sum_{z'} \pi(z') \times P(\mathrm{y}|\beta^*, \alpha^*, z')}$$

where the summation is over the class of $2^{T(T-1)/2}$ graphical models and $\pi(\cdot)$ is obtained from either the exchangeable or serial correlation prior of Section 3.3. β^*, α^* indicate current values of the parameters.

- Closed form expressions for the full conditionals of β's and α's do not exist. Furthermore, they are not known to be log-concave. Adaptive rejection metropolis sampling or ARMS (Gilks *et al.*, 1995) was used to draw a random sample from each conditional.

- We ran two samplers with over-dispersed starts for 7,500 iterations each. Convergence was assessed using the shrink factor diagnostic proposed in Gelman and Rubin (1992). This diagnostic is the factor by which the scale parameters of the marginal posterior distributions would be reduced if the chain were to run to infinity. In our analysis, the shrink factors were all close to 1 indicating convergence. Figure 3 below shows the trace plots for the regression coefficients in one of the runs. All other runs gave similar trace plots and give further visual evidence that convergence is achieved.

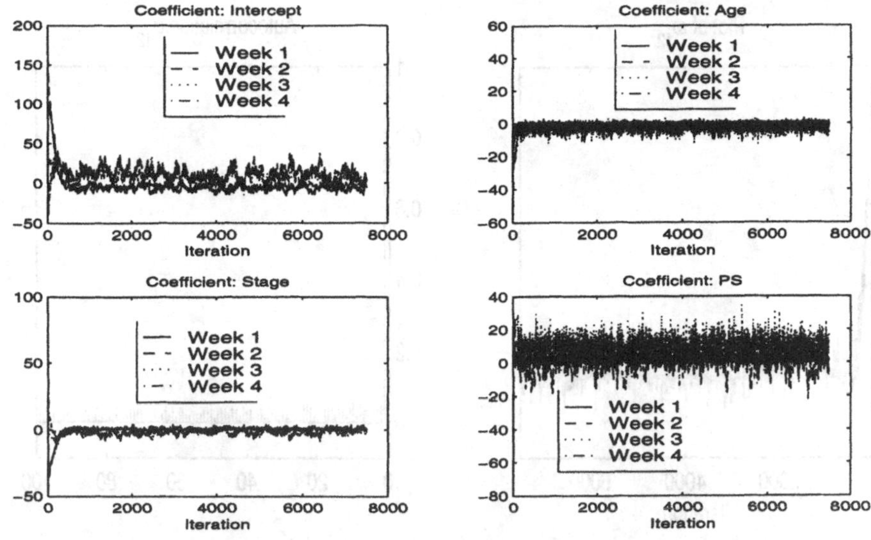

Exchangeable Prior on Graphs with $k_1 = k_2 = 100$
Figure 3: Trace of Posterior Samples of β

- We have also shown below the trace plot of the parameter ω_{12}. In Section 3.1, we defined ω_{12} and the other "interaction" parameters in terms of conditional log odds ratios (logits). Note that the posterior distribution has a mixture density with a positive mass at zero. Because of this feature, we provide the autocorrelation plot not of ω_{12} but rather of α_{12}. Recall from Section 4, that $\omega_{12} = Z_{12}\alpha_{12}$ and α_{12} has a continuous posterior. On a single run, the Raftery and Lewis (1992) convergence diagnostics all indicated convergence. Dependence factors were all below 3.0, suggesting that within-chain correlations did not impair convergence.

- The last 5,000 iterations were treated as samples from the posterior. The two runs gave almost identical results and the results were consolidated.

- Each iteration of the Gibbs sampler on an Ultrix machine required an average of 38 CPU seconds.

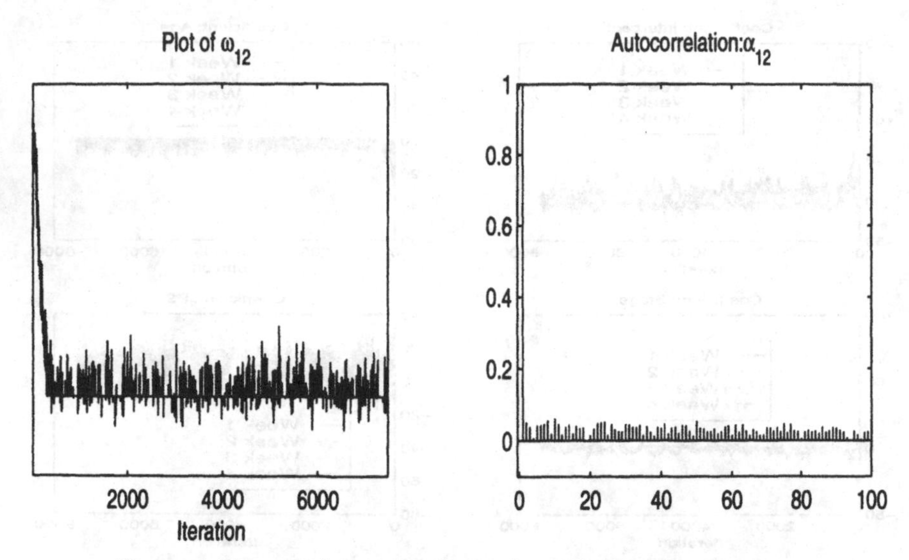

Figure 4: Trace and Autocorrelation for an interaction parameter

5 Results

5.1 Numerical results

This section presents some of our results. We used diffuse priors with $k_1 = k_2 = 100$. The hyperparameters k_1 and k_2 are as defined in Section 3. The variance for the ω's, σ^2 is also taken to be 100. It is interesting to compare the posterior probabilities of the edges for the two prior distributions on graphs: exchangeable and serial correlation. If a sample $(Z^{(k)}, k = 1, ..., K)$ is a sample from the posterior distribution of $[Z|y]$, then the posterior probability of edge e can be estimated by

$$\hat{P}(e) = \frac{1}{K} \sum_{k=1}^{K} \mathcal{I}\left(Z_e^{(k)} = 1\right) \tag{6}$$

where $\mathcal{I}$ is the indicator function.

Table 4: Posterior Probabilities of Edges
(Exchangeable prior on Graphical Models)

Edge	Prior Probability of Occurrence	Posterior Probability of Occurrence
12	.5	.11
13	.5	.43
23	.5	.49
14	.5	.38
24	.5	.38
34	.5	.46

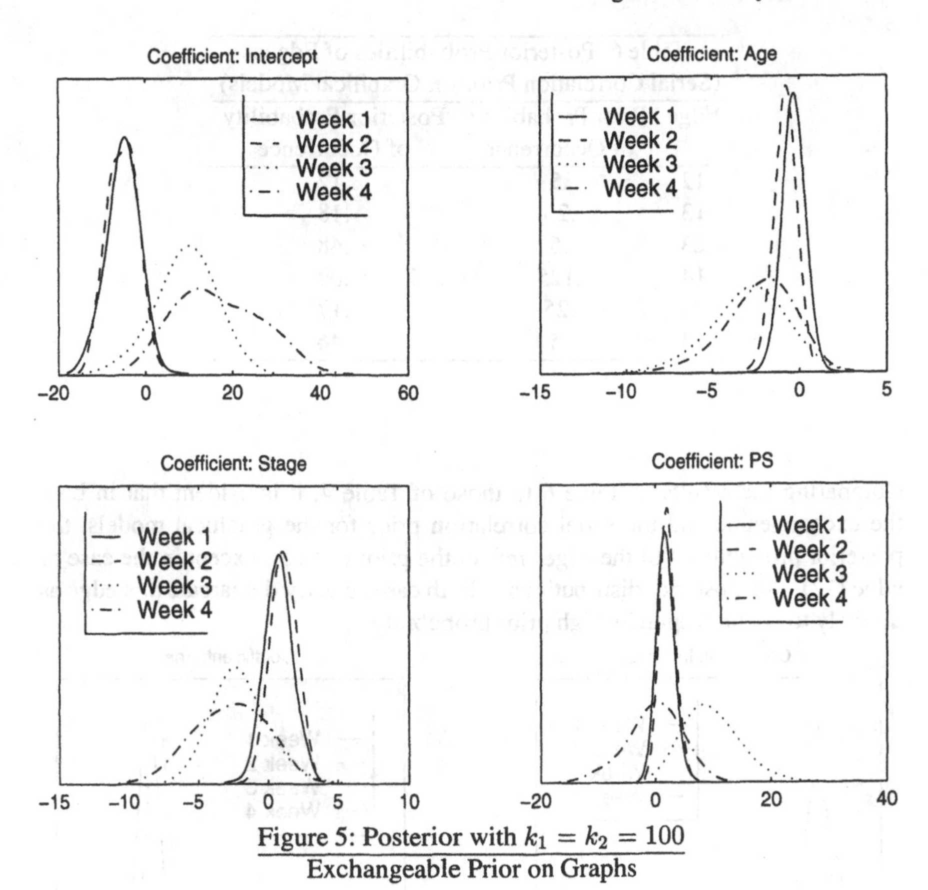

Figure 5: Posterior with $k_1 = k_2 = 100$
Exchangeable Prior on Graphs

Table 5: Summary Statistics of the Posterior Distribution
(Exchangeable Prior on Graphical Models)

Parameter	Description	Posterior Mean	95% credible Interval
β_{11}	Coefficient for Intercept at Week 1	-5.53	(-12.6, 0.82)
β_{12}	Coefficient for Age at Week 1	-0.36	(-1.60, 0.82)
β_{13}	Coefficient for Stage at Week 1	0.80	(-0.95, 2.65)
β_{14}	Coefficient for Ambulatory Status at Week 1	1.88	(-0.85, 5.01)
$\beta_{31} - \beta_{21}$	Difference in Intercept: Week 3 - Week 2	15.0	(2.1, 27.6)

All other differences, $\beta_t - \beta_{t-1}$ had a 95% credible interval containing 0

Table 6: Posterior Probabilities of Edges
(Serial Correlation Prior on Graphical Models)

Edge	Prior Probability of Occurrence	Posterior Probability of Occurrence
12	.5	.12
13	.25	.18
23	.5	.48
14	.125	.09
24	.25	.17
34	.5	.46

Comparing the results of Table 6 to those of Table 4, it is evident that in both the exchangeable and the serial correlation prior for the graphical models, the posterior probabilities of the edges reflect the prior choices, except in the case of edge (12). The posterior distributions in both cases clearly demarcate this edge as unlikely to occur despite its high prior probability.

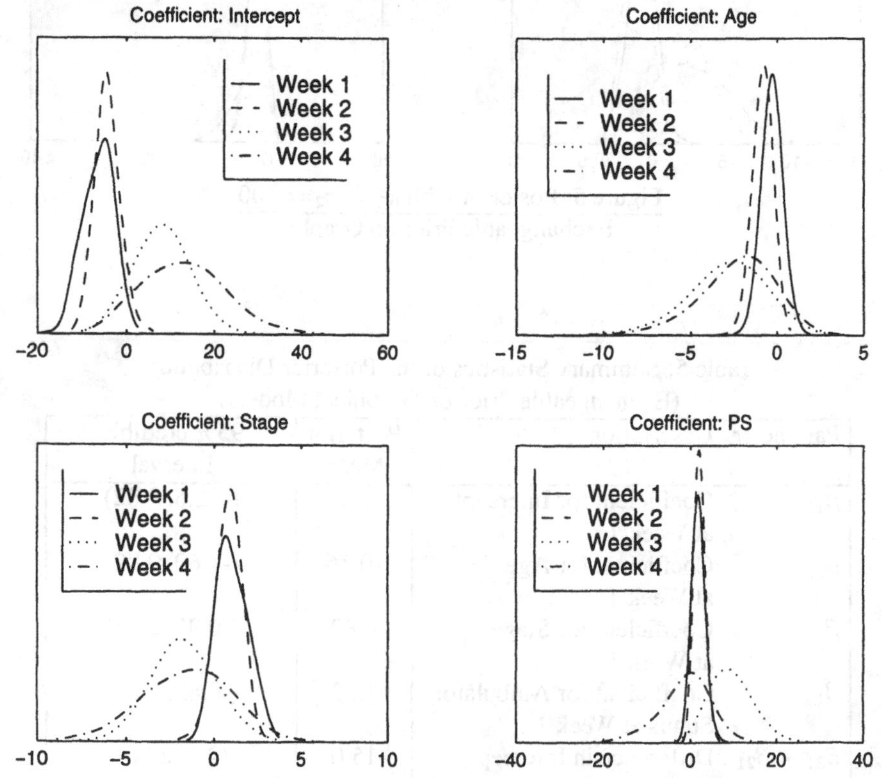

Figure 6: Posterior with $k_1 = k_2 = 100$
Serial Correlation Prior on Graphs

Table 7: Summary Statistics of the Posterior Distribution (Serial Correlation Prior on Graphical Models)

Parameter	Description	Posterior Mean	95% Credible Interval
β_{11}	Coefficient for Intercept at Week 1	-5.77	(-12.7, 0.02)
β_{12}	Coefficient for Age at Week 1	-0.33	(-1.51, 0.83)
β_{13}	Coefficient for Stage at Week 1	0.87	(-0.64, 2.72)
β_{14}	Coefficient for Ambulatory Status at Week 1	1.84	(-0.86, 5.13)
$\beta_{31} - \beta_{21}$	Difference in Intercept: Week 3 - Week 2	14.0	(0.9, 26.3)

All other differences, $\beta_t - \beta_{t-1}$ had a 95% credible interval containing 0

5.2 *Conclusions*

The numerical results obtained allow us to reach a few conclusions for ECOG trial 4593.

- The regression coefficients are not appreciably different from zero. This is partly a reflection of the relatively small sample size of 28 and a larger trial is needed to reach any reasonably reliable conclusion.

- The regression coefficients corresponding to the last two weeks are more dispersed than those corresponding to the first half of the study. Figures 5 and 6 show that there is a marked difference in the posterior distributions between these two sets. The credible intervals for the difference in the intercept coefficients, $(\beta_{31} - \beta_{21})$ given in Tables 5 and 7 reflect this. Recall that patients were given radiotherapy for the first two weeks only. Thus the statistical significance to the change in the regression parameters from week 2 to week 3 has a clinical interpretation, namely, toxicity sets in after treatment has stopped. This is also clear from Figure 1. The differences in the other coefficients (age, stage, and performance status) were not significant. This leads us to conclude that the longitudinal effect is more important than covariate effects as far as toxicity progression is concerned.

- The interaction between weeks 1 and 2 had very small posterior probabilities (about 0.11) for both graphical priors. The corresponding prior probabilities were 0.5. The other interactions between neighboring weeks did not show this feature. Allowing the data to find the most probable dependent structure allowed for this unexpected find. The observations for the first two weeks were only weakly dependent compared to the dependence between other neighboring weeks. Most patients developed toxicity some time during the treatment regimen (the first two weeks). The weak dependence suggests that there is little difference between the two weeks as far a time of onset is concerned. Once the symptoms are exhibited, they are usually present for the post-treatment period (the last two weeks) as well accounting for the stronger dependence between the later weeks.

- For the exchangeable prior case, the graphical model which allowed only for the interaction between week 3 and 4 was the most frequently visited model. However, the Bayes factor of this model relative to the independence model of no interactions was 1.05. By the interpretations for Bayes factors suggested in Kass and Raftery (1995) this is "not worth more than a bare mention". However, the Bayes factor of this model relative to the full model of all interactions was 54.5 ("strong evidence" against the full model). For the serial correlation prior, the graphical model with edges (23) and (14) had the highest Bayes factors relative to the independence and full models. In both cases, the full model allowing for all interactions fared very poorly.

- Preliminary survival and response-to-treatment data is reported in Mehta *et al.* (1997). Median survival for this patient population was 8.1 months. 23 of the 28 patients have died to date.

References

American Cancer Society (1996). *Cancer Facts and Figures*.

Bonney, G.E. (1987). Logistic regression for dependent binary observations, *Biometrics*, **43**, 951-973.

Carlin, B.P. and Chib, S. (1995). Bayesian model choice via Markov chain Monte Carlo methods, *Jour. Roy. Stat. Soc., Series B* **57**, 473-484.

Chib, S. and Greenberg, E. (1998). Analysis of Multivariate Probit Models *Biometrika*, **85**, 347-362.

Christensen, R. (1990). *Log-linear models*, Springer-Verlag.

Cowles, M.K., Carlin, B.P. and Connett, J.E. (1996). Bayesian Tobit modeling of longitudinal ordinal clinical trial compliance data with nonignorable missingness, *Jour. Amer. Stat. Assoc.* **91**, 86-98.

Cox, D.R. and Wermuth, N. (1993). Linear dependencies represented by chain graphs, em Statistical Science **8**, 204-218.

Cox, J.D. (1986). The role of radiotherapy in squamous, large cell, and adenocarcinoma of the lung, *Seminars in Oncology* **10**, 81-94.

Diggle, P.J., Liang, K.Y. and Zeger, S. L. (1994). *Analysis of longitudinal data*, Oxford Science Publications.

Fitzmaurice, G.M., Laird, N.M. and Rotnitzky, A.G. (1993). Regression models for discrete longitudinal responses, *Statistical Science* **8**, no. 3, 284-309.

Gelfand, A.E. and Smith, A.F.M. (1990). Sampling-based approaches to calculating marginal densities, *Jour. Amer. Stat. Assoc.* **85**, 398-409.

Gelman, A. and Rubin, D.R. (1992). A single series from the Gibbs sampler provides a false sense of security, In: *Bayesian Statistics* **4**, J.M. Bernardo et al.(ed.), 625-631.

Gilks, W.R., Best, N.G. and Tan, K.K.C. (1995). Adaptive rejection Metropolis sampling within Gibbs sampling, *Applied Statistics* **44**, 455-472.

Harrison, P.J. and Stevens, C.F. (1976). Bayesian forecasting (with disc.), *Jour. Roy. Stat. Soc., Series B*, **38**, 205-247.

Ibrahim, J.G. and Laud, P.W. (1991). On Bayesian analysis of generalized linear models using Jeffrey's prior, *Jour. Amer. Stat. Assoc.* **86**, no. 416, 981-986.

Kass, R.E. and Raftery, A.E. (1995). Bayes factors, *Jour. Amer. Stat. Assoc.* **90**, 773-795.

Mehta, M.P., Tannehill, S.P., Adak, S., Froseth, C., Martin L., Petereit, D.G., Wagner, H.W., Fowler, J.F. and Johnson, D. (1997). Phase II trial of hyperfractionated accelerated radiation therapy (HART) for unresectable non-small cell lung cancer: Preliminary results of ECOG 4593, *ECOG report.*

Murphy, G.P., Lawrence, W. and Lenhard, R.E. (1995). *Clinical oncology*, American Cancer Society.

Raftery, A.E. and Lewis, S. (1992). How many iterations in the Gibbs sampler? In *Bayesian Statistics 4*, (ed. J.M. Bernardo, J.O. Berger, A.P. Dawid and A.F.M. Smith), Oxford University Press, 763-774.

Soffer, E.E., Mitros, F., Doornbos, J.F., Friedland, J., Launspach, J. and Summers, R.W. (1994). Morphology and Pathology of radiation induced esophagitis. Double-blind study of naproxen vs placebo for prevention of radiation injury, *Dig Dis Sci* **39:3**, 655-660.

Withers, H.R., Taylor, J.M.G. and Maciejewski, B. (1988). The hazard of accelerated tumor clonogen repopulation during radiotherapy, *Acta Oncology* **27**, 131-146.

Analysis of hospital quality monitors using hierarchical time series models

Omar Aguilar and Mike West
ISDS, Duke University

ABSTRACT

The VA management services department invests considerably in the collection and assessment of data to inform on hospital and care-area specific levels of quality of care. Resulting time series of *quality monitors* provide information relevant to evaluating patterns of variability in hospital-specific quality of care over time and across care areas, and to compare and assess differences across hospitals. In collaboration with the VA management services group we have developed various models for evaluating such patterns of dependencies and combining data across the VA hospital system. This paper provides a brief overview of resulting models, some summary examples on three monitor time series, and discussion of data, modelling and inference issues. This work introduces new models for multivariate non-Gaussian time series. The framework combines cross-sectional, hierarchical models of the population of hospitals with time series structure to allow and measure time-variations in the associated hierarchical model parameters. In the VA study, the within-year components of the models describe patterns of heterogeneity across the population of hospitals and relationships among several such monitors, while the time series components describe patterns of variability through time in hospital-specific effects and their relationships across quality monitors. Additional model components isolate unpredictable aspects of variability in quality monitor outcomes, by hospital and care areas. We discuss model assessment, residual analysis and MCMC algorithms developed to fit these models, which will be of interest in related applications in other socio-economic areas.

1 Introduction

The performance monitoring system of the US Department of Veterans Affairs (VA) collects, reports and analyses data from over 170 hospitals. Policy interests lie in accurately estimating measures of hospital-level performance in key areas of health care provision, and in assessing changes over time in such measures to monitor impact of internal policy changes. Ultimately, these issues are related to the development of management and economic incentives designed to encourage and promote care provision at sustained and acceptable levels. As described in Burgess et al (1996), the quality monitor data are compiled annually and encompass a range of inpatient, outpatient and long term care activities at each of the VA medical centers. Each hospital records data on the total numbers of individuals who were exposed to a specific and well-defined outcome in each monitor area, and the number for whom that outcome occurred. There is a related covariate, referred to as the *DRG predictor*, based on exogenous information providing some correction for hospital/monitor specific case-mix and characteristics of patient population profiles. Further details appear in Burgess, Christiansen, Michalak and Morris (1996 and in related unpublished work), who discuss aspects of data analysis and hierarchical modelling (Christiansen and Morris 1997) in this context. Our study is concerned with evaluating

- patterns of variability over time, in hospital-monitor and area-specific performance measures across a selection of quality monitors, and
- patterns of dependencies between sets of monitors, in addition to and in combination with assessment of time-variations.

Christiansen and Morris have developed a variety of Bayesian hierarchical models for the observed outcomes, including regressions on the DRG predictor and hospital-specific parameters drawn from a hospital population prior (see references above). From this basis, we explore multiple-monitor time series models to address the above key questions. We focus on three specific monitors introduced in Section 2 where we provide some basic data description and perspective. Section 3 reviews our new models; these are multiple monitor, binomial/logit models in which hospital-specific random effects are related through time via a multivariate time series model. In addition to systematic patterns of variation over time, the models include components of unpredictable variability in outcome probabilities. In Section 4 we describe summary inferences for all hospitals and monitors, investigation of aspects of model fit, and examples of additional possible uses of the models. We conclude with summary comments about the study, and an appendix briefly summaries model theory and computation.

Our work relates closely to what are now essentially standard approaches in health care outcomes research and institutional comparisons – hierarchical Bayesian models that allow for various components of heterogeneity involving nested random effects. A recent contribution and overview appears in Normand et al (1997), for example. Our work is novel in several methodological respects, and draws on developments in Cargnoni, Müller and West (1997) related to both latent time series structure and computational algorithms. The methods will prove useful to workers dealing with longitudinal data structures in various socio-economic fields. Finally, more extensive details on the data analysis and modelling summarised here appears in an on-line report by West and Aguilar (1997).

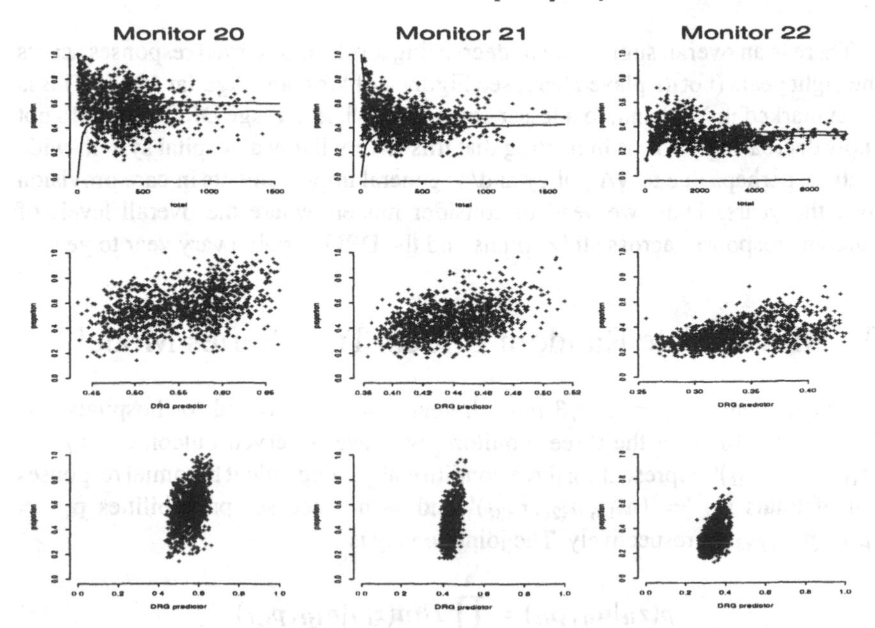

FIGURE 1. Observed and DRG-based predicted proportions on all three monitors and eight years of data (1988-95). The lower row is the DRG predictor on the probability scale.

2 Exploratory Data Analysis

The outcomes in the monitor areas represent annual numbers of individuals under a binary classification in an area of basic medical or psychiatric health care. The response recorded is the number of individuals who failed to return for an outpatient visit within 30 days of discharge out of the total number of annual discharges. Monitor M20 measures outcomes for *General Psychiatric*, M21 for *Substance Abuse Psychiatric*, and M22 for *Basic Medical and Surgical* care. Low return rates are indicative of low "quality" in these specific care areas. The data here covers years 1988 to 1995 for 152 hospitals having complete records.

Figure ?? displays the raw data on the three monitors separately, but combined over all eight years. There are $8 \times 152 = 1216$ observations per frame for the 8 years of data on $I = 152$ hospitals. The graphs plot the observed proportions of successes in each monitor against the total numbers of patients in each case, and then against the DRG-based predicted proportions. Super-imposed on each graph in the first row are approximate 99% intervals under marginal binomial distributions that assume "success" probabilities fixed at the overall average proportions for each monitor. Many observations lie outside these bands indicating considerable levels of over-dispersion relative to binomial models. This extra-binomial variation is to be explained by models that describe how the individual probabilities vary across hospitals and across years, using a combination of regression on the DRG predictor and random effects.

There is an overall suggestion of decreasing levels of observed responses across the eight years (not displayed here; see Figure 4 of West and Aguilar 1997). This is most marked in M20 and, to a lesser extent M21. The average DRG values do not show decreasing patterns indicating that this is very likely a hospital system-wide feature, perhaps due to VA policy and/or general improvements in care provision over the years. Thus, we need to consider models where the overall levels of outcome responses across all hospitals and the DRG variable vary year to year.

3 Multivariate Random Effects Time Series Model

Consider monitors $j = 1, 2, 3$ in each year $t = 1, \ldots, 8$ and for hospitals $i = 1, \ldots, I = 152$. On the three monitors, we have observed outcomes $\mathbf{z}_{it} = (z_{i1t}, z_{i2t}, z_{i3t})'$, representing three conditionally independent binomial responses out of totals $\mathbf{n}_{it} = (n_{i1t}, n_{i2t}, n_{i3t})'$ and with "success" probabilities $\mathbf{p}_{it} = (p_{i1t}, p_{i2t}, p_{i3t})'$, respectively. The joint density is

$$p(\mathbf{z}_{it}|\mathbf{n}_{it}, \mathbf{p}_{it}) = \prod_{j=1}^{3} Bin(z_{ijt}|n_{ijt}, p_{ijt}). \tag{1}$$

The p_{ijt} are hospital-specific parameters to be estimated, and the totals n_{ijt} are assumed uninformative about p_{ijt}. Our models for the p_{ijt} combine within-year random effects/hierarchical components with multivariate time series structure, now detailed.

3.1 Regression and hierarchical/random effects structure

For hospital i, the DRG-based predicted proportion of "successes" d_{ijt} is supposed to predict p_{ijt} on the basis of system-wide studies of patient case-mix profiles and historical data. Following Burgess et al (1996) we adopt a logistic regression as follows. Let $\mu_{ijt} = \log(p_{ijt}/(1-p_{ijt}))$ and $x^*_{ijt} = \log(d_{ijt}/(n_{ijt}-d_{ijt}))$, and define $x_{ijt} = x^*_{ijt} - x^*_{\cdot jt}$ where $x^*_{\cdot jt}$ is the arithmetic mean of the x^*_{ijt} across all hospitals $1 = 1, \ldots, I$. The logistic regression is $\mu_{ijt} = \beta_{0jt} + \beta_{1jt}x_{ijt}$ where the regression parameters β_{0jt} and β_{1jt} are unrestricted. In terms of the vectors $\boldsymbol{\mu}_{it} = (\mu_{i1t}, \mu_{i2t}, \mu_{i3t})'$, $\boldsymbol{\beta}_{0t} = (\beta_{01t}, \beta_{02t}, \beta_{03t})'$, $\boldsymbol{\beta}_{1t} = (\beta_{11t}, \beta_{12t}, \beta_{13t})'$, and matrices $\mathbf{X}_{it} = \text{diag}(x_{i1t}, x_{i2t}, x_{i3t})$, we have

$$\boldsymbol{\mu}_{it} = \boldsymbol{\alpha}_{it} + \mathbf{X}_{it}\boldsymbol{\beta}_{1t} + \boldsymbol{\nu}_{it} \tag{2}$$

where $\boldsymbol{\alpha}_{it} = (\alpha_{i1t}, \alpha_{i2t}, \alpha_{i3t})'$ and $\boldsymbol{\nu}_{it} = (\nu_{i1t}, \nu_{i2t}, \nu_{i3t})'$. For each monitor j and year t, the quantity α_{ijt} is an *absolute* hospital-specific random effect representing systematic variability that is related over time within each hospital. The ν_{ijt} represent residual, unpredictable variability, independent over time and across hospitals and monitors. The model assumes $\boldsymbol{\nu}_{it} \sim N(\boldsymbol{\nu}_{it}|\mathbf{0}, \mathbf{V})$ with monitor-specific variances v_1^2, v_2^2 and v_3^2 on the diagonal of the matrix $\mathbf{V}$, admitting cross-monitor dependencies through the covariances in $\mathbf{V}$.

Key to assessing quality levels are the *relative* random effects $\epsilon_{ijt} = \alpha_{ijt} - \beta_{0jt}$, i.e., hospital-specific deviations from the population level β_{0jt}. In terms of these quantities,

$$\mu_{it} = \beta_{0t} + \mathbf{X}_{it}\beta_{1t} + \epsilon_{it} + \nu_{it} \quad (3)$$

where $\epsilon_{it} = (\epsilon_{i1t}, \epsilon_{i2t}, \epsilon_{i3t})'$. This class of models accounts for variability over time in the hospital/monitor parameters β_{0t} and β_{1t} as well as the random effects α_{it} that together will account for the high levels of observed extra-binomial variability.

The parameter β_{0t} represents the hospital system-wide average in corrected responses on the logit scale. Management policies across the VA system, and improvements (or otherwise) in care provision impacting all hospitals in similar ways contribute to changes in β_{0t} from year to year. We do not currently impose structure on the hospital/monitor population parameters β_{0t} and β_{1t}. Predictive models, by contrast, would require evaluation of expert opinion about the reasons behind any inferred time evolution and the use of this in phrasing appropriate model extensions.

The ϵ_{it} terms represent hospital-specific departures from the system-wide underlying level β_{0t}. In Section ?? we model time series dependence over the years in the ϵ_{it} quantities to explain the structured variability over time. However, time series models introduce partial stochastic constraints so that some of the evident variation in the logit parameters μ_{it} will be unexplained by the regression and hospital-specific random effects ϵ_{it}. Hence the need for the residual random components ν_{it}.

3.2 *Time series structure of random effects*

Time series structure in the hospital-specific ϵ_{it} is modelled via a vector autoregression of order one – or VAR(1) model. This is a natural, interpretable model incorporating the view that there should be stability in the ϵ_{it} values within each hospital over such a short number of years. This stability represents true quality levels and any changes beyond this reflect unexplained random variations year to year due to the characteristics of the patient sample in each hospital. With such a short time span, more complex models are largely untenable. Moreover, the VAR(1) model has the desirable consequence that the annual marginal distributions of the hospital-specific effects are the same across years. The model structure is

$$\epsilon_{it} = \Phi\epsilon_{i,t-1} + \omega_{it} \quad (4)$$

over years t and independently across hospitals i within each year. Here $\Phi = \text{diag}(\phi_1, \phi_2, \phi_3)$ is the diagonal matrix of monitor-specific autoregressive coefficients. The ω_{it} terms are innovations vectors, with $\omega_{it} \sim N(\omega_{it}|\mathbf{0}, \mathbf{U})$ conditionally independent over time. In any year t, we have the implied marginal distribution $\epsilon_{it} \sim N(\epsilon_{it}|\mathbf{0}, \mathbf{W})$; the within-year relative random effects are a random sample from a zero-mean normal distribution. This is consistent with a view of no global changes in the hospital population makeup, i.e., with variability in expected levels being essentially constant over the short period of years once the DRG predictor and any system-wide changes are accounted for through β_{1t} and β_{0t}, respectively. Changes in relative performance of hospitals can therefore be assessed across years.

It follows that $\mathbf{W}$ satisfies $\mathbf{W} = \Phi\mathbf{W}\Phi + \mathbf{U}$, so that correlation patterns in $\mathbf{U}$ and $\mathbf{W}$, depend on the autoregressive parameters. In particular, for each monitor pair j, h we have covariance elements $\mathbf{W}_{jh} = \mathbf{U}_{jh}/(1 - \phi_j\phi_h)$. The matrix $\mathbf{W}$ represents the variability in the systematic components of corrected quality levels across the entire hospital population, the related variability in changes in relative quality levels year-to-year, and the dependencies between such quality measures across the three monitors. The autoregressive parameters ϕ_j will generally be close to one, lying in part of stationary region $0 < \phi_j < 1$. Large values of ϕ_j imply high positive correlations between the ϵ_{it} in a given hospital over the years. This is consistent with the view that a hospital that is generally "good" in a specific monitor/care in one year will have a high probability of remaining "good" the next year, and vice versa.

In terms of the absolute random effects α_{it} we have a centred VAR(1) model

$$\alpha_{it} = \beta_{0t} + \Phi(\alpha_{i,t-1} - \beta_{0,t-1}) + \omega_{it} \tag{5}$$

for $t > 1$, with yearly margins $N(\alpha_{it}|\beta_{0t}, \mathbf{W})$. Another feature to note concerns the time series structure of the combined hospital-specific random effects $\epsilon_{it} + \nu_{it}$ above. The addition of the residual/noise terms ν_{it} to the VAR(1) process ϵ_{it} modifies the correlation structure giving a VARMA(1,1) model with $N(\epsilon_{it} + \nu_{it}|\mathbf{0}, \mathbf{W} + \mathbf{V})$ yearly margins. Note that the overall levels of random effects variability, and the associated overall measures of cross-monitor dependencies, are represented through $\mathbf{W} + \mathbf{V}$. Our current model leaves $\mathbf{V}$ and $\mathbf{W}$ unrelated *a priori*, but the framework obviously permits the assessment of potential similarities in posterior inferences.

Finally, we assume constant values of Φ and $\mathbf{U}$ in the time series components. This assumption could be relaxed to allow for differing variances across hospitals and/or years as may be desirable for other applications.

3.3 Prior distributions

Inference is based on posterior distributions for all model parameters and random effects under essentially standard reference/uninformative priors for: (a) the annual population parameters β_{0t} and β_{1t}, (b) the population residual variance matrix $\mathbf{V}$, and (c) the variance-covariance matrix $\mathbf{U}$; the prior is completed with independent uniform priors for the autoregressive parameters ϕ_j on (0,1).

4 Results for the VA Data

Various marginal posterior distributions from the multiple monitor analysis are reported and discussed here. First, Figure ?? provides summaries of the marginal

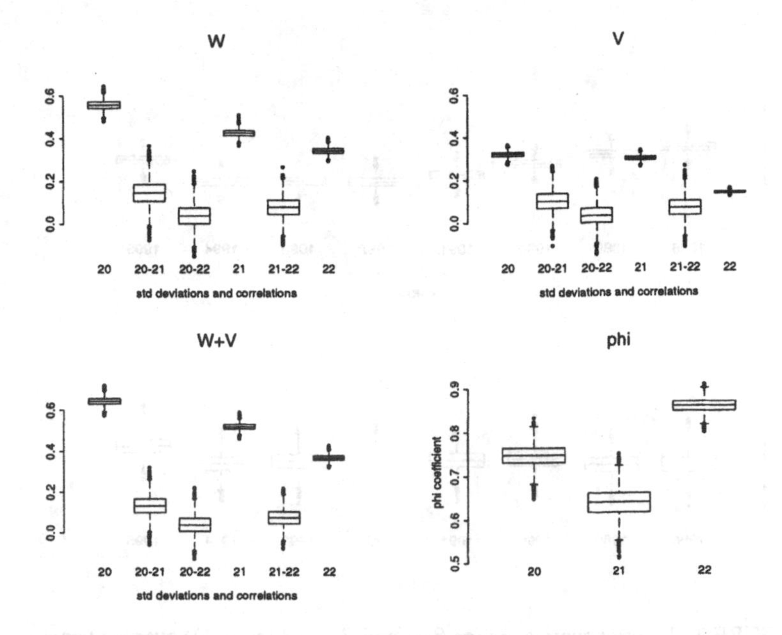

FIGURE 2. Posterior summaries for AR coefficients Φ and standard deviations and correlations of $\mathbf{V}$, $\mathbf{W}$ and $\mathbf{V}+\mathbf{W}$.

posteriors for correlations and standard deviations in $\mathbf{W}$, $\mathbf{V}$ and $\mathbf{W}+\mathbf{V}$. Here, and below, boxplots are centred at posterior medians, drawn out to posterior quartiles, and have notches at points 1.5 times the interquartile range beyond the edges of each box. These graphs indicate low overall correlations in each matrix. We focus on the key matrix $\mathbf{W}+\mathbf{V}$ that measures within-year, cross-monitor structure. Denoting posterior means by "hats" and writing $\mathbf{E}$ for the column eigenvector matrix of $\hat{\mathbf{W}}+\hat{\mathbf{V}}$, we have

$$\hat{\mathbf{W}}+\hat{\mathbf{V}} = \begin{pmatrix} 0.417 & 0.044 & 0.009 \\ 0.044 & 0.271 & 0.014 \\ 0.009 & 0.014 & 0.136 \end{pmatrix}, \ \mathbf{E} = \begin{pmatrix} 0.961 & -0.275 & -0.015 \\ 0.273 & 0.957 & -0.097 \\ 0.041 & 0.089 & 0.995 \end{pmatrix}.$$

This indicates correlations between M20 and M21 of around 0.13, between M20 and M22 of 0.04 and between M21 and M22 of 0.07, so supporting the suggestion that the correlation between M20 and M21 might be higher than any other combination, in view of the care areas of origination. The eigenvalues of $\hat{\mathbf{W}}+\hat{\mathbf{V}}$ are roughly 0.43, 0.26 and 0.13, so the principal components explain roughly 52%, 32% and 16% of variation; each of the eigenvectors is therefore relevant, and no data reduction seems appropriate. Posterior uncertainty about the variance matrices, and the eigen-structure, does not materially impact these qualitative conclusions. To exemplify this, the full posterior sample produces the following approximate posterior means and 95% intervals for the three eigenvalues of $\mathbf{W}+\mathbf{V}$:

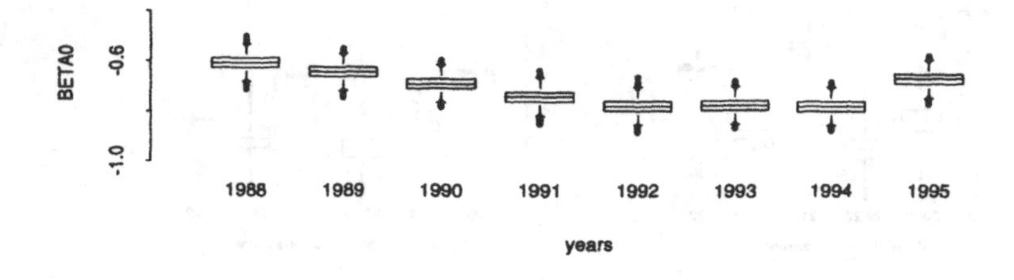

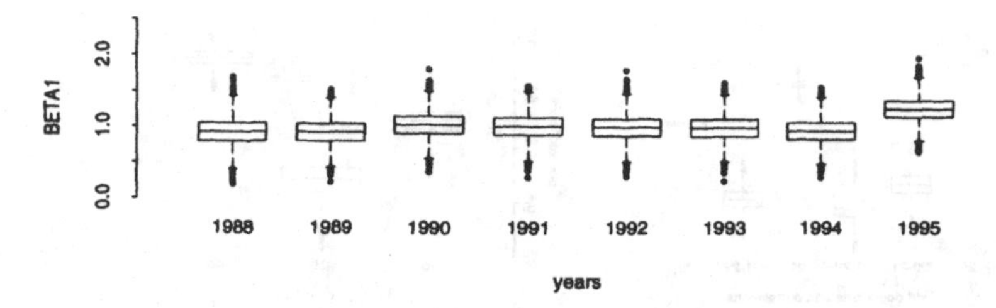

FIGURE 3. Posterior summaries for β_{02t} and β_{12t} (Monitor 22) over the years.

0.42 (0.38-0.48), 0.25 (0.22-0.29), 0.13 (0.11-0.16), closely comparable to the estimates quoted above. Evidently, the eigenvector matrix $\mathbf{E}$ is dominated by the diagonal terms, and all three are close to unity. Note that the eigenvector matrix would be the identity were the monitors uncorrelated. The first column represents an average of M20 and M21 dominated by the M20 psychiatric care component. The second column represents a contrast between M20 and M21 and the final column almost wholly represents M22 alone, and to the extent that the coefficients for M20 and M21 are non-ignorable, contrasts the two psychiatric care monitors with the general medical. The levels of correlation structure are clearly low for these specific monitors, perhaps surprisingly so for the first two in closely related care areas.

The lower right frame of Figure ?? provide summaries of the marginal posteriors for the three autoregressive parameters in Φ. These indicate highly significant dependence structures in each case, with inferred values of ϕ in the ranges $0.7-0.8$ for M20, $0.6-0.75$ for M21 and $0.8-0.9$ for M22. The dependence in the random effects time series is high in each case, but there are apparent differences between M22 and the other two monitors, perhaps associated different health care areas.

There are meaningful differences in the β_0 parameters across the eight years in each of the three monitors. The main feature is a general decreasing trend in β_0 over the years for all three monitors, more markedly so for Monitors M20 and M21. This corresponds to generally increased probabilities of return for outpatient visits within 30 days of discharge (i.e., increased "quality"), and the appar-

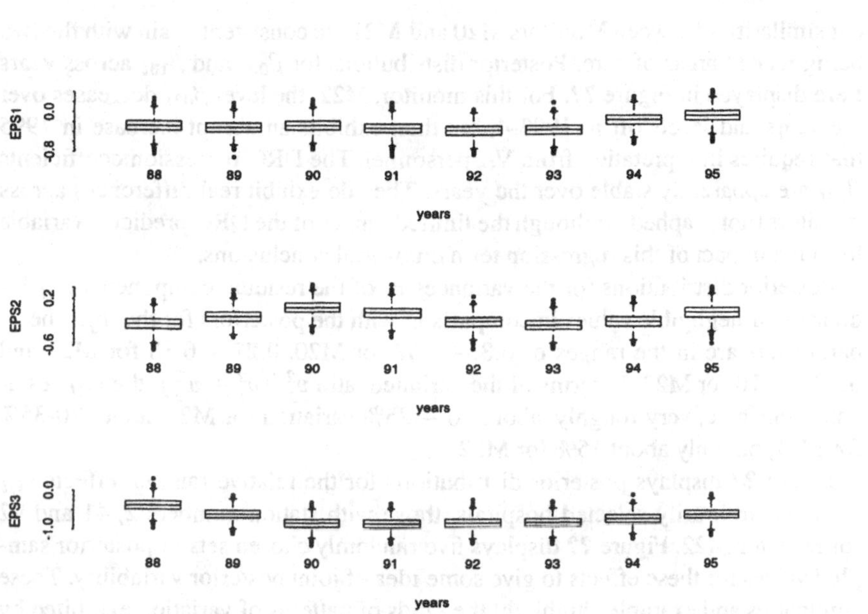

FIGURE 4. Posterior summaries for hospital-specific random effects ϵ_{ijt} on Monitor 22 over years t in hospitals $i = 2, 41$ and 92.

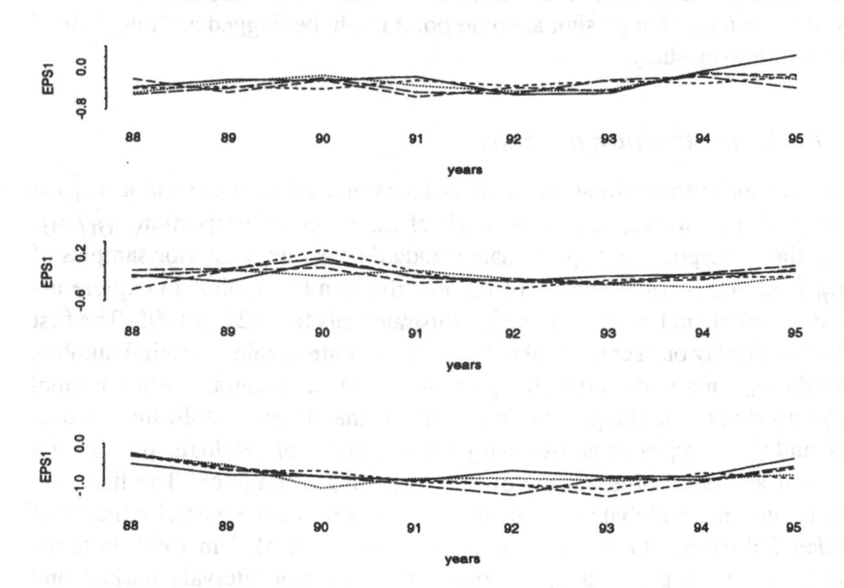

FIGURE 5. Selected posterior samples for hospital-specific random effects ϵ_{ijt} on Monitor 22 over years t in hospitals $i = 2, 41$ and 92.

ent similarities between Monitors M20 and M21 are consistent again with the two being related areas of care. Posterior distributions for β_{03t} and β_{13t} across years t are displayed in Figure ??. For this monitor, M22, the level β_{03t} decreases over the years and levels off in 1993-4, but then exhibits an abrupt increase in 1995 that requires interpretation from VA personnel. The DRG regression coefficients β_{13t} are apparently stable over the years. They do exhibit real differences across monitors (not graphed), although the limited ranges of the DRG predictor variable limit the impact of this regression term on overall conclusions.

Posterior distributions for the variances v_j of the residual components ν_{ijt} indicate non-negligible values in comparison with the posteriors for the w_j. The v_j parameters are in the ranges of $0.3 - 0.37$ for M20, $0.27 - 0.33$ for M21 and $0.12 - 0.16$ for M22. In terms of the variance ratio $v_j^2/(v_j^2 + w_j^2)$, the ϵ_{ijt} residuals contribute, very roughly, about $20 - 25\%$ variation for M20, about 30-35% for M21, but only about 15% for M22.

Figure ?? displays posterior distributions for the relative random effects ϵ_{ijt} for three arbitrarily selected hospitals, those with station numbers 2, 41 and 92 for Monitor M22. Figure ?? displays five randomly chosen sets of posterior sampled values for these effects to give some idea of joint posterior variability. These summaries and examples highlight the kinds of patterns of variation exhibited by the random effects within individual hospitals – the plots indicate the smooth, systematic dependence structure over time that is naturally expected. Hospitals that have tended to be below the population norm in terms of its proportions of outcomes in recent years will be expected to maintain its below average position this year, so that the ϵ parameters of this hospital will tend to be of the same sign. Hospitals whose effects change sign at some point might be flagged as "interesting" cases for follow-up study.

4.1 Residual structure analysis

The model implies approximate normality of the standardised data residuals $e_{ijt} = (y_{ijt} - \mu_{ijt})/s_{ijt}$ where, y_{ijt} is the logit of the observed proportion z_{ijt}/n_{ijt} and s_{ijt} the corresponding approximate standard deviation. Posterior samples of the μ_{ijt} lead to posterior samples of the e_{ijt} that can be graphed to explore aspects of model fit, and misfit. Figure ?? illustrates this for M21 in 1995. The first four frames display one such sample of residuals–plotted against hospital number, against the n_{it}, in a normal quantile plot, and finally a histogram with a normal density superimposed. The general impression is that of good conformity to normality, and this is repeated across many other samples of residuals, providing a measure of assurance of adequacy of this modelling assumption. The final two frames provide more global assessments. Here we explore the posterior means of the *ordered* observation residuals across all hospitals for M21 in 1995; in terms of a normal quantile plot with approximate 95% posterior intervals marked, and in terms of a histogram with the normal density superimposed. Again, adequacy of the normality assumption is indicated.

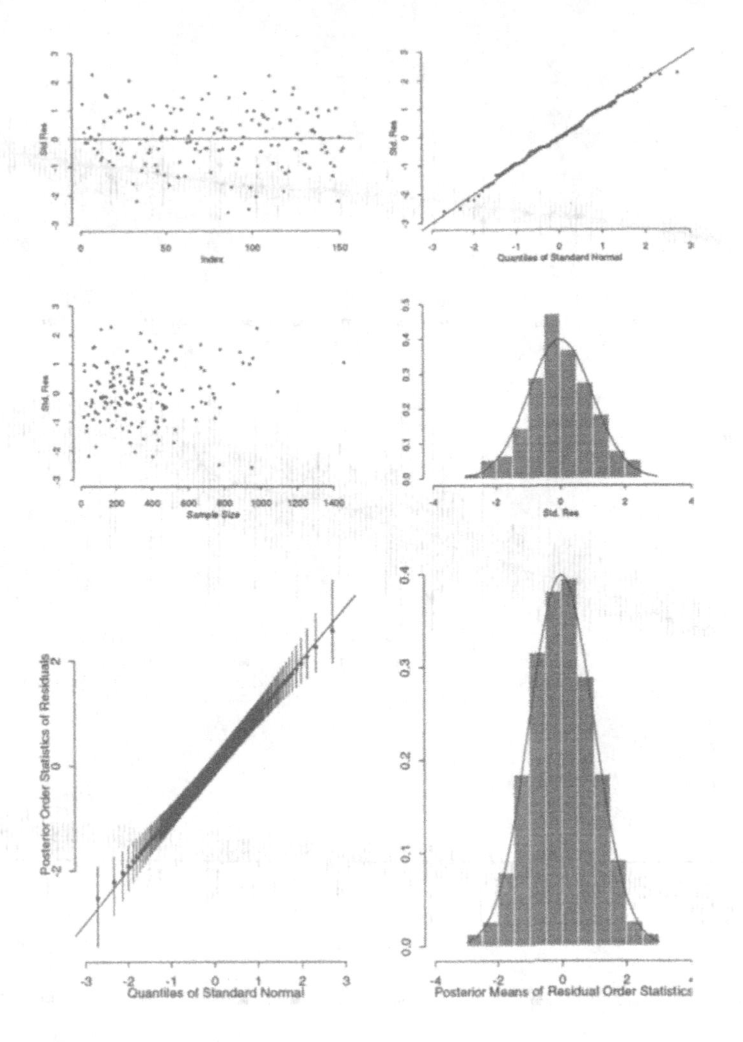

FIGURE 6. Draw from the posterior distribution of the observation residuals e_{ijt} across hospitals in M21 1995 (the four upper graphs). Posterior means of the ordered residuals e_{ijt} (the lower two graphs).

4.2 Summary Inferences for Monitor M21 in 1995

To illustrate additional uses of the model, we focus on M21 in 1995. Some summary posterior inferences appear in Figure ??, where a few specific hospitals are highlighted (with intervals drawn as dashed lines). Figure ??(a) displays approximate 95% intervals for the actual outcome probabilities p_{ijt}, ordering hospitals by posterior medians. Interval widths reflect posterior uncertainty which is a decreasing function of sample size. Hospitals with low n_{ijt} have wider intervals–

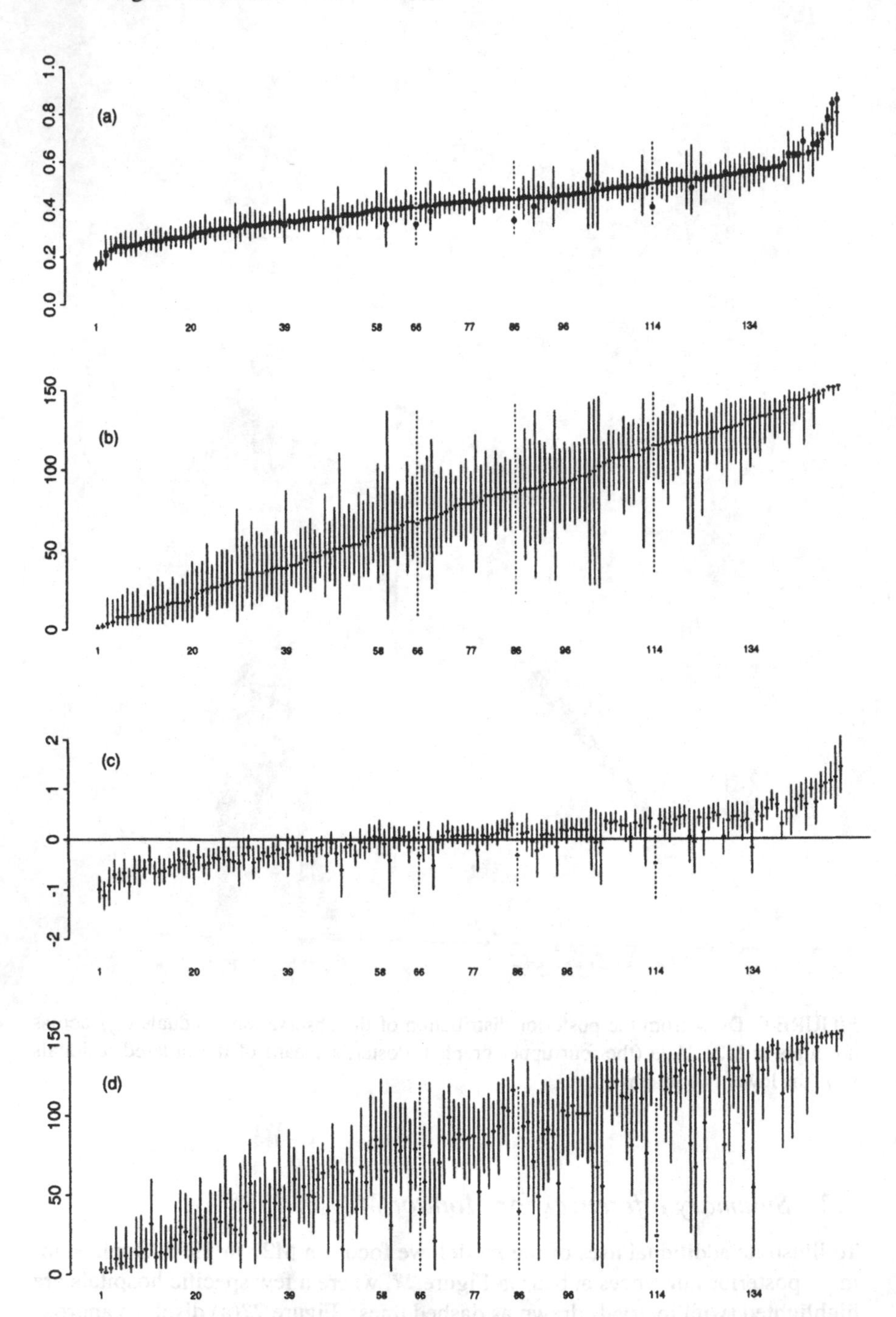

FIGURE 7. Posterior 95% intervals for all hospitals on M21 in 1995. (a) Outcome probabilities p_{ijt} (with dots marking the observed proportions), and (b) corresponding ranks of hospitals based on posterior for ordered p_{ijt}. (c) Composite random effects $\epsilon_{ijt} + \nu_{ijt}$, and (d) the corresponding ranks based on posterior for ordered random effects. The hospitals are graphed in order of posterior medians of p_{ijt} in each frame.

hospitals 66, 86 and 114, for example. Figure ??(c) displays corresponding intervals for the $\epsilon_{ijt} + \nu_{ijt}$. The "low" hospitals have random effects lower than average, indicating that the model has adapted to the extreme observations. Adaptation is constrained by the model form and also by the the high values of the DRG predictor. There is a general increasing trend in the random effects consistent with the ordering by outcome probabilities, though the pattern is not monotonic as the probabilities include the effects of the DRG predictor whereas the $\epsilon_{ijt} + \nu_{ijt}$ measure purely relative performance levels.

Figure ??(b) displays 95% posterior intervals for the *ranks* of the hospitals according to the p_{ijt}, and Figure ??(d) the intervals for ranks of the $\epsilon_{ijt} + \nu_{ijt}$. Evidently, the four or five hospitals with the highest (lowest) *estimated* outcome probabilities have very high (low) ranks, indicating that their *true* outcome probabilities are very likely to be among the largest (smallest) few across the system. Note that ranks based on p_{ijt} summarise absolute performance, impacted by patient-mix and other confounding factors, and ranks based on $\epsilon_{ijt} + \nu_{ijt}$ represent relative quality levels once these factors are accounted for via the model; the latter provide a firmer basis for assessing relative performance due to hospital-specific policies and practices. This is evident in the cases of hospitals 66, 86 and 114 noted above, for which appropriately lower rankings are indicated in Figure ??(d) than in the "unadjusted" rankings in Figure ??(b). Even then, there is high uncertainty about rankings for most hospitals, not only those with small sample sizes, reflecting the inherent difficulties in ranking now well understood in this and other areas (e.g., Normand et al 1997).

5 Summary Comments

We have presented a new class of multiple monitor, hierarchical random effects time series models to evaluate patterns of dependencies in series of annual measures of health care quality in the VA hospital system. A critical feature of our work has been the identification of several components of variability underlying within-year variation and across-year changes in observed quality levels. We split the hospital-specific variation in into two components: a partially systematic and positively dependent VAR component ϵ_{it}, and a purely unpredictable component ν_{it}. The latter component is non-negligible and contributes between 15-30% of the total random effects variance on the logit scale. Lower contributions in *general medical discharge* monitor than either of the psychiatric monitors. Hence hospital-specific levels of M22 are more stable over time and hence more predictable. Our multiple monitor time series models isolate changes over time and dependencies among such changes in the hospital-specific random effects across the three monitors. Though dependencies across monitors exist, they are apparently quite small. Summary graphs of posterior inferences for specific monitor:year choices provide useful insight into the distribution of outcome probabilities across the hospital system, about relative levels of performance, and about changes over time in such levels. There are evident changes in system-wide levels β_{0t} that require consideration and interpretation though such is beyond the data-analytic scope of our study. Display and assessment of posterior samples of model components provide insight in aspects of model fit and underpin the reported posterior inferences.

It should be clear that the models and computational methods (briefly detailed in the appendix below) may be applied in other contexts, and that the basic binomial sampling model may be replaced by other non-Gaussian forms as context demands. We expect that the work will be developed in such ways and that the models will find use in various other applications in the socio-economic arena.

Acknowledgements

This work was performed in collaboration with Jim Burgess and Ted Stefos of the VA Management Science Group, Bedford MA, and with important contributions and discussions with Cindy Christiansen, Carl Morris and Sarah Michalak. Our models and data exploration build on foundational contributions in hierarchical modelling of the hospital monitor data by Christiansen and Morris (see Burgess et al 1996). Corresponding author is Mike West, Institute of Statistics and Decision Sciences, Duke University, Durham, NC 27708-0251, USA; `http://www.stat.duke.edu`

Appendix: Model Theory and Computation

The full joint posteriors for all quantities $\{\mathbf{V}, \mathbf{U}, \Phi\}$ and $\{\beta_{0t}, \beta_{1t}, \alpha_{i,t}, \mu_{it}\}$ for all t was simulated via customised MCMC methods. The structure is related to that in Cargnoni, Müller and West (1997), although the methods involve substantial novelty as we are analysing new models. Results reported are based on a correlation-breaking subsample of 5,000 draws from a long chain of 100,000 iterates. The collection of conditional posteriors is briefly summarised here. Some are simulated easily, and are detailed without further comment; others require Metropolis-Hastings steps, which are noted as needed. By way of notation, for any set of parameters ξ, write ξ^- for the remaining parameters combined with the full data set $\mathbf{Z}$. The conditionals are as follows.

- For the β_{0t} we have conditional posteriors $N(\beta_{01}|\mathbf{b}_{01}, \mathbf{W}/I)$ and, for $t > 1$, $N(\beta_{0t}|\mathbf{b}_{0t}, \mathbf{U}/I)$ where

$$\mathbf{b}_{01} = \sum_{i=1}^{I} \alpha_{i1}/I \text{ and } \mathbf{b}_{0t} = \sum_{i=1}^{I} \{\alpha_{it} - \Phi(\alpha_{i,t-1} - \beta_{0,t-1})\}/I.$$

- For the β_{1t} we have conditionals $N(\beta_{1t}|\mathbf{b}_{1t}, \mathbf{B}_{1t})$ where

$$\beta_{1t} = \mathbf{B}_{1t}\sum_{i=1}^{I}\mathbf{X}'_{it}\mathbf{V}^{-1}(\mu_{it}-\alpha_{it}) \text{ and } \mathbf{B}_{1t}^{-1} = \sum_{i=1}^{I}\mathbf{X}'_{it}\mathbf{V}^{-1}\mathbf{X}_{it}.$$

- For $\mathbf{V}$ we have the conditional inverse Wishart $Wi(\mathbf{V}^{-1}|8I, \mathbf{H})$ where $\mathbf{H} = \sum_{i=1}^{I}\sum_{t=1}^{8}\nu_{it}\nu'_{it}$.

- For $\mathbf{U}^{-1}$ we have conditional density

$$p(\mathbf{U}^{-1}|\{\epsilon_{it}\}, \Phi) \propto a(\mathbf{U})Wi(\mathbf{U}^{-1}|7I, \mathbf{G})$$

with $\mathbf{G} = \sum_{i=1}^{I}\sum_{t=2}^{8}(\epsilon_{it}-\Phi\epsilon_{i,t-1})(\epsilon_{it}-\Phi\epsilon_{i,t-1})'$ and $\mathbf{W} = \Phi\mathbf{W}\Phi+\mathbf{U}$. With the inverse Wishart component as a Metropolis-Hastings proposal distribution, a candidate value $\mathbf{U}^*$ has acceptance probability $\min\{1, a(\mathbf{U}^*)/a(\mathbf{U})\}$ where $a(\mathbf{U}) = |\mathbf{W}|^{-I/2}\exp(-\text{tr}(\mathbf{W}^{-1}\mathbf{A})/2)$ and $\mathbf{A} = \sum_{i=1}^{I}\epsilon_{i1}\epsilon'_{i1}$.

- For Φ we have conditional posterior

$$p(\Phi|\{\epsilon_{it}\}, \mathbf{U}) \propto p(\Phi)N(\epsilon_{i1}|\mathbf{0}, \mathbf{W})\prod_{t=2}^{8}N(\epsilon_{it}|\Phi\epsilon_{i,t-1}, \mathbf{U})$$

where $\mathbf{W} = \Phi\mathbf{W}\Phi+\mathbf{U}$. Write $\phi = (\phi_1, \phi_2, \phi_3)'$ for the diagonal of Φ, and $\mathbf{E} = \text{diag}(\epsilon_{i,t-1})$. Then the density is proportional to $p(\Phi)c(\Phi)N(\phi|\mathbf{f}, \mathbf{F})$ where $\mathbf{f} = \mathbf{F}\sum_{i=1}^{I}\sum_{t=2}^{8}\mathbf{E}'\mathbf{U}^{-1}\epsilon_{it}$ and $\mathbf{F}^{-1} = \sum_{i=1}^{I}\sum_{t=2}^{8}\mathbf{E}'\mathbf{U}^{-1}\mathbf{E}$. A Metropolis-Hastings step generates a candidate Φ^* from the truncated multivariate normal here, and accepts it with probability $\min\{1, c(\Phi^*)/c(\Phi)\}$ where

$$c(\Phi) = |\mathbf{W}|^{-I/2}\exp(-\text{tr}(\mathbf{W}^{-1}\mathbf{A})/2)$$

with $\mathbf{A} = \sum_{i=1}^{I}\epsilon_{i1}\epsilon'_{i1}$ and $\mathbf{W} = \Phi\mathbf{W}\Phi+\mathbf{U}$.

- The conditional for the α_{it} is complicated but easily structured and sampled with dynamic linear modelling ideas. Write $\mathbf{y}_{it}$ for the vector of logit transforms of the observed outcome proportions. Then, for each t,

$$\begin{aligned}\tilde{\mathbf{y}}_{it} &= \alpha_{it}+\eta_{it} \text{ with } \eta_{it} \sim N(\eta_{it}|\mathbf{0}, \mathbf{V}+\mathbf{S}_{it}),\\ \alpha_{it} &= \beta_{0t}+\Phi(\alpha_{i,t-1}-\beta_{0,t-1})+\omega_{it}\end{aligned}$$

where $\tilde{\mathbf{y}}_{it} = \mathbf{y}_{it} - \mathbf{X}_{it}\beta_{1t}$ and $\mathbf{S}_{it} = \text{diag}(s_{i1t}, s_{i2t}, s_{i3t})$ is the diagonal matrix of approximate data variances in the normal-logit model. This is a multivariate dynamic linear model with known variance matrices and state vector sequence α_{it}. Standard results for simulation in DLMs now apply, as in West and Harrison (1997, chapter 15).

- For μ_{it} we have $p(\mu_{it}|\mu_{it}^-, \mathbf{z}_{it}) \propto p(\mathbf{z}_{it}|\mathbf{n}_{it}, \mu_{it})p(\mu_{it}|\mu_{it}^-)$ where the likelihood function $p(\mathbf{z}_{it}|\mathbf{n}_{it}, \mu_{it})$ is the product of the three binomial-logit functions, and $\mu_{it}|\mu_{it}^- \sim N(\mu_{it}|\alpha_{it} + \mathbf{X}_{it}\beta_{1t}, \mathbf{V})$. A Metropolis-Hastings step generates a candidate μ_{it}^* from the posterior based on the normal-logit approximation to the likelihood function. This delivers the proposal density $\mu_{it}|\mathbf{y}_{it} \approx N(\mu_{it}|\mathbf{m}_{it}, \mathbf{Q}_{it})$ where

$$\mathbf{Q}_{it} = (\mathbf{V}^{-1} + \mathbf{S}_{it}^{-1})^{-1} \text{ and } \mathbf{m}_{it} = \mathbf{Q}_{it}(\mathbf{V}^{-1}(\alpha_{it} + \mathbf{X}_{it}\beta_{1t}) + \mathbf{S}_{it}^{-1}\mathbf{y}_{it}).$$

The acceptance probability is $\min\{1, a(\mu_{it}^*)/a(\mu_{it})\}$ where $a(\cdot)$ is the ratio of the exact binomial to the approximate normal-logit likelihood.

References

Burgess, J.F., Christiansen, C.L., Michalak, S.E., and Morris, C.N. (1996) Risk adjustment and economic incentives in identifying extremes using hierarchical models: A profiling application using hospital monitors, *Manuscript*, Management Science Group, U S Department of Veterans Affairs, Bedford MA.

Cargnoni, C., Müller, P., and West, M. (1997) Bayesian forecasting of multinomial time series through conditionally Gaussian dynamic models, *Journal of the American Statistical Association*, **92**, 640-647.

Christiansen, C.L., and Morris, C.N. (1997) Hierarchical Poisson regression modeling, *Journal of the American Statistical Association*, **92**, 618-632.

Normand, S.T., Glickman, M.E., and Gatsonis, C.A. (1997) Statistics methods for profiling providers of medical care: Issues and applications. *Journal of the American Statistical Association*, **92**, 803-814.

West, M., and Harrison, P.J. (1997) *Bayesian Forecasting and Dynamic Models*, (2nd Edn.), New York: Springer Verlag.

West, M., and Aguilar, O. (1997) Studies of quality monitor time series: The V.A. hospital system, Report for the VA Management Science Group, Bedford, MA. *ISDS Discussion Paper #97-22*, Duke University. Available as `ftp://ftp.stat.duke.edu/pub/WorkingPapers/97-22a.ps`

Spatio-Temporal Hierarchical Models for Analyzing Atlanta Pediatric Asthma ER Visit Rates

Bradley P. Carlin
Hong Xia
Owen Devine
Paige Tolbert
James Mulholland

ABSTRACT Prevalence and mortality rates of asthma among children have been increasing over the past twenty years, particularly among African American children and children of lower socioeconomic status. In this paper we investigate the link between ambient ozone and pediatric ER visits for asthma in the Atlanta metro area during the summers of 1993, 1994, and 1995. Our statistical model allows for several demographic and meteorologic covariates, spatial and spatio-temporal autocorrelation, and errors in the ozone estimates (which are obtained from a kriging procedure to smooth ozone monitoring station data). As with most recent Bayesian analyses we employ a MCMC computing strategy, highlighting convergence problems we encountered due to the high collinearity of several predictors included in early versions of our model. After providing our choice of prior distributions, we present our results and consider the issues of model selection and adequacy. In particular, we offer graphical displays which suggest the presense of unobserved spatially varying covariates outside the city of Atlanta, and reveal the value of our errors in covariates approach, respectively. Finally, we summarize our findings, discuss limitations of (and possible remedies for) both our data set and analytic approach, and compare frequentist and Bayesian approaches in this case study.

1 Introduction

Prevalence and mortality rates of asthma among children have been increasing over the past twenty years (Evans et al., 1987), particularly among African American children and children of lower socioeconomic status (Centers for Disease Control, 1992). While a number of studies (see Tolbert et al., 1997, for a review) of emergency room (ER) visits for asthma have indicated that the disease's symptoms can be worsened by air pollution, there is less agreement with regard to the roles played by various pollutants. Specifically, ozone, particulate matter, acid aerosols, and nitrogen dioxide have each been suggested as having the strongest association with asthma exacerbation. A similar situation arises with respect to weather factors, with temperature, barometric pressure, humidity, precipitation, solar insolation, and wind speed all being potentially important covariates. Naturally arising allergens, such as pollen and mold counts, also play a role; these typically vary by year and by day within a particular year. Finally, certain sociodemographic variables, such as race, gender, age, and socioeconomic status (SES), clearly affect asthma prevalence, and are thought by some experts to interact with the pollution variables (say, due to a lower prevalence of air conditioners in socioeconomically deprived households); see for example Thurston et al. (1992).

In order to assess and quantify the impact of ozone and other air pollutants on asthma exacerbation, Tolbert et al. (1997) recently performed a spatio-temporal investigation of air quality in Atlanta in relation to pediatric emergency room (ER) visits for asthma. Visits to the major emergency care centers in the Atlanta metropolitan area during the summers of 1993, 1994, and 1995 were studied relative to ozone, particulate matter, nitrogen oxides, pollen, mold, and temperature. At the aggregate level, Poisson regression was employed to model the daily count of asthma presentations throughout Atlanta as a function of average Atlanta-wide exposures, as well as time covariates. A case-control treatment of the data was also employed, in which logistic regression was used to compare asthma cases to the ER patients presenting with non-asthma diagnoses with respect to exposures of interest, controlling for temporal and demographic covariates. In the case-control analysis, zip code of residence was used to link patients to spatially resolved ozone estimates obtained via a universal kriging procedure. Both analyses estimated the relative risk per standard deviation (20 ppb) increase in the maximum 8-hour ozone level as 1.04, with both findings significant at the 0.01 level. A similar exposure-response relationship was observed for particulate matter (PM-10). However, when both ozone and PM-10 were fit in a single model, the significance of both predictors was compromised, likely due to the high collinearity of the two pollutant variables.

The Poisson regression model used by Tolbert et al. (1997) accounted for long-term temporal trends, but not for temporal autocorrelation, nor spatial correlation in visit rates in nearby geographic regions. The logistic regression analysis incorporated spatially-resolved ozone observations (obtained via the kriging procedure), but did not account for the prediction error associated with these kriged estimates, nor the fact that they too are spatially correlated.

In this paper we use a spatio-temporal Poisson regression model which allows for both spatial and temporal correlation in the data, and accounts for error in the ozone covariate. Owing to the high dimensionality and complexity of the model, we adopt a Bayesian analytic approach, with computer implementation via Markov chain Monte Carlo (MCMC) methods. Our results indicate similar (though somewhat attenuated) estimates of relative risk due to ozone exposure as in Tolbert et al. (1997), but also provide a natural framework for drawing smoothed geographic information system (GIS) maps of the zip-specific relative risks on various days, and for investigating how changes in the covariates affect these maps. We offer tabular and graphical displays elucidating the impact of our errors in covariates model, as well as residual maps which suggest the presence of spatial variation as yet unaccounted for by the covariates in our data set. Finally, we discuss limitations of our data set and analytic approach, and contrast the frequentist and Bayesian approaches to this case study.

2 Spatio-temporal statistical modeling

2.1 *Description of data set*

Our study is an historic records-based investigation of pediatric asthma ER visits to the seven major emergency care centers in the Atlanta metropolitan statistical area (MSA) during the summers of 1993, 1994, and 1995. ER visit counts were recorded for each of the $I = 162$ zip codes in the MSA over each of 92 summer days (from June 1 through August 31) during 1993–1995, i.e., the total time period includes $T = 92 \times 3 = 276$ days. The ER data consist of C_{it}, the number of pediatric ER asthma visits in zip code i on day t, and n_{it}, the corresponding total number of pediatric ER visits. In addition, the sociodemographic variables age, gender, race, zip code of residence, and Medicaid payment indicator (a crude surrogate for SES) were obtained for each case from hospital billing records. Multiple ER visits by a single patient on a single day were counted only once.

The following air quality indices were assessed: ozone, particulate matter less than 10 μm in diameter (PM-10), total oxides of nitrogen (NO_x), pollen, mold, and various meteorological variables including temperature and humidity. With the exception of ozone, data on the exposure variables were insufficient for their spatial resolution. The daily ozone measurements at the various monitoring stations (eight in 1993 and 1994, and ten in 1995) were converted to zip-specific estimates in the following way. First, daily values for the maximum 1-hour average and maximum 8-hour (consecutive) average were calculated for each station. We then considered six different kriging methods available in the GIS ARC/INFO: four ordinary kriging methods (which modeled the variogram as linear, exponential, spherical, or Gaussian), and two universal kriging methods (which incorporated either a linear or quadratic structural component, and modeled the remaining variogram linearly). Based on meteorological factors, we would expect urban ozone distribution to have a structural component. Comparing the quality of the fitted variograms and plotted isopleths of the ozone predictions and variances, we selected the universal method that used the linear function to model drift (as well as a linear function for the rest of the variogram, hence one with no sill). Final kriged values and corresponding variance estimates were then obtained for each zip centroid via bilinear interpolation.

2.2 *Hierarchical model development*

In this subsection we describe a hierarchical modeling approach which accounts for spatio-temporal correlation in the fitted rates, as well as measurement error in the ozone covariate. The primary goal of the study is to evaluate a possible exposure-response relationship between childhood asthma and ambient ozone level. Recall from the previous subsection that only the ozone variable is both spatially and temporally resolved. The remaining exposure variables are only temporally resolved (i.e., available by day); conversely, the demographic covariates are only spatially resolved (i.e., available by zip; we would not expect these covariates to change much over our study period).

We model the observed pediatric asthma ER visit counts C_{it} using a Poisson likelihood,

$$C_{it} \sim Poisson(E_{it} \exp(\mu_{it})) ,$$

where E_{it} is the expected count from zip i on day t, and μ_{it} is the corresponding log-relative risk, where $i = 1, 2, \ldots, I$ and $t = 1, 2, \ldots, T$. The expected counts E_{it} are thought of as known constants. One way to obtain E_{it} is to use external standardization relying on an appropriate standard reference table for pediatric asthma visit rates. Since we lack such a reference table, we adopt the simpler alternative of *internal* standardization using the data themselves, i.e., we compute $E_{it} = n_{it}\hat{p}$, where $\hat{p} = \sum_{it} C_{it} / \sum_{it} n_{it}$, the observed overall visit rate. If reliable independent data are available that suggest that this rate varies widely by hospital (not just as a function of the demographic covariates), we may refine our estimates of E_{it} accordingly.

For the log-relative risks, we began with the following model:

$$\begin{aligned} \mu_{it} \;\; = \;\; & \mu + \beta_Z Z_{i,t-1} + \beta_P PM10_{t-1} + \beta_T Temp_{t-1} + \beta_L Pollen_{t-1} \\ & + \beta_M Mold_{t-1} + \beta_D Day_t + \beta_{D2} Day_t^2 + \beta_{Y94} Year94_t \\ & + \beta_{Y95} Year95_t + \alpha_A AvAge_i + \alpha_M PctMale_i \\ & + \alpha_S PctHiSES_i + \alpha_B PctBlack_i + \theta_i + \phi_i \end{aligned} \tag{1}$$

where Z_{it} is the true ozone level in zip i on day t, and Day_t indexes day of summer (i.e., $Day_t = t \bmod 92$). We include both linear and quadratic terms for day of summer, since our earlier exploratory data analysis suggested that asthma ER visit rates follow a rough U-shape over the course of each summer, with June and August higher than July. (We remark that while some previous studies have suggested that ER visits for asthma vary by day of the week, since the previous analysis of our data by Tolbert et al. found no significant differences in this regard, we do not pursue this option further.) $Year94_t$ and $Year95_t$ are indicators (0–1 dummy variables) for days in 1994 and 1995, respectively, so that 1993 is taken as the reference year. $PM10_t$, $Temp_t$, $Pollen_t$, and $Mold_t$ are the daily exposure values for PM-10, temperature, pollen, and mold, respectively, each observed at a single Atlanta location each day.

Note that all of these variables (and the ozone variable) are lagged one day in our model; that is, the level of each on a particular day is assumed to impact the asthma ER visit rate on the *following* day. (As such, there are now only $T = 91 \times 3 = 273$ days worth of usable data in our sample, instead of 276.) This decision was based in part on a review of the results for various lag times used in previous studies. Also, our main pollutant of interest, ozone, is an irritant that potentiates the effects of other exposures, leading to some expected time delay in producing a response. Furthermore we are not looking at onset of symptoms but the eventual behavior of deciding to go to the ER, which will further increase the likely time lag.

The zip-specific covariates, $AvAge_i$, $PctMale_i$, $PctHiSES_i$, and $PctBlack_i$, are average age, percent male, percent high socioeconomic status, and percent black race of those ER visitors from zip i, respectively. The parameter coefficients for these variables (in particular, whether each is significantly different from 0) will provide information about the role each variable plays in asthma ER visit relative risk.

The parameters corresponding to the above variables are fixed effects in the model, and we begin by assuming noninformative (flat) priors for them, hoping they will be well-identified by the data. The remaining two parameters, θ_i and ϕ_i, are random effects to capture overall heterogeneity and spatial clustering in the fitted rates, respectively. That is, we assume the following priors:

$$\boldsymbol{\theta} \mid \tau \sim N\left(\mathbf{0}, \frac{1}{\tau}\mathbf{I}\right) \quad \text{and} \quad \boldsymbol{\phi} \mid \lambda \sim CAR(\lambda) , \tag{2}$$

where τ and λ are hyperparameters that control the magnitude of the random effects, and CAR stands for a *conditionally autoregressive* structure (Besag et al., 1991), which encourages similarity of rates in adjacent zips. The most common form of this prior (Bernardinelli and Montomoli, 1992) has joint distribution proportional to

$$\lambda^{I/2} \exp\left[-\frac{\lambda}{2} \sum_{i \; adj \; j} (\phi_i - \phi_j)^2\right] \;\propto\; \lambda^{I/2} \exp\left[-\frac{\lambda}{2} \sum_{i=1}^{I} m_i \phi_i (\phi_i - \bar{\phi}_i)\right] ,$$

where i *adj* j denotes that regions i and j are adjacent, $\bar{\phi}_i$ is the average of the $\phi_{j \neq i}$ that are adjacent to ϕ_i, and m_i is the number of these adjacencies. This CAR prior is a member of the class of *pairwise difference priors* (Besag et al., 1995), which are identified only up to an additive constant. To permit the data to identify the grand mean μ, then, we also add the constraint $\sum_{i=1}^{I} \phi_i = 0$. A consequence of our prior specification is that

$$\phi_i \mid \phi_{j \neq i} \sim N(\bar{\phi}_i \, , \, 1/(\lambda m_i)) \, .$$

Vague (but proper) gamma hyperprior distributions for τ and λ complete this portion of the model specification.

An alternate model suggested by Waller et al. (1997) to accommodate spatio-temporal interactions not already accounted for by our exposure and demographic covariates would replace θ_i and ϕ_i by θ_{it} and ϕ_{it}. These new random effects can be viewed as surrogates for unobserved spatio-temporally varying covariates. If temporal dependence remains in the data even after fitting the time-varying parameters above, the definition of adjacency could be expanded to include temporal adjacency as well (e.g., also count $\phi_{i,t-1}$ and $\phi_{i,t+1}$ as being adjacent to ϕ_{it}). Alternatively, simple autoregressive models for the temporal effects might be considered. Since we found little evidence of excess temporal autocorrelation in our dataset, we do not pursue these options further.

Besides being the only spatio-temporally resolved covariate, the ozone variable is unique in that it is subject to prediction error introduced by the kriging process used to produce zip-specific estimates from the original point process observations. Ignoring the errors in this covariate may produce biased results and unrealistic interval estimates for various model quantities of interest. (In addition, for linear models with additive error structure, point estimates of the corresponding model coefficients will tend to be attenuated in the presence of measurement error, but no such general rule applies to our nonlinear, hierarchical model; see Carroll, Ruppert and Stefanski, p.23). This suggests adding an *errors in covariates*component to the model (Bernardinelli et al., 1997; Xia and Carlin, 1998). We introduce both prediction error and spatial correlation into the ozone covariate by letting

$$X_{it} \mid Z_{it} \overset{ind}{\sim} N(Z_{it}, \sigma^2_{it}), \text{ and } \mathbf{Z}_t \mid \nu_t \overset{ind}{\sim} CAR(\nu_t), \tag{3}$$

where X_{it} are the estimated ozone values (maximum eight hour average) obtained from our kriging procedure, which also produces the associated variance estimates σ^2_{it}. The ν_t are CAR scale parameters controlling the magnitude of spatial similarity in adjacent regions. The model specification is completed by assuming the ν_t arise exchangeably from a vague gamma hyperprior.

3 Computing

Our MCMC implementation used Gibbs sampling (Gelfand and Smith, 1990; Carlin and Louis, 1996, Sec. 5.4.2) updating steps for parameters with available conjugate full conditional distributions (i.e., τ, λ, and the ν_t), and Metropolis algorithm (Metropolis et al., 1953; Carlin and Louis, 1996, Sec. 5.4.3) steps for the remaining parameters. Convergence was monitored for a representative subset of the parameters using plots of the sampled values themselves, as well as sample autocorrelations and Gelman and Rubin (1992) convergence diagnostics.

Implementing this algorithm on the full model (1) resulted in convergence failure. This is not surprising in view of the lack of classical statistical significance previously found for many of the predictors in (1): there is insufficient information in the data to separately identify the effects of temperature, pollen, mold, average age, and percent male. This means that, given our noninformative prior structure, the MCMC algorithm will wander aimlessly forever, unable to decide amongst the infinite number of equally plausible values for these parameters. A primary culprit in this information paucity problem is the occasionally strong collinearity among the covariates. For example, the average age and percent male covariates exhibit little spatial variability (collinearity with the intercept) and are very similar overall (collinearity with each other). On the other hand, the distributions of SES and race appear to be more spatially diverse, though they are also fairly similar, both more or less distinguishing the city center from the suburbs.

We also found that the posterior medians of the heterogeneity parameters θ_i were tightly centered around zero, indicating no significant additional heterogeneity in the data beyond that explained by the CAR priors. As a result, in what follows we remove the above terms from our log relative risk model, resulting in what we refer to as the *basic* model,

$$\begin{aligned} \mu_{it} &= \mu + \beta_Z(Z_{i,t-1} - 50) + \beta_D Day_t + \beta_{D2} Day_t^2 \\ &\quad + \beta_{Y94} Year94_t + \beta_{Y95} Year95_t \\ &\quad + \alpha_S(PctHiSES_i - 0.5) + \alpha_B(PctBlack_i - 0.5) + \phi_i \,, \end{aligned} \tag{4}$$

which retains kriged ozone data, four temporal and two demographic covariates, and the space-varying clustering random effect. Also, the ozone, SES, and race covariates have been centered around rough midpoint values to ease collinearity with μ.

4 Results

4.1 Covariate and model selection

Using model (4) with vague $Gamma(1, 7)$ hyperpriors on λ and ν_t with the CAR priors in (2) and (3), we ran 5 independent, initially overdispersed MCMC chains for 1400 iterations each. Visual inspection of the sampling chains as well as the aforementioned convergence diagnostics suggested deletion of the first 400 iterations from each chain as pre-convergence "burn-in". The remaining $5 \times 1000 = 5000$ samples are summarized in Table 1, which shows the posterior medians, 95% central credible sets, and the fitted relative risks corresponding to the posterior medians for the parameters of interest in model (4). The sign and magnitude of β_Z answers our primary research question concerning the link between childhood asthma and ozone level. That is, there is a positive association between ozone level and asthma ER visits, with an increase of approximately 2.6% for every 20 ppb increase in ozone level (8 hour max). The lower endpoint of the associated 95% posterior credible set is zero to four decimal places, showing this finding to be "Bayesianly significant" (two-sided) at precisely the .05 level.

Turning to the other covariates, the significant temporal variables are the regression coefficients for Day, Day^2, and $Year95$, confirming the U-shape of asthma rates over any given summer and the increase in asthma rates during the final summer of observation. Regarding the demographic covariates, the association between pediatric asthma ER visits and the percent black in a zip is significant and positive; the fitted relative risk shows that a theoretical all-black zip

	posterior median	95% posterior credible set	fitted relative risk	comments
μ	−0.0618	(−0.1867, 0.0623)	—	
β_Z	0.0013	(0.0000, 0.0024)	1.026	per 20 ppb ozone increase
β_D	−0.0136	(−0.0177, −0.0088)	—	
β_{D2}	0.0002	(0.0001, 0.0002)	—	
β_{Y94}	0.0221	(−0.0581, 0.0923)	1.022	1994 vs 1993
β_{Y95}	0.2854	(0.2175, 0.3489)	1.330	1995 vs. 1993
α_S	−0.0638	(−0.5534, 0.2993)	0.938	all high SES vs. all low SES
α_B	1.0181	(0.7615, 1.2679)	2.768	all black vs. all non-black
λ	5.2185	(3.2526, 8.6324)	—	

TABLE 1. Fitted relative risks for the parameters of interest in the Atlanta pediatric asthma ER visit data, basic model.

would have relative risk nearly three times that of a comparable all-nonblack zip. Whereas in the report by Tolbert et al. (1997) the SES surrogate (paid by Medicaid) was significantly associated with asthma, in the current analysis the 95% credible set for this value broadly spans the null value, perhaps an indication of some instability in the model introduced by the correlated race variable.

Having settled on a satisfactory set of explanatory variables, we might now want to investigate the separate impacts of the spatial smoothing and errors in covariates aspects of our basic model. To do this, we re-ran our MCMC algorithm for three reduced models: one which eliminated the errors in covariates terms (by dropping equation (3) from the model and setting $X_{it} = Z_{it}$), one which eliminated the spatial smoothing terms (by setting $\phi_i = 0$), and one which eliminated both. To the order of accuracy in our Monte Carlo calculations, the point and interval estimates of the effect of ozone on log-relative risk, β_Z, remain constant across all four models, implying a constant increase in risk of about 2.5% per 20 ppb increase in ozone. The confidence interval for the spatial smoothing control parameter λ is a bit narrower when the errors in the ozone covariate are not acknowledged, suggesting slightly increased evidence of spatial clustering in this case.

Table 2 compares the fit of our basic model with that of the three reduced versions described above using the posterior loglikelihood score, computed as the sample median of the logged values of the Poisson likelihoods evaluated at the MCMC samples. While the usual chi-square asymptotics for differences in minus twice the loglikelihood score are not appropriate in our Bayesian hierarchical model setting, it seems clear from the table that including the spatial smoothing terms does lead to substantial improvement in fit. This in turn suggests the presence of some as yet unidentified spatially varying covariate which impacts pediatric ER asthma visit count. On the other hand, removal of the errors in covariates

model	loglikelihood score
basic	–5250
without errors in covariates	–5251
without spatial terms	–5359
without errors in covariates or spatial terms	–5359

TABLE 2. Median posterior loglikelihood scores, competing Bayesian models for the Atlanta pediatric asthma ER visit data.

portion of the model seems to have no appreciable impact on model fit. Apparently the relatively modest overall effect of ozone means that subtle changes in its modeling are not detected by the rather crude loglikelihood score statistic, though perhaps more sophisticated model choice criteria (e.g. Carlin and Louis, Chapter 6) might detect some benefit. In a related vein, the Poisson likelihood assumption itself might be examined via residual analysis; see Xia, Carlin and Waller (1997) for one possible method of comparing Poisson and normal likelihoods in a spatial hierarchical modeling context.

4.2 Mapping of fitted rates

Next, we obtain smoothed maps of pediatric asthma ER visit relative risk over time. These maps are obtained simply by replacing the parameters in equation (4) with the corresponding posterior medians from the MCMC run, and plotting the results for a given day t in `ARC/INFO`. Figures 1 through 3 present these fitted maps for three representative days during the summer of 1995: June 2, July 17, and August 31. We see that the late August date experiences the highest risk, followed by early June, with mid July having lowest risk, consistent with the U-shape of the asthma rate fit by our model. Corresponding maps of median posterior interquartile ranges (not shown) for these three days can be used to capture the variability in these maps. Not surprisingly, more heavily populated zip codes have less variable fitted relative risks; thus, the fitted risks for the central city of Atlanta and the northeast metro area are the most certain statistically.

These fitted log-relative risk maps (Figures 1–3) all have several features in common. First, the spatial clustering of high risk in the southern part of the city of Atlanta is quite apparent, with a second, less dramatic cluster evident in the northwestern corner of the metro area. Conversely, the southern metro area has persistently lower risk. The significant clustering of rates in adjacent zips suggests the impact of our spatial smoothing model, though of course the similar ozone measurements and sociodemographic makeup of adjacent zips also contributes to this clustering.

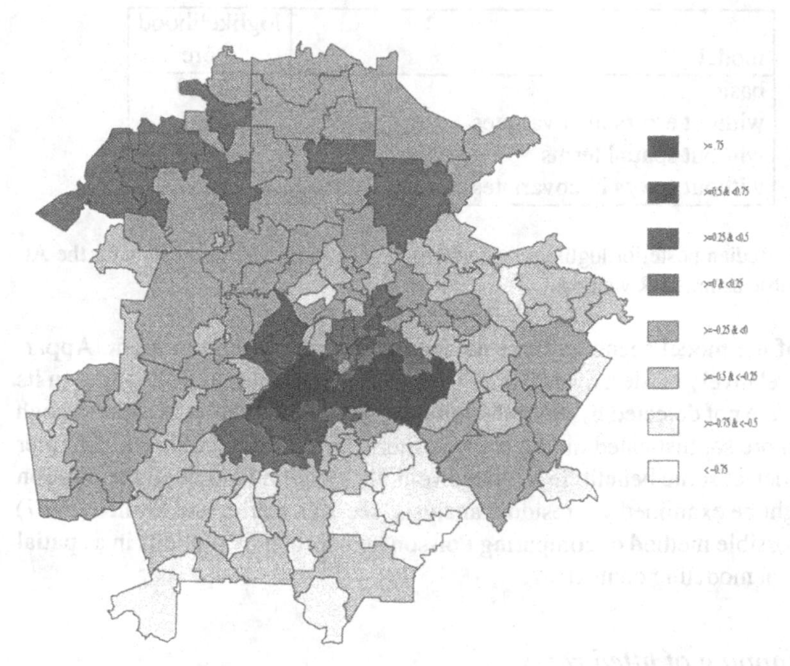

FIGURE 1. Fitted log-relative risk, 6/2/95, Atlanta pediatric asthma data

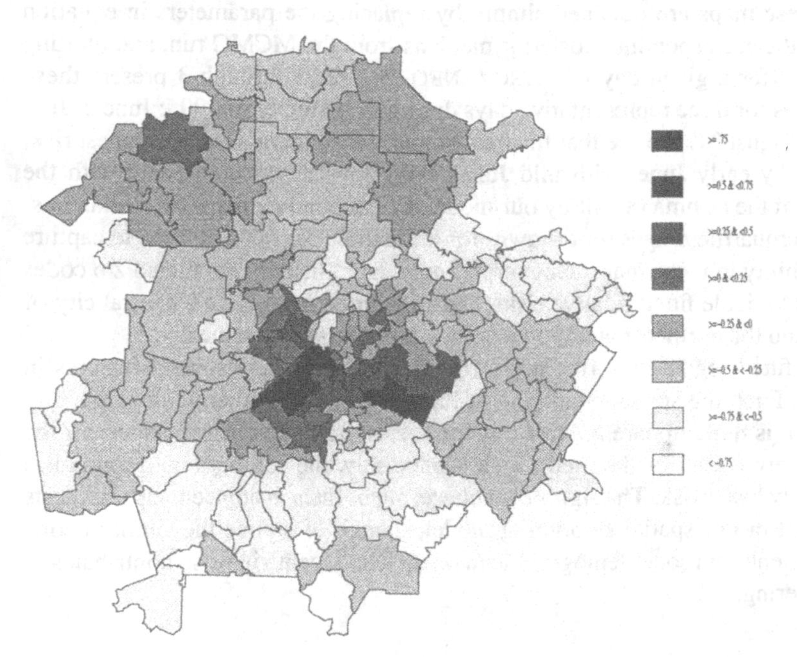

FIGURE 2. Fitted log-relative risk, 7/17/95, Atlanta pediatric asthma data

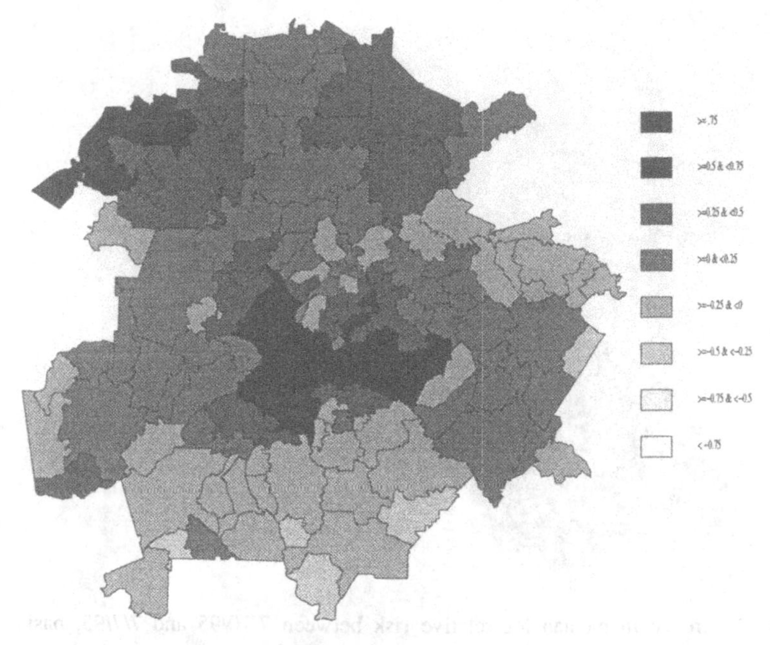

FIGURE 3. Fitted log-relative risk, 8/31/95, Atlanta pediatric asthma data

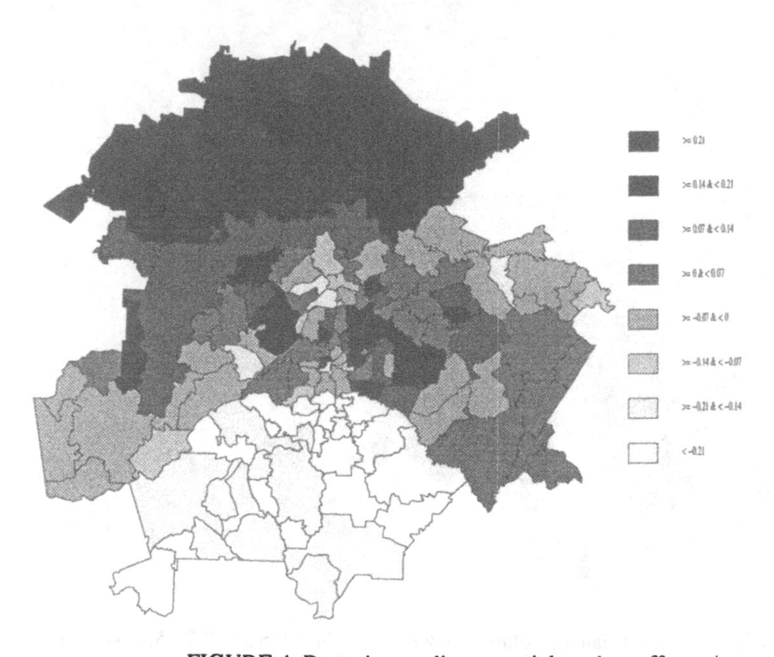

FIGURE 4. Posterior medians, spatial random effects ϕ_i.

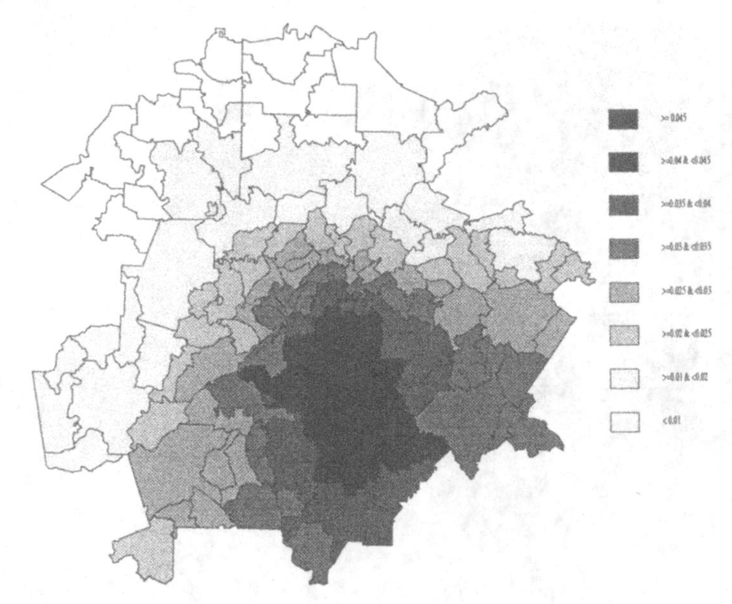

FIGURE 5. Difference in median log-relative risk between 7/10/95 and 7/7/95, basic model.

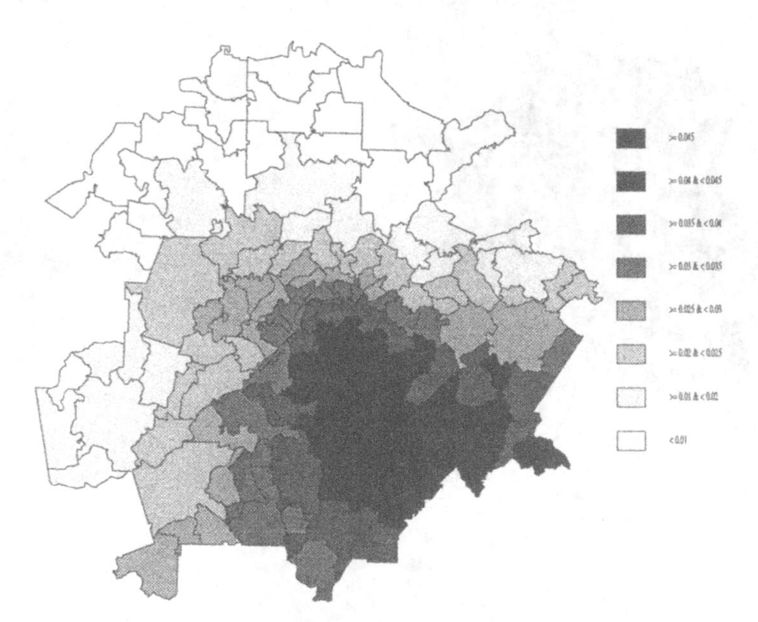

FIGURE 6. Difference in median log-relative risk between 7/10/95 and 7/7/95, basic model without errors in covariates.

The elevated risk clusters to the northwest and northeast seem primarily to reflect the spatial variation in baseline risk, independent of ozone and the various sociodemographic covariates measured in our study. To check this more formally, we consider the map of spatial residuals, namely, the posterior medians of the random effects ϕ_i in equation (4). The ϕ_i are residuals not so much in the usual model checking sense, but rather in that, when mapped, they can suggest spatial correlation not explained by the spatially varying covariates in the model (here, ozone, race, and SES). As such, they are often viewed as surrogates for underlying unobserved covariates. Figure 4 shows this median residual map, which reveals a fairly random pattern in the central city, but a very clear north-to-south pattern for the remainder of the metro area. Apparently after adjusting for our covariates, the northern metro has consistently higher risk than the south. Possible explanations for this pattern include local variations in ER usage patterns, inadequacy of our SES surrogate outside the metro area, or the presence of some other unmeasured spatially varying risk factor, such as PM-10 or pollen.

To check the impact of ozone on our fitted maps, we repeat this analysis for two days in the middle of this same summer having rather different ozone distribution patterns. July 6 precedes an ozone episode, with low ozone values (1 hour max around 60 ppb) throughout the metro area. On July 9, ozone levels were high and winds were blowing to the east, so that ozone readings in the northwest metro (upper 60's) were quite different from those just east of Atlanta (upper 120's); incidentally, this is the most typical spatial pattern for ozone in our data. We then were able to compute the log relative risks for July 7 and 10 (the two following days, which should reflect ozone's impact in our one-day-lagged model). Figure 5 shows the map of the differences in fitted log relative risk between July 10 and July 7. Because the terms other than ozone in equation (4) are either not time-varying or have the same impact on every geographical region, this map is able to reveal the estimated influence of ozone level alone on the relative risks. Figure 5 displays higher risks in the southeast metro and lower risks in the northwest, which is consistent with the aforementioned ozone episode on the previous day.

4.3 *Impact of errors in covariates*

We close this section with an effort to detect the impact of our errors in covariates assumption on our fitted maps. Figure 6 redraws the difference in fitted log-relative risk between July 7 and July 10 originally shown in Figure 5, but without the errors in covariates aspect of the model. Some differences between the two models are apparent. Figure 6 reflects less spatial smoothing than Figure 5, as we would expect since the errors in covariates portion of the model also allowed spatial correlation in the observed underlying ozone values. Generally higher rates in the southeast metro are also apparent.

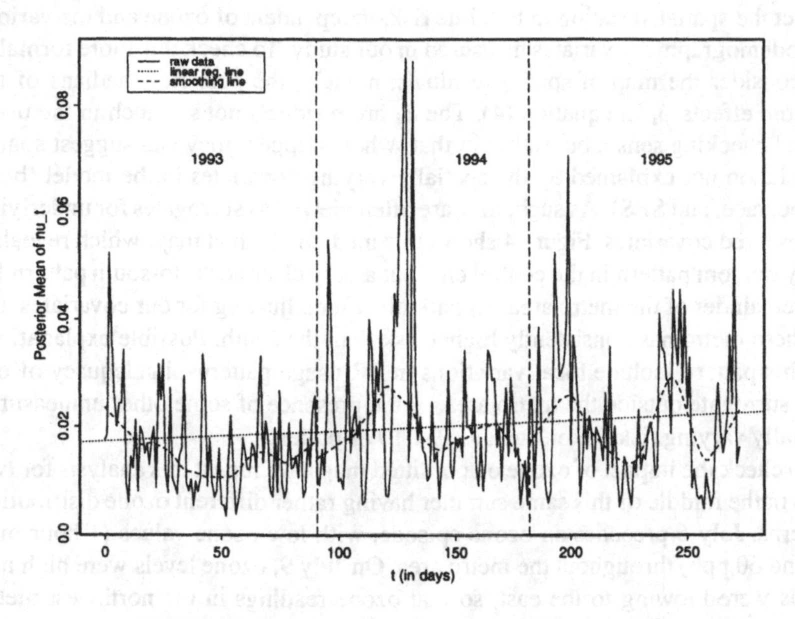

FIGURE 7. Posterior means of ν_t versus t, basic model.

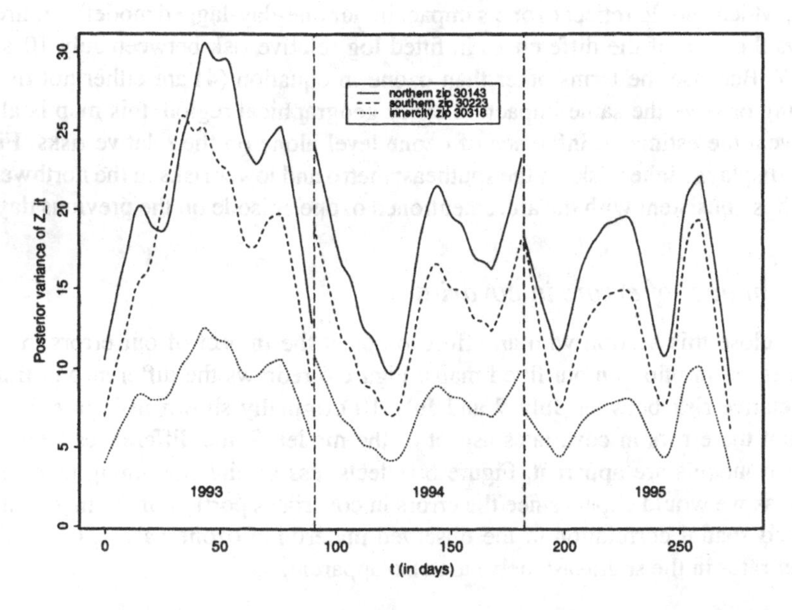

FIGURE 8. Posterior variances of Z_{it} versus t for 3 zip codes, basic model with year-specific lowess smoothing.

Figures 7 and 8 provide two final checks on the impact of the errors in covariates aspect of the model. The former plots the posterior means of the errors in covariates spatial smoothing terms ν_t versus t for the 273 days in our study. Recall large values of ν_t correspond to a high degree of spatial similarity in the ozone measurements; thus, a trend in this figure would suggest changing spatial structure in the true ozone values over time. While the overall trend (dotted straight line) is only slightly upwards, the shorter term trend captured by the lowess smoothing lines (three dashed lines, smoothed separately for each summer) reveals some cyclical behavior, especially in 1994 and 1995. Finally, Figure 8 plots the posterior variances of the Z_{it} versus t for three representative regions i; namely, a northern metro zip (30143), an innercity zip (30318), and a southern metro zip (30223). Here as in the previous figure we use the lowess smoother implemented in *S-Plus*, using a tri-cube weight function and a neighborhood size of 0.3 within a robust locally linear fitting algorithm (Cleveland, 1979). Even in the presence of such smoothing, the rather marked trends in these plots show changing error patterns in observed ozone levels over time. Further, the differences between the three lines indicate the differing ozone prediction reliability patterns for different zips, with the innercity zip (which is closest to the most ozone monitoring stations) having less variability overall. Both of these figures suggest interpretational advantages to our errors in covariates model which were not apparent in the strictly numerical calculations of Subsection 4.1, where acknowledging the (relatively small) variability in the Z_{it} failed to significantly improve the loglikelihood score.

5 Discussion

While the model described above has some advantages over those used in earlier attempts to model data such as ours, several difficulties remain. For instance, the variances used for the errors in covariates aspect of our model were those produced by the universal kriging procedure applied to the daily ozone measurements at the various monitoring sites. Error in the actual *measurement* of ozone was not included; however, this error is likely to be insignificant compared with the error inherent in the kriging procedure. Perhaps more worrisome is that the ozone measurements may or may not translate to the exposures of *individuals* living in the area. For example, indoor air that is air conditioned is likely to have very low ozone levels. Children who spend most of the day in air conditioned buildings would be expected to have much lower ozone exposure than children who do not. Unfortunately, no data on air conditioner usage or child activity patterns were available in our study area.

Other limitations of our data set include a lack of personal exposure to confounders such as cigarette smoke, cockroaches, and indoor mold. Also, our study population was limited to the seven ERs which agreed to participate, comprising only 80% of the total ER visits in the Atlanta MSA. Besides missing this

20% of ER visits, we are also missing all of the patients who present at facilities other than ERs. Patients with better healthcare coverage are likely to have family doctors who can help manage their asthma better, hence present less often in ER settings. It is for this reason that we defined our rate "denominators" n_{it} as the total number of pediatric ER visits in our sample, rather than the total number of children residing in the given zip on the given day. As such, we must emphasize that our results pertain only to children who present for asthma at ERs, not all pediatric asthma in the metro area.

Finally, regarding Bayes versus frequentist comparisons, our Bayesian findings generally reinforce the frequentist ones of Tolbert et al. (1997). In performing both analyses, many of the usual tradeoffs became apparent. Specifically, the Bayes approach enables a full space-time analysis acknowledging the errors in the ozone predictions, facilitates the drawing of smoothed relative risk maps, and provides honest variance estimates simultaneously reflecting all uncertainty in the model. Also, maps like Figure 4 of fitted spatial random effects ϕ_i can be helpful in identifying regions where important spatially-varying covariates may be missing from the model; patterns in a residual plot of heterogeneity parameters θ_i could similarly suggest missing *non*-spatially varying covariates. However, the frequentist analysis was far easier to carry out, using standard software (`Proc LOGISTIC` and the `GLIM` macro in `SAS`, instead of a complex "from scratch" C program) and requiring no expert input in making a convergence assessment for the algorithm. Also, the main results concerning the relative risk of ozone exposure were similar under the two models (1.039 versus 1.026), though the associated confidence interval was somewhat wider under the fully Bayesian models (regardless of whether they adjusted for errors in covariates and spatial autocorrelation or not). The high collinearity of some of the predictors was also easier to diagnose and handle using the traditional approach, since there it manifested itself in the usual way as "reduced significance" of the predictors, whereas the Bayesian algorithm, relying as it did on MCMC runs, could not even produce reliable point estimates in such cases due to convergence failure. Clearly further investigations and comparisons are warranted in similar challenging data settings to determine the overall utility of the fully Bayesian MCMC approach to such problems.

Acknowledgments

This research was supported by a grant from the Electric Power Research Institute and the Southern Company (#W09134–01). In addition, the work of the first two authors was supported in part by National Institute of Allergy and Infectious Diseases (NIAID) FIRST Award 1-R29-AI33466. The authors thank Fan Xu for substantial computing assistance.

References

Bernardinelli, L. and Montomoli, C. (1992). Empirical Bayes versus fully Bayesian analysis of geographical variation in disease risk. *Statistics in Medicine*, **11**, 983–1007.

Bernardinelli, L., Pascutto, C., Best, N.G. and Gilks, W.R. (1997). Disease mapping with errors in covariates. *Statistics in Medicine*, **16**, 741–752.

Besag, J., Green, P.J., Higdon, D. and Mengersen, K. (1995). Bayesian computation and stochastic systems (with discussion). *Statistical Science*, **10**, 3–66.

Besag, J., York, J.C., and Mollié, A. (1991). Bayesian image restoration, with two applications in spatial statistics (with discussion). *Annals of the Institute of Statistical Mathematics*, **43**, 1–59.

Carlin, B.P. and Louis, T.A. (1996). *Bayes and Empirical Bayes Methods for Data Analysis*. London: Chapman and Hall.

Carroll, R.J., Ruppert, D. and Stefanski, L.A. (1995). *Measurement Error in Nonlinear Models*. London: Chapman and Hall.

Centers for Disease Control (1992). Asthma – United States, 1980–1990. *Morbidity and Mortality Weekly Report*, **41**, 733–735.

Cleveland, W. S. (1979). Robust locally weighted regression and smoothing scatterplots. *Journal of the American Statistical Association*, **74**, 829–836.

Evans, R., Mullally, D.I., Wilson, R.W., Gergen, P.J., Rosenberg, H.M., Grauman, J.S., Chevarly, F.M., and Feinleib, M. (1987). National trends in the morbidity and mortality of asthma in the US. *Chest*, **91**, 65S–74S.

Gelfand, A.E. and Smith, A.F.M. (1990). Sampling based approaches to calculating marginal densities. *Journal of the American Statistical Association*, **85**, 398-409.

Gelman, A. and Rubin, D.B. (1992). Inference from iterative simulation using multiple sequences (with discussion), *Statistical Science*, **7**, 457–511.

Metropolis, N., Rosenbluth, A.W., Rosenbluth, M.N., Teller, A.H., and Teller, E. (1953). Equations of state calculations by fast computing machines. *J. Chemical Physics*, **21**, 1087–1091.

Thurston, G.D., Ito, K., Kinney, P.L., and Lippmann, M. (1992). A multi-year study of air pollution and respiratory hospital admissions in three New York State metropolitan areas: Results for the 1988 and 1989 summers. *J. Expos. Anal. Environ. Epidemiol.*, **2**, 429–450.

Tolbert, P., Mulholland, J., MacIntosh, D., Xu, F., Daniels, D., Devine, O., Carlin, B.P., Butler, A., Wilkinson, J., Russell, A., Nordenberg, D., Frumkin, H., Ryan, B., Manatunga, A., and White, M. (1997). Spatio-temporal analysis of air quality and pediatric asthma emergency room visits. To appear *Proc. A.S.A. Section on Statistics and the Environment*, Alexandria, VA: American Statistical Association.

Waller, L.A., Carlin, B.P., Xia, H., and Gelfand, A.E. (1997). Hierarchical spatio-temporal mapping of disease rates. *J. Amer. Statist. Assoc.*, **92**, 607–617.

Xia, H. and Carlin, B.P. (1998). Spatio-temporal models with errors in covariates: Mapping Ohio lung cancer mortality. To appear *Statistics in Medicine*.

Xia, H., Carlin, B.P., and Waller, L.A. (1997). Hierarchical models for mapping Ohio lung cancer rates. *Environmetrics*, **8**, 107–120.

Validating Bayesian Prediction Models: a Case Study in Genetic Susceptibility to Breast Cancer

Edwin Iversen Jr.
Giovanni Parmigiani
Donald Berry

ABSTRACT A family history of breast cancer has long been recognized to be associated with predisposition to the disease, but only recently have susceptibility genes, BRCA1 and BRCA2, been identified. Though rare, mutation of a gene at either locus is associated with a much increased risk of developing breast as well as ovarian cancer. Understanding this risk is an important element of medical counseling in clinics that serve women who present with a family history. In this paper we discuss validation of a probability model for risk of mutation at BRCA1 or BRCA2. Genetic status is unknown, but of interest, for a sample of individuals. Family histories of breast and ovarian cancer in 1st and 2nd degree relatives are available and enable calculation, via the model of Berry et al. (1997) and Parmigiani et al. (1998b), of a carrier probability score. Results of genetic tests with unknown error rates are available with which to validate carrier probability scores. A model is developed which allows joint assessment of test sensitivity and specificity and carrier score error, treating genetic status as a latent variable. Estimating risk and using receiver operating characteristic (ROC) curves for communicating results to practitioners are discussed.

1 Introduction

Accurate assessment of breast cancer risk helps women plan their lives. Individuals with a family history of breast cancer may seek counseling regarding genetic testing and prophylactic measures whose benefits depend on risk and admission to chemoprevention trials may depend on risk (see Berry and Parmigiani 1997 and Parmigiani et al. 1998a for a more complete discussion of the use of breast cancer risk models in the context of medical decision making).

The etiology of breast cancer is complex and a variety of risk modifiers are currently associated with incidence, the most significant of which is a family history of breast and ovarian cancer. A family history is often the result of a genetic proclivity to cancer, such as an inherited mutation of a susceptibility gene

(Garber 1997). Current research has focused on two loci (locations, or sites, on the human genome), BRCA1 (chromosome 17q) and BRCA2 (chromosome 13q). While there is evidence that other genetic factors (Ford and Easton 1995) are at work, it is thought that BRCA1 and BRCA2 are associated with a preponderance of heritable breast cancer cases and that hereditary breast cancer accounts for 5% to 10% of *all* breast cancer cases (Claus et al. 1991; Newman et al. 1988).

Among the general population of U.S. women, lifetime risk of breast cancer is approximately 1/8 and lifetime risk of ovarian cancer is approximately 1/70. While mutations of BRCA1 and BRCA2 genes are rare, incidence of breast and ovarian cancers among women who carry such mutations is high. By age 70 it is estimated (Ford and Easton 1995) that 85% of BRCA1 and 63% of BRCA2 carriers will experience breast and 63% of BRCA1 and less than 10% of BRCA2 carriers will experience ovarian cancer. Women of Ashkenazi Jewish heritage are more likely to carry a mutation at BRCA1 or BRCA2 (Oddoux et al. 1996; Roa et al. 1996) and may have different age–specific incidence functions (Struewing et al. 1997). Berry et al. 1997 and Parmigiani et al. 1998b have developed a prediction model for the probability that an individual carries a germ–line mutation at BRCA1 or BRCA2 based on his or her family's history of breast and ovarian cancer. This model assumes an autosomal dominant mode of inheritance for the genes and employs Bayes' theorem to derive the required carrier probability score conditional on family history. The model uses published estimates of mutation frequency and cancer rates among carriers and non-carriers.

Data for the current validation consists of a sample of individuals for whom family histories of breast and ovarian cancer in first- and second-degree relatives are available, enabling us to calculate carrier probabilities. In addition, results of genetic tests are available for each individual. These tests have unknown error rates. The joint likelihood of test and probability scores is multinomial conditional on genetic status. The unconditional likelihood requires population specific prevalence parameters which are unknown. Prior beliefs regarding test sensitivity and specificity are incorporated in the model.

Validation is important to clinicians and genetic counselors. A standard tool for summarizing the predictive performance of threshold-based classifiers is the receiver operating characteristic (ROC) curve. We present practical ways of adapting ROC curves to communicating the predictive accuracy of probability scores.

The remainder of this paper is organized as follows. Section 2 addresses two sources of evidence regarding an individual's mutation status: a probability score and genetic testing. Here we briefly describe Parmigiani et al. 1998b's model for the probability that an individual carries a BRCA1 or BRCA2 mutation, and outline properties of genetic tests. Section 3 presents a preliminary analysis, one that ignores uncertainty inherent in genetic tests, of 121 tested individuals in which we introduce ROC curves as a measure of a probability score's calibration. Finally, in Section 4, we detail a latent variable analysis for validating the probability score that accounts for uncertainty in genetic test results. Conclusions and a discussion are presented in Section 5.

2 Probability Score

The probability that an individual carries a germ-line mutation at either BRCA1 or BRCA2 is derived conditionally on family history of breast and ovarian cancer via the calculation detailed in Parmigiani et al. 1998b (see also Parmigiani et al. 1998a). We give a brief overview of the model here. Individual i presents with a family history F_i consisting of a pedigree of first- and second-degree relatives with breast and ovarian cancer status and age(s) at diagnosis, if applicable, and current age, or age at death indicated for each member.

The BRCA1 and BRCA2 loci are on autosomes (chromosomes other than the sex chromosomes) and two genes are present at each loci. If one or both genes at a given locus are mutated variants then they will be phenotypically expressed (a higher risk of breast and ovarian cancer will be observed as a result of this expression), and hence the genes are dominant. The probability model assumes inheritance of BRCA1 and BRCA2 genes is autosomal dominant and independent, assumptions supported in the literature (Claus et al. 1994). More than 100 risk-increasing mutations have been discovered, but we make a distinction only between mutations prevalent in the Ashkenazi Jewish population as opposed to all others. Within each subpopulation we consider only two types of alleles (gene variants): normal and mutated. Hence an individual's BRCA1 and BRCA2 genotype status can be summarized in a 3 by 3 table according to the number (0, 1 or 2) of alleles classified as mutated at BRCA1 and BRCA2. For the purposes of this analysis define

$$M_{mi} = \begin{cases} 1 & \text{if one or more BRCA\{m\} gene is mutated,} \\ 0 & \text{if neither BRCA\{m\} gene is mutated,} \end{cases}$$

for individual i, where $m = 1$ or 2 according to the locus in question (BRCA1 or BRCA2).

The likelihood $Pr(F_i \mid M_{1i}, M_{2i})$ depends on the above inheritance assumptions and published age- and genetic status-specific breast and ovarian cancer rates. The posterior distribution of genetic status given family history $Pr(M_{1i}, M_{2i} \mid F_i)$ is obtained via Bayes theorem using published prevalence estimates as prior probabilities. The purpose of our analysis is to validate the predictive accuracy of the probability score $Pr(M_{1i}, M_{2i} \mid F_i)$.

Genetic tests for mutations at BRCA1 and BRCA2 are available. These tests involve a search of the genes for known mutations and have unknown error rates. Tests are expensive as there are over 100 risk-increasing mutations known to exist, and an uncertain number to be discovered. For individual i define

$$T_{mi} = \begin{cases} 1 & \text{a mutation detected on one or both BRCA\{m\} genes,} \\ 0 & \text{no mutation detected on either BRCA\{m\} gene,} \end{cases}$$

where $m = 1$ or 2.

Test sensitivity is the probability of identifying a mutation given that one exists, $Pr(T_{mi} = 1 \mid M_{mi} = 1)$ where $m = 1$ or 2 depending on the gene.

Sensitivity is less than one for both genes due to missed and unknown mutations. Test specificity is the probability that the test correctly identifies non-mutated individuals as such, $Pr(T_{mi} = 0 \mid M_{mi} = 0)$, again where m is the gene identifier. Expert opinion suggests that sensitivity is nearly one. Sensitivity and specificity may be mutation-specific, but we assume they do not depend on the mutation.

3 Preliminary Evaluation of Probability Score

For this analysis we confine attention to mutation status at *either* locus. Hence, for the ith individual we observe scores $S_i = Pr(M_{\bullet i} = 1 \mid F_i)$ where

$$M_{\bullet i} = \begin{cases} 1 & \text{if } M_{1i} = 1 \text{ or } M_{2i} = 1, \\ 0 & \text{otherwise,} \end{cases}$$

summarizes the individual's genetic status. We also observe test results

$$T_{\bullet i} = \begin{cases} 1 & \text{if } T_{1i} = 1 \text{ or } T_{2i} = 1, \\ 0 & \text{otherwise.} \end{cases}$$

Scores were calculated for 121 genetically tested individuals whose pedigrees were provided by the Dana Farber Cancer Institute for our model validation. While this is not a population-based sample, it is representative of individuals who seek genetic testing and hence an appropriate sample for model validation. Using $T_{\bullet i}$ as a proxy for unknown genetic status $M_{\bullet i}$, we assess model performance. Figure 1 plots results of a logistic regression of $T_{\bullet i}$ against S_i. The fitted line is solid, dashed lines are plotted at the fitted line plus and minus one standard error. Scores perfectly calibrated for predicting $T_{\bullet i}$ have the property that $Pr(T_{\bullet i} \mid S_i = s) = s$, hence the logistic regression should be close to the diagonal (shown as a dashed line) if genetic testing is calibrated. The logistic regression indicates that the probability model under-predicts the likelihood of a positive test result in the range 0 to 0.8 and over predicts in 0.8 to 1.s

Figure 1 also plots frequencies of $T_{\bullet i} = 1$ within categories of width 1/10 running from zero to one as X's positioned at category midpoints overlaying a step function. These frequencies should also roughly follow the diagonal, but tend to be high for scores below 0.6 and low for scores above 0.6. Clusters of scores are found just below 0.5 and 1.0, scores for individuals in the cluster below 0.5 tend to be under-predictions. Many of these individuals are young and cancer-free and have a family history strongly indicative of a mutation in one parent.

A graphical summary, familiar in medical decision making and applicable here, is the receiver operating characteristic (ROC) curve (McNeil et al. 1975). ROC curves are used to summarize the power of diagnostic tests that threshold a continuous measurement S for predicting a binary outcome D, usually presence or absence of disease. Choose a threshold s and define the test $T^*(s) = I_{S>s}$, the

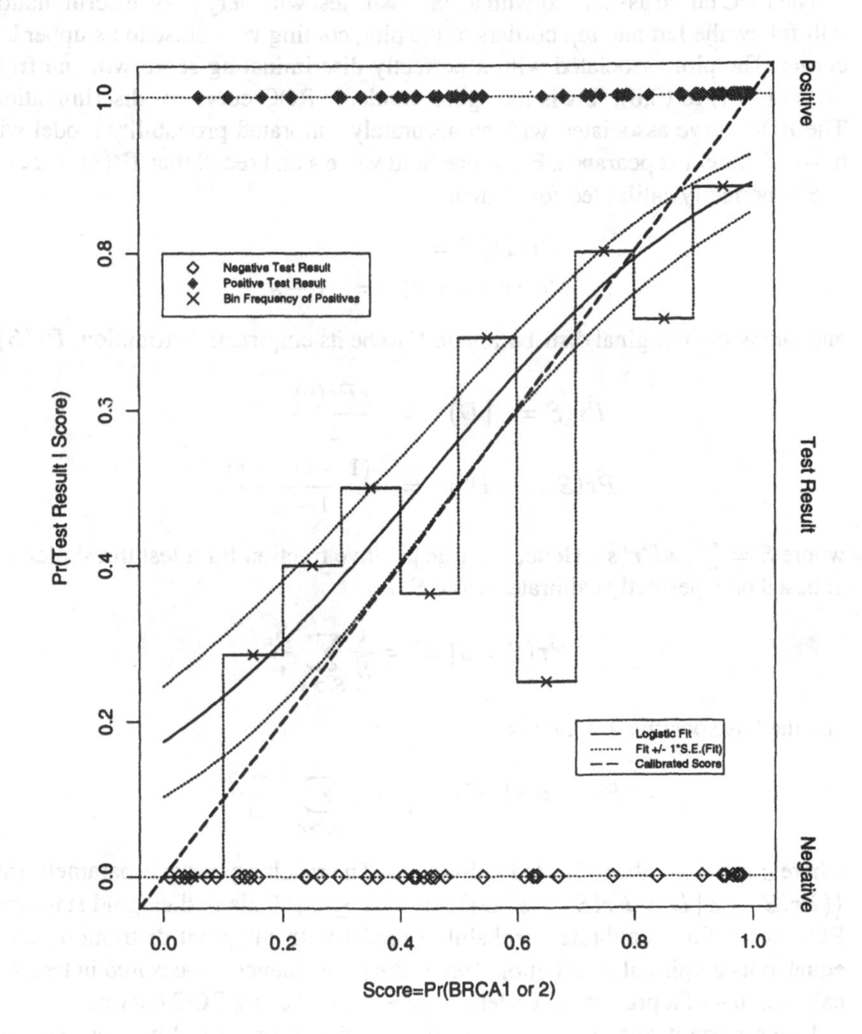

FIGURE 1. Observed data and logistic regression of genetic test result on probability score. Fitted regression line and one standard error bars are plotted as solid and dashed lines, respectively. Individuals with positive tests are plotted as solid diamonds; those with negative tests are plotted as empty diamonds. The frequency of positive tests within categories of width 1/10 running from zero to one are plotted as X's overlaying a step function.

ROC curve for T^* is a plot of the fraction true positives $(T^*(s) = 1 \mid D)$ against the fraction of false positives $(T^*(s) = 1 \mid D^c)$ over possible values of s. Figure 2 plots the ROC curve for thresholding S to predict T.

The ROC curve associated with a diagnostic test with very good discrimination will follow the left and top borders of the plot, coming very close to its upper left corner. The plot associated with a perfectly discriminating score will run from (1,1) to (0,1) to (0,0); this is the "gold standard" ROC curve for discrimination. The ROC curve associated with an accurately calibrated probability model will have a different appearance. Fix a threshold value s and recall that $T^*(s) = I_{S>s}$. If S is perfectly calibrated for D then

$$\begin{aligned} Pr(D \mid S = s) &= s \\ Pr(D^c \mid S = s) &= 1 - s, \end{aligned}$$

and taking the marginal distribution on S to be its empirical distribution, $\hat{Pr}(S)$,

$$\begin{aligned} \hat{Pr}(S = s \mid D) &= \frac{s\hat{Pr}(s)}{\bar{S}} \\ \hat{Pr}(S = s \mid D^c) &= \frac{(1-s)\hat{Pr}(s)}{1-\bar{S}}, \end{aligned}$$

where $\bar{S} = \sum_s s\hat{Pr}(s)$. Hence the true positive fraction for a test thresholded at s, based on a perfectly calibrated score S, is

$$\hat{Pr}(S > s \mid D) = \frac{1}{\bar{S}} \sum_{S_i > s} \frac{S_i}{n}$$

and the false positive fraction is

$$\hat{Pr}(S > s \mid D^c) = \frac{1}{1-\bar{S}} \sum_{S_i > s} \frac{1-S_i}{n},$$

where n is the number of tested individuals. The resulting curve is parameterized $\{(\hat{Pr}(S > s \mid D^c), \hat{Pr}(S > s \mid D)) : 0 \le s \le 1\}$. This is the "gold standard" ROC curve for a calibrated probability model with marginal distribution on S equal to its empirical distribution. Hence the performance – measured in terms of calibration – of a predictive model can be evaluated using ROC curves.

In the current analysis, S_i is the ith individual's probability score and the individual's test result, $T_{\bullet i}$, is the binary outcome. The associated ROC curve, depicted in Figure 2, is below the reference curve; for each threshold value s, the corresponding point on the reference curve is northwest of the corresponding point on the observed ROC curve, indicating that the probability score over predicts the likelihood of a positive test result among those who test negative and under predicts the likelihood of a positive test result among those who test positive. The probability score, however, predicts an individual's genotype $M_{\bullet i}$ not their test result $T_{\bullet i}$. Hence this analysis does not provide an honest appraisal of the probability model.

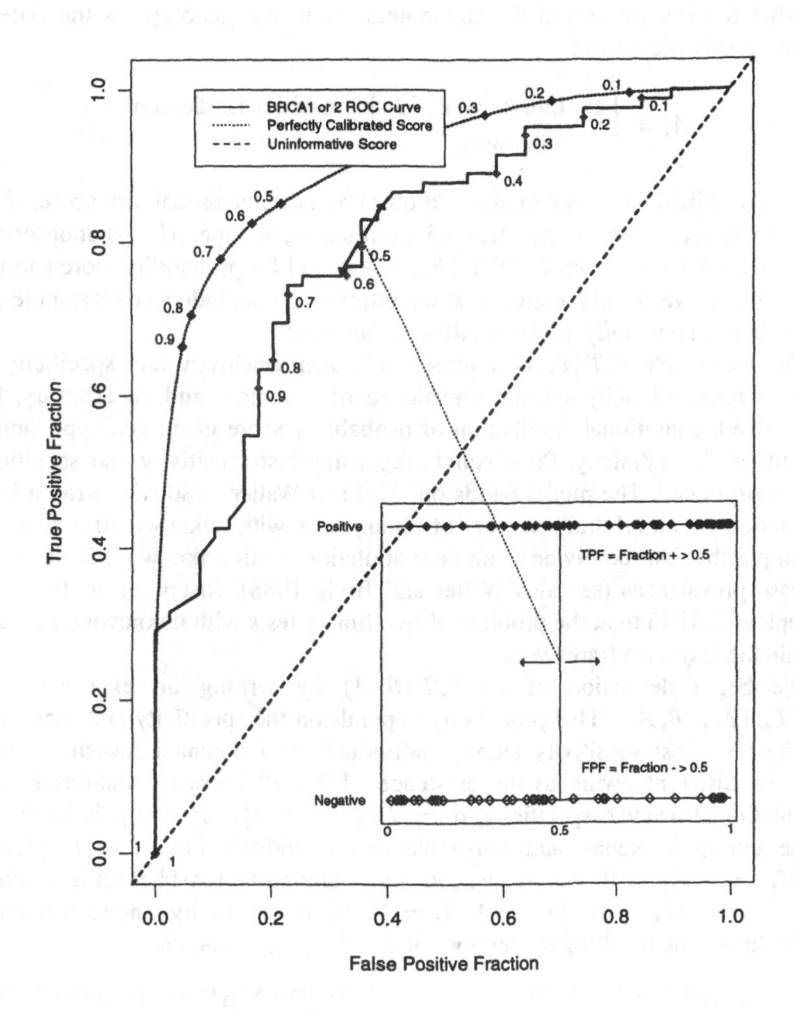

FIGURE 2. Receiver Operating Characteristic (ROC) curve for classifying genetic test results using the probability score. The ROC curve is a plot of the True Positive Fraction (TPF) against the False Positive Fraction (FPF) for all possible threshold values of the probability score. Threshold values that are multiples of 0.10 are plotted as diamonds and labeled, and the point associated with a threshold value of 0.5 is explicated in the plot's lower right corner. The fraction of 'Positive' ('Negative') points to the right of the vertical at 0.5 is the TPF (FPF); as the vertical is moved from 0 to 1, the ROC curve is traced from (1,1) to (0,0). The solid arc is the ROC curve associated with a perfectly calibrated score where the marginal distribution on S is taken to be its empirical distribution. The diagonal is the ROC curve associated with an uninformative test (e.g. a uniform random deviate).

4 Latent Variable Analysis

In what follows we extend the above analysis to use genotype as the (latent) response variable. Define

$$A_i = \begin{cases} 1 & \text{Individual } i \text{ of Ashkenazi Jewish descent,} \\ 0 & \text{Otherwise.} \end{cases}$$

For each individual in our sample we observe a carrier probability score, S_i, a test result, $T_{\bullet i}$, and ethnicity, A_i; the individual's genotype, $M_{\bullet i}$, is unobserved. Our analysis derives from $Pr(S, T \mid \theta, A)$, a model for probability score and test result given a vector of parameters θ and ethnicity for an independent sample (all individuals in our study are from different families).

The model $Pr(S, T \mid \theta, A)$ assumes unknown sensitivity and specificity of genetic tests, ethnicity-specific prevalence of mutation, and an arbitrary, but discretized, conditional distribution of probability score given genotype, and is conditional on ethnicity. Prior beliefs regarding test sensitivity and specificity are incorporated. The model builds on Hui and Walter 1980 who write a joint likelihood for an arbitrary number of binary tests with unknown error rates for a sample that can be divided into subpopulations with unknown, but different, disease prevalences (see also Walter and Irwig 1988). Joseph et al. 1995 and Joseph et al 1996 treat the problem of two binary tests with unknown error rates within the Bayesian framework.

We begin derivation of $Pr(S, T \mid \theta, A)$ by writing an expression for $Pr(T_{\bullet i} \mid M_{\bullet i}, \theta, A_i)$. This probability depends on the specificity and sensitivity of the test. Test sensitivity among individuals of Ashkenazi Jewish ancestry may be different owing to the existence of 3 well known mutations in this population. However, specificity, $\alpha = Pr(T_{\bullet i} = 0 \mid M_{\bullet i} = 0)$, is likely the same among Ashkenazi and non-Ashkenazi individuals. Let $\beta_0 = Pr(T_{\bullet i} = 1 \mid M_{\bullet i} = 1, A_i = 0)$ denote test sensitivity among non-Ashkenazi individuals and $\beta_1 = Pr(T_{\bullet i} = 1 \mid M_{\bullet i} = 1, A_i = 1)$ denote sensitivity among Ashkenazi individuals. The resulting expression for $Pr(T_{\bullet i} \mid M_{\bullet i}, \theta, A_i)$ is

$$[\beta_0^{T_{\bullet i}}(1-\beta_0)^{1-T_{\bullet i}}]^{M_{\bullet i}(1-A_i)}[\beta_1^{T_{\bullet i}}(1-\beta_1)^{1-T_{\bullet i}}]^{M_{\bullet i}A_i}[\alpha^{1-T_{\bullet i}}(1-\alpha)^{T_{\bullet i}}]^{1-M_{\bullet i}},$$

a multinomial.

Next, we write an expression for $Pr(S_i|M_{\bullet i}, \theta, A_i)$. To avoid assuming a functional form for it, we discretize the likelihood of S_i given $M_{\bullet i}$, by dividing the range of S_i into N intervals of equal width, with cell probabilities

$$\phi_{0b} = Pr\left(\frac{b-1}{N} < S_i \leq \frac{b}{N} \;\middle|\; M_{\bullet i} = 0\right)$$

for mutation–free individuals, where $b = 1 \ldots N$, and

$$\phi_{1b} = Pr\left(\frac{b-1}{N} < S_i \leq \frac{b}{N} \;\middle|\; M_{\bullet i} = 1\right)$$

for mutation carriers, where $b = 1 \dots N$. In our analysis we will take $N = 10$. Hence, $Pr(S_i|M_{\bullet i}, \theta, A_i)$ is

$$[\Pi_{b=1}^{N} \phi_{1b}^{I_{ib}}]^{M_{\bullet i}} [\Pi_{b=1}^{N} \phi_{0b}^{I_{ib}}]^{1-M_{\bullet i}}$$

where

$$I_{ib} = \begin{cases} 1 & \text{if } \frac{b-1}{N} < S_i < \frac{b}{N}, \\ 0 & \text{otherwise,} \end{cases}$$

indicates whether individual i has score S_i in cell b. Note that $Pr(S_i|M_{\bullet i}, \theta, A_i)$ does not depend on ethnicity.

We assume that, conditional on genetic status, an individual's test result and probability score are independent. This is reasonable as test results do not depend on phenotypic expression of the gene. Hence

$$\begin{aligned} Pr(S_i, T_{\bullet i} \,|\, M_{\bullet i}, \theta, A_i) &= Pr(S_i \,|\, M_{\bullet i}, \theta, A_i)\, Pr(T_{\bullet i} \,|\, M_{\bullet i}, \theta, A_i) = \\ & [\beta_0^{T_{\bullet i}(1-A_i)} (1-\beta_0)^{(1-T_{\bullet i})(1-A_i)} \ \beta_1^{T_{\bullet i} A_i} (1-\beta_1)^{(1-T_{\bullet i})A_i} \ \Pi_{b=1}^{N} \phi_{1b}^{I_{ib}}]^{M_{\bullet i}} \\ & [\alpha^{1-T_{\bullet i}} (1-\alpha)^{T_{\bullet i}} \ \Pi_{b=1}^{N} \phi_{0b}^{I_{ib}}]^{1-M_{\bullet i}}, \end{aligned}$$

where $\theta = (\alpha, \beta_0, \beta_1, \phi_0, \phi_1, \pi_0, \pi_1)$.

$Pr(S_i, T_{\bullet i} \,|\, \theta, A_i)$ is obtained by integrating over unknown genetic status $M_{\bullet i}$:

$$Pr(S_i, T_{\bullet i} \,|\, \theta, A_i) = \sum_{M_{\bullet i}=0}^{1} Pr(S_i, T_{\bullet i} \,|\, \theta, M_{\bullet i}, A_i) Pr(M_{\bullet i} \,|\, \theta, A_i).$$

This sum requires mutation prevalence, $Pr(M_{\bullet i} \,|\, \theta, A_i)$, a parameter that is known to be different among individuals of Ashkenazi Jewish (AJ) descent. It is likely that this difference persists among individuals that present to high–risk breast cancer clinics like Dana Farber, the source of our data, so we introduce ethnicity-specific prevalence parameters. Let $\pi_1 = P(M_{\bullet i} = 1 \,|\, A = 1)$ denote mutation prevalence among individuals of Ashkenazi descent and $\pi_0 = P(M_{\bullet i} = 1 \,|\, A = 0)$ denote mutation prevalence among those not of Ashkenzi descent presenting to the clinic.

Rewriting the sum,

$$\begin{aligned} Pr(S_i, T_{\bullet i} \,|\, \theta, A_i) = \sum_{M_{\bullet i}=0}^{1} & [\beta_0^{T_{\bullet i}(1-A_i)} (1-\beta_0)^{(1-T_{\bullet i})(1-A_i)}]^{M_{\bullet i}} \times \\ & [\beta_1^{T_{\bullet i} A_i} (1-\beta_1)^{(1-T_{\bullet i})A_i} \Pi_{b=1}^{N} \phi_{1b}^{I_{ib}}]^{M_{\bullet i}} [\alpha^{1-T_{\bullet i}} (1-\alpha)^{T_{\bullet i}} \ \Pi_{b=1}^{N} \phi_{0b}^{I_{ib}}]^{1-M_{\bullet i}} \\ & \times [\pi_0^{M_{\bullet i}} (1-\pi_0)^{1-M_{\bullet i}}]^{1-A_i} [\pi_1^{M_{\bullet i}} (1-\pi_1)^{1-M_{\bullet i}}]^{A_i}. \end{aligned}$$

Viewed as the likelihood of $\theta = (\alpha, \beta_0, \beta_1, \pi_0, \pi_1, \phi_0, \phi_1)$, this expression assigns the same value to $(\alpha, \beta, \beta, \pi_0, \pi_1, \phi_0, \phi_1)$ and $(1-\beta, 1-\alpha, 1-\alpha, 1-\pi_0, 1-\pi_1, \phi_1, \phi_0)$, hence parameters are not identifiable without constraints. This corresponds to the fact that a test that identifies carriers is equivalent to one that identifies non–carriers. We restrict attention to $\alpha > 0.5$, $\beta_0 > 0.5$ and $\beta_1 > 0.5$ as we know genetic testing to have high specificity and sensitivity and $\phi_{1N} > \phi_{0N}$ as experience has shown very high carrier-scores to be informative. The resulting joint likelihood for a sample of n independent patients is

$$Pr(S,T \mid \theta, A) = \Pi_{i=1}^{n} Pr(S_i, T_{\bullet i} \mid \theta, A_i)\; I_{\alpha>0.5}\; I_{\beta_0>0.5}\; I_{\beta_1>0.5}\; I_{\phi_{1N}>\phi_{0N}}.$$

Priors on likelihood parameters complete model specification. We place an informative prior on only one parameter, α, the genetic test specificity. We take $\alpha \sim$ Beta(500, 1), as it is *very unlikely* to classify an individual as a mutation carrier when they are not. Specificity is effectively restricted to equal one. Priors on $\beta_0, \beta_1, \pi_0, \pi_1, \phi_0$, and ϕ_1 are taken to be uniform.

Gibbs sampling (Gelfand and Smith 1990) is employed to sample the posterior $Pr(\alpha, \beta_0, \beta_1, \phi_0, \phi_1, \pi_0, \pi_1 \mid S, T, A)$ with genetic status, M, treated as a latent variable. The resulting full conditional distributions are

$$\begin{aligned}
\alpha &\sim \text{Beta}\left(\sum (1 - M_{\bullet i})(1 - T_{\bullet i}) + 500,\ \sum (1 - M_{\bullet i})T_{\bullet i} + 1\right),\\
\beta_0 &\sim \text{Beta}\left(\sum M_{\bullet i}T_{\bullet i}(1 - A_i) + 1,\ \sum M_{\bullet i}(1 - T_{\bullet i})(1 - A_i) + 1\right),\\
\beta_1 &\sim \text{Beta}\left(\sum (M_{\bullet i}T_{\bullet i}A_i) + 1,\ \sum (M_{\bullet i}(1 - T_{\bullet i})A_i) + 1\right),\\
\pi_0 &\sim \text{Beta}\left(\sum M_{\bullet i}(1 - A_i) + 1,\ \sum (1 - M_{\bullet i})(1 - A_i) + 1\right),\\
\pi_1 &\sim \text{Beta}\left(\sum M_{\bullet i}A_i + 1,\ \sum (1 - M_{\bullet i})A_i + 1\right),\\
\phi_0 &\sim \text{Dir}\left(\sum (1 - M_{\bullet i})I_{i1} + 1, \ldots, \sum (1 - M_{\bullet i})I_{iN} + 1\right),\\
\phi_1 &\sim \text{Dir}\left(\sum M_{\bullet i}I_{i1} + 1, \ldots, \sum M_{\bullet i}I_{iN} + 1\right),\\
M_{\bullet i} &\sim \text{Bern}\,(P_i),
\end{aligned}$$

where $P_i = a_{1i}/(a_{1i} + c_i)$ for individuals of Ashkenazi descent and $P_i = a_{0i}/(a_{0i} + c_i)$ otherwise, where

$$\begin{aligned}
a_{0i} &= \beta_0^{T_{\bullet i}}(1 - \beta_0)^{(1 - T_{\bullet i})}\ \pi_0\ \Pi_{b=1}^{N}\phi_{1b}^{I_{ib}}\\
a_{1i} &= \beta_1^{T_{\bullet i}}(1 - \beta_1)^{(1 - T_{\bullet i})}\ \pi_1\ \Pi_{b=1}^{N}\phi_{1b}^{I_{ib}}\\
c_i &= (1 - \alpha)^{T_{\bullet i}}\alpha^{1 - T_{\bullet i}}(1 - \pi_0)^{1 - A_i}(1 - \pi_1)^{A_i}\ \Pi_{b=1}^{N}\phi_{0b}^{I_{ib}}
\end{aligned}$$

Twenty-five thousand samples were drawn following this prescription. Raftery and Lewis' convergence diagnostics (Raftery and Lewis 1996) were calculated using CODA (Best et al. 1995) to determine the number of iterations, burn-in iterations and thinning interval needed to estimate the 2.5^{th} and 97.5^{th} percentiles to within an accuracy of ± 0.005 with probability 0.95. The first 21 iterations were discarded to allow for stabilization of the chain and every third iteration was kept for inference, leaving a correlated sample of size 8327.

Figure 3 summarizes the 8327 realizations of the MCMC sample. Frame a) is a histogram of the test sensitivity parameter for non-Ashkenazi Jewish individuals, β_0. Our estimate of the posterior mean of β_0 is 0.922; an equal–tailed 95% interval estimate is (0.772, 0.996). Frame c) is a histogram of the test sensitivity parameter for Ashkenazi Jewish individuals, β_1. Our estimate of the posterior mean of β_1 is 0.807; an equal–tailed 95% interval estimate is (0.660, 0.951). The lower test sensitivity among the Ashkenazi may be due to a different testing protocol for these individuals; there are 3 predominant mutations among the Ashkenazi and this fact is often used to simplify the testing procedure.

Frames b) and d) of Figure 3 summarize the sample of prevalence parameters for non-Ashkenazi and Ashkenazi individuals. Sample means are 0.742 and 0.801 and interval estimates are (0.586, 0.881) and (0.652, 0.934), respectively. Thirty of the 42 non-Ashkenazi and 52 of the 79 Ashkenazi women in our sample *tested positive* for a mutation at BRCA1 or BRCA2. However, due to the test's imperfect sensitivity, about 75% of non-Ashkenazi and about 80% of Ashkenazi individuals tested are estimated to be *mutation carriers.* Frame f) provides a histogram of the total number of BRCA1 and BRCA2 *mutation carriers* in the sample. Our estimate of the posterior mean of this margin is 94, an equal–tailed 95% interval estimate is (86, 105). Eighty-two, or 68% of individuals in the sample *tested positive* for the mutation.

In our analysis, test sensitivity is restricted to be nearly one: the prior on this parameter is a beta density with parameters 500 and 1. Frame e) of Figure 3 is a histogram of sampled test specificity α. The posterior of this parameter closely matches the prior; we estimate its mean to be 0.998 (coincides with the prior mean), a 95% interval estimate for this parameter is (0.993, 1.00).

Finally, frames g) and h) of Figure 3 give box plots of sampled cell probabilities for the histogram of probability score given no mutation (frame e)) and mutation (frame f)). Among non-carriers, the probability score is below 1/2 about 65% of the time, while among carriers, the probability score is above 1/2 about 81% of the time with about 55% of carriers receiving a score of 9/10 or above (95% interval estimate ranges from 45% to 65%). If one were to define a test by thresholding the probability score at 9/10, the resulting test would have specificity about 93.5% and sensitivity about 55.3%. This defines one point (false positive fraction = 1 - 0.935, true positive fraction = 0.553) on the posterior mean ROC curve given in Figure 5 described below. Among individuals who test positive using this test, 94.7% are estimated to be mutation carriers.

In Section 3 the analysis proceeded using the genetic test result as a proxy for genotype. Using the Markov chain Monte Carlo sample summarized above we can repeat that analysis with unknown genotype M taken as the response variable. From the MCMC sample, we derive a sample of regressions of M on S and a sample of ROC curves from sampled cell probabilities ϕ_0 and ϕ_1.

Figure 4 plots the sample of non-parametric posterior regressions of (latent) genotype M against discretized probability score. This regression is presented as

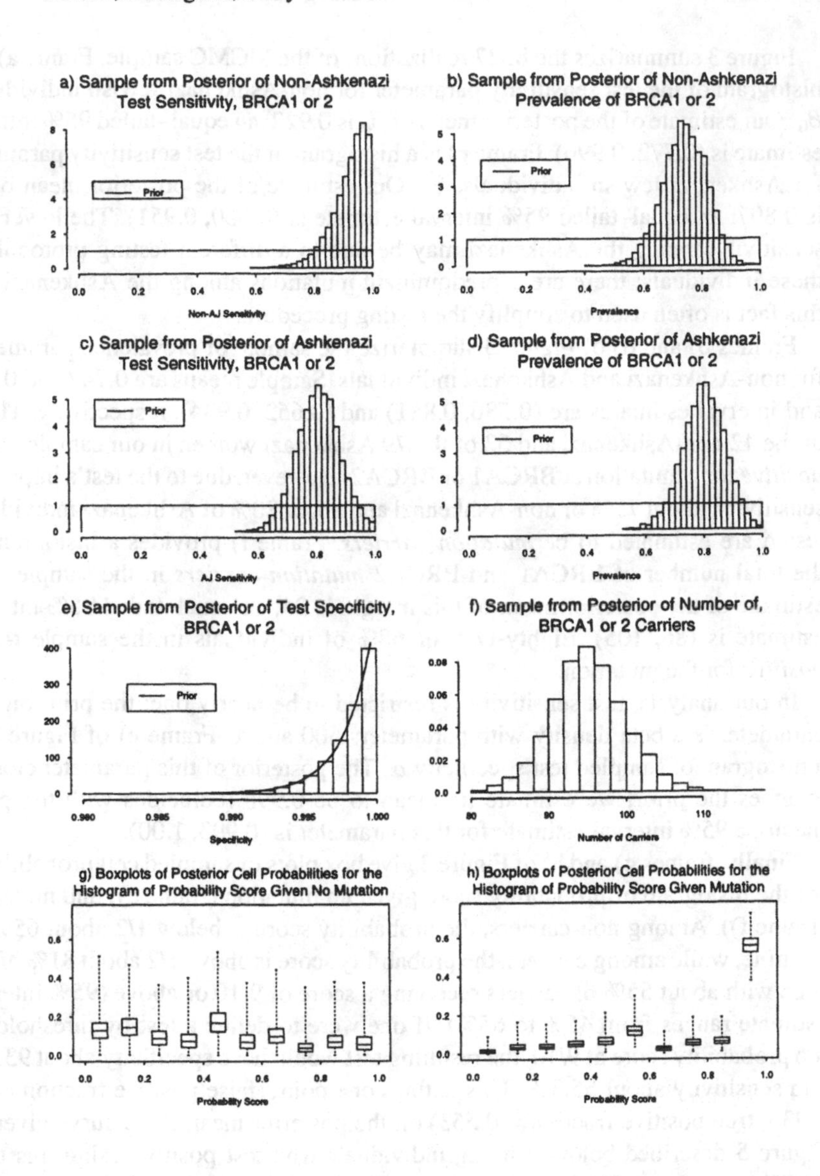

FIGURE 3. Summary of a Markov chain Monte Carlo sample from the posterior. Histograms of test sensitivity among respondents of Ashkenazi (frame c) and non-Ashkenazi (frame a) heritage, specificity (frame e) and prevalence of mutation among respondents of Ashkenazi (frame d) and non–Ashkenazi (frame b) heritage are given; prevalence, sensitivity and specificity are shown with prior overlayed. Specificity is restricted to be nearly equal to one. Frame f) is a histogram of the number of BRCA1 and BRCA2 carriers in the sample. Frames g) and h) give box plots of posterior cell probabilities for the histogram (10 cells) of probability score for individuals with no (frame e) mutations and those with one or more (frame f) mutation.

a series of box plots of

$$Pr\left(M_{\bullet i} = 1 \mid \frac{b-1}{N} < S_i \leq \frac{b}{N}\right)$$

derived from sampled ϕ_0, ϕ_1, and sampled prevalence parameters. The posterior regression overlays the logistic regression and cell frequencies of Figure 1. The posterior regression of M on S is uniformly above the regression of T on S due to the imperfect sensitivity and near perfect specificity of genetic tests. The posterior regression indicates that the probability model under predicts the likelihood of a positive test result in the ranges 0 to 0.4 and 0.5 to 0.6 and over predicts in the range 0.6 to 0.7, observations consistent with our earlier analysis. This analysis also shows that the model is roughly calibrated in the range 0.7 to 1.0, a fact not obvious from the earlier analysis. Probability scores in the range 0.6 to 0.7 appear to be out of line with scores in adjacent ranges. Explanations for this feature are currently under investigation.

Figure 5 plots 20 receiver operating characteristic (ROC) curves for predicting genotype using the probability score. These curves are derived from the MCMC sample paths of ϕ_0, the histogram of probability score given no mutation, and ϕ_1, the histogram of probability score given mutation, as described above. These curves differ from the curve in Figure 2 as they evaluate the probability score's ability to predict genotype (its designed purpose) not the test result. Uncertainty in genotype is reflected in the sample of ROC curves given. The solid arc is the ROC curve associated with a hypothetical perfectly calibrated score and the diagonal dashed line is the ROC curve associated with a hypothetical perfectly uninformative score. An estimate of the posterior mean ROC curve is plotted as a dark dashed line among the lighter sampled curves. Sampled curves are quite variable and mostly below the perfectly calibrated curve. The later observation is consistent with the probability score over predicting the likelihood of mutation among those free of mutation and under predicting the likelihood of mutation among mutation carriers.

5 Discussion

We have provided a method for estimating the accuracy of a probability prediction when the predicted event is observed with unknown error. The analysis provides estimates of this measurement error, together with estimates of prediction error and event prevalence in the sample population. An important aspect of this work is in communicating results to consumers of the probability scoring model and genetic tests. Receiver operating characteristic curves are one good way to accomplish this. Our analysis allows us to present an estimate of the ROC curve for the probability score together with a measure (a sample of curves) of variability in the curve due to uncertainty regarding genetic status, prevalence, and uncertainty in the calculated score.

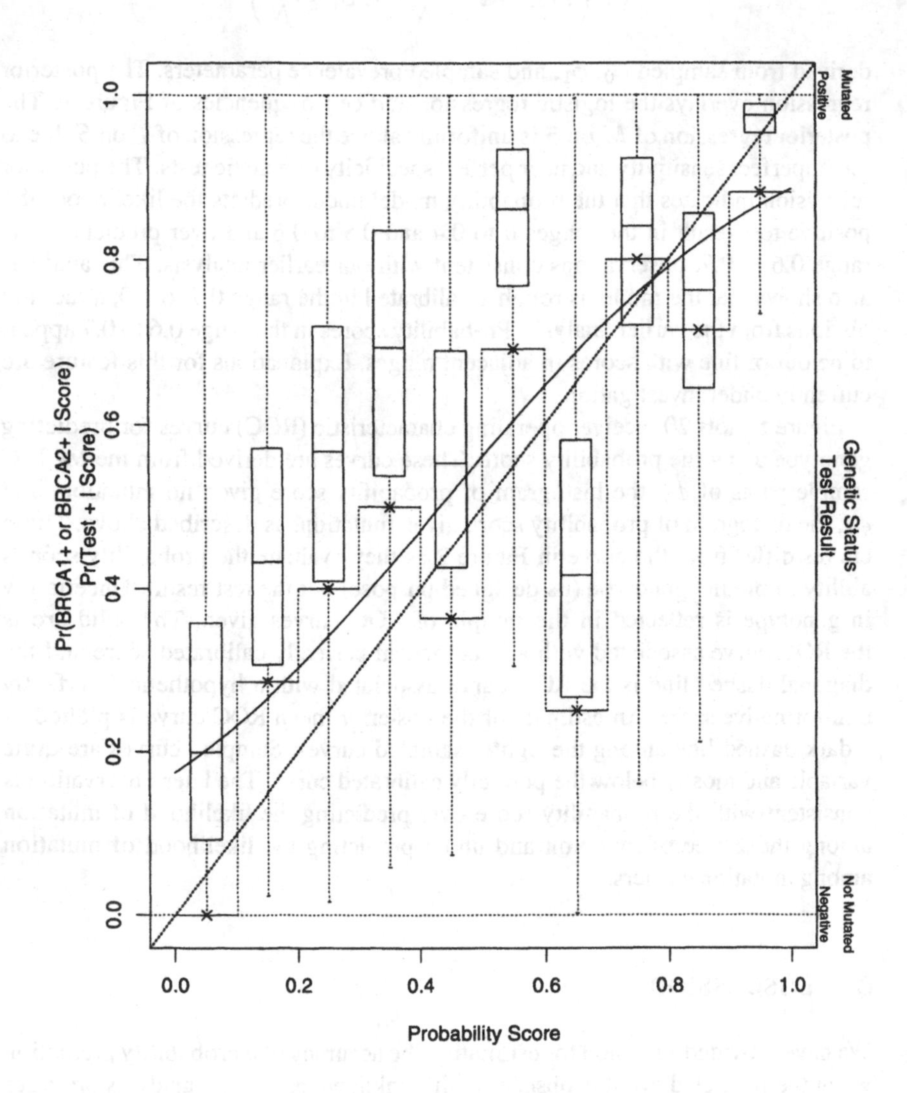

FIGURE 4. Summary of sampled regressions of (unknown) genetic status on probability score. The regression is non-parametric and discrete: probability score is divided into 10 categories of equal width and the probability of M given S is obtained via Bayes rule. Box plots of 8.327 samples from the posterior are plotted together with the naive logistic regression of T on S. Positive test frequencies are plotted by score category.

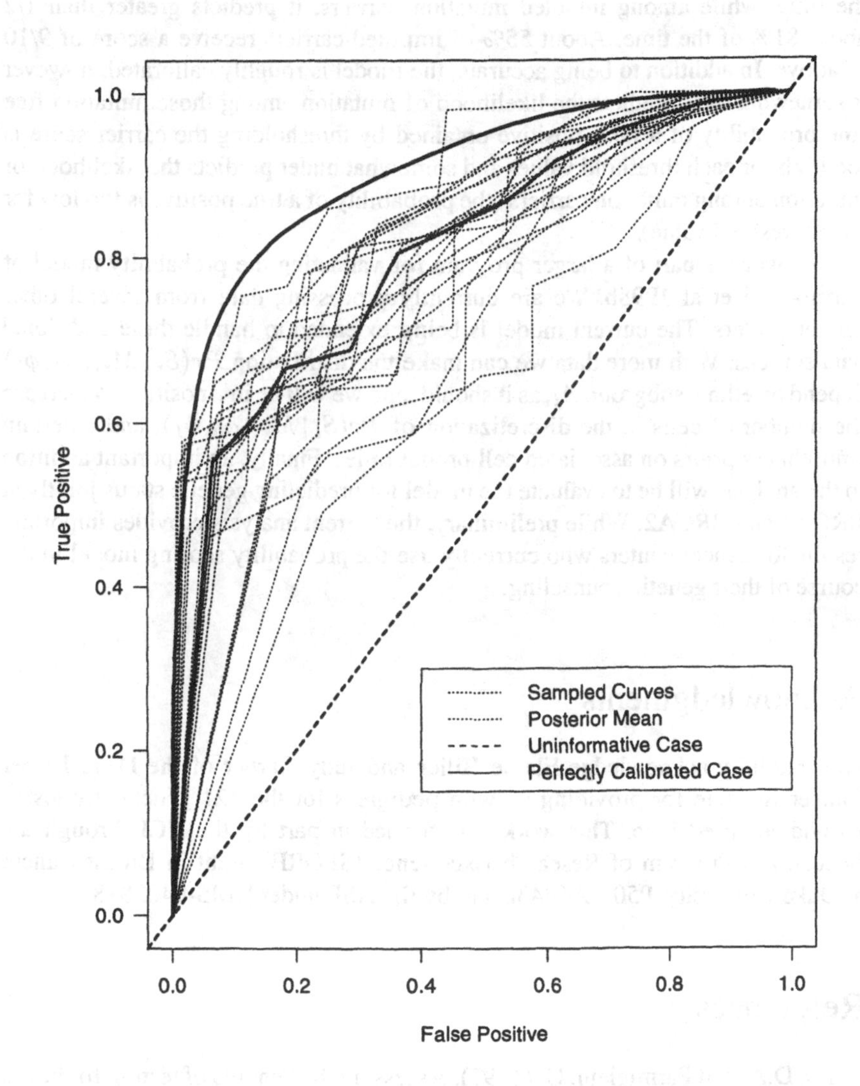

FIGURE 5. A sample of 20 ROC curves from the posterior are derived from the sample path of cell probabilities and are plotted together with the posterior mean ROC curve (heavy dashed line), the theoretical ROC curve associated with perfectly calibrated scores where the marginal distribution on S is taken to be its empirical distribution (solid arc) and the ROC curve associated with perfectly uninformative scores (dashed diagonal).

Our analysis shows that the probability model of Parmigiani et al 1998b is effective in predicting the presence of a mutated BRCA1 or BRCA2 gene. Among imputed non-carriers, the probability model predicts less than 1/2 about 65% of the time, while among imputed mutation carriers, it predicts greater than 1/2 about 81% of the time. About 55% of imputed carriers receive a score of 9/10 or above. In addition to being accurate, the model is roughly calibrated, however it somewhat over predicts the likelihood of mutation among those mutation free (the probability of a false positive obtained by thresholding the carrier score is too high for each threshold value) and somewhat under predicts the likelihood of mutation among mutation carriers (the probability of a true positive is too low for each threshold value).

This work is part of a larger program for validating the probability model of Parmigiaini et al 1998b. We are currently processing data from several other cancer centers. The current model is being extended to handle these additional data sources. With more data we can make the distribution $Pr(S_i \mid M_{\bullet i}, \phi_0, \phi_1)$ depend on ethnic subgroup A_i, as it should, and we will be in a position to increase the number of cells in the discretization of $Pr(S_i \mid M_{\bullet i}, \phi_0, \phi_1)$, and entertain smoothness priors on associated cell probabilities. Finally, an important addition to the analysis will be to evaluate the model for predicting genetic status jointly at BRCA1 and BRCA2. While preliminary, the current analysis provides important results for cancer centers who currently use the probability scoring model in the course of their genetic counseling.

Acknowledgments

We gratefully acknowledge Elaine Hiller and Judy Garber of the Dana Farber Cancer Institute for providing us with pedigrees for the 121 genetically tested individuals used here. This work was funded in part by the NCI through the Specialized Program of Research Excellence (SPORE) grant in Breast Cancer at Duke University, P50 CA68438 and by the NSF under DMS-9403818.

References

Berry, D.A. and Parmigiani, G. (1997). Assessing the benefits of testing for breast cancer susceptibility genes: A decision analysis. *Breast Disease*.

Berry, D.A., Parmigiani, G., Sanchez, J., Schildkraut, J.M., and Winer, E.P. (1997). Probability of carrying a mutation of breast-ovarian cancer gene BRCA1 based on family history. *J Natl Cancer Inst*, 89:9–20.

Best, N.G., Cowles, M.K., and Vines, K. (1995). CODA: Convergence diagnostics and output analysis software for Gibbs sampling output, version 0.30 Technical report, MRC Biostatistics Unit, University of Cambridge.

Claus, E.B., Risch, N., and Thompson, W.D. (1991). Genetic analysis of breast cancer in the cancer and steroid hormone study. *American Journal of Human Genetics*, 48:232–242.

Claus, E.B., Risch, N., and Thompson, W.D. (1994). Autosomal dominant inheritance of early-onset breast cancer: Implications of risk prediction. *Cancer*, 73: 643–651.

Ford, D. and Easton, D.F. (1995). The genetics of breast and ovarian cancer. *Br J Cancer*, 72:805–812.

Garber, J.E. (1997). Breast cancer markers: genetic markers of breast cancer predispositoin. In *Proceedings of the Thirty-third nnual Meeting of ASCO*, pp 213–216. ASCO.

Gelfand, A.E., and Smith, A.F.M. (1990). Sampling-based approaches to calculating marginal densities. *Journal of the American Statistical Association*, 85(410):398–409.

Hui, S.L. and Walter, S.D. (1980). Estimating the error rates of diagnostic tests. *Biometrics*, 36:167–171.

Joseph, L., Gyorkos, T.W., and Coupal, L. (1995). Bayesian estimation of disease prevalence and the parameters of diagnostic tests in the absence of a gold standard. *American Journal of Epidemiology*, 141:263–272.

Joseph, L., Gyorkos, T.W., and Coupal, L. (1996). Inferences for likelihood ratios int he absence of a gold standard. *Medical Decision Making*, 16:412–417.

McNeil, B.J., Keeler, E.K., and Adelstein, S.J. (1975). Primer on certain elements of medical decision making. *N Engl J Med*, 293:211–215.

Newman, B., Austin, M.A., Lee, M., *et al.* (1988). Inheritance of human breast cancer: evidence for autosomal dominant transmission of high-risk families. *Proc Natl Acad Sci USA*, 85:3044-3048.

Oddoux, C., Struewing, J.P., Clayton, C.M., Neuhausen, S., Brody, L.C., Kaback, M., Haas, B., Norton, L., Borgen, P., Jhanwar, S., Goldgar, D.E., Ostrer, H., and Offit, K. (1996). The carrier frequency of the BRCA2 6174delT mutation among Ashkenazi Jewish individuals is approximately 1%. *Nature Genetics*, 14:188–190.

Parmigiani, G., Berry, D., Iversen, E.S., Jr., Müller, P., Schildkraut, J., and Winer, E.P. (1998a). Modeling risk of breast cancer and decisions about genetic testing. In *Case Studies in Bayesian Statistics*, Volume IV, pages 133-203, Springer-Verlag.

Parmigiani, G., Berry, D.A., and Aguilar, O. (1998b). Determining carrier probabilities for breast cancer susceptibility genes BRCA1 and BRCA2. *American Journal of Human Genetics*, 62:145–158.

Raftery, A.E. and Lewis, S.M. (1996). Implementing MCMC. In Gilks, W.R., Richardson, S., and Spiegelhalter, D.J., editors, *Markov Chain Monte Carlo in Practice*, pages 115–127, London. Chapman and Hall.

Roa, B.B., Boyd, A.A., Volcik, K., and Richards, C.S. (1996). Ashkenazi Jewish population frequencies for common mutations in BRCA1 and BRCA2. *Nature Genetics*, 14:185–187.

Struewing, J.P., Hartge, P., Wacholder, S., Baker, S.M., Berlin, M., McAdams, M., Timmerman, M.M., Brody, L.C., and Tucker, M.A. (1997). The risk of cancer associated with specific mutations of BRCA1 and BRCA2 among Ashkenazi Jews. *New England Journal of Medicine*, 336:1401.

Walter, S.D. and Irwig, L.M. (1988). Estimation of test error rates, disease prevalence and relative risk from misclassified data: a review. *Journal of Clinical Epidemiology*, 41:923–937.

Mixture Models in the Exploration of Structure-Activity Relationships in Drug Design

Susan Paddock
Mike West
S. Stanley Young
Merlise Clyde

ABSTRACT We report on a study of mixture modeling problems arising in the assessment of chemical structure-activity relationships in drug design and discovery. Pharmaceutical research laboratories developing test compounds for screening synthesize many related candidate compounds by linking together collections of basic molecular building blocks, known as monomers. These compounds are tested for biological activity, feeding in to screening for further analysis and drug design. The tests also provide data relating compound activity to chemical properties and aspects of the structure of associated monomers, and our focus here is studying such relationships as an aid to future monomer selection. The level of chemical activity of compounds is based on the geometry of chemical binding of test compounds to target binding sites on receptor compounds, but the screening tests are unable to identify binding configurations. Hence potentially critical covariate information is missing as a natural latent variable. Resulting statistical models are then mixed with respect to such missing information, so complicating data analysis and inference. This paper reports on a study of a two-monomer, two-binding site framework and associated data. We build structured mixture models that mix linear regression models, predicting chemical effectiveness, with respect to site-binding selection mechanisms. We discuss aspects of modeling and analysis, including problems and pitfalls, and describe results of analyses of a simulated and real data set. In modeling real data, we are led into critical model extensions that introduce hierarchical random effects components to adequately capture heterogeneities in both the site binding mechanisms and in the resulting levels of effectiveness of compounds once bound. Comments on current and potential future directions conclude the report.

1 Introduction and Background

Medicinal chemists involved in drug design and discovery synthesize large numbers of pharmacological compounds for initial testing to screen for chemical effectiveness in a laboratory setting. This high-throughput screening process assesses many related candidate compounds, each being created by combining basic building blocks drawn from separate sets of complex molecules, or *monomers.* This activity is called combinatorial synthesis. The screening process aims to identify biologically active compounds using aspects of molecular structure associated with biological activity and effectiveness. Our focus here is on statistical modeling and inferential issues arising in exploring such *structure - activity relationships.*

We consider the simplest case of test compounds created by linking together two monomers drawn from two separate test sets. This produces a *library* of two-monomer compounds; the notation $A + B \rightarrow AB$ identifies monomer A from the first set and monomer B from the second set. Various properties of A and B, such as `molecular weight`, partly characterize the resulting compound AB. Chemical activity is measured through experiments that estimate the extent to which the compound inhibits certain chemical reactions; the resulting outcomes are measures of *potency* of the test compound. The synthesized compounds are designed to bind to a structure of interest; for example, synthesized compounds could bind to healthy body tissue to block natural but harmful compounds from binding to the tissue. The site at which the compounds bind to a cell in the tissue is called a receptor. In addition to the effects of constituent monomers, the outcome potency depends on the *binding configuration* of the compound; that is, the compound may bind to a receptor in such a way that individual binding sites are matched by monomers. Hence an effective bind of a two-monomer compound involves two binding sites at the receptor, with one monomer binding to each site. Given a binding configuration, the chemical/physical monomer characteristics are determinants of potency. In practice, it is impossible to observe or measure binding configuration, though it is understood that the various monomer characteristics may play a role in determining configuration.

Though the perspective of the medicinal chemist is the synthesis of monomers, the relationship of a compound to its binding sites depends little upon how the compound was or will be made. Rather, the binding of a compound is governed by the properties of the receptor sites and the compound itself. With a receptor having two binding sites, S_1 and S_2, the binding can be AB to S_1S_2 or BA to S_1S_2; see Figure 1. A data set with n compounds can then be arranged in 2^n configurations so that, as n increases, the number of possible configurations is enormous. For more complicated problems with more than two sites/two monomers, the situation becomes even more complex, of course. Here we develop and fully explore the two sites/two monomers setup, and discuss models we have developed that combine regression models for both binding configurations and potency levels given binding configurations. This introduces two novelties from the chemists' viewpoints: the use of descriptors to model both orientation and binding via

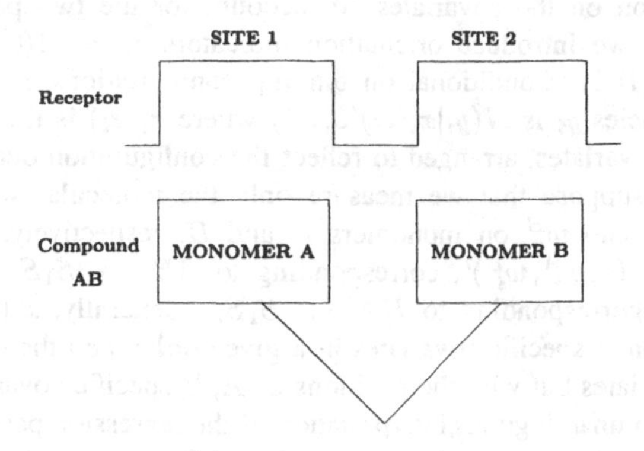

FIGURE 1. Two-monomer, two-site configuration framework. Which orientation is correct: $AB \to S_1S_2$ or $AB \to S_2S_1$?

separate model components, and the accounting for differences in potency due to binding orientations. This leads to structured mixture models in which linear regressions for potencies are mixed with respect to binary regressions predicting binding configurations. Our work on the development of MCMC methods to implement these models, problems encountered en route, benchmark assessments on simulated data and some preliminary results with real data are described and illustrated here. Our basic mixture model performs well on simulated test data, but is found to be less adequate when applied to real data sets from experiments at Glaxo Wellcome Inc. This leads us into practically critical model extensions that introduce hierarchical random effects components to adequately capture two kinds of heterogeneity in the compound population: significant over-dispersion relative to standard probit predictors in the site binding/orientation mechanism, coupled with a lesser degree of over-dispersion relative to normal linear models in the potency outcome mechanism. We discuss this, explaining the model extensions and developments involved in analysis, with illustration in analysis of a Glaxo Wellcome Inc. data set.

2 Basic Mixture Model and Analysis

2.1 Model Structure

In a context with a two-site receptor S_1S_2, consider a sample of n two-monomer compounds A_jB_k, $(j = 1,\dots,n_A, k = 1,\dots,n_B)$, where n_A and n_B are the number of monomers in sets A and B, respectively, and the resulting $n_A \times n_B = n$ compounds are indexed by i $(i = 1,\dots,n)$. Assume that compound binding configurations and potency outcomes are independent across compounds. Covariates are measured on each monomer and our potency model assumes that, given a particular binding configuration, outcomes follow a normal linear regression on the covariates. To account for the two possible binding configurations, we introduce orientation indicators $z_i = 1(0)$ if binding to S_1S_2 is AB (BA). Conditional on binding configurations z_i, the model for outcome potencies y_i is $N(y_i|x_i(z_i)'\beta, \sigma^2)$ where $x_i(z_i)$ is the column vector of monomer covariates, arranged to reflect the configuration determined by z_i. For example, suppose that we measure only the molecular weights of each monomer, w_i^A and w_i^B on monomers A and B, respectively. Then we will take $x_i(1) = (1, w_i^A, w_i^B)'$, corresponding to $AB \rightarrow S_1S_2$, and $x_i(0) = (1, w_i^B, w_i^A)'$, corresponding to $BA \rightarrow S_1S_2$. Generally, $x_i(1)$ is specified with the monomer-specific covariates in a given order, and then $x_i(0)$ contains the same covariates but with the positions of A, B-specific covariates switched. This ensures an unambiguous interpretation of the regression parameters as *site-specific*. In the above example, $\beta = (\beta_0, \beta_1, \beta_2)'$ where β_0 is an intercept term and, for $j = 1, 2$, β_j is the regression coefficient on molecular weight of the monomer that binds to S_j, irrespective of whether that is monomer A or B. The quantity σ measures unexplained dispersion, including assay variability attributable to biological variability present in the receptor and compounds during experimentation and the calibration stage. Nominal levels of assay variability alone are in the 10-20% range on the standard "percent inhibition" scale for potency levels.

Configurations are assumed related to monomer characteristics via a binary regression model. We have adopted a probit link in our work to date, so that the indicators z_i are governed by conditional binding probabilities $\pi_i = Pr(z_i = 1) = \Phi(h_i'\theta)$ where h_i is a column vector of covariates drawn from the same set as those in $x_i(\cdot)$, and $\Phi(\cdot)$ is the standard normal cdf. Hence the model for observed potency outcomes is a mixture over the unobserved configuration, given by

$$y_i \sim \pi_i N(y_i|x_i(1)'\beta, \sigma^2) + (1 - \pi_i) N(y_i|x_i(0)'\beta, \sigma^2)$$

Naturally we deal with this mixture model by explicitly including the latent indicators z_i. Thus the full model is defined by the joint distribution

$$p(y, z|\beta, \sigma, \theta) = \prod_{i=1}^{n} N(y_i|x_i(z_i)'\beta, \sigma^2)\Phi(h_i'\theta)^{z_i}(1 - \Phi(h_i'\theta))^{1-z_i}$$

where y and z are the vectors of y_i and z_i respectively. We are interested in inference about the full set of uncertain quantities $(z, \beta, \sigma, \theta)$ based on observing y. Our analyses use standard, conditionally conjugate prior distributions for the regression model parameters, i.e., independent normal priors for β and θ, and inverse gamma priors for σ^2.

We note that there are symmetries in the model induced by the arbitrary labeling of receptor sites. This is an example of identification features common to most mixture models (Titterington et al. 1985; West 1997). Note that

$$p(y, z|\beta, \sigma, \theta) = p(y, 1 - z|\beta^*, \sigma, -\theta)$$

where β^* is a permutation of the elements of β obtained by switching the labels of receptor sites S_1 and S_2. While predictions are unaffected by non-identifiability, synthesis of future compounds based upon posterior inferences of properties of A and B is problematic in such a non-identified model. Below we note this and describe how to handle the issue *a posteriori* in the context of MCMC analysis.

2.2 *Posterior computations*

We exploit the standard latent variable augmentation method for probit models (Albert and Chib 1993) to enable direct Gibbs sampling (Gelfand and Smith 1990) for posterior computation. Introduce latent variables ω_i such that $\omega_i|\theta \sim N(h_i'\theta, 1)$, implying that $\pi_i = Pr[\omega_i \geq 0]$. Then the set of full conditional posterior distributions has the following structure:

- The posterior for $\beta|z, y, \sigma$ is a normal distribution resulting from the conditional linear regression of the y_i on $x_i(z_i)$.
- The posterior for $\sigma^2|z, y, \beta$ is inverse gamma, also a direct derivation from the conditional linear regression of the y_i on $x_i(z_i)$.
- The posterior for $\theta|w$ is a normal distribution resulting from linear regression of the ω_i on h_i.
- The ω_i are conditionally independent and $\omega_i|z_i, \theta$ has a truncated normal posterior (following Albert and Chib 1993).
- The z_i are conditionally independent with posterior odds on $z_i = 1$ versus $z_i = 0$ given by

$$\Phi(h_i'\theta)N(y_i|x_i(1)'\beta, \sigma^2)/\{(1 - \Phi(h_i'\theta))N(y_i|x_i(0)'\beta, \sigma^2)\}$$

The calculation and simulation of each of these distributions, in turn, is standard and easy, and so enables direct Gibbs sampling. The question of parameter identification is dealt with in the posterior computations by examining MCMC samples from an identified region of the parameter space, rather than by attempting to restrict the simulation analysis to such a region by imposing constraints in the prior. This follows what is now essentially standard practice in other kinds of mixture models (e.g., West 1997). Finally, in order to explore predictive questions, for both model evaluation (as in Gelman et al. 1996) and for use in predicting potency of new compounds, we simulate predictive distributions based on the posterior parameters and latent variable samples.

3 A Simulated Dataset

We report an analysis of a test dataset simulated to explore and validate the model and our implementation, which is illuminating in its own right. Here we generated $n = 200$ potency values under a model with one covariate on each monomer. The covariate values are sampled from the $U(-5, 5)$ distribution for monomer A, and from the standard normal distribution for monomer B. Binding orientations z_i are generated from the probit regression on the two covariates with regression vector $\theta = (-0.5, -0.25, -0.5)'$. Then, given orientations, potency levels are drawn from the normal linear regressions with parameter vector $\beta = (30, 3, 8)'$ and variance $\sigma^2 = 9$. The values here were chosen to produce observed data configurations resembling real data sets, although the residual variance σ^2 is lower than typical levels of assay variability. Our analysis is based on quite vague proper priors in the conditionally conjugate class, namely $\sigma^{-2} \sim Ga(\sigma^{-2}|.01, 1)$, $\beta \sim N(\beta|0, 10000I)$, and $\theta \sim N(\theta|0, I)$. Our MCMC analyses have involved a range of experiments with Monte Carlo sample sizes and starting values, and MCMC diagnostics. Following this, our summary numerical and graphical inferences here are based on post burn-in samples of size 20,000. The resulting histogram approximations to marginal posteriors for parameters appear in Figure 2, with related posterior inferences in Figure 3. The top row in this latter graph displays posteriors of the (posterior) classification probabilities of orientation for four selected observations. The actual orientation probabilities and indicators are $\{.67, .68, .45, .86\}$ and $\{0, 1, 0, 1\}$, respectively. Note the concordance of posteriors and true values. To provide insight into aspects of model adequacy, posterior predictive checks are examined. Approximate posterior predictive distributions for potency levels of the four example compounds are displayed in Figure 3 (bottom row). The distributions are centered close to the actual observations, which are marked by crosses along the axes, supportive of model adequacy. A further posterior predictive check involves graphical comparison of the actual data with sampled predictives for the full data set. For a set of draws from the posterior, we sampled 430 observations, one at each design point, producing replicates from the posterior predictive distribution of the data set. Smoothed versions of ten of the resulting histograms are overlaid on the real data histogram in Figure 4. These evidence a comforting degree of conformity to the data histogram.

To the best of our knowledge, we do not encounter the identifiability problem in this analysis. To explore this, we ran another analysis of this simulated data in which the prior for σ^2 was inappropriately biased and concentrated on very large values, namely $\sigma^{-2} \sim Ga(\sigma^{-2}|5, 30000)$. The analysis details were otherwise the same as above. The MCMC trajectories (not shown) from this analysis indicate the lack of identifiability induced by arbitrary labeling of sites; the trajectories for β_1 and β_2 are interchanged two or three times, and the values of the θ_j change sign at the same points. To handle the identification issue we can simply examine a subset of iterations corresponding to one identified region of the parameter space, or map the MCMC samples to a single identified region of the parameter space. The latter strategy is easily implemented in these models by changing the signs of the θ values corresponding to $\theta_1 < 0$, permuting the site-specific β terms at the corresponding iterates, and reflecting the corresponding indicators z_i to $1 - z_i$.

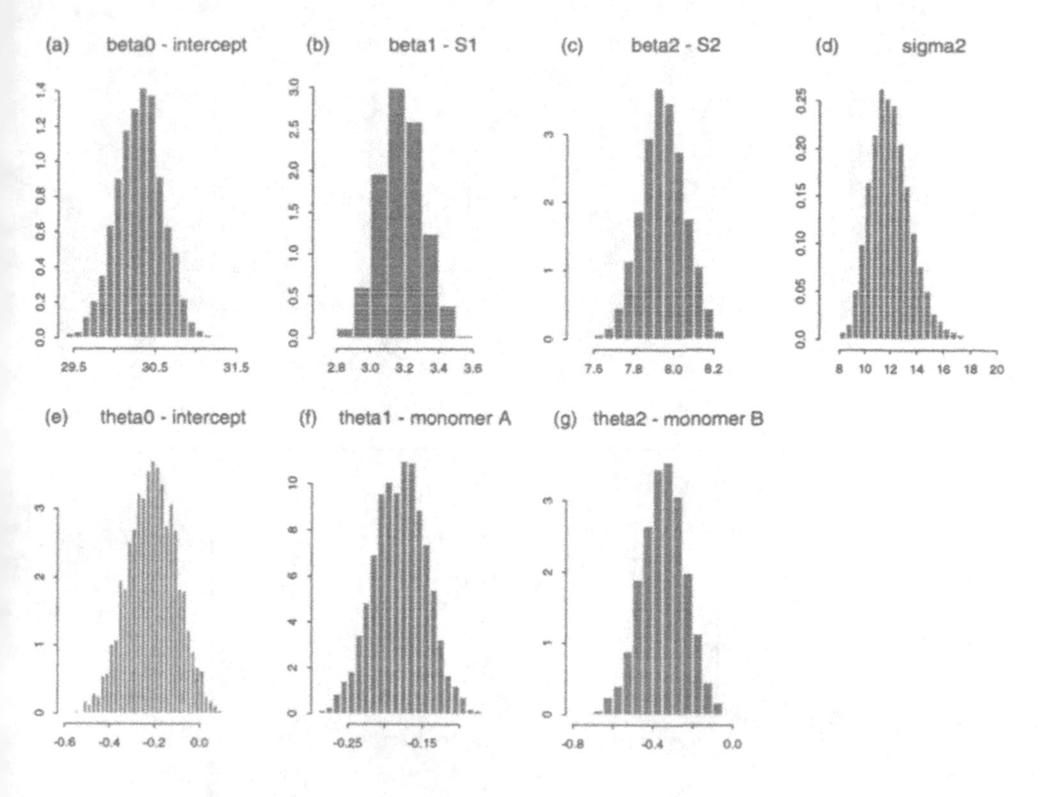

FIGURE 2. (a)-(c) Posteriors for $(\beta_0, \beta_1, \beta_2)$, (d) σ^2, and (e)-(g) $(\theta_0, \theta_1, \theta_2)$ in analysis of simulated data.

In summary, this simulated test data analysis indicates that the complicated mixture structure of these models can be adequately unwrapped to identify the effects of differences between monomer characteristics on both binding orientation and potency outcomes.

4 Critical Model Elaborations

Initial exploration of the modeling approach on real data sets from laboratories at Glaxo Wellcome Inc. indicated lack of model fit that we address here through model extensions. Deficiencies in the basic model are apparent in that posterior predictive distributions are rather overdispersed, and also obviously biased in some respects, relative to the observed data histogram. Related to this, the corresponding posterior for σ favors values very significantly larger than the ranges of 10-20% for assay standard deviations, which are normally expected by the chemists for assays of this type. We address this via model elaborations that introduce compound-specific random effects to account for and isolate departures from the basic model. The first model extension adds random effects to the

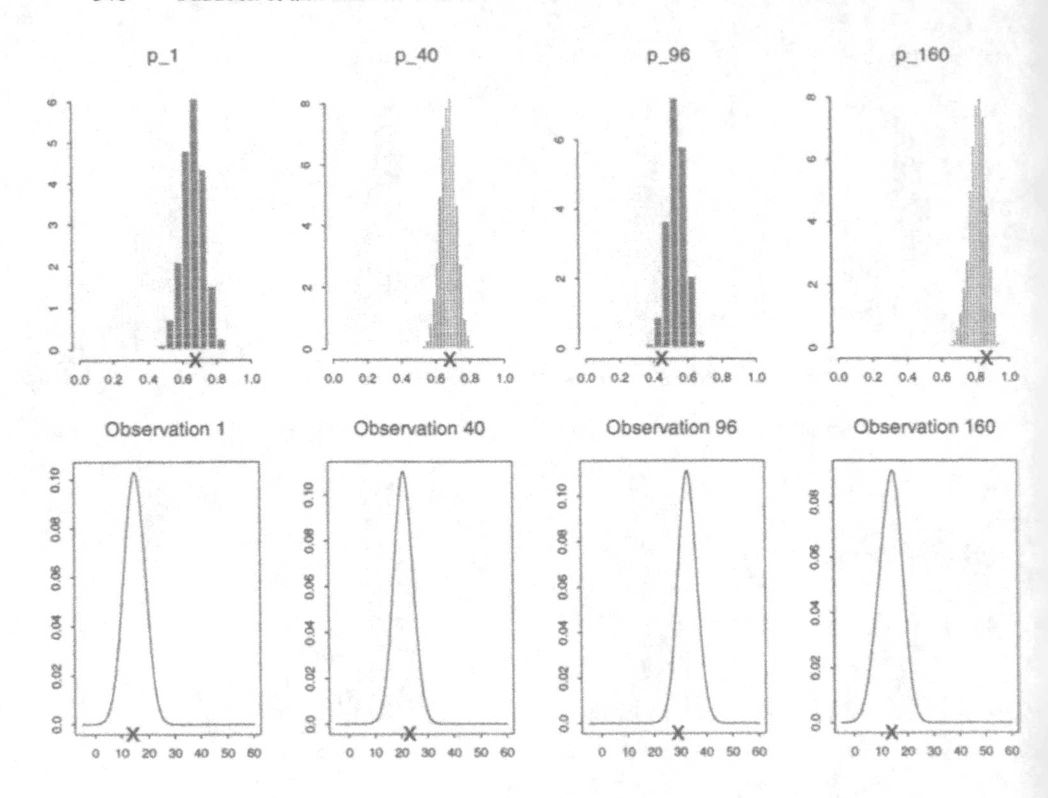

FIGURE 3. *Top row:* Posteriors for classification probabilities of four selected observations in the simulated dataset. *Bottom row:* Posterior predictive densities for the four cases. True values are marked as crosses on the axes in each row.

orientation/site binding mechanism to relax the strict probit regression on the specific linear predictor chosen. This results in a hierarchical model for over-dispersion in binary regression relative to the probit model. The second extension similarly modifies the strict normality assumption in the linear regressions, inducing heavier tails in the potency outcome distribution. Again this is done by adding hierarchical random effects to allow for additional, compound-specific variability. These two model elaborations very substantially improve model fit to the real data sets, with the resulting interpretations that: (a) there are a number of potency outcomes that are extreme relative to the basic normal regression (conditional on orientation), and (b) there is a significant degree of additional variability in the selection of binding orientations that is simply not captured by the specified linear predictor in the specific probit regressions for binding.

The extended model has the following structure. First, the binding indicators z_i are remodeled via $\pi_i = Pr(z_i = 1) = \Phi(h_i'\theta + \alpha_i)$ where we introduce conditionally independent random effects $\alpha_i \sim N(\alpha_i|0, \tau^2)$. These α_i represent compound-specific effects on binding that are unexplained by the chosen regressors h_i. Inference on these and τ will measure the levels and nature of the extra-probit variability. Second, conditional on binding configurations, the potency outcome levels are remodeled as $y_i \sim N(y_i|x_i(z_i)'\beta, \sigma^2/\lambda_i)$ where we introduce conditionally independent random effects $\lambda_i \sim Ga(\lambda_i|k/2, k/2)$. These λ_i represent compound-specific effects on potency that are unexplained by the chosen regressors $x_i(\cdot)$. This converts the normal error structure of the regression to the more heavy-tailed Student T_k forms induced by integrating out the λ_i (e.g., West 1984).

Application now includes the sets of random effects $\alpha = \{\alpha_i\}$ and $\lambda = \{\lambda_i\}$, together with the hyperparameter τ^2, in the posterior analysis. We choose to specify the degrees of freedom parameter k but require inference on τ^2. Details of model modifications are straightforward. Conditional upon $\{\alpha, \lambda\}$, the analysis is changed in only minor ways, adjusting the probit linear predictor and latent variables by the specific α_i, and weighting observations in the linear models by the λ_i. Then, at each stage of the MCMC analysis, all original model parameters, latent configuration indicators and so forth are sampled from these modified conditional posteriors. This is coupled with resampling of the new random effects and hyperparameters from their corresponding conditionals. These are trivial: the conditional posteriors for the α_i are independent normals, those for the λ_i are independent gammas, and that for τ^2 is an inverse gamma under an inverse gamma prior. Further details can be easily derived by the reader, or can be obtained on request from the authors.

5 A Glaxo Wellcome Dataset

A real data set from Glaxo Wellcome Inc. was selected from an original symmetric three-position library of compounds – that is, a library of three-monomer compounds. We selected 430 compounds with a common monomer and treat this smaller data set as a two-position library. The specific library under study is comprised of compounds similar to those synthesized by Whitten et al. (1996), who examine inhibition for a set of compounds consisting of a central triazine ring with two distinct monomers. Three covariates named `clogP`, `flexibility` and `molecular weight` are given for each monomer; `clogP` is a continuous-

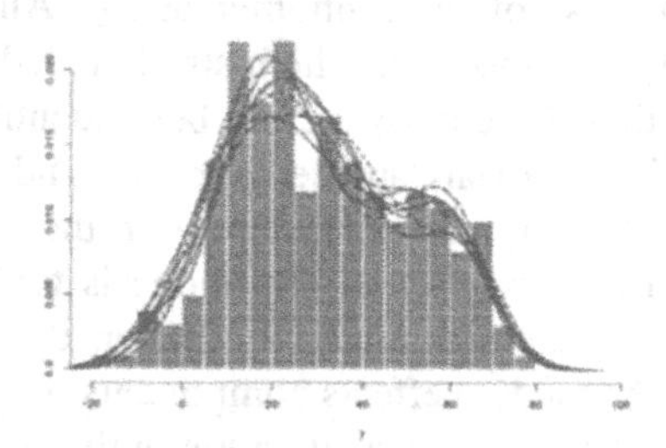

FIGURE 4. Posterior predictive model check for the simulated dataset.

valued measure of lipophilicity and `flexibility` and `molecular weight` are integer-valued binned variables. Potency levels y_i measure percent inhibition, which is a measure of how well the synthesized compounds perform with respect to a standard; for example, the standard can be a natural compound for which the medicinal chemists hope to discover a synthetic substitute with similar chemical behavior. At this stage of experimentation, the chemists search for highly potent compounds that will be retested in subsequent experiments to further assess their potential usefulness in drug design.

All three covariates are used in the linear regression model for potency outcomes, so that each x_i is a 7-vector, including the intercept constant. For the orientation model we difference the observed covariate values so that each h_i is a 3-vector representing relative covariates between the two monomers, plus an intercept. Our analysis is based on quite vague proper priors in the conditionally conjugate class. The prior for the illustrated analysis has the following independent margins: $N(\theta|0, I)$, $N(\beta|0, 10000I)$, $Ga(\sigma^{-2}|0.01, 1)$, and $Ga(\tau^{-2}|2, 1)$; we additionally set $k = 10$. Again following experiments with ranges of MC sample sizes and starting values for the MCMC analysis, summary numerical and graphical inferences here are based on a post burn-in sample of 5,000 iterates subsampled from a longer run, and deemed adequate based on repeat runs. We note that there is no appearance of the switching phenomenon in the MCMC trajectories induced by the model identification problem illustrated with our modified simulated data analysis earlier. Though this does not fully assure us that the MCMC run has remained in a single identified region of the parameter space, it does give some reason to believe that to be the case. Resulting sampled values are therefore assumed to be drawn from a single identified region, and are presented as histograms.

Figure 5(a-c) presents posteriors for the potency outcome regression parameter β. Note the clearly identified differences between the estimated effects of monomer characteristics when binding at different sites. Consider the variable `clogP`, for example, and the posterior in Figure 5(a). This indicates that a unit increase in `clogP` of the monomer at S_1 is associated with an average increase of about 3 units in potency. By contrast, a unit `clogP` increase of the monomer at S_2 has an associated average potency increase of around 17 units. From Figure 5(b) and (c), it appears that the coefficients of `flexibility` and `molecular weight` for the monomer at S_1 are apparently small or negligible, in distinct contrast to those of the monomer at S_2. All variables are clearly relevant, as is expected by the medicinal chemists, though these differential effects due to binding orientations have never before been identified. The posterior for σ^2 in Figure 5(g) indicates a standard deviation around 15-19, which is large on this outcome scale though quite consistent with the experimental chemistry involved. As noted earlier, assay variability alone is typically benchmarked at levels consistent with σ values around 10-20% on the inhibition scale. This supports the view that the random effects components α_i and λ_i have adequately catered for over-dispersion and mis-specification in the regression components.

Graphs in Figure 5(d-f,h) display the posteriors for the components of θ.

It is apparent that the relative values of `clogP` and `molecular weight`, describing the difference between the two monomers, are strongly associated with orientation outcomes, whereas `flexibility` is evidently much less important if not quite irrelevant. As mentioned previously, `clogP` is continuous, whereas the other covariates were computed as binned values; such binning could induce a loss of information resulting in the suppression of a relationship. Nevertheless, these preliminary conclusions are quite consistent with the consensus of chemists, who agree that `clogP` is generally one of the best chemical descriptors available for these data. Our isolation and estimation of the differential effects of `clogP` and `molecular weight` on binding versus potency outcomes is quite novel, and provides more incisive information about the relationships.

The posterior for the probit random effects variance τ^2, Figure 5(i), supports standard deviation values in the 0.6-0.7 range. This translates to the random effects alone accounting for a substantial range of variability on the orientation probability scale, potentially determining the orientation in some instances. Though the probit regression does isolate a meaningful explanatory predictor, there is evidently a fair degree of residual, compound-specific variation unexplained and that has here been simply estimated, compound by compound, via the α_i random effects. Further work with chemists should consider examination of compounds with significant random effects to enquire about possible additional explanatory variables and the reliability of assays.

Some global aspects of model adequacy are explored via posterior predictive checks, as illustrated in Figure 5(j,k). For each observed compound, we compute approximate posterior predictive distributions at the observed design points and evaluate the resulting set of 430 predictive quantiles at the actually observed potency outcomes. Assuming the effects of dependencies induced by posterior parameter uncertainties to be ignorable, departures of this set of quantiles from an appearance of approximate uniformity will indicate global model deficiencies. These figures display a uniform quantile plot and a histogram of the values, and the concordance with uniformity is comforting, suggestive of an adequate model fit.

We further explore model fit and interpretation of the random effects parameters in Figure 6. Potency predictions for eight randomly selected compounds appear in the top two rows. Predictions for compounds with extreme posterior means for their α_i and λ_i values appear in rows 3 and 4, respectively. In each case, the posterior means of λ_i and α_i are given, and the observed potency values are marked on the axis. In the randomly selected cases, predictions are made apparently adequately. The compounds in row 3 have apparently extreme α_i values but moderate λ_i values, while the situation is reversed for compounds in row 4. The observed values for the compounds in rows 3 and 4 consistently fall in the tails of their predictive distributions. To explore the implications of accounting for heterogeneity of the compounds with respect to activity, consider the first histogram in row 4. The mean of the λ_i here is 0.292. The observed potency outcome is -29, which is initially puzzling since the data measure percent inhibition. Despite the assumption of inhibition for the compounds under

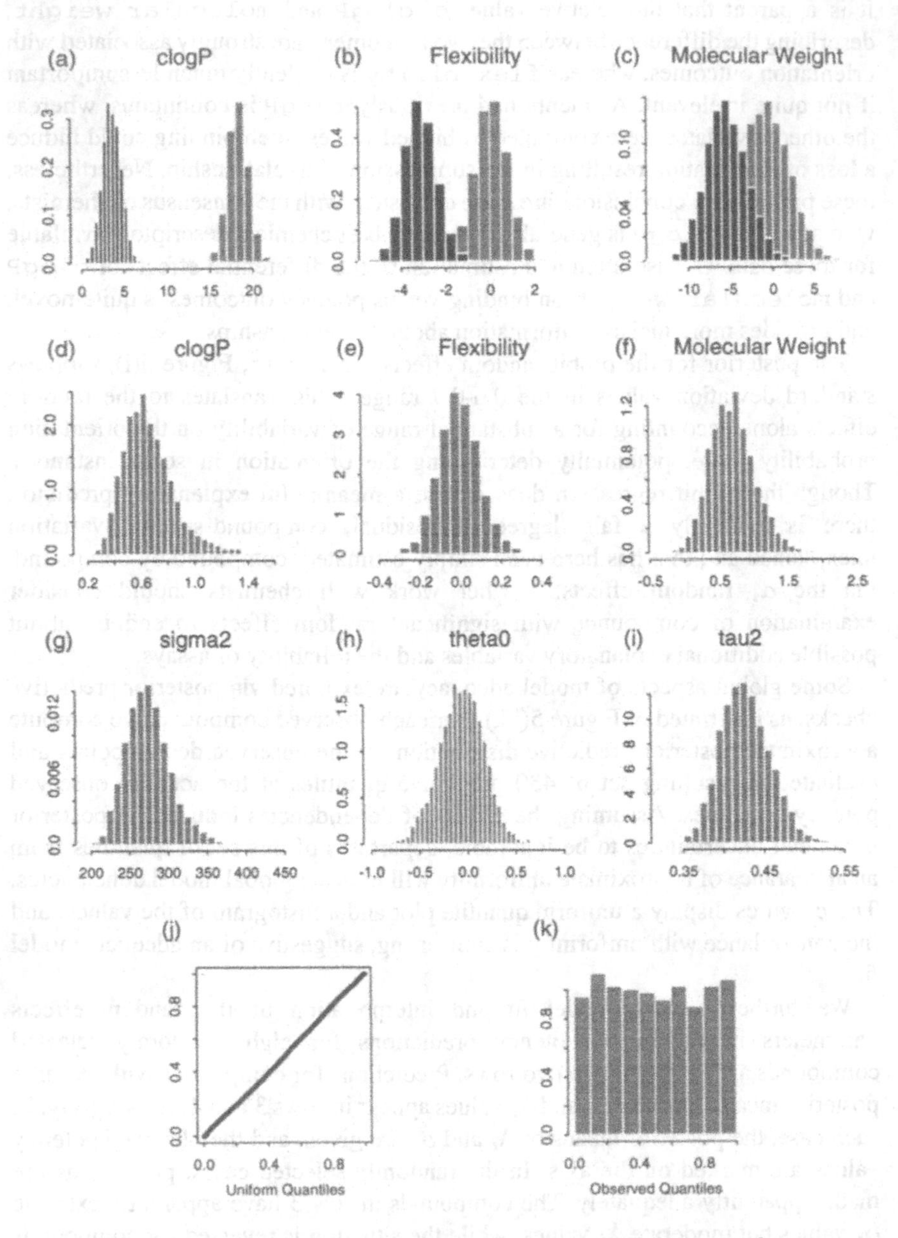

FIGURE 5. Posteriors in analysis of Glaxo Wellcome dataset. (a-c): β parameters in potency regressions. The right-most histograms are for regression coefficients for monomers binding to S_1, and the left-most for monomers binding to S_2. (d-f,h): Probit regression parameters θ. (g): σ^2. (i): Binding probability variance parameter τ^2. (j,k): Predictive quantiles of observed data in uniform quantile plot and histogram.

experimentation, it is the case that excitation occurs occasionally due to biological variability, explaining this outcome that is quite outside the model for inhibition we have constructed. Hence the posterior indicates a very low λ_i value, isolating this case as an apparent outlier. Here, as in the other cases, it is interesting to note that lack of fit of the original model is isolated via extreme values of just one of the two random effects, indicating that compounds poorly represented by the basic model depart in terms of either extreme potency outcomes, or in terms of an unusual binding orientation, but, at least for the selected compounds, not both simultaneously.

6 Discussion

Our analysis, a new approach to assessing chemical structure-activity relationships, promises to be useful for examining libraries of compounds through exploration of monomer space. Our results also provide motivation to enhance computation of chemical descriptors of monomers, so providing additional covariates for analysis. Evidently, the real data analysis here isolates some meaningful relationships between covariates and both responses – binding and potency – but much more is needed to improve predictive ability for future compounds. Our discussion indicates that, with the extensions to incorporate hierarchical model components describing heterogeneity in both the binding/orientation selection mechanism and the potency outcome distributions, our model appears to adequately represent this particular real data set, in spite of a relatively high degree of assay and experimental variability. Posterior distributions on model parameters indicate subsets of the chosen covariates are informative about the structure-activity relationship, and we have contributed to the chemistry in novel ways by isolating and estimating the differential effects of covariates on binding orientation and potency scales separately. The high levels of assay variability limit predictive accuracy, though our derived predictions are generally accurate. Our random effects models are critical in accounting for compound-specific characteristics that are not measured or understood, and this may prove useful in future studies of similar compounds whose monomers have identical covariate values but, when analyzed, apparently distinct random effects. The framework may also be modified to include random effects for the monomers, rather than just the overall compounds, in support of such experimental developments.

Current research on the statistical side includes exploration of the identifiability problem, especially when the number of covariates is small and/or when the assay variance is large. We have also begun the development of related models for more complex experiments involving more than two monomers and receptor binding sites. Additionally, we need to explore more monomer covariates, perhaps many more than the simple few here. What has been learned in this simpler case is naturally guiding these challenging but potentially rewarding investigations.

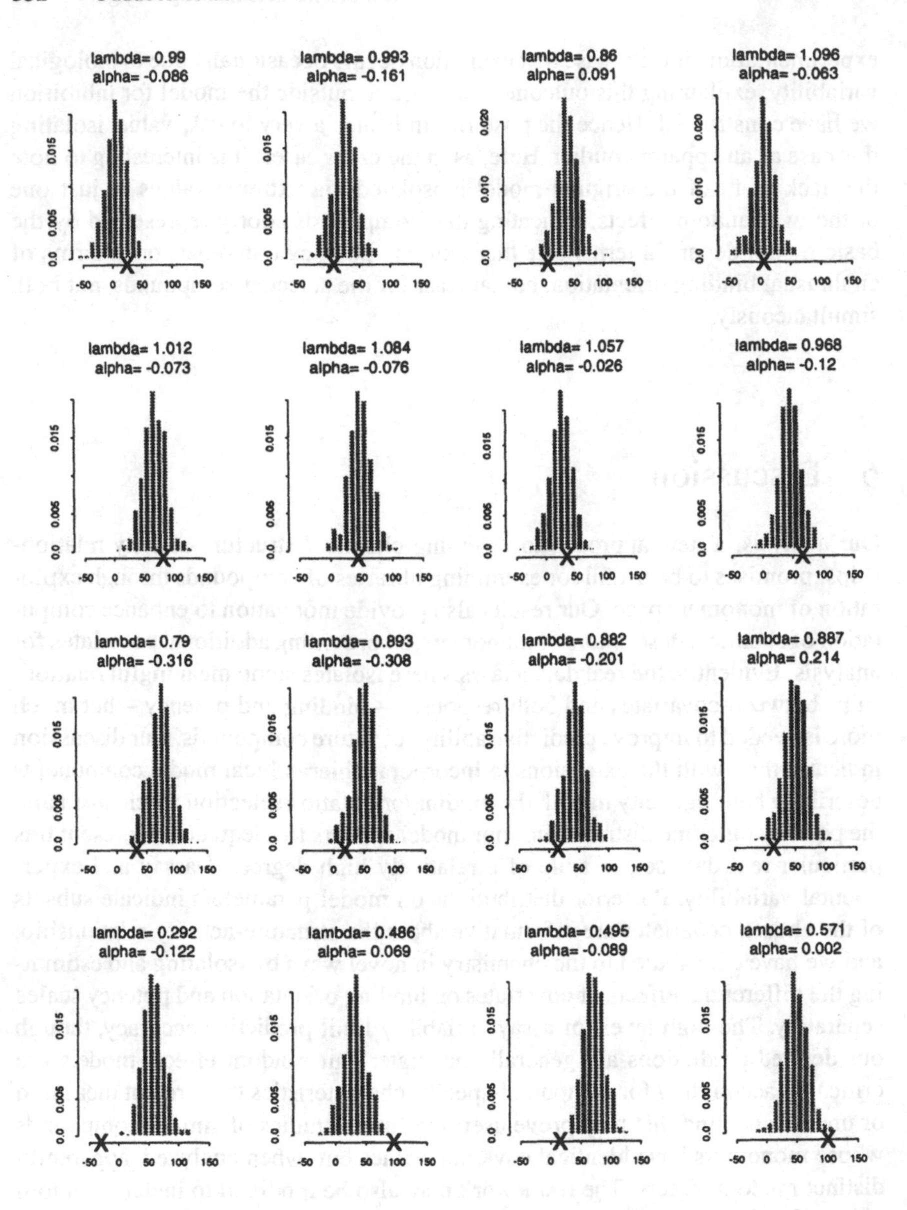

FIGURE 6. Predictive densities of eight randomly-selected compounds (rows 1 and 2), four compounds with extreme α_i values (row 3), and four compounds with extreme λ_i value (row 4). Data values appear as crosses on the axes.

Acknowledgments

This research is partially supported by Glaxo Wellcome, 5 Moore Drive, RTP, NC 27709, and the National Institute of Statistical Sciences (NISS), `http://www.niss.org`. Corresponding author is Susan Paddock, Institute of Statistics and Decision Sciences, Duke University, Durham, NC 27708-0251, USA, `http://www.stat.duke.edu`

References

Albert, J.H. and Chib, S. (1993) Bayesian analysis of binary and polychotomous response data, *Journal of the American Statistical Association*, **88**, 669-679.

Gelfand, A.E. and Smith, A.F.M. (1990) Sampling-based approaches to calculating marginal densities, *Journal of the American Statistical Association*, **85**, 398-409.

Gelman, A., Carlin, J.B., Stern, H., and Rubin, D.B. (1996) *Bayesian Data Analysis*, London: Chapman and Hall.

Titterington, D.M. and Smith, A.F.M., and Makov, U.E. (1985) *Statistical Analysis of Finite Mixture Distributions*, London: Wiley.

Whitten, J.P., Xie, Y.F., Erickson, P.E., Webb, T.R., De Souza, E.B., Grigoriadis, D.E., and McCarthy, J.R. (1996) Rapid microscale synthesis, a new method of lead optimization using robotics and solution phase chemistry: Application to the synthesis and optimization of Corticotropin-releasing factor$_1$ receptor antagonists, *Journal of Medicinal Chemistry*, **39** 4354-4357.

West, M. (1984) Outlier models and prior distributions in Bayesian linear regression. *Journal of the Royal Statistical Society (Ser. B)*, **46**, 431-439.

West, M. (1997) Hierarchical mixture models in neurological transmission analysis. *Journal of the American Statistical Association*, **92**, 587-606.

Population Models for Hematologic Data

J. Lynn Palmer
Peter Müller

ABSTRACT This manuscript was motivated by the need to adequately model and predict the distribution of blood stem cells collected from cancer patients. These blood stem cells are collected from patients over multiple days prior to their undergoing high-dose chemoradiotherapy as a treatment for cancer. The blood stem cells are returned to the patient after the high-dose chemoradiotherapy in order to enable permanent reconstitution of the white blood cell components and the patient's healthy blood system. Maximizing the number of blood stem cells collected in as few aphereses as possible is desirable. We use population models to model these collections and to design optimal apheresis schedules to collect blood stem cells from cancer patients.

1 Introduction and Background

Reconstitution of blood cell components after high-dose chemoradiotherapy is an essential need of patients who undergo this type of treatment for cancer. This high-dose treatment may be used for the treatment of, for example, acute myeloblastic leukemia or metastatic breast cancer. High-dose chemoradiotherapy in this setting (also called a myeloablative treatment) is designed to completely eradicate the blood cell systems. The reconstitution of blood cell components can be achieved either by bone marrow transplant or by transfusion of blood stem cells collected from the circulating blood. Stem cell transfusion has the advantage of having a more rapid reconstitution of white blood cells, thus reducing the length of time the patient is at high risk for a potentially lethal infection. It is also an option for older and other patients considered ineligible for bone marrow transplant. Patients can also be discharged earlier than those undergoing bone marrow transplantation. In order to maximize the usefulness of blood-derived stem cells, two issues must be resolved: the timing and number of collection procedures for these stem cells. These issues are especially important because the number of circulating stem cells at any given time differs according to the type and timing of pre-treatments.

Blood stem cells, the precursors of mature blood cells, reside in bone marrow and also routinely circulate to a limited degree in the peripheral blood system. The concentration of blood stem cells in the circulating blood is about 1/10 to 1/20 of that in the marrow (from McCarthy and Goldman, 1984). Pre-treatment regimens (regimens incorporating cyclophosphamide at sufficient dosages to cause brief, reversible neutropenia) have been designed for patients scheduled for high-dose chemoradiotherapy which increase the number of blood stem cells circulating in the peripheral blood system. Often, these pre-treatments result in a transient ten-fold or more increase in the concentration of circulating blood stem cells, which can then be collected from the patient's blood. These brief rises of rebound blood stem cells "flood" into the circulatory system and usually abate within seven days. It is during this temporary increase in the concentration of peripheral blood stem cells that the opportunity for selecting optimal times for blood stem cell collection occurs.

During the blood stem cell collection process (which is also known as apheresis) blood is drawn from the patients, then centrifuged and its components separated. The white blood cell components which contain the blood stem cells are collected and the remaining cells are returned to the patient. The collected blood stem cells are frozen for later use by the patient after high-dose chemoradiotherapy treatment for cancer. Each apheresis procedure requires about three hours, although its duration may vary, based on the fraction of the patient's blood volume processed. Usually three to five apheresis procedures are performed. Although in some types of cancers eight or more aphereses would be required to achieve a minimum number of blood stem cells, this type of procedure is not often used in those circumstances. The initial apheresis for this study was scheduled to occur on the fifth or the seventh day after an initial pre-treatment regimen (depending on which pre-treatment was given) with additional blood stem cell collections then scheduled each day for five days unless the target number of cells was collected earlier.

The two pre-treatments used in this study were G-CSF (Granulocyte Colony-Stimulating Factor) alone and IL-3 (Interleukin-3, an immunomodulatory agent) followed by G-CSF. G-CSF acts to mobilize blood stem cells into the peripheral blood. Results suggest that IL-3 expands an early blood stem cell population *in vivo* that subsequently requires the action of a later factor such as G-CSF to complete its development. See Donahue et al. (1988) or Robinson et al. (1987) for further information.

Blood stem cells are cells capable of self-renewal and differentiation into one or all of the mature cells found in the peripheral blood. To et al. (1989) recommended a minimum safe dose of blood stem cells for permanent engraftment to be 50×10^4 per kg body weight for peripheral blood stem cell transplants.

Another important consideration is whether the removal of such large numbers of stem cells is harmful to the patient. Results of the study by To et al. (1989) showed that peripheral blood stem cell aphereses carried out during very early remission of acute myeloid leukemia have no deleterious effects on hematopoietic (blood stem cell) recovery, as measured by the rates of rise of neutrophils, platelets, and lymphocytes during and following apheresis. There was also no difference in the median duration of first remission or the apparent plateau in the survival curve between the apheresed and the non-apheresed patients.

Methods to isolate blood stem cells must rely on physical properties that can distinguish these cells from other cells contained in marrow or blood. Investigators have subsequently discovered that a monoclonal antibody raised against the leukemia cell line KG-1A reacted with 1-2% of mononuclear cells in human marrow that gave rise to virtually all hematopoietic colonies and their precursors detectable *in vitro*. The antigen defined by this antibody was designated $CD34$. Several other monoclonal antibodies have been raised against the $CD34$ antigen, greatly facilitating efforts to purify stem cells. For more information see Gale, Juttner and Henon (1994). There are also several subsets of $CD34^+$ cells used in these type of studies. It is not known which of these subsets (or combination of several subsets, or proportions of the relevant subsets) are needed to prove that these cells alone are necessary and sufficient for hematopoietic recovery after high-dose chemoradiotherapy (marrow ablative therapy). This study uses the total number of $CD34^+$ cells in its predictions and, for convenience and consistency, refers to them as blood stem cells or as stem cells.

The total blood stem cell yield per apheresis may depend on variables such as: blood volume processed, separation efficiency of the device used, centrifuge speed, hematocrit and white blood cell counts before apheresis, disease status at time of apheresis and response to pre-treatment. The highest concentration of circulating blood stem cells is most often observed in patients who have received minimal or no prior chemoradiotherapy.

2 Population Models

2.1 Background

The optimal design for the timing of scheduled stem cell collections, or aphereses, will be made by inference in a Bayesian population model using longitudinal data. The model will be estimated by Markov chain Monte Carlo simulation techniques. Using these techniques, we can find optimal designs for future patients by minimizing a specified loss function over the posterior predictive distribution for the new patient.

This method is generalized from other methods recently proposed for inference in population models, such as Wakefield, Smith, Racine-Poon and Gelfand (1994), Rosner and Müller (1994, 1997), Wakefield (1996), and Müller and Rosner (1997). In general, population models are concerned with inference as follows. Let y_{ij}, $i = 1, \ldots, I$, $j = 1, \ldots, n_i$ denote n_i repeated observations for each of I individuals (in this case, stem cell counts collected at each of several days). For each individual i we introduce a parameter vector θ_i to describe the joint distribution of $y_i = (y_{i1}, \ldots, y_{in_i})$. Then θ_i parametrizes a mean curve $E(y_{ij}|t_{ij})$ over time and describes the joint sampling distribution of y_i. The model is then completed by a second stage prior $p(\theta_i|\phi)$, which could involve a regression on some patient specific covariates x_i.

We use the posterior predictive distribution to solve an optimal design problem. Given the posterior predictive distributions for future patients we find an apheresis schedule that minimizes some pre-specified loss function. The loss function includes a penalty for failing to collect a certain target number of stem cells and a cost for each stem cell collection scheduled. Wakefield (1994) has proposed a similar approach with the specific aim of determining an optimal dose in a pharmacokinetic study. For a review of Bayesian methods for optimal design problems see, for example, Chaloner and Verdinelli (1995).

2.2 *Estimating Blood Stem Cell Profiles*

The magnitude and length of time of elevated stem cell levels in the peripheral blood after an initial pre-treatment may vary considerably depending on variables such as the pre-treatment regimen received and the patient's health status. As a result, a high increase in the number of stem cells may occur quickly but drop off rapidly, or a less extreme peak may occur but continue with an elevated number of stem cells over a longer period of time. Figure 1 shows typical blood stem cell ($CD34$) counts of four patients with data points connected by dark or dashed straight lines. Palmer and Müller (1998), provide further details for the mathematical modeling of this application. The approach is briefly summarized below.

The variety of shapes that are possible for blood stem cell level profiles will be modeled by using a scaled Gamma probability density function as a nonlinear regression function. Let $g(t; e, s)$ denote a Gamma probability density function with shape and scale parameters chosen such that mean and standard deviation are e and s, respectively, and rescaled such that $\sup_t g(t; e, s) = 1.0$, i.e.,

$$g(t; e, s) = \left[(e^2 - s^2)/e\right]^{-e^2/s^2+1} \exp\left[-e/s^2(t-e) - 1\right] t^{e^2/s^2-1}, \qquad e > s.$$

The hierarchical model specified below includes for each patient a nonlinear regression with mean function $f(t) = z_i\, g(t; e_i, s_i)$.

A hierarchical prior distribution on the profile parameters $\theta_i = (e_i, s_i, z_i)$ is defined by $\theta_i \sim N(\eta_k, V)$, where $k = 1, 2$ is the pre-treatment administered to patient i and η_k is the mean profile parameter for patients subjected to pre-treatment k. The model is completed by a hyperprior distribution $\eta_k \sim N(\mu, \Sigma)$, $k = 1, 2$ and conjugate Wishart and Gamma hyperpriors on V^{-1} and σ^{-2}. Specifically,

$$
\begin{aligned}
y_{ij} &= z_i\, g(t_{ij}; e_i, s_i) + \epsilon_{ij},\ i = 1, \ldots, I,\ j = 1, \ldots, n_i \\
&\quad \epsilon_{ij} \sim N(0, \sigma^2), \\
\theta_i &\sim N(\eta_k, V), k = 1, 2 \\
&\quad \theta_i = (e_i, s_i, z_i), \\
\eta_k &\sim N(\mu, \Sigma),\ k = 1, 2 \\
V^{-1} &\sim W(q, (qQ)^{-1}) \\
\sigma^{-2} &\sim Gamma(a_0/2, b_0/2).
\end{aligned} \tag{1}
$$

The hierarchical model allows for information to be used from different sources, borrowing strength across different patients and different pre-treatments. The model includes a nonlinear regression as a submodel for each patient. Instead of estimating each submodel separately as an independent probability model we assume common prior parameters η_k for all patients who undergo the same pre-treatment.

The model is estimated by implementing a Markov chain Monte Carlo scheme using methods similar to those found in, for example, Gilks, et al. (1993). In our example we will let $\omega = (\theta_1, \ldots, \theta_I, \eta_1, \eta_2, V, \sigma^2)$ denote the full parameter vector. Using MCMC posterior simulation it is possible to generate by computer simulation a Monte Carlo sample $M = \{\omega^1, \ldots, \omega^T\}$. Here, ω^t is the imputed parameter values after t iterations of the Monte Carlo simulation. For details of our setup of the Monte Carlo simulation used, see Palmer and Müller (1998). The fitted curves in Figure 1 are posterior means of fitted values, $f(t) = z_i\, g(t; e_i, s_i)$, estimated by averages over the Monte Carlo sample M.

2.3 *Determining Optimal Apheresis Design*

The decision problem is to find the optimal apheresis schedule for a new patient, $h = I + 1$, who is administered a specified pre-treatment. Given that the first day of collection is as previously determined (5 or 7 days, depending on the pre-treatment), which of the following days should be selected for blood stem cell collections? The process of finding the optimal apheresis schedule can be described as minimizing in expectation the value of some pre-specified loss function. The loss function in this problem is given by a sampling cost, p_1, for each stem cell collection and a penalty, p_2, for failing to collect a pre-specified minimum number of blood stem cells. The decision maker must specify values for each of these penalties. Let d denote the decision of which days an apheresis should be scheduled. Let $h = I + 1$ index a new future patient. Let D denote the event of failing to achieve the minimum number of blood stem cells. The optimal design is determined by maximizing in expectation the utility function

$$u(d, y_h) = -p_1 n_d - p_2 I(D) \tag{2}$$

where $I(D)$ is an indicator function of event D.

The computed utilities, and thus eventually the optimal design, depend critically on the relative trade-off p_1/p_2 between sampling cost and the penalty for not achieving the target. The solution reported in the next section is based on a relative penalty of 1 to 10. In this application, it is most critical that the target number of blood stems cells is reached. However, it is also important not to burden the patient with unnecessary aphereses. Each collection requires about three hours, and is not a trivial process. Besides these general considerations, and a reasonable assumption that the number of aphereses be penalized linearly, the specific choice of the functional form for the utility function $u(\cdot)$ is arbitrary. Alternatives could, for example, consider a penalty for not achieving the target different from the step function $I(D)$ used here. Compare also with the discussion at the end of Section 3.2 about sensitivity of the final results to alternative choices of the utility function.

Since the apheresis schedule d for the new patient h has to be chosen before y_h is observed, we need to maximize the expectation of $u(\cdot)$ with respect to y_h. Let $p(y_h|y)$ denote the posterior predictive distribution for the profile of a future patient $h = I + 1$ in model (1), given the observed data $y = (y_{ij}, i = 1, \ldots, I, j = 1, \ldots, n_i)$. We can formally state the design problem as

$$\max_d \underbrace{\int u(d, y_h)p(y_h|y)dy_h}_{U(d)} \tag{3}$$

Due to the medical limitation of one blood stem cell collection per day, there is a limited, countable number of possible schedules for blood stem cell collections. We therefore enumerated all possible designs and evaluated the loss associated with each possible schedule. A thorough discussion of this process which uses Monte Carlo integral approximation can be found in Palmer and Müller (1998).

3 Results

3.1 *Population Models*

The model as described can be used to fit individual profiles to each patient. Figure 1 shows results of fitted profiles for four typical patients. The two patients on the left side of the figure were given pre-treatment 1, G-CSF (granulocyte colony-stimulating factor) alone. The two patients on the right side of the figure were given pre-treatment 2, IL-3 (interleukin-3) followed by G-CSF. The dark or dashed straight lines connect actual blood stem cell counts for patients given pre-treatments 1 or 2 and the solid curve plots the fitted profile using the model as described previously. It can be seen that pre-treatment 1 (G-CSF only) achieves a peak earlier than pre-treatment 2 (IL-3 followed by G-CSF). However, the two fitted curves for each pre-treatment appear to share some common characteristics. The hierarchical nature of the prior (allowing for more than one pre-treatment group), allows the model to make use of these shared characteristics and takes into account the different mean profile parameters for patients in each pre-treatment group.

A second feature of the model is to fit predicted profiles for new patients who were given a specified pre-treatment. Figure 2 displays predicted profiles for future patients given one of the two pre-treatments. Pre-treatment 1 (G-CSF) is again shown in the left panel and pre-treatment 2 (IL-3 followed by G-CSF) in the right. The dashed lines represent $\pm$ one posterior standard deviation. A quick glance at the plots would suggest that earlier blood stem collections should be made for pre-treatment 1 as compared to pre-treatment 2.

Another feature of the model is to fit an individual patient's predicted profile, given their first blood stem cell collection results (the number of blood stem cells collected) and knowledge of the pre-treatment they were given. Figure 3 shows the predicted profiles conditional on a patient's first blood stem cell collection and their pre-treatment. As in the previous figure, pre-treatment 1 (G-CSF) is shown on the left and pre-treatment 2 (IL-3 followed by G-CSF) on the right. The dashed lines plot $\pm$ one posterior standard deviation. The star in each panel identifies the first blood stem cell count for that patient. Also again, it appears that earlier blood stem cell collections would be suggested for pre-treatment 1 as compared to pre-treatment 2.

3.2 *Optimal Design Results*

The expected utilities were computed by evaluating Monte Carlo averages. Almost no additional computations were required beyond the simulation of the Monte Carlo sample to solve the optimal design problem.

The results of the optimal design problem, which designates the optimal blood stem cell collection schedule for an individual patient given their initial blood stem cell collection count and their pre-treatment, are shown in Table 1. This table summarizes optimal designs for two future patients, one in each pre-treatment group, in addition to their second- and third-best blood stem cell collection designs. The first column displays the estimated value $-U(d)$. The smallest values for $-U(d)$ minimize this loss function. The initial day of pre-treatment, d_0, differs for the two pre-treatments. For the first patient (pre-treatment 1) the first observation was taken on day 5, for the second patient (pre-treatment 2), on day 7. Even given this shift in initial days, it can be seen that the two pre-treatments differ in estimated optimal apheresis schedules.

For example, for the first patient the optimal design would be to schedule a blood stem cell collection on day 5 and on the following day. It appears that in most cases, the minimum number of stem cells either was achieved in two collections for these patients, or it was not achieved regardless of the number of days of collection. There is some indication that the first day of blood stem cell collection should have been scheduled earlier, on day 3 or 4. The next optimal schedule for this patient would be to begin on day 5, also collect on day 6 and again on day 7. This design closely matches the procedure that was followed at

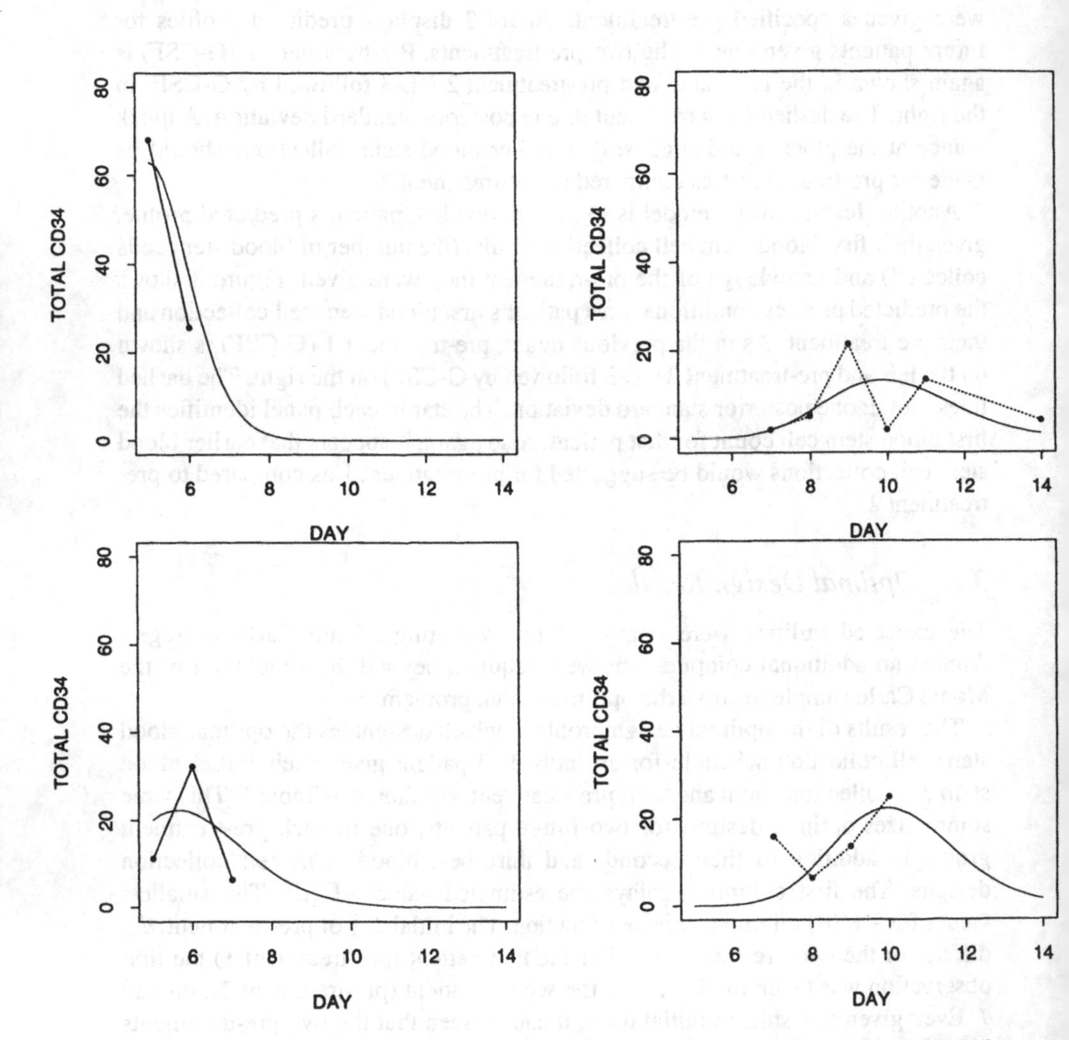

FIGURE 1. Four typical patients' blood stem cell (CD34) counts. The dark or dashed straight lines connect the counts for pre-treatments 1 or 2. The solid curve plots the fitted profile for the individual patient.

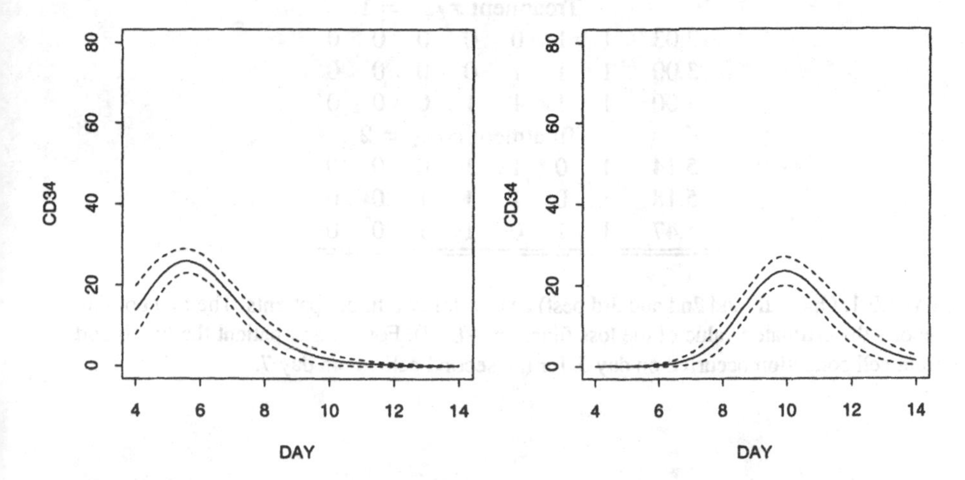

FIGURE 2. Predicted profiles of blood stem cell (CD34) collection counts plotted against days for future patients subjected to pre-treatment 1 (on the left) and pre-treatment 2 (on the right). The dashed lines plot ± one posterior standard deviation.

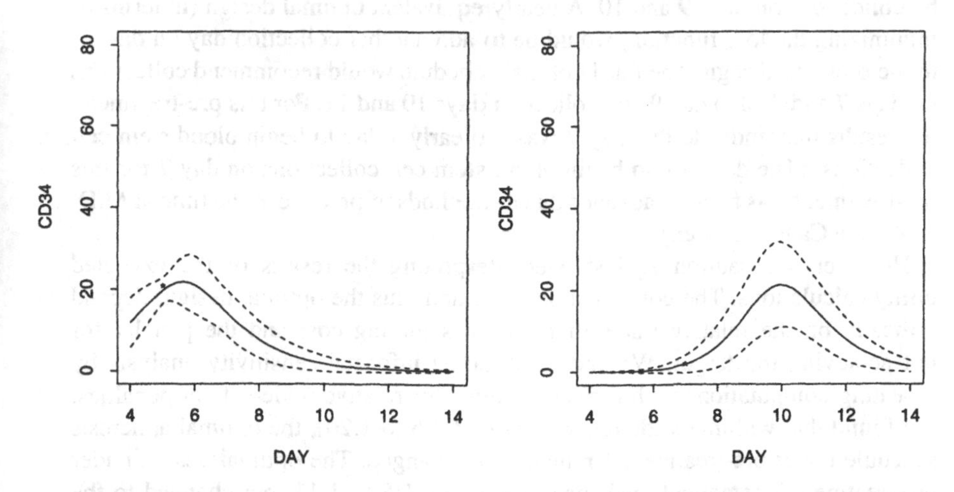

FIGURE 3. Predicted profiles for future patients conditional on a first blood stem cell (CD34) collection count plotted against days. The left figure shows a future patient subjected to pre-treatment 1, and the right shows a patient subjected to pre-treatment 2. The dashed lines plot ± one posterior standard deviation. The asterisk marks the first observation.

Value	d_0	d_1	d_2	d_3	d_4	d_5	d_6
	Treatment $x_{I+1} = 1$						
2.03	1	1	0	0	0	0	0
3.00	1	1	1	0	0	0	0
4.00	1	1	1	1	0	0	0
	Treatment $x_{I+1} = 2$						
5.14	1	0	1	1	0	0	0
5.18	1	0	1	1	1	0	0
5.47	1	1	0	1	1	0	0

TABLE 1. Optimal (and 2nd and 3rd best) design for two future patients. The first column reports the estimated value of the loss function $-U(d)$. For the first patient the first blood stem cell collection occurred on day 5, for the second patient, on day 7.

M.D. Anderson Cancer Center at the time that this data was collected.

The second patient (in pre-treatment 2) has a differing optimal design for the first-, second-, and third-best schedules. In this patient's case, the first collection would be made on day 7, but then no collection would be made on day 8, followed by collections on days 9 and 10. A nearly equivalent optimal design (in terms of minimizing the loss function) would be to add another collection day on day 11 to the previous design. The third optimal schedule would recommend collections on days 7 and 8, skip day 9 and collect on days 10 and 11. For this pre-treatment, the results may indicate that day 7 was too early a day to begin blood stem cell collections. (The decision to begin blood stem cell collections on day 7 for this pre-treatment was fixed in advance due to methods of practice at the time at M.D. Anderson Cancer Center.)

However, we caution against over-interpreting the results of the expected utility calculations. The computed utilities, and thus the optimal design, depend critically on the relative trade-off between sampling cost and the penalty for not achieving the target. We undertook some informal sensitivity analysis by repeating computations with different values for relative trade-offs in penalties. We found that within a wide range (ratios of 1:5 to 1:20), the optimal apheresis schedule under pre-treatment 1 remained unchanged. The optimal design under pre-treatment 2 remained unchanged for ratios 1:5 to 1:11, but changed to the design of the previous second-best optimal schedule when ratios of 1:11 to 1:20 were used. In general, the optimal design selected for use in cancer patients may be based on other considerations as well. Such as, if the minimum number of cells have been collected early, a later collection may not be necessary. Also, it may not always be possible to conduct a blood stem cell collection if the patient's health or other scheduling problems preclude it.

4 Discussion

The practical issues of determining optimal blood stem cell collections for cancer patients have been addressed using population model methods. We have used this method to profile blood stem cell counts from individual patients over time, to model and predict the profile of blood stem cell counts for patients given different pre-treatments, and also to predict a profile for an individual patient given their initial blood stem cell count and their initial pre-treatment. The methods of this paper can also be generalized to include results from more than two pre-treatments and to predictions for a specific patient based on more than one day of blood stem cell counts. We have also summarized the results and interpretation of optimal apheresis schedules that can be made with these population model methods.

The benefits of the Bayesian approach in this application are that the above results were accomplished. Non-Bayesian methodology is not available to replicate these models or predictions, or for optimizing apheresis scheduling. Also, the methodology could be used even though only a small amount of data was available. The difficulties in the Bayesian approach in this application include a high level of mathematical modeling to arrive at a solution. However, once the theory is in place and related programs have been written, it is relatively easy to apply this method to other data sets and other similar situations. In summary, the methodology may be complex, but it simplifies the understanding of the application (the profiles of blood stem cell counts) and clarifies the benefits of different apheresis schedules.

Acknowledgments

The authors thank Dr. Martin Körbling, Department of Hematology, The University of Texas M.D. Anderson Cancer Center, for the use of the data presented in this paper, and for discussions on this topic of applied research.

J. Lynn Palmer was partially supported by the National Institutes of Health Cancer Center Support Grant CA-16672.

Peter Müller was partially supported by the National Science Foundation under grant DMS-9704934.

References

Chaloner, K. and Verdinelli, I. (1995). Bayesian experimental design: a review, *Statistical Science*, **10**, 273–304.

Donahue, R.E., Seehra, J., Metzger, M., Lefebvre, D., Rock, D., Carbone, S., Nathan, D.G., Garnick, M., Sehgal, P.K., Laston, D., LaVallie, E., McCoy, J., Schendel, P.F., Norton, C., Turner, K., Yang, Y.C. and Clark, S. (1988).

Human IL-3 and GM-CSF act synergistically in stimulating hematopoiesis in primates, *Science*, **241**, 1820–1823.

Gale, R.P., Juttner, C.A. and Henon, P. (1994). *Blood Stem Cell Transplants*, Cambridge University Press, Cambridge, Great Britain.

Gelfand, A.E. and Smith A.F.M. (1990). Sampling based approaches to calculating marginal densities, *Journal of the American Statistical Association*, **85**, 398–409.

Gilks, W.R., Clayton, D.G., Spiegelhalter, D.J., Best, N.G., McNeil, A.J., Sharples, L.D. and Kirby, A.J. (1993). Modelling complexity: applications of Gibbs sampling in medicine, *Journal of the Royal Statistical Society Ser. B*, **55**, 39–52.

McCarthy, D.M. and Goldman, J.M. (1984). Transfusion of circulating stem cells: A review of clinical results, *CRC Critical Reviews in Clinical Laboratory Sciences*, **20**, 1–24.

Müller, P. and Rosner, G. (1997). A semiparametric Bayesian population model with hierarchical mixture priors, *Journal of the American Statistical Association*, in press.

Palmer, J.L. and Müller, P. (1998). Bayesian Optimal Design in Population Models of Hematologic Data, *Statistics in Medicine*, in press.

Robinson, B.E., McGrath, H.E. and Quesenberry, P.J. (1987). Recombinant murine granulocyte-macrophage colony stimulating factor has megakaryocyte colony stimulating activity and augments megakaryocyte colony stimulation by Interleukin-3, *J Clin Invest*, **79**, 1648–1652.

Rosner, G. and Müller, P. (1994). Pharmacodynamic analysis of hematologic profiles, *Journal of Pharmacokinetics and Biopharmaceutics*, **22**, 499–524.

Rosner, G. and Müller, P. (1997). Bayesian population pharmacokinetics and pharmacodynamic analyses using mixture models, *Journal of Pharmacokinetics and Biopharmaceutics*, **25**, 209–233.

To, L.B., Haylock, D.N., Thorp, D., Dyson, P.G., Branford, A.L., Ho, J.Q.K., Dart, G.D., Roberts, M.M., Horvath, N., Bardy, P., Russel, J.A., Millar, J.L., Kimber, R.J. and Juttner, C.A. (1989). The optimization of collection of peripheral blood stem cells for autotransplantation in myeloid leukaemia, *Bone Marrow Transplantation*, **4**, 41–47.

Wakefield, J. (1994). An expected loss approach to the design of dosage regimens via sampling-based methods, *The Statistician*, **43**, 13–29.

Wakefield, J.C., Smith, A.F.M., Racine-Poon, A. and Gelfand, A. (1994). Bayesian analysis of linear and nonlinear population models using the Gibbs sample, *Applied Statistics*, **43**, 201–221.

Wakefield, J.C. (1996). The Bayesian analysis of population pharmacokinetic models, *Journal of the American Statistical Association*, **91**, 62–75.

A Hierarchical Spatial Model for Constructing Wind Fields from Scatterometer Data in the Labrador Sea

J. A. Royle
L. M. Berliner
C. K. Wikle
R. Milliff

ABSTRACT Wind fields are important for many geophysical reasons, but high resolution wind data over ocean regions are scarce and difficult to collect. A satellite-borne scatterometer produces high resolution wind data. We constructed a hierarchical spatial model for estimating wind fields over the Labrador sea region based on scatterometer data. The model incorporates spatial structure via a model of the u and v components of wind *conditional* on an unobserved pressure field. The conditional dependence is parameterized in this model through the physically based assumption of geostrophy. The pressure field is parameterized as a Gaussian random field with a stationary correlation function. The model produces realistic wind fields, but more importantly it appears to be able to reproduce the true pressure field, suggesting that the parameterization of geostrophy is useful. This further suggests that the model should be able to produce reasonable predictions outside of the data domain.

1 Introduction

The momentum and thermodynamic balances of the ocean depend critically upon the wind field forcing at the surface. The implied air-sea interaction is an important factor in processes that occur on scales from very local, to those of relevance to the global climate. Numerical models of regional and global ocean circulation are sensitive to the temporal (Large et al. 1991), and spatial (Milliff et al. 1996) resolution of the surface winds imposed as boundary conditions. Satellite-borne instruments ("scatterometers") have been developed, and the technology is evolving to meet the need for high resolution surface wind information over the global ocean, including remote and harsh environments at high-latitudes, such as the Labrador Sea region (290° E - 320° E, 50° N - 65° N).

Surface wind field data (at 10 m above the surface) derived from present-day scatterometer observations occur in swaths that follow the satellite ground track in a polar orbit sampling configuration. For the NASA Scatterometer (NSCAT), orbits occur at 100 minute intervals, and precess westward to cover the global oceans on the order of a few days. Different portions of the Labrador Sea region are covered by successive orbits twice in a 24-hour period (near 0 and 1400 UTC, Coordinated Universal Time). Within the sub-region of the scatterometer swath, the wind estimates are dense such that large data sets are produced. However, ocean models and scientific analyses often require fields of surface winds that extend beyond the sub-region sampled by the scatterometer at a given time. This poses the statistical problem to be addressed in this paper; how can large volumes of surface wind information covering a variable-sized sub-region be used to generate realistic estimates of a regional wind field?

The primary focus of this paper is to develop a hierarchical spatial approach to constructing wind fields from scatterometer data. Analyses are complicated by the fact that the data arise from a bivariate spatial field. That is, the winds are vectors, typically represented by orthogonal components: u represents the component in the "x-direction" (east); v is the component in the "y-direction" (north). In using standard cokriging techniques (e.g. Cressie, 1993, sec. 3.2.3), one would model both marginal covariance functions and a cross-covariance function between u and v. We pursue a hierarchical Bayesian formulation involving a model which posits a simple covariance structure for the wind fields *conditional* on a pressure field. In this analysis, actual pressure fields are not observed. However, we base this model on (approximate) physical theory. We model the pressure field as a spatially correlated random field, thus imposing the spatial structure at this simpler (univariate) stage of the model. A heuristic view is that (i) we employ observed winds to estimate pressures at observation sites; (ii) this information is then used to estimate pressures at other locations; (iii) the wind-pressure relationship then provides information about winds at sites with no data. This is accomplished rigorously in our Bayesian calculations. For the Labrador Sea region, important small-scale, intense features of the surface pressure field (i.e. "polar lows") might be deduced from our model. While this is appealing, at least two issues arise. First, the Bayesian calculations may be numerically intensive due to the size and complexity of the model. Second, the analysis is strongly dependent on specification of the model components, the validity of which are not rigorously studied. This paper primarily offers a test of concept. After beginning this work, estimated pressure fields became available. We offer a brief comparison of our results to these data as a rough confirmation of our approach. (In work in progress, we consider the incorporation of pressure data into the model.)

The proposed model permits comparison of scatterometer-based regional wind fields with existing estimates from the European Center for Medium-Range Weather Forecasts (ECMWF). The ECMWF wind fields of interest here occur on a regular 1° grid for the Labrador Sea region. This is the target grid for the regional surface wind estimate to be generated from the scatterometer data. The ECMWF surface field estimates are known to be too smooth at sub-regional scales (Milliff et al. 1996). Therefore, a useful outcome of this research, which is currently under investigation, is a comparison of deduced regional surface pressure and surface wind fields with ECMWF analysis fields. Ideally, the scatterometer-based fields possess more realistic sub-regional variability (i.e. not too smooth).

In this paper we only treat the purely spatial problem. The extension to space-time will be pursued elsewhere, but a general formulation of a space-time hierarchical model is given by Wikle et al. (1998).

2 The Scatterometer Data

Scatterometer instruments do not measure wind speed and direction directly. Instead, wind speed and wind direction are inferred, via a geophysical model function, from multiple observations of the change imparted on radar pulses (i.e. "backscatter") by the sea surface (e.g., see Naderi et al, 1991; Freilich and Dunbar 1993; Stoffelen and Anderson, 1997). Individual backscatter observations occur with point support. Typically, 4 to 16 backscatter observations are averaged over a 50×50 km wind vector cell (WVC) before the geophysical model function is applied to obtain estimates of wind speed and direction. The number of backscatter observations averaged in each WVC is an important piece of statistical information that can enter into the likelihood of the data.

The NSCAT wind data from a typical set of 2 swaths from November 19, 1996 are plotted as a *vector* plot in the top panel of Figure 1. This plot is a representation of the wind field where the length of the vector is proportional to the magnitude (wind speed) and the angle of the vector indicates wind direction. There are 289 ECMWF analysis grid points located over the Labrador Sea, which are shown as the dots in both panels of Figure 1. Note that the empty regions of Figure 1 represent the Coast of Labrador and Greenland (i.e. land as opposed to ocean) and so no scatterometer data are available over these regions. The bottom panel of Figure 1 shows a contour plot of the ECMWF pressure field which will be discussed in Section 5.

3 Hierarchical Model

3.1 *Motivation*

Let $\mathbf{U}$ and $\mathbf{V}$ each be N-vectors of the u and v components of the wind field at N grid locations. For our problem, these N locations are chosen to correspond to the ECMWF analysis grid of Figure 1. Let $\mathbf{P}$ be the pressure field at these locations.

The geostrophic approximation is a well-known concept in geophysical fluid dynamics (e.g. Holton, 1992, p. 59) that relates geopotential height fields (denoted by Z) and the u and v components of wind. (Geopotential heights are proportional to the elevation above sea level of a constant pressure surface.) The geostrophic relationship assumes a balance between the gradient of geopotential height and the Coriolis acceleration (in response to the earth's rotation). The relationship holds fairly well in the extratropics, above the planetary boundary layer (approximately 1 km). Below the planetary boundary layer, the effects of friction are significant and must be included in the force balance. Thus, we would not expect a model based solely on this relationship to be entirely adequate for surface wind fields in the Labrador Sea. Nevertheless, the relationship does provide some physically based information that can be built into a hierarchical model.

Thiebaux (1985) derived geostrophic covariance relationships under an additional assumption that the Z field is a realization of a second-order (spatial) autoregressive stochastic process. Although a low-order autoregressive model is somewhat simplistic, the covariance model from such a process has been shown to adequately model many atmospheric phenomena, both in the spatial and spectral domains. For s and s' any two spatial coordinates, the stationary spatial autoregressive covariance model is:

$$Cov(P(s), P(s')) = [\cos(\alpha_2||s - s'||) + \frac{\alpha_1}{\alpha_2}\sin(\alpha_2||s - s'||)]e^{-\alpha_1||s-s'||}. \tag{1}$$

This exponentially damped sinusoid produces good empirical fits to data which exhibit quasi-periodic correlation structure and which can have negative correlations. Although Thiebaux's work was concerned with the geostrophic relationship on a constant pressure surface, our interest is with a constant height surface (i.e. 10 meters). The geostrophic relationship on such a surface then relates the pressure gradient field to the Coriolis force (e.g. Holton, 1992, p. 40).

3.2 *The Model*

The data, denoted as $\mathbf{D}_u$ and $\mathbf{D}_v$ are not, in general, observed at grid locations. There are typically 2-3 observations within each grid cell (see Figure 1). We relate these observations to the u and v components of wind at grid sites in the measurement stage of the model:

Stage 1:

$$[\mathbf{D}_u, \mathbf{D}_v | \mathbf{U}, \mathbf{V}, \mathbf{H}_u, \mathbf{H}_v, \sigma_e^2] \text{ is } Gau(\begin{pmatrix} \mathbf{H}_u\mathbf{U} \\ \mathbf{H}_v\mathbf{V} \end{pmatrix}, \begin{pmatrix} \sigma_e^2\mathbf{I} & \mathbf{0} \\ \mathbf{0} & \sigma_e^2\mathbf{I} \end{pmatrix}). \tag{1}$$

Here, $\mathbf{H}_u$ and $\mathbf{H}_v$ are matrices which map the u and v components at the grid locations to the data locations. If u and v are always observed together (which they are in the dataset analyzed here), then we specify $\mathbf{H}_u = \mathbf{H}_v = \mathbf{H}$. A simple form for these matrices, and the one that we will use here, maps observation sites to the nearest grid location (Wikle, Berliner, and Cressie, 1998). That is, it is assumed that observations within the grid box are observations of u and v at the grid node *plus* independent, mean zero measurement error. Formally, the rows of $\mathbf{H}$ are all 0's with a single 1, indicating the element of $\mathbf{U}$ with which the particular observation is associated. One could produce a more complicated $\mathbf{H}$ matrix by taking rows to be interpolating weights, mapping the gridded process to data locations, for example. In some situations this may be more reasonable, particularly if the objective is off-site prediction. Since the data arise as averages (of the backscatter measurements) within small rectangles (e.g. $50 \times 50\ km$), it is likely that the measurement error variance is small. However, the fact that observations from a typical "time-slice" come from two swaths separated by 100 minutes does induce measurement error. In continuing work, we will incorporate the number of observations in each WVC into the model at this stage. In this article the measurement error variances are assumed constant. It is suspected that a systematic measurement bias exists in the scatterometer data which is proportional to u and v (Freilich, 1997). Future development of this model will attempt to parameterize this bias.

Stage 2: We now concentrate on a model for the u and v processes. In the second stage of the model, we specify the distribution of $\mathbf{U}$ and $\mathbf{V}$, *conditional* on $\mathbf{P}$:

$$[\mathbf{U}, \mathbf{V}|\mathbf{P}, \mathbf{B}_u, \mathbf{B}_v, \mu_u, \mu_v, \mathbf{\Sigma}] \text{ is } Gau\left(\begin{pmatrix} \mu_u \mathbf{1} + \mathbf{B}_u \mathbf{P} \\ \mu_v \mathbf{1} + \mathbf{B}_v \mathbf{P} \end{pmatrix}, \mathbf{\Sigma}\right) \tag{2}$$

Some flexibility can be permitted in specification of $\mathbf{\Sigma}$. We discuss this in Section 3.3. One must also choose a parameterization of the matrices $\mathbf{B}_u$ and $\mathbf{B}_v$. These could conceivably be very complicated. However, in this problem a natural choice arises based on physical arguments, and we discuss this parameterization in Section 3.4.

Stage 3: Next, we formulate a stochastic model for the pressure field, $\mathbf{P}$. At this stage of the model, we incorporate information derived from the second-order spatial autoregressive model previously discussed. Although $\mathbf{P}$ is defined on a lattice, we can base its distribution on a model generating a pressure field with continuous spatial index. In particular, we assume that the pressure field is a Gaussian random process with (prior) constant mean, μ_p and covariance kernel $\sigma_p^2 r(s, s')$. That is, the covariance of the pressure at arbitrary locations, s, s' is given by $\sigma_p^2 r(s, s')$ (note that we are assuming *a priori* that the variance σ_p^2 of the pressure is the same at all sites). This model implies that the gridded values $\mathbf{P}$ have a multivariate Gaussian distribution with correlation matrix $\mathbf{R}_\alpha$, where the entries in $\mathbf{R}$ are the appropriate values of r for the gridded sites. That is,

$$[\mathbf{P}|\mu_p, \sigma_p^2, \mathbf{R}_\alpha] \text{ is } Gau(\mu_p \mathbf{1}, \sigma_p^2 \mathbf{R}_\alpha). \tag{3}$$

Here, $\boldsymbol{\alpha} = (\alpha_1, \alpha_2)$ are the parameters in the covariance function given by (1).

The basic idea of this model for the pressure field is to place very strong spatial structure in $\mathbf{P}$ and use this information to produce (primarily geostrophic) predictions of u and v away from the data locations. Moreover, it is hoped that by putting (meaningful) covariance structure at this stage, the remaining structure (i.e. the spatial variation in u and v) can be explained more parsimoniously in the mean of the stage 2 conditional distribution, thus permitting a simplified parameterization of $\mathbf{\Sigma}$.

Note that some parameters of the full model are not identifiable. In principle, a valid posterior exists as long as all priors are proper. However, some care must be taken in the formulation of these priors as well as in the interpretation of results. For example, posteriors on unidentified parameters are not necessarily of value, though posterior inferences on other unknowns and some functions of unidentified parameters can be useful. In our context, recall that we parameterize the dependence of u and v on the *gradient* of $\mathbf{P}$ (discussed in Section 3.4). Hence, the mean of $\mathbf{P}$ cannot be meaningfully updated without pressure data. Moreover, σ_p^2 and the parameters of $\mathbf{B}_u$ and $\mathbf{B}_v$ are related to one another; multiplying $\mathbf{P}$ by a constant reduces the magnitude of the parameters in $\mathbf{B}_u$ and $\mathbf{B}_v$ by the inverse of the constant. To overcome these issues, we viewed $\mathbf{P}$ as a *standardized* field. The idea is to view the latent pressure field as unitless, with mean zero and variance one. We wanted to do this in a fashion which still used the correlation structure in Section 3.1 and was reasonably flexible. Hence, we assigned priors to μ_p and σ_p^2 that favor μ_p near 0 and σ_p^2 near 1 (see Section 3.5). We then only make use of the gradient of the pressure field in other calculations and only present results concerning standardized pressures. Further, since the the gradient of the pressure field is unitless, the regression parameters (β's in (2)) are measured in meters per second (as are the wind vectors). Having established these units, we can meaningfully develop priors on all other parameters in the Stage 2 model.

3.3 *Parameterization of* $\mathbf{\Sigma}$

The matrix $\mathbf{\Sigma}$ is interpreted as the ageostrophic covariance (i.e. that not accounted for by the gradient relationship with $\mathbf{P}$). One of the motivating factors of the hierarchical modeling approach is to simplify the structure of $\mathbf{\Sigma}$ from a large and complicated joint variance-covariance matrix to a simpler model, letting the more complicated structure be taken up in the univariate random field model for $\mathbf{P}$.

The parameterization that we use here assumes that u and v are correlated at the same site, but not across sites, so that $\mathbf{\Sigma} = \mathbf{K}_{2\times 2} \otimes \mathbf{I}_{N\times N}$ where

$$\mathbf{K} = \begin{pmatrix} \sigma_u^2 & \sigma_{uv} \\ \sigma_{uv} & \sigma_v^2 \end{pmatrix} \tag{4}$$

is the variance-covariance matrix of $(U(s), V(s))$. More generally, $\mathbf{\Sigma}$ may be spatially heterogeneous, and so one might consider placing additional spatial structure on, for example, the elements of $\mathbf{K}$.

3.4 *Parameterization of* $\mathbf{B}_u$ *and* $\mathbf{B}_v$

Physical arguments suggest that, to a first-order approximation, the u and v components should be proportional to the gradient of the pressure field. More precisely, the geostrophic approximation yields:

$$v_g \propto \frac{1}{f}\frac{dP}{dx} \text{ and } u_g \propto -\frac{1}{f}\frac{dP}{dy}$$

where x and y are the east-west and north-south directions, respectively and f is the Coriolis parameter (e.g. Holton, 1992; p. 40).

Consider the following first-order neighborhood of the point s_3:

$$\begin{array}{ccc} & s_2 & \\ s_1 & s_3 & s_5 \\ & s_4 & \end{array}$$

The east-west gradient in P is proportional to

$$([P(s_5) - P(s_3)] + [P(s_3) - P(s_1)])/2 = ([P(s_5) - P(s_1)])/2.$$

The north-south gradient in P is proportional to

$$([P(s_2) - P(s_3)] + [P(s_3) - P(s_4)])/2 = ([P(s_2) - P(s_4)])/2.$$

Thus, under geostrophy, we might expect that $U(s_3) = \beta_1([P(s_5) - P(s_1)]/2) + \beta_2([P(s_2) - P(s_4)]/2)$. Under geostrophy, we further expect $\beta_1 = 0$ for the u component (Holton, p.40). However, for now we retain this as a parameter to be estimated. Similarly, we might expect $V(s_3) = \beta_3([P(s_5) - P(s_1)]/2) + \beta_4([P(s_2) - P(s_4)]/2)$, and $\beta_4 = 0$ under geostrophy (Holton, p.40). These relationships imply a simple anisotropic nearest-neighbor structure on the matrices $\mathbf{B}_u$ and $\mathbf{B}_v$.

One must take care in dealing with points on the boundary of the region. In such cases, there will not be observations on all sides of a given point. We used the available observations to the east or west and north or south, and modified the pressure derivative, re-expressing the relationship between wind and pressure as, for example, $U(s_3) = \beta_1[P(s_5) - P(s_3)] + \beta_2[P(s_3) - P(s_4)]$ in the case where there does not exist a point to the west and north of s_3.

It may be computationally convenient to re-express $\mathbf{B}_u\mathbf{P}$ as $\mathbf{B}_u\mathbf{P} = \boldsymbol{\Psi}_p\boldsymbol{\beta}_u$ where $\boldsymbol{\Psi}_p$ is an $N \times 2$ matrix containing appropriate elements of differences in $\mathbf{P}$ and $\boldsymbol{\beta}_u = (\beta_1, \beta_2)$. For example, in the above example, for the u component, $\mathbf{B}_u$ and $\mathbf{P}$ look like:

$$\mathbf{B}_u\mathbf{P} = \begin{pmatrix} -\beta_1 & 0 & \beta_1 & 0 & 0 \\ 0 & \beta_2 & -\beta_2 & 0 & 0 \\ -\beta_1/2 & \beta_2/2 & 0 & -\beta_2/2 & \beta_1/2 \\ 0 & 0 & \beta_2 & -\beta_2 & 0 \\ 0 & 0 & -\beta_1 & 0 & +\beta_1 \end{pmatrix} \begin{pmatrix} P_1 \\ P_2 \\ P_3 \\ P_4 \\ P_5 \end{pmatrix}$$

Note that the rows of the matrices $\mathbf{B}_u$ and $\mathbf{B}_v$ necessarily sum to 0 due to the discrete definition of the pressure gradient. We can express this in terms of $\boldsymbol{\Psi}_p$

and $\boldsymbol{\beta}_u$ as:

$$\boldsymbol{\Psi}_p\boldsymbol{\beta}_u = \begin{pmatrix} -P_1 + P_3 & 0 \\ 0 & P_2 - P_3 \\ \frac{1}{2}(P_5 - P_1) & \frac{1}{2}(P_2 - P_4) \\ 0 & P_3 - P_4 \\ P_5 - P_3 & 0 \end{pmatrix} \begin{pmatrix} \beta_1 \\ \beta_2 \end{pmatrix} \tag{5}$$

where the first column in $\boldsymbol{\Psi}_p$ represents the east-west gradient, and the second column represents the north-south gradient for each point.

3.5 *Prior distributions on parameters*

The final stage of the model requires specification of prior distributions on the variance parameters σ_e^2, σ_u^2, σ_v^2, and σ_p^2, the mean parameters μ_u, μ_v and μ_p, the parameters of $\boldsymbol{\Sigma}$, and of $\mathbf{B}_u$ and $\mathbf{B}_v$ and finally $\boldsymbol{\alpha}$, the parameters of the correlation matrix of $\mathbf{P}$. We assume that all of these parameters are independent. For the measurement error variance we use the conjugate inverse gamma prior (denoted as IG):

$$\sigma_e^2 \sim IG(a_e, b_e). \tag{6}$$

We fix these hyperparameters at $a_e = 6$ and $b_e = .2$ to give the prior mean 1 and standard deviation .5. We fix the prior of σ_p^2 to have a mean of 1 and variance 1 so that

$$\sigma_p^2 \sim IG(a_p, b_p) \tag{7}$$

where $a_p = 3$ and $b_p = .05$.

Under the separable parameterization for $\boldsymbol{\Sigma}$ given in (3.3), we specify an inverted-Wishart on $\mathbf{K}_{2\times 2}$:

$$\mathbf{K}^{-1} \sim W((\nu\mathbf{K}_o)^{-1}, \nu). \tag{8}$$

The means (diagonal elements of $\mathbf{K}_o$) for σ_u^2 and σ_v^2 were set at 10, and the off-diagonal element was set at 0 to reflect a prior belief in 0 correlation between u and v (conditional on $\mathbf{P}$). The parameter ν was set at 2 to express little prior information in $\mathbf{K}_o$ relative to the data.

For the mean parameters we assume

$$\mu_u \sim Gau(\lambda_u, \gamma_u^2), \mu_v \sim Gau(\lambda_v, \gamma_v^2), \mu_p \sim Gau(\lambda_p, \gamma_p^2), \tag{9}$$

The actual values used here are $\lambda_u = 0$ and $\lambda_v = 0$, both with standard deviations of 4, suggesting little prior information. We set $\lambda_p = 0$ and $\gamma_p^2 = 1$ (recall that the actual value of μ_p is irrelevant lacking real data).

The parameters $\beta_i : i = 1, 2, 3, 4$ of $\mathbf{B}_u$ and $\mathbf{B}_v$ are assumed to be independent Gaussians with means b_i and variances η_i^2:

$$\beta_i \sim Gau(b_i, \eta_i^2) \tag{10}$$

Physics suggests that the means of the prior distributions should be $b_1 = 0$, $b_2 > 0$, $b_3 > 0$ and $b_4 = 0$, respectively, to reflect geostrophy. But, specification of precise values for the parameters of these prior distributions is difficult.

We approach this problem by choosing values to reflect prior beliefs on the *percent of variance* that should be explained by geostrophy (i.e. the R^2 of the regression of u on $\mathbf{B}_u\mathbf{P}$, for example). Empirical analyses of extratropical near surface wind observations indicate that it is reasonable to assume that geostrophy explains roughly 70% of the variance in wind fields (Deacon, 1973). Thus, we choose prior values for the gradient parameters $\beta_1, \beta_2, \beta_3$ and β_4 accordingly. The standard deviation of these parameters will be such that about 95% of the mass of these distributions falls between values implying an R^2 of 60% and 80%. Assume that $\sigma_p^2 = 1$ (the following argument applies for any value, but we choose this for convenience). We can derive prior parameter values for the gradient parameters by noting that (for u, and similarly for v) $Var(u) = Var(E(u|P)) + E(Var(u|P))$ which is:

$$Var(\mathbf{U}_i) = \mathbf{b}_i \mathbf{R}_\alpha \mathbf{b}_i' + \sigma_u^2 \tag{11}$$

where $\mathbf{b}_i$ is a the ith row of $\mathbf{B}_u$. We then use an estimate of $Var(\mathbf{U}_i)$ based on the data at hand and solve for a variety of values of β_2, holding β_1 fixed at 0. The prior mean of β_2 is selected so that the first term in (11) is equal to 70% of the total. This suggests a value for b_2 of 7.3 with the 60% and 80% values being 6.7 and 7.9, respectively. Thus, we chose $\eta_2^2 = .3^2$. Similarly, $\eta_1^2 = .3^2$, with $b_1 = 0$. This exercise was repeated to get prior values on the v gradient parameters, leading to $b_3 = 7.7$ with $\eta_3^2 = .25^2$ and $b_4 = 0$ with $\eta_4^2 = .25^2$ (recall that the 0 values for b_1 and b_4 reflect geostrophy).

Finally, we need a distribution for $[\alpha] = [\alpha_1][\alpha_2]$. As parameterized by (1), α_1 is a parameter describing the correlation "range", and must be positive. Similarly, α_2 is a parameter describing the frequency of the correlation function (in essence, the distance between the correlation maximum and minimum, which is related to the spatial scales of major weather disturbances). As parameterized here, $\alpha_2/(2\pi)$ is the frequency *per* 1000 km. Thus, prior distributions on these parameters would need to honor this restricted range. Therefore, we assumed fairly uninformative truncated Gaussian distributions for both of these parameters:

$$\begin{aligned} \alpha_1 &\sim Gau(a_1, \sigma_{a1}^2) : \alpha_1 > 0 \\ \alpha_2 &\sim Gau(a_2, \sigma_{a2}^2) : \alpha_2 > 0 \end{aligned} \tag{12}$$

The hyperparameters were set at $a_1 = 2.5$, $a_2 = 5$, $\sigma_{a1} = .5$ and $\sigma_{a2} = 1$, which were loosely based on estimates fit to the raw wind data.

4 Gibbs Sampling

Here we give the various full conditionals required for carrying out a Gibbs sampling scheme. We avoid the details of the derivation. Since all priors are conjugate, these are straight forward with the exception of the full conditionals for the covariance parameters in $\mathbf{R}_\alpha$.

4.1 *Full conditionals for* U *and* V

We can compute $[\mathbf{U}|\cdot]$ and $[\mathbf{V}|\cdot]$ as $[\mathbf{U}|\cdot] \propto [\mathbf{U}|\mathbf{V},\mathbf{P}][\mathbf{D}_u|\mathbf{U}]$ and similarly for $[\mathbf{V}|\cdot]$. Define $\boldsymbol{\theta}_u = \mu_u\mathbf{1} + \mathbf{B}_u\mathbf{P}$ and $\boldsymbol{\theta}_v = \mu_v\mathbf{1} + \mathbf{B}_v\mathbf{P}$. It is easy to show that $[\mathbf{U}|\cdot]$ is Gaussian with mean:

$$(\frac{1}{\sigma_e^2}\mathbf{H}'\mathbf{H} + \frac{1}{\sigma_u^2 - \frac{\sigma_{uv}^2}{\sigma_v^2}}\mathbf{I})^{-1}(\frac{1}{\sigma_e^2}\mathbf{D}_u'\mathbf{H} + \frac{1}{\sigma_u^2 - \frac{\sigma_{uv}^2}{\sigma_v^2}}[\boldsymbol{\theta}_u + \frac{\sigma_{uv}}{\sigma_v^2}(\mathbf{V} - \boldsymbol{\theta}_v)])$$

and variance:

$$(\frac{1}{\sigma_e^2}\mathbf{H}'\mathbf{H} + \frac{1}{\sigma_u^2 - \frac{\sigma_{uv}^2}{\sigma_v^2}}\mathbf{I})^{-1}$$

Similarly, the full conditional $[\mathbf{V}|\cdot]$ is Gaussian with mean:

$$(\frac{1}{\sigma_e^2}\mathbf{H}'\mathbf{H} + \frac{1}{\sigma_v^2 - \frac{\sigma_{uv}^2}{\sigma_u^2}}\mathbf{I})^{-1}(\frac{1}{\sigma_e^2}\mathbf{D}_v'\mathbf{H} + \frac{1}{\sigma_v^2 - \frac{\sigma_{uv}^2}{\sigma_u^2}}[\boldsymbol{\theta}_v + \frac{\sigma_{uv}}{\sigma_u^2}(\mathbf{U} - \boldsymbol{\theta}_u)])$$

and variance:

$$(\frac{1}{\sigma_e^2}\mathbf{H}'\mathbf{H} + \frac{1}{\sigma_v^2 - \frac{\sigma_{uv}^2}{\sigma_u^2}}\mathbf{I})^{-1}$$

4.2 *Full conditional for* P

For $[\mathbf{P}|.]$ we need to analyze $[\mathbf{P}|.] \propto [\mathbf{U},\mathbf{V}|\mathbf{P}][\mathbf{P}]$. Define $\mathbf{B} = (\mathbf{B}_u, \mathbf{B}_v)'$. It can be shown that

$$[\mathbf{P}|.] \text{ is } Gau(\mathbf{Q}^{-1}\mathbf{A}, \mathbf{Q}^{-1}) \tag{1}$$

with $\mathbf{Q} = \mathbf{B}'\boldsymbol{\Sigma}^{-1}\mathbf{B} + (1/\sigma_p^2)\mathbf{R}^{-1}$ and $\mathbf{A} = [(\mathbf{U},\mathbf{V})\boldsymbol{\Sigma}^{-1}\mathbf{B} + (\mu_p/\sigma_p^2)\mathbf{1}'\mathbf{R}^{-1}]'$.

4.3 *Full conditionals for* μ_u, μ_v *and* μ_p

The prior on $\boldsymbol{\mu} = (\mu_u, \mu_v)$ is $Gau(\boldsymbol{\lambda}, \mathbf{A})$. Define $\hat{\boldsymbol{\mu}} = (\mathbf{X}'\boldsymbol{\Sigma}^{-1}\mathbf{X})^{-1}\mathbf{X}'\boldsymbol{\Sigma}^{-1}\mathbf{d}$ for

$$\mathbf{X} = \begin{pmatrix} 1 & 0 \\ 0 & 1 \end{pmatrix} \text{ and } \mathbf{d} = \begin{pmatrix} \mathbf{U} - \mathbf{B}_u\mathbf{P} \\ \mathbf{V} - \mathbf{B}_v\mathbf{P} \end{pmatrix}.$$

Note that $\hat{\boldsymbol{\mu}} \sim Gau(\boldsymbol{\mu}, \mathbf{I})$. Therefore, we have:

$$[\boldsymbol{\mu}|\cdot] \text{ is } Gau(\hat{\boldsymbol{\mu}} - (\mathbf{I} + \mathbf{A})^{-1}(\hat{\boldsymbol{\mu}} - \boldsymbol{\lambda}), (\mathbf{I} + \mathbf{A}^{-1})^{-1}) \tag{2}$$

Similarly, the full conditional for μ_p is a standard Gaussian update of the prior.

4.4 Full conditionals for β_u and β_v

Following (5), we can express the mean of $\mathbf{U}$, $\mu_u\mathbf{1} + \mathbf{B}_u\mathbf{P}$, as $\mu_u\mathbf{1} + \mathbf{\Psi}_p\boldsymbol{\beta}_u$ (and similarly for the mean of $\mathbf{V}$) where $\mathbf{\Psi}_p$ is an $N \times 2$ matrix with elements that depend on $\mathbf{P}$. Then

$$\begin{pmatrix} \mathbf{U} - \mu_u\mathbf{1} \\ \mathbf{V} - \mu_v\mathbf{1} \end{pmatrix} \sim Gau(\begin{pmatrix} \mathbf{\Psi}_p & \mathbf{0} \\ \mathbf{0} & \mathbf{\Psi}_p \end{pmatrix} \begin{pmatrix} \boldsymbol{\beta}_u \\ \boldsymbol{\beta}_v \end{pmatrix}, \mathbf{\Sigma})$$

Computing $\hat{\boldsymbol{\beta}}$ from this, and using the priors given in (13), we see that

$$[\boldsymbol{\beta}|\cdot] \text{ is } Gau(\hat{\boldsymbol{\beta}} - (\mathbf{I} + \mathbf{A}_\eta)^{-1}(\hat{\boldsymbol{\beta}} - \boldsymbol{\beta}_o), (\mathbf{A}_\eta^{-1} + \mathbf{I})^{-1})$$

where $\mathbf{A}_\eta = diag(\eta_1^2, \eta_2^2, \eta_3^2, \eta_4^2)$ and $\boldsymbol{\beta}_o = (b_1, b_2, b_3, b_4)$.

4.5 Full conditional for $\mathbf{K}$

The prior on $\mathbf{K}^{-1}$ is the conjugate inverted-Wishart, $[\mathbf{K}^{-1}]$ is $W((\nu\mathbf{K}_o)^{-1}, \nu)$, and so it follows that

$$[\mathbf{K}^{-1}|\cdot] \text{ is } W((\hat{\mathbf{K}} + \nu\mathbf{K}_o)^{-1}, \nu + N)$$

where $\hat{\mathbf{K}} = (\mathbf{U} - \boldsymbol{\theta}_u, \mathbf{V} - \boldsymbol{\theta}_v)(\mathbf{U} - \boldsymbol{\theta}_u, \mathbf{V} - \boldsymbol{\theta}_v)'$.

4.6 Full conditionals for σ_e^2 and σ_p^2

For n the number of NSCAT data observations (recall, N is the number of grid points), define

$$2n\hat{\sigma}_e^2 = (\mathbf{D}_u - \mathbf{HU})'(\mathbf{D}_u - \mathbf{HU}) + (\mathbf{D}_v - \mathbf{HV})'(\mathbf{D}_v - \mathbf{HV}).$$

Then,

$$[\sigma_e^2|\cdot] \text{ is } IG(a_e + n, (\frac{1}{b_e} + \frac{1}{2}(2n\hat{\sigma}_e^2))^{-1}). \tag{3}$$

Note that there is an n here instead of $n/2$ because $\mathbf{D}_u$ and $\mathbf{D}_v$ have common measurement error variance. Similarly,

$$[\sigma_p^2|\cdot] \text{ is } IG(a_p + N/2, (\frac{1}{b_p} + \frac{1}{2}(\mathbf{P} - \mu_p\mathbf{1})'\mathbf{R}_\alpha^{-1}(\mathbf{P} - \mu_p\mathbf{1}))^{-1}) \tag{4}$$

4.7 Full conditionals for α_1 and α_2

The full conditionals for the correlation parameters of $\mathbf{R}_\alpha$ are the only non-standard distributions for this particular problem. We note that the correlation function given by (1) is not strictly positive definite for all values of α_1 and α_2, even for strictly positive values of these parameters and thus one must use rejection sampling to to ensure that the parameter space is restricted to positive definite covariance functions. In general, the full conditionals for these parameters are not tractable. A common solution is to rely on a Metropolis-Hastings algorithm for sampling from these distributions (Gilks, 1996). To do this, we sampled from approximating distributions that are truncated Gaussian with means $\hat{\alpha}_1$ and $\hat{\alpha}_2$, the least-squares estimates of the covariance parameters. In addition, draws of α_1 and α_2 were checked to ensure that they implied a positive definite covariance matrix.

5 Application

The hierarchical model of Section 3.2 was applied to the data described in Section 2. The Gibbs sampler of the previous section was run for 2000 iterations, and we used the first 250 samples for burn-in. Graphical inspection indicated that convergence occurred reapidly. Here we concentrate on summarizing the results of the posterior mean wind and pressure fields. The posterior pressure field is shown as a contour plot in both panels of Figure 2, and the posterior wind field is depicted as a vector plot in the top panel, as in Figure 1. We notice a slight departure from geostrophy (Figure 2, bottom panel) in these predictions (i.e. the winds are not parallel to the pressure contour lines).

The estimates of the mean parameters are $\hat{\mu}_u = -1.22$, $\hat{\mu}_v = -4.18$ and $\hat{\mu}_p = -0.05$. The gradient parameter estimates are $\hat{\boldsymbol{\beta}}_u = (0.07, 4.23)$ and $\hat{\boldsymbol{\beta}}_v = (4.47, 0.008)$. The u and v variance estimates are $\hat{\sigma}_u^2 = 11.53$, $\hat{\sigma}_v^2 = 6.45$, $\hat{\sigma}_{uv}^2 = -1.12$, $\hat{\sigma}_p^2 = 0.84$ and $\hat{\sigma}_e^2 = 1.25$. Finally, the covariance parameter estimates are $\hat{\alpha}_1 = 2.55$ and $\hat{\alpha}_2 = 4.71$. We note that the estimates of $\hat{\boldsymbol{\beta}}_u$ and $\hat{\boldsymbol{\beta}}_v$ are consistent with geostrophy since $\hat{\beta}_1$ and $\hat{\beta}_4$ are essentially 0.

Since we are interested in the validity of our parameterization of geostrophy, we can examine the predicted component of the wind field due to geostrophy (i.e. the $\mathbf{B}_u\mathbf{P}$ and $\mathbf{B}_v\mathbf{P}$ component of the model) by producing the wind field from residuals from the mean. That is, $\hat{\mathbf{U}} - \mu_u$ and $\hat{\mathbf{V}} - \mu_v$. Note that the direction and speed are not linear operations, and so subtracting the constant mean will generally impact the results. The predicted wind field due to geostrophy is shown in the bottom panel of Figure 2. Recall the ECMWF pressure field ("analysis" field) given in the bottom panel of Figure 1. As shown in Figure 1, this is actually a standardized field to facilitate comparison with the posterior mean pressure field from Figure 2. The ECMWF field can be viewed as the current "best guess" based primarily on numerical models, rather than the "true" pressure field. In any case it provides a source of validation of our model, since the ECMWF fields are believed to be reasonable accurate over large scales. We note that other than slightly more variability appearing in the predicted field, the broad structure is similar between the two which adds further support to our parameterization of geostrophy.

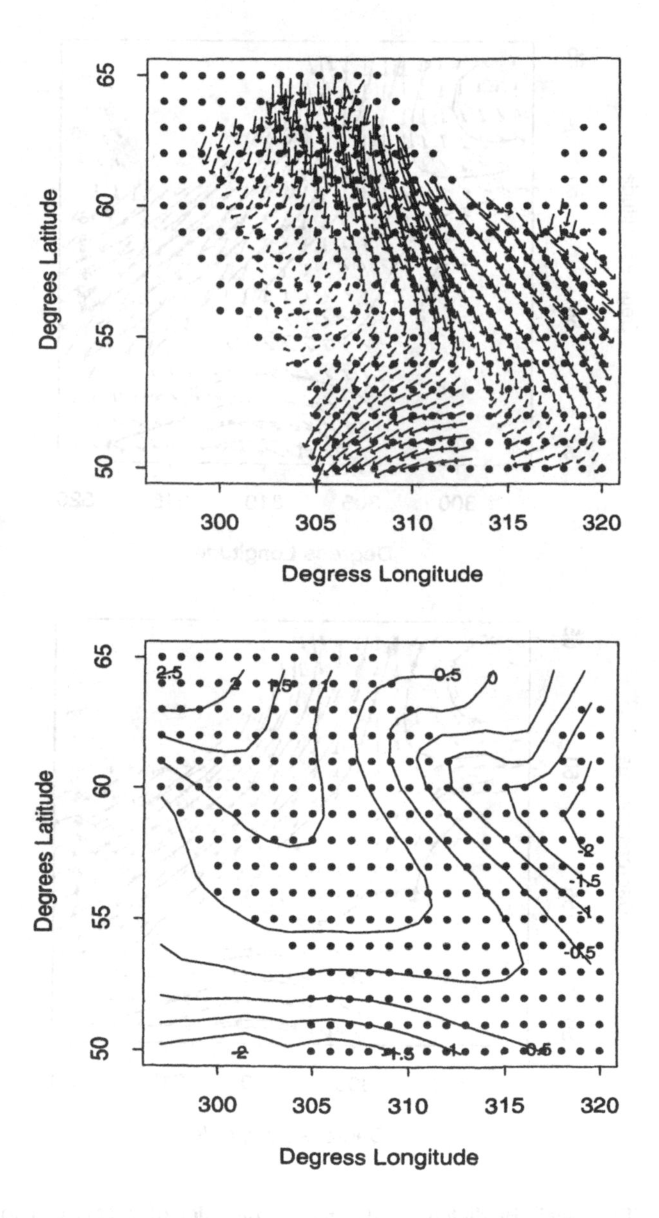

FIGURE 1. Top panel: Scatterometer observations and ECMWF grid locations; Bottom: Contour plot of ECMWF pressure field and grid locations.

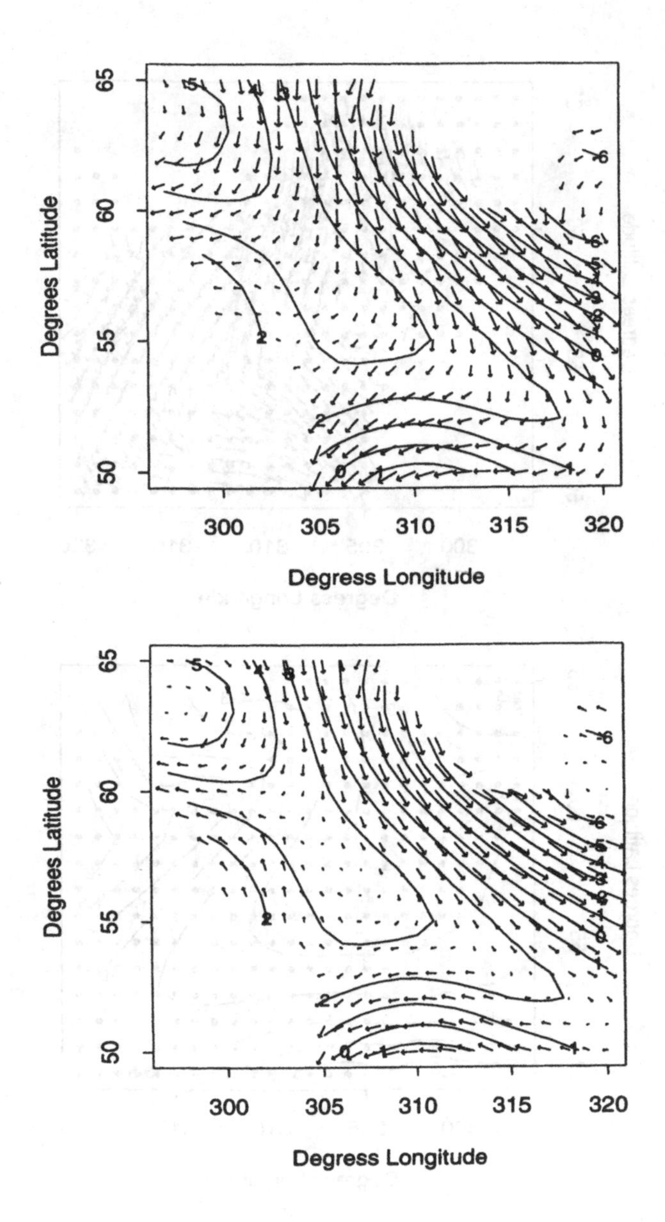

FIGURE 2. Top panel: Predicted winds and contour plot of pressure field (posterior means); Bottom: Predicted wind field component due to pressure gradient and contour plot of pressure field.

6 Conclusions

We proposed a hierarchical spatial model for constructing scatterometer-derived wind fields, which incorporates spatial dependence through an unobserved pressure field. The dependence of the wind field on pressure was parameterized by geostrophy. The important conclusion from this model applied to scatterometer data is that the geostrophic parameterization provides a coherent physical basis for model parameterization and prior formulation that can be used to perform multivariate spatial prediction. The posterior predicted pressure field adequately represents the true pressure field, and the geostrophic parameter estimates are consistent with geostrophy. One implication of this is that reasonable wind field predictions outside of the data domain should at least capture the variation due to geostrophy. Use of this model to construct finer resolution and more realistic wind fields is promising. Important directions of future work include incorporation of ECMWF pressure fields into the model, and investigation of more complicated ageostrophic covariance structures.

Acknowledgments

This research was supported by the National Center for Atmospheric Research, Geophysical Statistics Project, sponsored by the National Science Foundation under grant #DMS93-12686. In addition, we acknowledge the sustained support of the NSCAT Science Working Team agreement with the NCAR Climate and Global Dynamics Division administered by NASA/JPL.

References

Cressie, N.A.C., 1993, *Statistics for Spatial Data*, John Wiley, New York, 900 pp.

Freilich, M.H., 1997, "Validation of vector magnitude datasets: Effects of random component errors," *J. of Atmospheric and Oceanic Technology*, 14, 695-703.

Freilich, M.H. and Dunbar, R.S., 1993, "A preliminary C-band scatterometer model function for the ERS-1 AMI instrument," *Proceeding of the First ERS-1 Symposium: Space at the Service of Our Environment*, November 1992, Cannes, FRANCE, ESA SP-359 volume 1 (March 1993), 79-83.

Gilks, W.R., 1996, "Full conditional distributions". In *Markov Chain Monte Carlo in Practice*. W.R. Gilks, S. Richardson, and D.J. Spiegelhalter (Eds.), Chapman and Hall, London, pp.75-88.

Holton, J.R., 1992, *An Introduction to Dynamic Meteorology*, 3rd Ed., Academic Press, San Diego, 507 pp.

Large, W.G., Holland, W.R. and Evans, J.C., 1991, "Quasigeostrophic ocean response to real wind forcing: The effects of temporal smoothing," *J. Physical Oceanography*, 21, 998-1017.

Milliff, R.F., Large, W.G., Holland, W.R. and McWilliams, J.C., 1996, "The general circulation responses of high-resolution North Atlantic Ocean models to synthetic scatterometer winds," *J. Physical Oceanography*, 26, 1747-1768.

Naderi, F.M., Freilich, M.H. and Long, D.G., 1991, "Spaceborne radar measurement of wind velocity over the ocean- an overview of the NSCAT scatterometer system," *Proceedings IEEE*, 79, 850-866.

Stoffelen, A. and D. Anderson, 1997, "Scatterometer data interpretation; measurement space and inversion," *J. Atmos. and Ocean. Tech.*, **14**, 1298 - 1313.

Thiebaux, H.J., 1985, "On approximations to geopotential and wind-field correlation structures," *Tellus*, 37A, 126-131.

Wikle, C. K., Berliner, L.M., and Cressie, N., 1998, "Hierarchical Bayesian Space-Time Models," *Environmental and Ecological Statistics*, to appear.

Redesigning a Network of Rainfall Stations

Bruno Sansó
Peter Müller

ABSTRACT We consider the problem of reducing a network of existing rainfall monitoring stations. The aim is to reduce a current system of 80 stations in the state of Guárico, Venezuela, to around 40 stations. The problem is to choose these 40 stations out of the original 80 in a way which least reduces our ability to make inference about local rainfall and minimise cost. We formulate the problem as an optimal design problem with decision variables d, a probability model $p(y, \theta)$ on data y and parameter θ, and a utility function $u(d, \theta, y)$ which formalizes the decision criterion. We approach the optimal design problem by simulation. Specifically, we define an auxiliary probability model $h(d, \theta, y)$ such that simulation from $h(\cdot)$ can be shown to be equivalent to solving the original optimal design problem.

1 Introduction

We address the problem of redesigning a network of rainfall stations in the state of Guárico in the Venezuelan central plains, north of the Orinoco and Apure rivers. Currently $D = 80$ stations cover the region. Because of budget constraints the Ministry of the Environment (MARNR) seeks to reduce the network to roughly 40 stations.

The aim is to choose the subset of 40 remaining stations in a way which minimally compromises inference about local rainfall in the regions of interest. As a proxy for defining the localities where inference is important we use the locations of the exisiting 80 stations, which were historically at locations considered important, mainly because of interests related to agricultaral use.

For most of the current stations we have rainfall data available for the past $T = 16$ years. A formal solution of the decision problem posed above will involve a probability model for these annual data. The implied probability model on future years will be instrumental in formalizing the decision criterion.

Similar problems related to decisions about monitoring stations are discussed in Caselton *et al.* (1992), Guttorp *et al.* (1993) and Nychka and Saltzman (1997). In the first two papers the authors consider an approach based on minimising entropy and a normal-inverted-Wishart model. Nychka and Saltzman (1997) use a kriging model to estimate the response surface of interest and show how the problem of designing the grid is equivalent to a variable selection problem, and proceed from there. None of the papers mentioned above consider the problem of explicitly taking into account the cost of running the network. Bras and Rodríguez-Iturbe (1985) consider the design problem over a grid with a small number of possible locations and study a cost function similar to the one we consider in this paper.

The focus of the discussion in this paper is on the solution of the decision problem. The proposed approach is based on a formal description of the optimal design problem in a decision theoretic setup (DeGroot, 1970, chapter 8). Many approaches to implementing Bayesian optimal design have been focused on experimental design in normal linear models, see for example Verdinelli (1992), Chaloner and Larntz (1992), Toman (1994), Clyde and Chaloner (1996); Chaloner and Verdinelli (1995) present a recent review of work in Bayesian optimal design. Recent references concerned with numerical solutions of more general optimal design problems are Verdinelli and Kadane (1992) who consider problems arising with multiple competing loss functions, Müller and Parmigiani (1996) and Bielza, Müller and Ríos Insua (1996) who consider Monte Carlo simulation to evaluate expected utilities for specific instantiations of the decision variables.

Since in this paper we are mainly interested in the decision problem, we constrain the discussion of the underlying probability model to commonly used spatial models. We use a kriging model as probability model for the annual rainfalls at the 80 locations. The model is based on Bayesian kriging models discussed in Handkock and Stein (1993); the literature on statistical models for rainfall is considerably large, see Sansó and Guenni (1996) for discussions and references.

In Section 2 we define the utility function formalizing the decision criterion and a probability model for the annual rainfall data. In Section 3 we introduce the equivalent simulation problem which allows to solve the optimal design problem by simulation. In Section 4 we discuss results and conclude with a critical discussion in Section 5.

2 The decision problem

2.1 *The utility function*

We want to reduce the number of rainfall monitoring stations to decrease cost but mantain accuracy. We want to predict the amount of rainfall in important locations with as much accuracy as possible and as cheaply as possible. As a proxy to identify locations of interest in Guárico we use the locations of the currently exisiting 80 stations which were placed historically at localities deemed important for monitoring rainfall data.

Let y_i denote the rainfall, on a logarithmic scale, in a given year at station

i, $i = 1, \ldots, n$. We will denote with $d = (d_1, \ldots, d_n)$ a vector of indicators specifying which stations remain in the network ($d_i = 1$), and which do not ($d_i = 0$). We will use y_d and $y_{\not d}$ to denote the subvector of $y = (y_1, \cdots, y_n)$ corresponding to the stations in the network and out of the network, respectively. We will use $D = \{i : d_i = 1\}$ for the index set of all stations included in the design, and D^c for the index set of not included stations. Finally, let $\hat{y}_i(y_d)$ denote the prediction for station i based on the observed stations y_d, with the understanding that for any station $i \in D$ the prediction is the observed value itself, i.e., $\hat{y}_i = y_i$.

For a formal description of the decision problem we require a utility function $u(d, y)$ which specifies the utility of decision d under outcome y, i.e., the payoff (or negative loss) which we incur if under decision d data y is observed. As in most case studies, translating the vague decision criterion into a formal utility function requires some arbitrary specifications. To start, we consider the following utility function. Let $\mathbf{1}(A)$ denote the indicator function for event A, then

$$u(d, y) = C \sum_{i \in D^c} \underbrace{\mathbf{1}(y_i \in \hat{y}_i(y_d) \pm \delta)}_{S_i} - \sum_{i \in D} c_i + C_0. \qquad (1)$$

In words, for every station i which is not included in the network we obtain a payoff C if rainfall y_i is predicted within an accuracy of δ. And for every station which is included we incur a cost c_i. We will later discuss reasonable choices for δ and the trade-off parameters C and c_i. The rationale behind this utility function is that a collection d of stations has high utility if it is such that stations in the complement D^c can be accurately predicted by the model and, on the other hand, the sum of the costs associated with stations in D is not too high. A constant shift C_0 is added to keep the utility function non-negative, a technical constraint required for the simulation based approach discussed in the next section.

Of course, several choices in the specification of the utility function are arbitrary. For example, alternative functions could be used instead of the step function $\mathbf{1}(S_i)$ which rewards any prediction within the interval $(y_i \pm \delta)$ equally, and penalizes any prediction outside the interval equally. Or the summation in the first term could be extended over all stations. For $i \in D$ the indicator S_i would always evaluate to 1, and we could combine S_i, $i \in D$ with the corresponding term c_i in the second summation, leading to: $u_2(d, y) = C \sum_{i \in D^c} \mathbf{1}(S_i) + \sum_{i \in D}(C - c_i) + C_0$. The factor C has an interpretation as utility of one individual measurement. We will assume $(C - c_i) > 0$, since rainfall monitoring stations would not have been built if the utility C of a measurement were not greater than the cost c_i of maintaining a station.

The utility function u assigns a value to each pair of decision d and data y. A rational decision maker chooses a decision by maximizing the utility $u(d, y)$. Of course, at the time of decision making the future rainfalls y are still unknown. Thus we replace $u(d, y)$ by its expectation with respect to y, and the formal optimal design problem becomes:

$$\text{maximize } U(d) = E[u(d, y)] \qquad (2)$$

In the next section we define the relevant probability model on y with respect to which the exepected value in (2) is defined.

2.2 *The probability model*

As a probability model for the data y we consider a spatial kriging model as defined in Handkock and Stein (1993) which has been widely used for interpolating isotropic fields. The isotropy asumption is justified by the lack geographical diversity in the area. Exploration of the historical data sustains this asumption.

The network consists of rainfall stations scattered over a region, their latitude x_{i1}, longitude x_{i2} and elevation x_{i3}, are known, and no other measurements that could be used as covariates are available. We assume that y is multivariate normal with a regression on x_1 and x_2. The elevation has been dropped to simplify the model. Again, this asumption is justified by the analysis of the historical data. We then have

$$y \sim N[x\beta, \sigma^2 V(\lambda)], \tag{3}$$

where $V(\lambda)$ is a correlation matrix that depends on a parameter λ and is consistent with the assumption that the observations correspond to an isotropic field, that is, the covariance between any two locations only depends on the distance between them. For the purpose of our analysis we assumed $V_{ij}(\lambda) = \exp(-\lambda d_{ij})$, where $\lambda > 0$ and d_{ij} is the distance between stations i and j.

In order to solve the decision problem we need to define an informative prior distribution on the unknown model parameters $\theta = (\beta, \lambda, \sigma^2)$ based on available historical data. We proceeded as follows: for each station we have a set of observations consisting of annual rainfall recorded in millimeters of water over a period of $T = 16$ years. Let z_{ij} be the log of the annual rainfall recorded at station i during year j, where $i = 1, \ldots, n$ and $j = 1, \ldots, T$. Let $z_j = (z_{1j}, \ldots, z_{nj})$. Since the data are annual rainfall, serial correlations are neglible and we will suppose that $z_j, j = 1, \ldots, T$ are exchangable and

$$z_j \overset{iid}{\sim} N(x\gamma, \tau^2 V(\kappa)), \; j = 1, \ldots, T. \tag{4}$$

We complete the model with a non-informative prior $p(\gamma, \kappa, \tau^2) \propto 1/\tau^2 p(\kappa)$ where $p(\kappa)$ is a gamma density with a large variance and a mean that is compatible with the empirical estimation of the covariogram of z.

We run a Markov chain Monte Carlo (MCMC) simulation to obtain samples from the posterior distribution of (γ, κ, τ) given z and use the samples to specify an informative prior for (3). The posterior variance for κ was reasonably small. Thus we fixed λ at the posterior mean $E(\kappa|z)$. For $p(\beta, \sigma)$ we assume a normal-inverse-gamma prior $\sigma^{-2} \sim \text{Gamma}(a, b)$, $\beta|\sigma^2 \sim N(m, \sigma^2 S)$ with hyperparameters a, b, m and S chosen to have prior means matching the posterior means for γ and τ^2, and prior variances equal to ten times the posterior variances for γ and τ^2. The inflated variances account for possible changes over time.

We determined a value for δ by considering for each station the residuals from fitting the historical data z using model (4). We set δ equal to twice the sample standard deviation of these residuals.

3 A dual simulation problem

Solving the optimal design problem of redesigning the network of rainfall monitoring stations amounts to finding the design d which in expectation maximizes the given utility function $u(\cdot)$, i.e., we need to find d^* which maximizes expected utility

$$U(d) = \int u(d,y)p(\theta)p(y|\theta)d\theta dy.$$

A straightforward implementation could use conventional optimization techniques, evaluating the target function $U(d)$ for every considered design d by large scale Monte Carlo, i.e.,

$$U(d) \approx \tilde{U}(d) = 1/M \sum u(d, y^m), \tag{5}$$

where $(\theta^m, y^m) \sim p(\theta)p(y|\theta), m = 1, ..., M$, are M simulated experiments under design d. In general, this is impracticable since each evaluation of $u(d, y)$ requires a posterior integration and a costly maximization has to be performed.

Bielza, Mueller and Ríos Insua (1996) propose to consider instead of the optimization problem a related simulation problem. They show how to set up an artificial probability model $h(d, \theta, y)$ such that $h(d) \propto U(d)$. The optimal design problem can then be translated into a problem of simulating from a given distribution, $h(d, \theta, y)$. If one can obtain by computer simulation a Monte Carlo sample from $h(\cdot)$, then one can solve the optimal design problem by finding the mode of the empirical distribution of the simulated values for d. Roughly speaking, the design d which is most frequently generated is the solution of the original design problem. The attractiveness of setting up this equivalent simulation problem is that in recent years a barrage of literature related to simulating from a given, possibly high dimensional, distribution has appeared in the statistical literature. See, for example, Gilks, Richardson and Spiegelhalter (1996) for a review. As a result of these developments, algorithms are now available to simulate from essentially any distribution as long as it can be evaluated pointwise. Evaluation up to a constant factor suffices. The methods are geared towards simulating from posterior distributions, but are equally applicable to any other distribution which can be pointwise evaluated, for example the artificial probability model $h(d, \theta, y)$.

One possible choice for $h(\cdot)$ is

$$h(d, \theta, y) \propto p(\theta)p(y|\theta)u(d, \theta, y). \tag{6}$$

Subject to some technical constraints on the utility function, expression (6) defines a probability density function on (d, θ, y). Specifically, we need $u(d, \theta, y) > 0$. Note that a utility function can be shifted by an arbitrary constant without changing the solution of an optimal design problem. It is easily verified that the implied marginal in d is proportional to $U(d) = \int u(d, \theta, y)p(\theta)p(y|\theta)d\theta\, dy$, as desired. MCMC simulation from the artificial probability density function $h(\cdot)$ can be used to solve optimal design problems. The starting point is the following algorithm. We use d^t to denote the currently imputed design after t iterations of the Markov chain simulation (we use superindices to distinguish d^t from the indicator d_i^t for the i-th station).

1. Start with a design d^0. Simulate $(\theta, y) \sim p(\theta)p(y|\theta)$, i.e., simulate an experiment under design d^0. Evaluate $u_0 = u(d, y)$.

2. Generate $\tilde{d}$ from a probing distribution $g(\tilde{d}|d^0)$. Choice of the probing distribution will be discussed below.

3. Simulate $(\tilde{\theta}, \tilde{y})$, as in 1, under design $\tilde{d}$. Evaluate $\tilde{u} = u(\tilde{d}, \tilde{y})$.

4. Compute

$$a = \min\left\{1, \frac{h(\tilde{d}, \tilde{\theta}, \tilde{y})}{h(d^0, \theta, y)} \frac{g(d^0|\tilde{d})p(\theta, y)}{g(\tilde{d}|d^0)p(\tilde{\theta}, \tilde{y})}\right\} = \min\left\{1, \frac{\tilde{u}}{u_0} \frac{g(d^0|\tilde{d})}{g(\tilde{d}|d^0)}\right\}$$

5. Set

$$(d^1, u_1) = \begin{cases} (\tilde{d}, \tilde{u}) & \text{with probability } a \\ (d^0, u_0) & \text{with probability } 1 - a \end{cases}$$

6. Repeat steps 2 through 5 to generate $d^2, u_2, \ldots, d^T, u_T$, until the chain is judged to have practically converged after T iterations.

The algorithm defines an MCMC scheme to simulate an approximate Monte Carlo sample from $h(d, \theta, y)$, and thus marginally from $h(d) \propto U(d)$. It follows the general format of what Tierney (1994) refers to as Metropolis-Hastings chains with independence proposal. Convergence to the desired $h(\cdot)$ can be established using results from Tierney (1994).

There still remains the specification of the probing distribution $g(\tilde{d}|d)$. Choosing a symmetric probing distribution, that is, one for which $g(\tilde{d}|d) = g(d|\tilde{d})$, leads to a simple expression for the acceptance probability, $a = \min(1, \tilde{u}/u_t)$. In this application we used a probing distribution that flipped one d_j at a time, i.e., $g(\tilde{d}|d)$ is defined by first uniformly sampling an index $j \in \{1, \ldots, n\}$ and then setting $\tilde{d}_j = (1 - d_j)$ and $\tilde{d}_h = d_h, h \neq j$.

There is a major limitation of this scheme. With N stations we have 2^N possible designs. Probing new designs with a random walk probing distribution can not be expected to cover the design space. Thus we consider the algorithm as a stochastic exploration for interesting designs, but not necessarily as a search for the optimal design. It is worth mentioning, though, that none of the approaches that we are aware of for redesigning networks of stations produces an optimal solution when dealing with a realistically large number of stations.

4 Results

We implemented the described algorithm to find promising designs. We constrain n_d, the number of stations, to be between 35 and 45. Alternatively, enforcing a constraint to exactly 40 stations would be easily possible by replacing the probing distribution $g(\tilde{d}|d)$ given in the previous section with a probing distribution which first randomly selects a pair of indicators (i, j) with $(d_i, d_j) = (0, 1)$ and then proposes a new design $\tilde{d}$ with $(\tilde{d}_i, \tilde{d}_j) = (1, 0)$ and $\tilde{d}_h = d_h, h \notin \{i, j\}$.

There remains the issue of choosing C and $c_i, i = 1, \ldots, n$. We used $C = 1.5$ and $c_1 = \ldots = c_n = 1$, based on the argument that the exisiting stations would not have been built if the payoff of recording rainfall at a site would not be greater than the cost of maintaining the site, i.e., $C > c_i$. The smaller C/c_i is, the fewer stations we would consider. For a first exploration we chose $C/c_i = 1.5$.

Following a suggestion by Bielza *et al.* (1996) we used cluster analysis to estimate the mode of $h(d)$, i.e., the optimal design. Denote with d^m, $m = 1, \ldots, M$ the designs generated in the MCMC scheme described earlier. For each design d^m we consider the distance from all other simulated designs. We formally define the distance between two designs d and d' as the number of mismatches between d and d', i.e., $\sum_{i=1}^{80} \mathbf{1}(d_i \neq d'_i)$. We obtain a hierarchical clustering tree based on this binary distance and cut it at a certain height. We thus get clusters of simulated designs which differ by less than a given number of stations. The largest cluster corresponds to designs which are around the mode of $h(d)$. Figures 1 and 2 show some summaries of these clusters.

The approach is simulation based and in the end hinges upon identifying the mode of the distribution $h(d)$ based on the simulation output. Because of the involved approximations, the finally proposed set of designs does not necessarily include the optimal design. Rather, the proposed designs should be considered "interesting" solutions to the challenging problem. To verify that the proposed designs are in fact improvements over initial guesses we can for a few identified designs evaluate the large scale Monte Carlo approximation $\tilde{U}(d)$ given in (5). Table 1 reports such approximations for some selected designs.

5 Discussion

We have approached a challenging optimal design problem by considering an equivalent simulation problem. We consider the scheme as a method of searching the design space for interesting designs, not as a final solution to the original design problem. There are two remaining obstacles to a complete solution. First

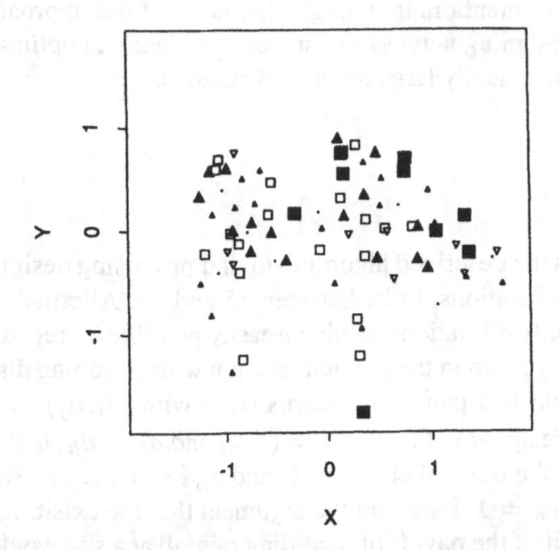

FIGURE 1. Frequency of occurence of individual stations in designs included in the top cluster. Each plotting symbol corresponds to one station, centered at a location corresponding to longitude and latitude of the respective station. Dots, triangles, inverted triangles, squares, filled triangles, and filled squares indicate stations which are included 0,1,2,3,4 and 5 times, respectively. The size of the plotting symbol varies from small to large according to the frequency of the station. The figure has to be interpreted with care since it provides only a marginal summary of the multivariate problem.

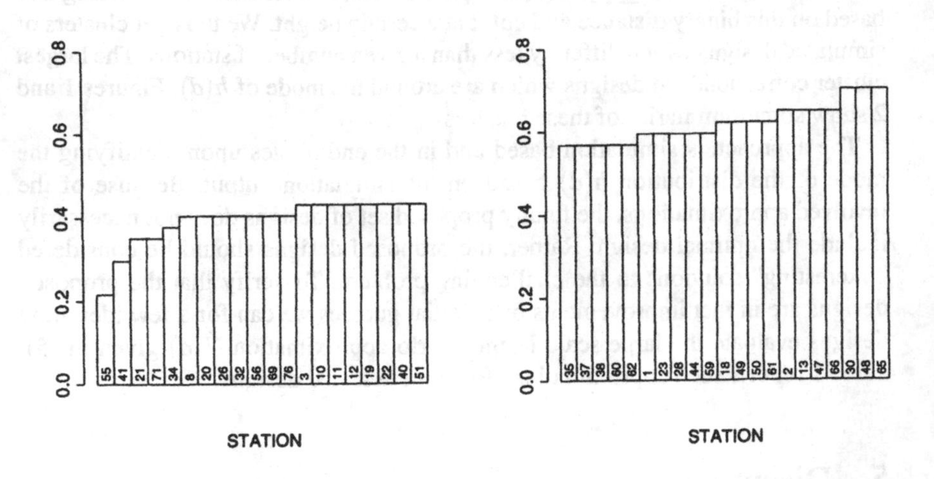

FIGURE 2. Histogram of frequencies of all stations in clusters of sizes 4 and higher. The left panel shows the 20 least frequently included stations. The right panel shows the 20 most frequently included stations. Again, the figure has to be interpreted with care since it provides only a marginal summary of the multivariate problem.

best			worst						west/east		
d	$\tilde{U}(d)$	n_d	d	$\tilde{U}(d)$	n_d	d	$\tilde{U}(d)$	n_d	d	$\tilde{U}(d)$	n_d
d_1	109.8	44	d_6	106.2	36	d_{11}	108.0	35	d_{16}	104.3	39
d_2	115.3	42	d_7	104.8	35	d_{12}	109.8	36	d_{17}	91.3	41
d_3	114.0	39	d_8	106.7	39	d_{13}	106.2	35			
d_4	114.1	44	d_9	109.4	35	d_{14}	105.9	35			
d_5	117.6	40	d_{10}	107.6	39	d_{15}	104.3	35			

TABLE 1. Approximate expected utilities $\tilde{U}(d)$ using $M = 200$ independent simulations. The table reports $\tilde{U}(d)$ for the five designs in the top cluster $(d_1, \ldots, d_5)$, the 10 designs in the smallest (singleton) clusters $(d_6, \ldots, d_{15})$, and for comparsion, for two designs which choose all stations on the western half of the region (d_{16}) and on the eastern half of the state (d_{17}), respectively. The column labeled n_d reports the number of stations included in each design. The numerical standard deviation of the reported Monte Carlo sample means $\tilde{U}(d)$ is less than 0.5.

the problem of defining a good proposal distribution $g(\cdot)$ for generating new designs; and second, the problem of summarizing the high dimensional output of simulated designs, trying to uncover the mode of the underlying distribution $h(d)$, i.e., essentially the problem of estimating the mode in a high dimensional distribution.

In current research we consider solving the first problem by changing the algorithm to a Gibbs sampler on blocks of design parameters. For $N = 80$, we partition the design vector d in, say, eight subvectors $d = (d_1, \ldots, d_8)$. We replace step 2. by

Step 2'. Generate a new proposal $\tilde{d}$ from $h(d_j | d_i, i \neq j, y, \theta)$, scanning over $j = 1, \ldots, 8$ in subsequent iterations.

Step 4/5'. Replace (d, u) by $(\tilde{d}, \tilde{u})$.

The modified step 2' generates a new value of d_j given currently imputed values for d_i, $i \neq j$, the parameters θ and future data y, using the conditional distribution under $h(\cdot)$. Each d_j has $2^{10} = 1024$ possible values. Thus generation from $h(d_j | \ldots)$ requires evaluation of $u_m = u(d, \theta, y)$ for each of the 1024 possible values $d_j = m$, and a multinomial sampling with probabilities proporitional to u_m, $m = 1, \ldots, 1024$. Using complete conditinals under $h(\cdot)$ as probing distribution the acceptance probability in step 4. reduces to a constant 1.0, i.e., the MCMC becomes a Gibbs sampler.

Acknowledgments

We would like to thank Dr Lelys Guenni for posing the problem to us and MARNR of Venezuela, for providing the data. The first author was partially supported by grants BID-CONICIT I-06 and USB GID-01. Peter Müller was partially supported by the National Science Foundation under grant DMS-9704934.

References

Bielza, C., Müller, P., and Ríos-Insúa, D. (1996). 'Monte Carlo Methods for Decision Analysis with Applications to Influence Diagrams,', Discusssion paper 96-07, ISDS, Duke University.

Bras, R. L. and Rodríguez-Iturbe, I. (1985). *Random Function and Hydrology*. Addison-Wesley.

Caselton, W.F., Kan, L. and Zidek, J.V. (1992). 'Quality data networks that minimize entropy'. In Statistics in the Environment and Earth Sciences, (eds. P. Guttorp and A. Walden), London: Griffin.

Chaloner, K. and Verdinelli, I. (1995).'Bayesian experimental design: a review', *Statistical Science*, **10**, 273–304 .

Chaloner, K. and Larntz, K. (1992). 'Bayesian design for accelerated life testing', *Journal of Statistical Planning and Inference*, **33**, 245–259.

Clyde, M and Chaloner, K. (1996). 'The equivalence of contrained and weighted designs in multuple objective design problems', *Journal of the American Statistical Association*, **89**, 1236–1244.

De Groot, M.H. (1970). *Optimal Statistical Decisions*, McGraw-Hill, New York.

Gelfand, A.E. and Smith A.F.M. (1990). 'Sampling based approaches to calculating marginal densities', *Journal of the American Statistical Association*, **85**, 398–409.

Gilks, W.R., Clayton, D.G., Spiegelhalter, D.J., Best, N.G., McNeil, A.J., Sharples, L.D., and Kirby, A.J. (1993).'Modelling complexity: applications of Gibbs sampling in medicine', *Journal of the Royal Statistical Society Ser. B*, **55**, 39–52.

Guttorp, P., Le, N., Sampson, P.D. and Zidek, J.V. (1993). 'Using entropy in the redesign of an environmental monitoring network'. In *Multivariate Environmental Statistics*, (G.P. Patil and C.R. Rao eds.), Elsevier Science Publisher, 175–202.

Handcock, M. and Stein, M. (1993). 'A Bayesian analysis of kriging', *Technometrics*, **35**, 403–410.

Palmer, J. and Müller, P. (1998). 'Bayesian Optimal Design in Population Models of Hematologic Data,' *Statistics in Medicine*, to appear.

Nychka, D. and Saltzman, N. (1997). 'Design of air quality monitoring networks'. Tech. report, NISS.

Sansó, B. and Guenni, L. (1996). 'A Bayesian estimation of the parameters of a space-time model for rainfall', Tech. Report 96-01, CESMa, Universidad Simón Bolívar.

Toman, B. (1994). 'Bayes optimal designs fo two and three level factorial experiments', *Journal of the American Statistical Association*, **89**, 937–946.

Tierney, L.(1994). 'Markov chains for exploring posterior distributions', *Annals of Statistics*, **22**, 1701–1728 .

Verdinelli, I. (1992).'Advances in Bayesina experimental design', In *Bayesian Statistics 4*, (eds. J. M. Bernardo, J. O. Berger, A. P. Dawid and A. F. M. Smith eds.), Oxford University Press, 467–48.

Using PSA to detect prostate cancer onset: An application of Bayesian retrospective and prospective changepoint identification

Elizabeth H. Slate
Larry C. Clark

ABSTRACT Serum prostate-specific antigen (PSA) concentrations are now widely used to aid in the detection of prostate cancer. When prostate cancer is present, PSA levels typically increase. But a number of benign conditions will also cause elevated PSA levels and, conversely, prostate cancer has been diagnosed in the absence of raised PSA.

We analyze one of the most extensive data sets currently available for longitudinal PSA readings, obtained by an historical prospective study of frozen serum samples from the Nutritional Prevention of Cancer Trial (Clark *et al.*, 1996). These data consist of serial readings for over 1200 men taken at approximate six-month intervals over an 11 year period.

We fit a fully Bayesian hierarchical changepoint model to the longitudinal PSA readings. Our ojectives include better understanding the natural history of PSA levels in patients who have and have not been diagnosed with prostate cancer and identifying subject-specific changepoints that are indicative of cancer onset. With the goal of accurate early detection, we perform a prospective sequential analysis to compare several diagnostic rules, including a rule based on the posterior distribution of individual changepoints.

1 Introduction

According the American Cancer Society Facts & Figures, prostate cancer is the most common cancer, excluding skin cancer, and the second leading cause of cancer death in American men. Prostate cancer incidence has tripled in the past decade, to an estimated 334,500 new cases in 1997, largely due to increased detection by the use of serum prostate-specific antigen (PSA) measurements. PSA is a glycoprotein produced by the prostate gland that increases with the volume of the prostate. Many studies have demonstrated the utility of PSA for detecting cancer. Catalona *et al.* (1991, 1993) found that combining PSA levels with ultrasound and digital rectal exam (DRE) significantly improved the detection of prostate cancer. They interpreted PSA levels of 4 ng/ml (on the monoclonal scale) as arousing suspicion of cancer. Oesterling *et al.* (1993) concluded from a prospective study that PSA increases gradually among healthy men and, consequently, suggested age-specific normal ranges for PSA. Recently, Carter *et al.* (1997) supported the cutoff of 4 ng/ml for maintaining the detection of curable prostate cancer. However, a single PSA measurement is an imprecise indicator of disease status. The proportion of men without prostate cancer who have PSA levels over 4 ng/ml is 6–41% and even higher in some subpopulations, while 18–35% of men with prostate cancer have PSA readings less than 4 ng/ml (Carter *et al.* 1992a,Catalona *et al.* 1991,Pearson *et al.* 1996,Oesterling *et al.* 1993). Similarly Gann *et al.* (1995) andWhittemore *et al.* (1995) reported that the cutoff of 4 ng/ml has estimated sensitivities of 73–75% and specificities of 88–91% for detection of prostate cancer in the next 4–7 years.

A longitudinal series of PSA measurements taken periodically from a subject may lead to the development of diagnostic criteria with much higher sensitivities and specificities. Few published studies have investigated the behavior of longitudinal PSA measurements. Carter *et al.* (1992a, b),Pearson *et al.* (1991, 1994) andMorrell *et al.* (1995) analyzed data obtained from the Baltimore Longitudinal Study of Aging (BLSA). These data consisted of series of PSA readings obtained from frozen blood samples collected approximately bi-annually for 54 men. The series spanned the 7–25 years before determination of prostate disease status, with a median number of approximately 10 readings per person. They fit linear mixed effects models and concluded that the exponential growth rate of PSA is significantly greater for cases prior to diagnosis than for controls, indicating that the rate of change of PSA may be a more sensitive and specific marker for prostate cancer than PSA levels. These investigators also used nonlinear mixed effects models to estimate changepoints in the rate of increase of the PSA for cases. They found that the transition from slow to rapid increase typically occurred 7.3 and 9.2 years before diagnosis for local and metastatic cancers, respectively.

A second longitudinal study was described by Whittemore *et al.* (1995), based on data available from a screening program run by the Kaiser Permanente Medical Care Program (KPMC). They analyzed PSA trajectories spanning 1–5 years for 320 men, with a median number of readings of approximately 4 per person. These investigators also fit linear and nonlinear mixed effects models and found significant differences in the growth rate of PSA between cases and controls, reporting that the transition from slow to rapid growth occurred about 13–14.5 years before diagnosis. Moreover, Whittemore *et al.* evaluated diagnostic rules based on absolute levels and rates of change of PSA and concluded, surprisingly, that a single PSA measurement was a more sensitive indicator of cancer within the next 7 years than any index of change based on the entire trajectory. This result may be partially explained by the lack of readings near diagnosis in their data.

In this paper we investigate a much larger set of longitudinal PSA readings obtained from frozen blood samples available for patients in the Nutritional

Prevention of Cancer Trial (NPCT). The NPCT is a randomized double-blind cancer prevention trial using a nutritional dose of the essential trace element selenium as the intervention agent. The results of a ten-year follow up were described by Clark *et al.* (1996). The PSA data that we analyze here consist of serial readings for 1210 men spanning up to 11 years with a median number of readings of 4 per person; additional details are given in Section 2. For comparison with the BLSA and KPMC studies, we briefly describe the results of fitting linear and nonlinear mixed effects models to our data in Section 3. However, the focus of this paper is the fully Bayesian hierarchical changepoint model described in Section 4. Our model is similar to one studied by Cronin (1995), which was motivated by this PSA context, and also to that used by Slate and Cronin (1997) to analyze serial PSA readings obtained from patients following radiation treatment for prostate cancer. The model is like that described by Carlin *et al.* (1992) and used by Lange *et al.* (1992) for CD4 T-cell counts, but with a continuous changepoint as in Stephens (1994). As does Stephens, we use our model to retrospectively estimate changepoints, but we also consider our data as arising from an historical prospective study (see, for example, Carter *et al.*, 1997) and dynamically estimate the changepoints. This dynamic implementation enables us to evaluate and compare various diagnostic rules using receiver operating characteristic (ROC) curves adapted for the longitudinal tests; we do this in Section 5. Additional discussion appears in Section 6.

2 The NPCT PSA data

The NPCT investigated the utility of a nutritional dose of selenium for preventing cancer. A total of 1736 patients at high risk of skin cancer were randomized into efficacy and safety arms of the trial, with 1255 of these being males with no prior history of prostate cancer. Blood samples were collected and frozen semi-annually from trial participants. Our PSA data were obtained using the Abbott IMx assay (Abbott Park, Ill.; Vessella *et al.*, 1992) on the frozen samples from the men in the trial. After removing PSA readings from samples taken after diagnosis of prostate cancer, after the start of finasteride, which is a medication used to treat symptomatic BPH and known to reduce PSA levels, and after the trial unblinding date of Feb. 1, 1996, our data consisted of 6659 readings for 1210 case-control study men. Of these men, 85 had biopsy-confirmed diagnoses of prostate cancer and will be called the cases, while the remainder will be called the controls. Thus it is possible that a man we categorize as a control in fact has undiagnosed prostate cancer. Note that because of the historical prospective nature of our data, the PSA readings were typically not used to aid diagnosis.

Figure 1 shows representative PSA trajectories from a sample of cases and controls from NPCT data. The horizontal scale in these plots is labeled as "Years Before Reference Date." The reference date is defined as the date of diagnosis for cases and the date of the last recorded reading for controls (as in Pearson *et*

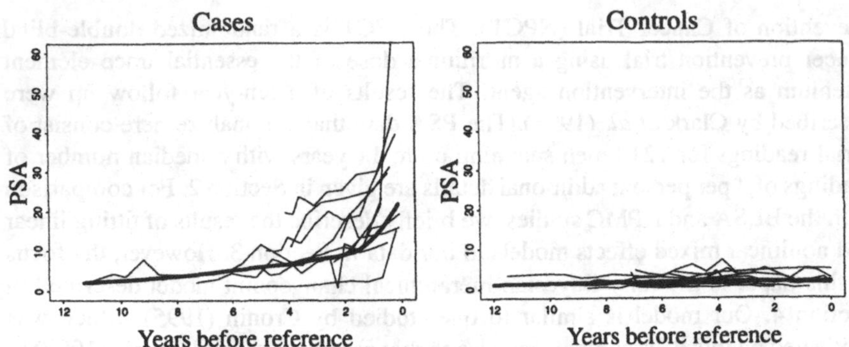

FIGURE 1. Typical PSA trajectories and for cases and controls. The thin lines are linear interpolations of selected trajectories and the thick lines, one for each graph, are the smoothed trajectories for all cases and controls.

al., 1994).Also shown are loess (Cleveland and Devlin, 1988) smooth fits to all 1125 controls and 85 cases. Similar graphs can be found in Figures 1, 2 and 3 of Carter *et al.* (1992a), Figure 1 of Pearson *et al.* (1994) and in Figure 1 of Whittemore *et al.* (1995). It is important to note that there does appear to be a difference in the behavior of the PSA trajectories for the two groups: those for the controls remain relatively flat, whereas the trajectories for the cases appear to show more of an increasing trend. These (apparent) differences in the PSA trajectories according to prostate condition suggest that, indeed, longitudinal PSA readings can be informative about prostate disease status.

Table 1 summarizes key characteristics of the NPCT data. Separately for cases and controls, the median, average and range are given for the number of readings per person, time span of the readings per person, age at the reference date, and median PSA per person. Also summarized across all readings for cases and controls is the years prior to the reference date (YPR) at which the blood samples were taken. Although the NPCT data contain readings for many more subjects, the number of readings per subject and the time span of those readings fall between these values for the BLSA and KPMC data. Median PSA readings appear to be a little lower in the NPCT data than in the BLSA data. For example, the median PSA readings at the reference date in the NPCT data are 7.2 and 1.2 ng/ml for cases and controls, respectively, while these values are approximately 15.9 and 2.0 ng/ml in the BLSA data. Similarly, the medians of the PSA readings about five years before the reference date are 3.4 and 1.2 ng/ml for the cases and controls here, while in the BLSA data these medians are about 4.7 and 1.7 ng/ml. These differences, particularly among the control group, might be explained by the younger subjects in the NPCT. Nearly 6% of the men in the control group were less than age 50 years at the reference date; there are no men under age 50 at this time in the BLSA and KPMC studies.

	Cases (N = 85)			Controls (N = 1125)		
	Med.	Av.	Range	Med.	Av.	Range
No. Readings	5.0	6.0	(1–19)	4.0	4.5	(1–20)
Time Span (yrs)	2.5	3.1	(0–10.8)	4.5	4.2	(0–11.6)
Age at Ref.	72.0	72.5	(56.7–84.6)	70.5	68.9	(20.7–89.5)
Median PSA	5.3	10.0	(0.9–150.4)	1.1	1.6	(0–25.1)
YPR	3.0	3.3	(0–11.3)	2.5	3.2	(0–12.3)

TABLE 1. Summary of the NPCT PSA data. The median (Med.), average (Av.) and range are given for key characteristics as described in the text.

3 Mixed effects models

The analyses of both the BLSA and KPMC studies included mixed effects models for the PSA trajectories. These models incorporate both population and subject-specific effects and can capture the serial correlation expected in the marker measurements recorded within a subject. The linear mixed effects model is useful for describing the apparent differences in the PSA trajectories for the cases and controls. The nonlinear mixed effects model additionally permits the estimation of changepoints representing the time at which cancer first affects the trajectories for the cases. Here we provide an illustration of each type of model, its application to the NPCT data and a brief comparison to the BLSA and KPMC results.

3.1 Linear mixed effects model

Whittemore *et al.* (1995) and Pearson *et al.* (1994) consider models similar to the following:

$$\begin{aligned}\ln(\text{PSA}_{ij}+1) &= (\beta_0+b_{0i})+\beta_1\,\text{Case}_i+(\beta_2+b_{2i})\,\text{PriorYears}_{ij}\\ &+ \beta_3\,\text{Case}_i\times\text{PriorYears}_{ij}+\beta_4\,\text{RefAge}_i+\epsilon_{ij} \qquad (1)\end{aligned}$$

Here, i indexes the subject and j indexes the observation within that subject. "Case_i" equals 1 if subject i has been diagnosed with prostate cancer and 0 otherwise, and "RefAge_i" is the age of subject i at the reference date (recall that the reference date is the date of diagnosis for cases and the date of the last reading for controls). Also "PriorYears" denotes years prior to the reference date with a negative sign attached. (This sign convention is needed so that the slope coefficients β_2 and β_3 will be nonnegative in situations such as those depicted in Figure 1.) The population or fixed effects are denoted by βs and are unknown constants to be estimated. The random or subject-specific effects b_{0i} and b_{2i} permit the intercept (β_0+b_{0i}) and slope (β_2+b_{2i}) to vary across subjects. The vectors $\{(b_{0i},b_{2i})\}$ are assumed to be independently distributed as multivariate normal with mean $\mathbf{0}$ and arbitrary variance-covariance matrix. The errors $\{\epsilon_{ij}\}$ are modeled as independent and identically distributed normal random variables with constant (but unknown) variance σ^2. The inclusion of the predictor RefAge adjusts for the effect of age at the reference date on the

PSA levels (see Oesterling *et al.*, 1993, for results concerning the effect of age on PSA levels). The transformation ln(PSA + 1) is used by Whittemore *et al.* (1995) and implies linear growth on the log scale, which might be justified both biologically by the exponential growth of malignancies and statistically as an approximate variance stabilizing transformation. (The addition of one is to diminish the influence of extremely small PSA readings.)

We fit model (1) to the NPCT data using restricted maximum likelihood (Lindstrom and Bates, 1988, 1990). The population median fits are shown in Figure 2. The approximately horizontal fits are for control subjects and the increasing curves are for case subjects. The effect of age at diagnosis in this model is to shift the fitted curves upwards for greater ages: for both cases and controls the three lines correspond to ages at the reference date of 50, 65 and 80 years as the PSA levels increase. All estimated coefficients are highly statistically significant.

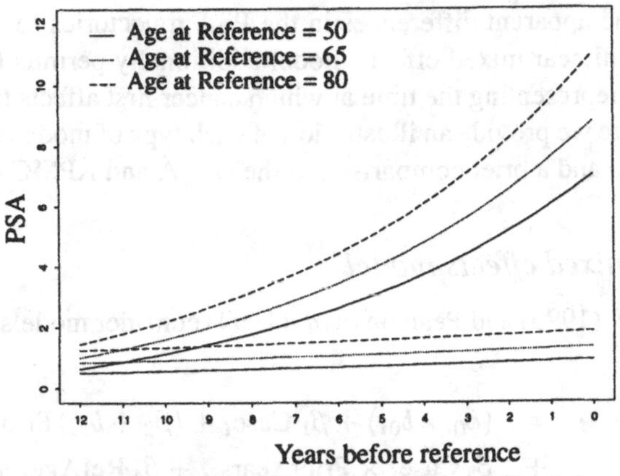

FIGURE 2. The population median PSA trajectories predicted by model (1) for the preliminary NPCT data. The approximately horizontal fits are for control subjects and the increasing curves are for case subjects.

The slope of the ln(PSA + 1) curves, here either β_2 for controls or $\beta_2 + \beta_3$ for cases, is termed the exponential growth rate or "velocity" of the PSA and is of primary interest. For these data the estimated exponential growth rate is 0.02 (se = 0.002) for controls and 0.13 (0.009) for cases. Using models similar to (1), approximate exponential growth rates for the KPMC data are 0.03 (0.004) and 0.16 (0.012) for the controls and cases, respectively. Pearson *et al.* (1994) obtain a slope of approximately 0.04 for their control subjects and, including a quadratic term for the cases, obtain slopes of 0.10, 0.15 and 0.19 at 10, 5, and one year before diagnosis (se not available from published reports). All three studies show a significant difference between the growth rates for PSA for the cases and controls.

3.2 *Nonlinear mixed effects model*

Assuming that at one time the cases were cancer-free and hence controls, it is reasonable to seek a changepoint in the PSA trajectories for the cases that indicates the time when a malignancy first affects the PSA readings. Similar to Whittemore *et al.* (1995), one model that we have fit is

$$\begin{aligned}\ln(\mathrm{PSA}_{ij}+1) = &\\ &\theta_{0i} + \theta_1\,(\mathrm{PriorYears}_{ij} - t_i) + \\ &\theta_{2i}\,(\mathrm{PriorYears}_{ij} - t_i)\,\mathrm{sgn}(\mathrm{PriorYears}_{ij} - t_i) + \epsilon_{ij}, \qquad (2)\end{aligned}$$

where sgn is the sign function, θ_{0i}, θ_{2i} and t_i are subject-specific parameters, *i.e.* $\theta_{0i} = \delta_0 + d_{0i}$, $\theta_{2i} = \delta_2 + d_{2i}$ and $t_i = \alpha + a_i$, $\{(d_{0i}, d_{2i}, a_i)\}$ are independent multivariate normal vectors with mean zero, and the errors are independent and normally distributed. The interpretation of the parameters for the trajectory for subject i is that θ_{0i} is the overall level, θ_1 is the average slope, θ_{2i} is half the difference in slopes before and after the changepoint, and t_i is the changepoint. (This particular parameterization was chosen for its numerical properties for detecting changes in slope, see Seber and Wild, 1989, Sec. 9.4.1.) Thus the level, changepoint, and slopes before and after the changepoint are modeled as random. We smoothed the transition between the linear regimes by replacing $\mathrm{sgn}(z)$ with $h(z,\gamma) = (z^2+\gamma)^{1/2}/z$, where γ is a small positive smoothing parameter (see Seber and Wild, 1989, Sec. 9.4.1).

Figure 3 shows the median maximum likelihood fit of this model to the PSA trajectories for 84 cases from the NPCT data set. (The two PSA readings of 167.4 and 133.3 for one case were omitted.) This fit was obtained by using $\gamma = 0.5$ in the smoothing function $h(\cdot,\gamma)$, maximum likelihood estimates for all population parameters and medians of the modal estimates of the subject-specific effects. The population parameters are highly significant in this model. The fixed effect pre- and post-changepoint slope estimates are 0.03 (se = 0.03) and 0.17 (se = 0.02), respectively, and the medians of the subject-specific slope estimates are 0.03 for controls and 0.16 for cases. The median changepoint is 4.25 years before diagnosis, and all subjects' estimated changepoints (*i.e.* $\hat{t}_i = \hat{\alpha} + \hat{a}_i$) are before diagnosis. The transition time of 4.25 years before diagnosis is considerably less than the values of 7–9 and 13–14 years obtained by Pearson *et al.* (1994) and Whittemore *et al.* (1995). This difference might be explained by the fact that these data involved a selected population actively participating in a clinical trial with frequent physician visits. Residual plots (not shown) confirm that this model provides an adequate fit to these data.

4 Bayesian hierarchical changepoint model

Although the nonlinear mixed effects model accomodates serial correlation and subject-specific changepoints in the PSA trajectories, there are difficulties

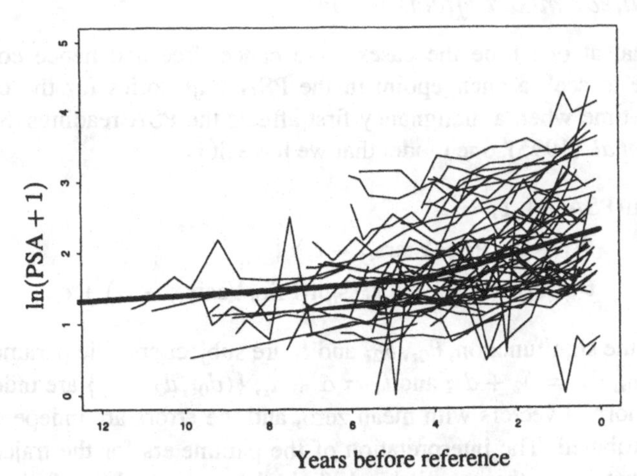

FIGURE 3. The ln(PSA + 1) trajectories for 84 cases from the NPCT data set and the median fit from the model (2).

associated with these models. For most standard fitting routines, model (2) can be fit to the cases only and cannot be used to predict when in the future the changepoint may be encountered by controls. Also, these fitting routines tend to rely on smoothness of the model function and hence require smoothing at the transition point. Furthermore, linear approximations in fitting may create biases in the estimates of the random effects (Breslow and Lin, 1995). A Bayesian model is appealing because it has the benefits of the mixed effects model and enables prediction of changepoints for controls, is easily fit using Monte Carlo Markov chain (MCMC) techniques, and, most importantly, provides an immediate answer to the question "What is the probability that this subject has encountered his changepoint?" For example, if PSA levels react immediately to the underlying prostate condition, then this question is equivalent to "What is the probability that prostate cancer has initiated in this subject?" The answer that the Bayesian model provides is the probability assigned prior to the current time by the posterior distribution of the particular subject's changepoint.

We use a segmented linear regression model for the transformed PSA:

$$\ln(\mathrm{PSA}_{ij} + 1) = a_{0i} + a_{1i}\, x_{ij} + (b_i - a_{1i})(x_{ij} - t_i)^+ + \epsilon_{ij} \tag{3}$$

where x_{ij} is the age of subject i at j-th reading, and $z^+ = z$ if $z > 0$ and 0 otherwise. Note that our predictor here is age, rather than years prior to the reference date. This allows incorporation of the dependence on age and obviates the need for the concept of a "reference date" as used in Section 3. The intercept a_{0i}, slope before the changepoint a_{1i}, slope after the changepoint b_i and the changepoint t_i are all random effects.

The full model is (3) with the prior distributions given below:

$$
\begin{aligned}
\begin{pmatrix} a_{0i} \\ a_{1i} \end{pmatrix} \Big| \begin{pmatrix} \alpha_0 \\ \alpha_1 \end{pmatrix}, \Omega_a &\sim \mathrm{MVN}\left\{ \begin{pmatrix} \alpha_0 \\ \alpha_1 \end{pmatrix}, \Omega_a \right\} \\
\begin{pmatrix} \alpha_0 \\ \alpha_1 \end{pmatrix} &\sim \mathrm{MVN}\left\{ \begin{pmatrix} 1 \\ .02 \end{pmatrix}, \begin{pmatrix} 100 & 0 \\ 0 & 10000 \end{pmatrix} \right\} \\
\Omega_a &\sim W\left\{ \left[5 \begin{pmatrix} .1 & 0 \\ 0 & .0001 \end{pmatrix} \right]^{-1}, 5 \right\} \\
b_i \mid \beta, \tau_b &\propto \mathrm{N}(\beta, \tau_b)\ I(b_i > .08) \\
\beta &\sim \mathrm{N}(.15, 3600) \\
\tau_b &\sim \mathrm{Gamma}(48.0, .0133) \\
t_i | \mu_t, \tau_t &\sim \mathrm{N}(\mu_t, \tau_t) \\
\mu_t &\sim \mathrm{N}(80, 0.10) \\
\tau_t &\sim \mathrm{Gamma}(47.0, 4700) \\
\epsilon_{ij} | \tau_i &\sim \mathrm{N}(0, \tau_i) \\
\tau_i &\sim \mathrm{Gamma}(5.0, 0.25).
\end{aligned} \tag{4}
$$

The subject-specific parameters are conditionally independent, as are the within-subject errors. All normal distributions are parameterized in terms of a mean and a precision; thus Ω_a is a 2×2 precision matrix, and τ_b, τ_t and τ_i are precisions. The precision matrix Ω_a follows a Wishart distribution, as parameterized in Press (1982, Chapter 5), for example. The Gamma(a, b) distributions have mean a/b. The slope after the changepoint, b_i, is constrained to be larger than 0.08, a conservative value consistent with previous estimates of the exponential growth rate for cases. This restriction facilitates the model's distinction between a_{1i} and b_i. The model specifies that all men will eventually encounter their changepoint if they live long enough. Thus, because it appears that a large fraction of men do not encounter this changepoint, the parameter μ_t, the mean age at which the changepoint occurs, tends to be quite large. Because of the conjugate structure, it is straightforward to fit this model using the Gibbs sampler (Geman and Geman, 1984;Gelfandand Smith, 1990), as describedin Slate and Cronin (1997),Cronin *et al.* (1994) andCronin (1995). The"hierarchical centering"of the random effects (see Gelfand, Sahu, and Carlin, 1995) generally aids the MCMC convergence.

The prior information used for this analysis is drawn from the PSA literature, particularly Carter *et al.%* (1992a, b), Pearson*et al.* (1994), Oesterling *et al.* (1993) andWhittemore *et al.* (1995). This prior information about the PSA trajectories is best conveyed via the plots in Figure 4, which show the .1, .5, and .9 quantiles of the $\ln(\text{PSA} + 1)$ and PSA trajectories implied by the chosen prior distributions. Note in particular, the changepoint is typically at about age 80.

Figure 5 shows the prior and posterior distributions based on the data for the 1210 men for selected population parameters after 100,000, 900,000 and one million iterations of the Gibbs sampler. These kernel density estimates are based

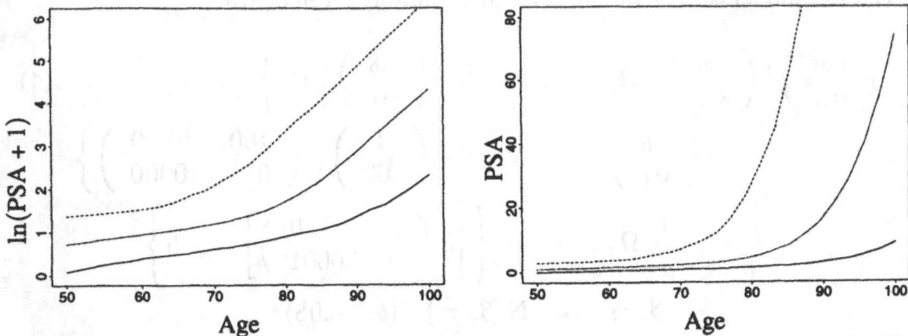

FIGURE 4. The 0.1, 0.5 and 0.9 quantiles of the ln(PSA + 1) and PSA trajectories implied by the prior distributions given in equations (4).

on 1000 sampled points obtained after lagging successive iterates by up to 75. For all population parameters, the posterior distributions are much less diffuse than the prior distributions. The characteristics of these distributions depend on the mix of cases and controls in the data set. The NPCT data is nearly 93% controls. The posterior mode of the mean of the slopes before the changepoint, α_1, is approximately 0.015, whereas the mean of the slopes after the changepoint, β, has mode about 0.14. The parameter μ_t is the mean age at the changepoint (cancer onset) and has mode approximately 87, substantially higher than the prior mean of 80 years. This shift in the distribution of μ_t is not surprising given that controls dominate the data set. Because there are few PSA readings taken at ages above 80 (less than 5%), the posterior for μ_t indicates that many subjects have yet to experience their changepoint.

4.1 *Retrospective changepoint identification*

The posterior distribution of the changepoint for each case subject can be used to estimate the age of cancer onset. Similarly, the posterior distribution of the changepoint for a control subject summarizes all current information about when onset is likely to occur for this subject (and this distribution may indicate that onset has already occurred despite the lack of diagnosis). Figure 6 shows two sample PSA trajectories, one for a case subject and one for a control, and the corresponding posterior distributions for the changepoints. The variability in the posterior distribution for the changepoint for the control subject is nearly the same as the variability in the prior distribution. The posterior exhibits some right skewness and has mode at about age 90, which is well beyond the range of the PSA data available for this subject. The posterior distribution of the changepoint for the case subject, however, is quite peaked and centered at about age 64. Thus the model fit indicates that this case experienced his changepoint before entering the trial. Note that his first PSA is above 4 ng/ml.

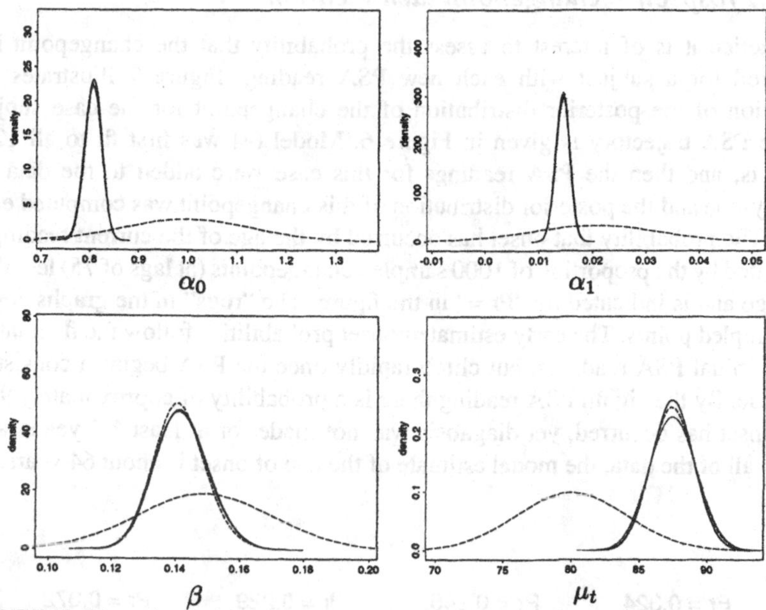

FIGURE 5. The prior (wide dashes) and posterior distributions of selected population parameters after 100,000 (dotted), 900,000 (dashed) and one million (solid) iterations of the Gibbs sampler.

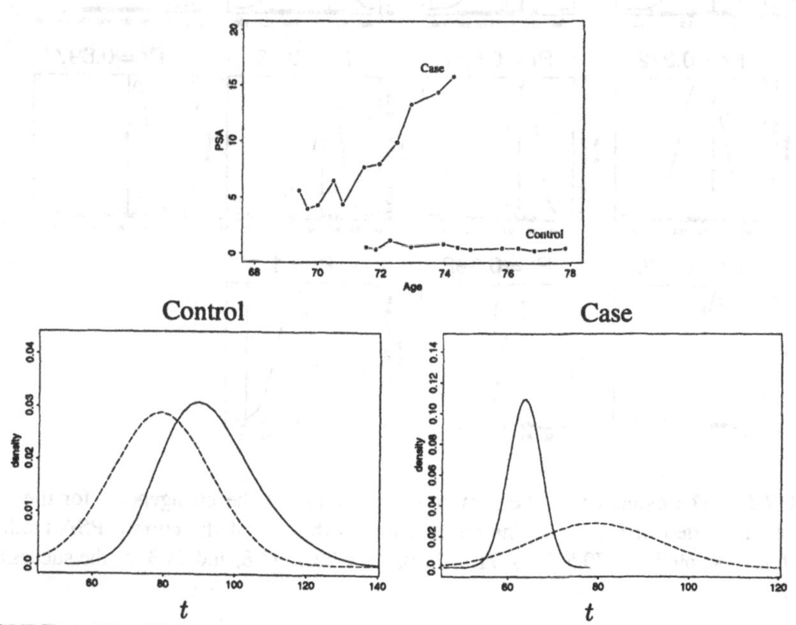

FIGURE 6. The PSA trajectories (top panel) and prior (dashed) and posterior (solid) distributions of the changepoint for selected case and control subjects.

4.2 Prospective changepoint identification

In practice it is of interest to assess the probability that the changepoint has occurred for a subject with each new PSA reading. Figure 7 illustrates the evolution of the posterior distribution of the changepoint for the case subject whose PSA trajectory is given in Figure 6. Model (4) was first fit to all 1210 subjects, and then the PSA readings for this case were added to the data set one-by-one and the posterior distribution of this changepoint was computed each time. The probability that onset has occurred by the age of the current reading is estimated by the proportion of 1000 sampled changepoints (at lags of 75) less than this age and is indicated by "Pr =" in the figure. The "rugs" in the graphs depict the sampled points. The early estimated onset probabilities follow the fluctuation of the initial PSA readings, but climb rapidly once the PSA begins a consistent increase. By the eighth PSA reading there is a probability of approximately 90% that onset has occurred, yet diagnosis was not made for at least 1.5 years later. Using all of the data, the modal estimate of the age of onset is about 64 years.

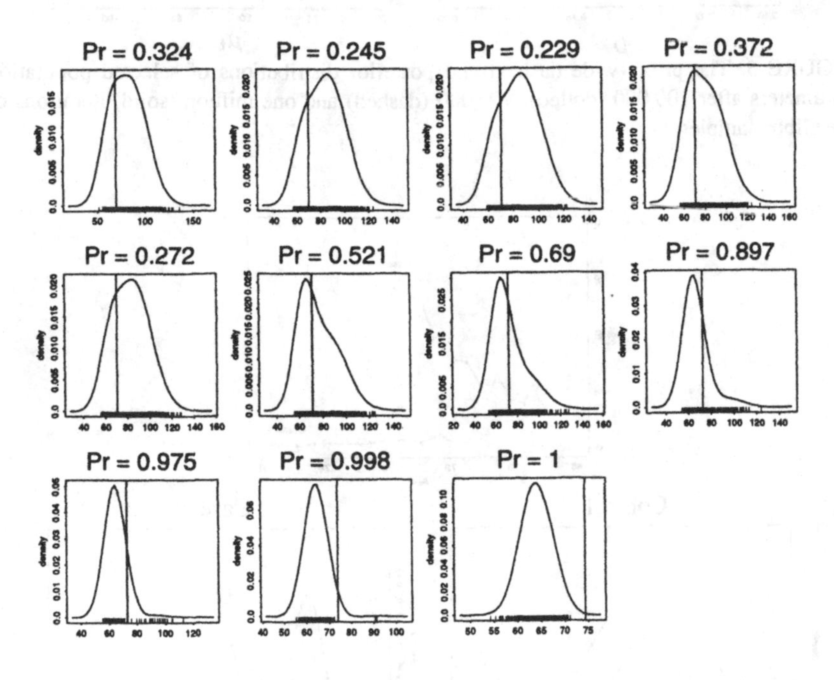

FIGURE 7. The evolution of the posterior distribution of the changepoint for the case subject illustrated in Figure 6. The vertical line is the age of the current PSA reading, which is 69.4, 69.7, 70, 70.5, 70.8, 71.5, 71.9, 72.5, 72.9, 73.8, and 74.3 for the successive panels.

5 ROC Methodology

Figure 7 suggests a diagnostic rule that declares that the changepoint has occurred once the cumulative posterior probability of onset exceeds a specified threshold. The performance of this rule can be compared to that of other diagnostic rules by examining the ROC curves. To do so, the notions of sensitivity and specificity must be extended from the usual context of a one-time test to a series of longitudinal tests. Murtagh and Schaid(1991), Murtagh (1995) andEmir *et al.* (1995) have discussed longitudinal ROC curves. We describe and apply a method similar to that of Slate and Cronin (1997).

Specificity is the probability of a negative test given the subject is disease-free. In the longitudinal setting, we define specificity as an average of individual specificity rates, where the specificity rate for a subject is the probability of a negative test while the subject is free of disease. To estimate specificity, we restrict attention to the control subjects and, for each of these, estimate the specificity rate as the proportion of negative tests. Then the estimated specificity is the average of the estimated rates:

$$\begin{aligned} \widehat{\text{spec}}_i &= \text{proportion of negative tests for control subject } i \\ \widehat{\text{spec}} &= \text{average of the } \widehat{\text{spec}}_i. \end{aligned}$$

The estimated specificity rates are equally weighted when computing the estimate of the overall specificity.

The sensitivity of a test is the probability that the test is positive (indicates that disease is present) given that disease is indeed present. In the longitudinal context, sensitivity depends on the proximity of the test to disease onset. For example, a negative test one month after onset is not comparable to a negative test five years after onset. Moreover, a positive test result ends the series of tests. To account for this time dependence, we use *K-period sensitivity*, which is the probability that a test based on data available K time periods from an origin is positive given that disease is (ultimately) present. The appropriate choice of origin depends on the context and will often vary across individuals. In our PSA setting, we use the time of diagnosis as the origin with K extending backward in time. Thus our estimate of K-period sensitivity is the proportion of case subjects who test positive according to the most recent test that can be formed using (potentially) all readings taken prior to K years before diagnosis.

We compare three diagnostic rules here: the threshold rule, a one-year slope rule, and the posterior probability. The threshold rule gives a positive result if the most recent PSA reading exceeds a cutoff. The one-year slope rule gives a positive result if the increase per year in PSA, as determined from the two most recent readings (typically 6-12 months apart for the NPCT data) exceeds a cutoff. The posterior probability rule gives a positive result for subject i at time t if the posterior probability that t_i is less than t exceeds a cutoff.

Using these definitions of sensitivity and specificity for the longitudinal setting, the ROC curves shown in Figure 8 result for a subset of 54 cases and 54 controls

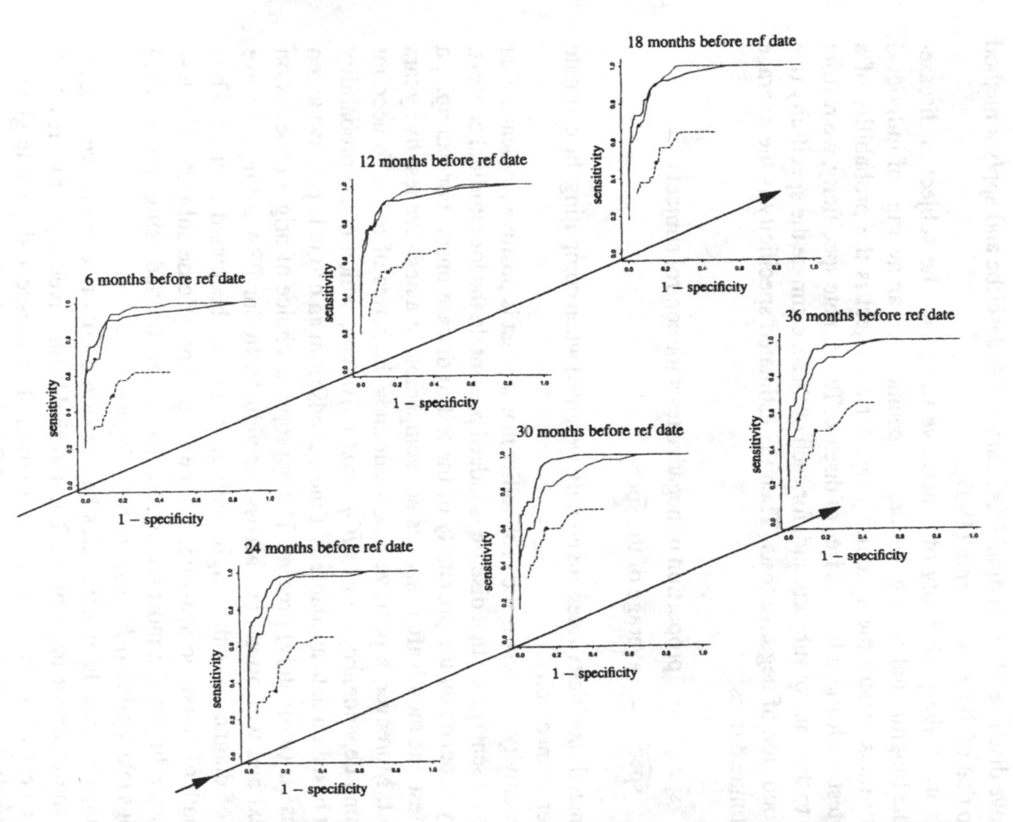

FIGURE 8. The ROC curves for the threshold (dotted line), slope (dashed line) and posterior probability (solid line) rules computed using a sample of 108 subjects from the NPCT data set. The circle on the threshold line corresponds to the usual cutoff of 4 ng/ml. The circle on the slope line corresponds to the cutoff of 1 ng/ml/year.

matched on various characteristics for the NPCT data. The curves for the posterior probability rule were obtained by first fitting the model to all 1210 subjects, and then separately adding each of the 108 subjects in the matched data, one reading at a time, and computing the requisite posterior probability. In this data set, the slope rule emerges as markedly inferior. The posterior probability rule performs at least as well as the threshold rule for all specificities of serious interest ($> 80\%$), and the superiority appears greater for the large values of K.

6 Discussion

For detecting cancer onset, we have emphasized the posterior distribution of the subject-specific changepoint as it evolves in time. The Bayesian framework is ideal here because this posterior distribution allows us to answer, for each subject, the question "What is the probability that cancer is present now?" Furthermore, upon diagnosis, the posterior distribution of the slope after the changepoint, b_i, may be valuable for the selection of appropriate therapy. The size of this slope may distinguish aggressive from slow-growing tumors and hence aid the decision of whether to remove the cancer or to pursue watchful waiting. We are currently compiling tumor grade and stage information for the prostate cancers that have been detected in the NPCT participants. By including these variables as covariates in the Bayesian model, we will be able to assess differences in the PSA trajectories for aggressive and slow-growing tumors.

One potentially confounding factor in our data is the presence of benign prostatic hyperplasia (BPH), an enlargement of the prostate that will also cause elevated PSA levels. Carter *et al.*(1992a, b) showed that PSA levels increase at a faster rate among BPH cases than among controls, but that this rate of increase is nonetheless significantly less than that among cancer cases. Diagnoses of BPH are available for some subjects in the NPCT data, but it is believed much of the incidence has not been reported and that among men in the age group of our cohort, BPH is the rule rather than the exception. In our analysis, we categorized the men diagnosed with BPH as controls.

It is also of interest to use PSA to monitor for recurrence among men who have undergone radiotherapy for prostate cancer. Slate and Cronin (1997) investigated one- and two-changepoint Bayesian models in this context.

Acknowledgments

We thank Beth Jacobs and Dawn Houghton for performing the PSA assays. We also thank Bruce Turnbull for many valuable discussions. This work was supported by a grant from the National Cancer Institute and NSF grant DMS 9505065.

References

Breslow, N.E. and Lin, X. (1995). Bias correction in generalised linear mixed models with a single component of dispersion, *Biometrika*, **82**, 81–91.

Carlin, B.P., Gelfand, A.E. and Smith, A.F.M. (1992). Hierarchical Bayesian analysis of changepoint problems. *Appl. Statist*, **41**, No. 2, 389–405.

Carter, H.B., Pearson, J.D., Metter, E.J., Brant, L.J., Chan, D.W., Andres, R., Fozard, J.L. and Walsh, P.C. (1992a). Longitudinal evaluation of prostate-specific antigen levels in men with and without prostate disease. *J. Amer. Med. Assoc.*, **267**, No. 16, 2215–2220.

Carter, H.B., Morrell, C.H., Pearson, J.D., Brant, L.J., Plato, C.C., Metter, E.J., Chan, E.W., Fozard, J.L. and Walsh, P.C. (1992b). Estimation of prostatic growth using serial prostate-specific antigen measurements in men with and without prostate disease, *Cancer Research*, **52**, 2232–3328.

Carter, H.B., Epstein, J.I., Chan, D.W., Fozard, J.L. and Pearson, J.D. (1997). Recommended prostat-specific antigen testing intervals for the detection of curable prostate cancer, *J. Amer. Med. Assoc.*, **277**, 1456–1460.

Catalona, W.J., Smith, D.S., Ratliff, T.L. and Basler, J.W. (1993). Detection of organ-confined prostate cancer is increased through prostate-specific antigen-based screening. *J. Amer. Med. Assoc.*, **270**, No. 8, 948–954.

Catalona, W.J., Smith, D.S., Ratliff, T.L., Dodds, K.M., Coplen, D.E., Yuan, J.J., Petros, J.A. and Andriole, G.L. (1991). Measurement of prostate-specific antigen in serum as a screening test for prostate cancer. *New Engl. J. Med.*, **324**, No. 17, 1156–1161.

Clark, L.C., Combs, G. F., Turnbull, B.W., Slate, E.H. . Chalker, D.K., Chow, J., Davis, L.S., Glover, R.A., Graham, G.F., Gross, E.G., Krongrad, A., Lesher, J.L., Park, H.K., Sanders, B.B., Smith, C.L., Taylor, J.R. and the Nutritional Prevention of Cancer Study Group (1996). Effects of selenium supplementation for cancer prevention in patients with carcinoma of the skin: A randomized clinical trial, *J. Amer. Med. Assoc.*, **276**, No. 24, 1957–1963.

Cleveland, W.S., and Devlin, S.J. (1988). Locally-weighted Regression: An Approach to Regression Analysis by Local Fitting, *J. Amer. Statist. Assoc.*, **83**, 596–610.

Cronin, K.A. (1995). *Detection of Changepoints in Longitudinal Data*, Ph.D. Thesis, School of Operations Research, Cornell University.

Cronin, K.A., Slate, E.H., Turnbull, B.W. and Wells, M.T. (1994). Using the Gibbs sampler to detect changepoints: application to PSA as a longitudinal marker for prostate cancer, *Computing Science and Statistics*, **26**, eds. J. Sall and A. Lehman, Interface Foundation of North America, pp. 314–318.

Emir, B., Wieand, S., Su, J.Q. and Cha, S. (1995). The analysis of repeated markers used to predict progression of cancer. *Proceedings of the International Biometric Society (ENAR).*

Gann, P.H., Hennekens, C.H. and Stampfer, J.J. (1995). A prospective evaluation of plasma prostate-specific antigen for detection of prostatic cancer, *J. Amer. Med. Assoc.*, **273**, 289–294.

Gelfand, A.E. and Smith, A.F.M. (1990). Sampling-based approaches to calculating marginal densities, *J. Amer. Statist. Assoc.*, **85**, 398–409.

Gelfand, A.E., Sahu, S.K. and Carlin, B.P. (1995). Efficient parametrisations for normal linear mixed models, *Biometrika*, **82**, 479–88.

Geman, S. and Geman, D. (1984). Stochastic relaxation, Gibbs distributions, and the Bayesian restoration of images, *IEEE Transactions on Pattern Analysis and Machine Intelligence*, **6**, 721–741.

Lange, N., Carlin, B.P. and Gelfand, A.E. (1992). Hierarchical Bayes models for the progression of HIV infection using longitudinal CD4 T-cell numbers, *J. Amer. Statist. Assoc.*, **87**, No. 419, 615-632.

Lindstrom, M.J. and Bates, D.M. (1988). Newton-Raphson and EM algorithms for linear mixed-effects models for repeated-measures data, *J. Amer. Statist. Assoc.*, **83**, 1014–1022.

Lindstrom, M.J. and Bates, D.M. (1990). Nonlinear Mixed Effects Models for Repeated Measures Data, *Biometrics*, **46**, 673–687.

Morrell, C.H., Pearson, J.D., Carter, H.B. and Brant, L.J. (1995). Estimating unknown transition times using a piecewise nonlinear mixed-effects model in men with prostate cancer, *J. Amer. Statist. Assoc.*, **90**, 45–53.

Murtagh, P.A. (1995). ROC curves with multiple marker measurements. *Biometrics*, **51**, No. 4, 1514–1522.

Murtagh, P.A. and Schaid, D.J. (1991). Application of R.O.C. curve methodology when markers are repeated measures. *Proc. of the International Biometric Society (ENAR)*, Houston, TX, March 24–27.

Oesterling, J.E., Jacobsen, S.J., Chute, C.G., Guess, H.A., Girman, C.J., Panser, L.A. and Lieber, M.M. (1993). Serum prostate-specific antigen in a community-based population of healthy men. *J. Amer. Med. Assoc.*, **270**, No. 5, 660–664.

Pearson, J.D., Luderer, A.A., Metter, E.J., Partin, A.W., Chan, D.W., Fozard, J.L. and Carter, H.B. (1996). Longitudinal analysis of serial measurements of free and total PSA among men with and without prostatic cancer, *Urology*, **48** (6A), 4–9.

Pearson, J.D., Morrell, C.H., Landis, P.K., Carter, H.B. and Brant, L.J. (1994). Mixed-effects regression models for studying the natural history of prostate disease, *Statist. Med.*, **13**, 587–601.

Pearson, J.D., Kaminski, P., Metter, E.J., Fozard, J.L., Brant, L.J., Morrell, C.H. and Carter, H.B. (1991). Modeling longitudinal rates of change in prostate specific antigen during aging, *Proc. Social Statistics Section of the Amer. Statist. Assoc.*

Press, S.J. (1982). *Applied Multivariate Analysis: Using Bayesian and Frequentist Methods of Inference*, Robert E. Krieger Publishing, New York.

Seber, G.A.F. and Wild, C.J. (1989). *Nonlinear Regression,* Wiley, New York.

Slate, E.H. and Cronin, K.A. (1997). Changepoint modeling of longitudinal PSA as a biomarker for prostate cancer, in C. Gatsonis, J.S. Hodges, R.E. Kass, R. McCulloch, P. Rossi and N.D. Singpurwalla, (eds), *Case Studies in Bayesian Statistics III*, Springer-Verlag, New York, pp. 444–456.

Stephens, D.A. (1994). Bayesian retrospective multiple-changepoint identification. *Appl. Statist*, **43**, No. 1, 159–178.

Vessella, R.L., Noteboom, J. and Lange, P.H. (1992). Evaluation of the Abbott IMx automated immunoassay of prostate-specific antigen, *Clinical Chemistry*, **38**, 2044–2054.

Whittemore, A.S., Lele, C., Friedman, G.C., Stamey, T. Vogelman, J.J., and Orentreich, N. (1995). Prostate-specific antigen as predictor of prostate cancer in black man and white men, *J. Nat. Cancer Inst.*, **87**, No. 5, 354–360.

Author Index

Subject Index

Lecture Notes in Statistics

For information about Volumes 1 to 66 please contact Springer-Verlag

Vol. 105: Constantine Gatsonis, James S. Hodges, Robert E. Kass, Nozer D. Singpurwalla(Editors), Case Studies in Bayesian Statistics, Volume II. x, 354 pages, 1995.

Vol. 106: Harald Niederreiter, Peter Jau-Shyong Shiue (Editors), Monte Carlo and Quasi-Monte Carlo Methods in Scientific Computing. xiv, 372 pages, 1995.

Vol. 107: Masafumi Akahira, Kei Takeuchi, Non-Regular Statistical Estimation. vii, 183 pages, 1995.

Vol. 108: Wesley L. Schaible (Editor), Indirect Estimators in U.S. Federal Programs. viii, 195 pages, 1995.

Vol. 109: Helmut Rieder (Editor), Robust Statistics, Data Analysis, and Computer Intensive Methods. xiv, 427 pages, 1996.

Vol. 110: D. Bosq, Nonparametric Statistics for Stochastic Processes. xii, 169 pages, 1996.

Vol. 111: Leon Willenborg, Ton de Waal, Statistical Disclosure Control in Practice. xiv, 152 pages, 1996.

Vol. 112: Doug Fischer, Hans-J. Lenz (Editors), Learning from Data. xii, 450 pages, 1996.

Vol. 113: Rainer Schwabe, Optimum Designs for Multi-Factor Models. viii, 124 pages, 1996.

Vol. 114: C.C. Heyde, Yu. V. Prohorov, R. Pyke, and S. T. Rachev (Editors), Athens Conference on Applied Probability and Time Series Analysis Volume I: Applied Probability In Honor of J.M. Gani. viii, 424 pages, 1996.

Vol. 115: P.M. Robinson, M. Rosenblatt (Editors), Athens Conference on Applied Probability and Time Series Analysis Volume II: Time Series Analysis In Memory of E.J. Hannan. viii, 448 pages, 1996.

Vol. 116: Genshiro Kitagawa and Will Gersch, Smoothness Priors Analysis of Time Series. x, 261 pages, 1996.

Vol. 117: Paul Glasserman, Karl Sigman, David D. Yao (Editors), Stochastic Networks. xii, 298, 1996.

Vol. 118: Radford M. Neal, Bayesian Learning for Neural Networks. xv, 183, 1996.

Vol. 119: Masanao Aoki, Arthur M. Havenner, Applications of Computer Aided Time Series Modeling. ix, 329 pages, 1997.

Vol. 120: Maia Berkane, Latent Variable Modeling and Applications to Causality. vi, 288 pages, 1997.

Vol. 121: Constantine Gatsonis, James S. Hodges, Robert E. Kass, Robert McCulloch, Peter Rossi, Nozer D. Singpurwalla (Editors), Case Studies in Bayesian Statistics, Volume III. xvi, 487 pages, 1997.

Vol. 122: Timothy G. Gregoire, David R. Brillinger, Peter J. Diggle, Estelle Russek-Cohen, William G. Warren, Russell D. Wolfinger (Editors), Modeling Longitudinal and Spatially Correlated Data. x, 402 pages, 1997.

Vol. 123: D. Y. Lin and T. R. Fleming (Editors), Proceedings of the First Seattle Symposium in Biostatistics: Survival Analysis. xiii, 308 pages, 1997.

Vol. 124: Christine H. Müller, Robust Planning and Analysis of Experiments. x, 234 pages, 1997.

Vol. 125: Valerii V. Fedorov and Peter Hackl, Model-oriented Design of Experiments. viii, 117 pages, 1997.

Vol. 126: Geert Verbeke and Geert Molenberghs, Linear Mixed Models in Practice: A SAS-Oriented Approach. xiii, 306 pages, 1997.

Vol. 127: Harald Niederreiter, Peter Hellekalek, Gerhard Larcher, and Peter Zinterhof (Editors), Monte Carlo and Quasi-Monte Carlo Methods 1996, xii, 448 pages, 1997.

Vol. 128: L. Accardi and C.C. Heyde (Editors), Probability Towards 2000, x, 356 pages, 1998.

Vol. 129: Wolfgang Härdle, Gerard Kerkyacharian, Dominique Picard, and Alexander Tsybakov, Wavelets, Approximation, and Statistical Applications, xvi, 265 pages, 1998.

Vol. 130: Bo-Cheng Wei, Exponential Family Nonlinear Models, ix, 240 pages, 1998.

Vol. 131: Joel L. Horowitz, Semiparametric Methods in Econometrics, ix, 204 pages, 1998.

Vol. 132: Douglas Nychka, Walter W. Piegorsch, and Lawrence H. Cox (Editors), Case Studies in Environmental Statistics, viii, 200 pages, 1998.

Vol. 133: Dipak Dey, Peter Müller, and Debajyoti Sinha (Editors), Practical Nonparametric and Semiparametric Bayesian Statistics, xv, 408 pages, 1998.

Vol. 134: Yu. A. Kutoyants, Statistical Inference For Spatial Poisson Processes, vii, 284 pages, 1998.

Vol. 135: Christian P. Robert, Discretization and MCMC Convergence Assessment, x, 192 pages, 1998.

Vol. 136: Gregory C. Reinsel, Raja P. Velu, Multivariate Reduced-Rank Regression, xiii, 272 pages, 1998.

Vol. 137: V. Seshadri, The Inverse Gaussian Distribution: Statistical Theory and Applications, xi, 360 pages, 1998.

Vol. 138: Peter Hellekalek, Gerhard Larcher (Editors), Random and Quasi-Random Point Sets, xi, 352 pages, 1998.

Vol. 139: Roger B. Nelsen, An Introduction to Copulas, xi, 232 pages, 1999.

Vol. 140: Constantine Gatsonis, Robert E. Kass, Bradley Carlin, Alicia Carriquiry, Andrew Gelman, Isabella Verdinelli, Mike West (Editors), Case Studies in Bayesian Statistics, Volume IV, xvi, 456 pages, 1999.